ISBN 978-1-331-91814-1
PIBN 10253823

OF THE

SURVEY AND EXPLORATION

OF

Northern Ontario

1900

PRINTED BY ORDER OF
THE LEGISLATIVE ASSEMBLY OF ONTARIO

TORONTO:
PRINTED AND PUBLISHED BY L. K. CAMERON,
Printer to the King's Most Excellent Majesty.
1901.

WARWICK BRO'S & RUTTER, PRINTERS,

TORONTO.

REPORT OF THE SURVEY AND EXPLORATION OF NORTHERN ONTARIO, 1900.

To His Honor the Honorable Sir Oliver Mowat, G.C.M.G., Lieutenant-Governor of the Province of Ontario:

MAY IT PLEASE YOUR HONOR:—

I have the honor to transmit herewith for presentation to the Legislative Assembly the Report of the Surveys and Explorations carried on under the supervision of my Department in Northern Ontario during the summer and fall of 1900.

E. J. DAVIS,
Commissioner Crown Lands.

DEPARTMENT OF CROWN LANDS,
TORONTO, March 15th, 1901.

TABLE OF CONTENTS

ERRATA IN ILLUSTRATIONS.

Page 8. In title of centre picture read "west" instead of "north."
Pages 24 and 25 Picture in centre of page 24 and picture at bottom of page 25 are transposed, but titles are in proper place.
Page 30. In title of bottom picture on page "Moose R." should read "Mouse R."
Page 76 Transpose the titles of the pictures.

INTRODUCTORY.

The Government of the Province having determined upon a policy of opening up and exploiting the resources of New Ontario, with a view of increasing the industrial wealth and population of the Province as a whole, it became necessary to take stock of the various kinds of natural wealth which existed there in order to ascertain of what they consisted and in what quantities, their situation and location and their possibility of development. When this information was had the people would know the value of their heritage, and would be able to approve with intelligence the expenditure of whatever money was necessary to render available the latent wealth of that enormous region, and the Government would be informed as to the sections which would best repay immediate development, and from what points now accessible by rail, railways or colonization roads should be built to open up the territory for settlement, lumbering and mining.

Having concluded to examine and explore the region, the next question was where and how the exploration should be made. Speaking roughly all Ontario north of Lake Nipissing and the French river, and indeed parts of the districts of Parry Sound and southern Nipissing might be included in the term "New Ontario", but sufficient information was at hand in the reports of surveyors, etc., to establish what are the resources and characteristics of the country south of the French river and Lake Nipissing, as well as that part of the district of Algoma lying between Sudbury and Sault Ste Marie. It was accordingly considered expedient to draw some well-defined base line from the eastern to the western boundary of the Province from which the explorations should start north, and it was thought that the Canadian Pacific Railway track might well be taken as the southern boundary of the region to be explored, all the way from the Ottawa river to the western boundary of the Province. It is true that south of the C. P. R. track there are large regions not yet opened up or surveyed, notably in the districts of Algoma, Thunder Bay and Rainy River, but as to these there is some good general idea of their resources and characteristics, and as railway construction is going on through the parts of Thunder Bay and Rainy River not yet opened up which will have the effect of bringing their resources to light, and as the main line of the Algoma Central Railway and its Michipicoton branch will do the same in the unexplored parts of Algoma, it was considered best to confine the exploration to the territory north of the C. P. R. track, which was practically a *terra incognita* for there, after all, must lie our greatest hopes for future expansion, as, if there is little or no good land in that immense country eight hundred miles in length from the Ottawa to the western boundary of the Province and in places two hundred and sixty miles in width between the C. P. R. and James Bay, then our area of good agricultural land is limited. If we have no forests of merchantable timber north of the C. P. R. then our pulpwood resources are also limited; and so on with other things valuable or to be made valuable by the opening up of that region.

Of course some information was at hand, but it was very vague. Parties had penetrated to James Bay from time to time, and these had seen good land and quantities of spruce along the streams which they had travelled, but at what distance from any known point and in what quantities they could not say. Mr. E. B. Borron as stipendiary magistrate of northern Nipissing, who had that immense region under his charge, had made several trips to James Bay by different canoe routes; and he, too, had seen and reported good land and quantities of spruce and other timber upon the different streams he had travelled, but as to quality and locality he was necessarily very indefinite, having only observed as he navigated or portaged in an entirely unsurveyed territory. The Geological Survey of Canada, too, had examined the geology and characteristics of the land adjacent to certain streams which they had ascended or descended, and some of our own surveyors in running a few base lines and the boundary line between the Districts of Nipissing and Algoma, had passed through good land and spruce timber; but after all, this sort of information was disjointed as well as limited, although of sufficiently favorable character to warrant the belief that when systematic exploration was made great sources of natural wealth would be discovered in that little known region. Accordingly at the session of 1900 the Government asked the Legislature to appropriate the sum of

$40,000 for the exploration of that part of the Province which it had been determined to examine, and it was voted without demur or objection.

As already stated, the southern base of the exploration was the C.P.R., and the northern base the boundary of the Province. The magnitude of the immense region to be explored will be better understood when it is stated that the area of the whole Province is about 126,000,000 acres, and the old and settled part contains about 23,000,000 acres, while the area that was to be explored north of the C. P. R. contains 60,000,000 acres. The boundaries of this territory having been fixed, the system to be pursued, and the selection of the surveyors and other officers necessary to obtain and bring back intelligent information was entrusted to the Commissioner of Crown Lands. Under his direction the region was divided into such convenient areas as might be explored in one season by the parties selected, and in order to permit this the territory was divided into ten separate districts and a party assigned to each.

It was necessary that the explorations should be so performed that the country could be mapped out and the situation of the good land and timber accurately laid down on the plans of the Crown Lands Department. Therefore a surveyor was placed at the head of and in charge of each party, with the exception of one, and an expert timber man who was also a judge of land was attached to each party as an estimator and explorer. The geological and topographical features of the region were also of importance, and therefore a geologist of good training was sent with each party. Upon these three officers devolved the responsibility of obtaining the information desired by the Department. There was necessarily a number of packers, canoemen, etc., as each party had to take in sufficient supplies for about five months.

The work of selecting, instructing and equipping the several exploration parties at so short a notice was one which involved much diligence and care, as, upon the personnel of the explorers and the objects to which their attention was specially directed depended in a very large degree the accomplishment of the ends aimed at and the obtaining of value for the money expended. The net results which were achieved are set out in detail below. On the whole the operations were carried out with skill, energy and carefulness, and there is ground for congratulation in the fact that so much tangible information has been secured in so short a time and at the expenditure of what is, comparatively speaking, a small sum.

The general outline of the system of exploration to be pursued was prepared and printed, and the heads of each party were supplied with copies of the same. In addition to this they were given specific instructions as to the details of the particular work which they were to perform, and all the information in the possession of the Crown Lands Department was furnished to them.

GENERAL SCHEME OF EXPLORATION.

As the objects of these explorations and the manner in which they were to be attained can best be gathered from this printed scheme it is included herein. Following it will be found the particular instructions directed to a surveyor in charge which will give a general idea of the instructions to the heads of the different parties.

Districts to be Explored.

No. 1. A base line due east from the 198th mile of the Algoma-Nipissing line across the District of Nipissing to the boundary between Ontario and Quebec, a distance of 70 or 75 miles. To be explored 50 miles on each side of the base line.

No. 2. A base line due west from the 198th mile of the Algoma-Nipissing line into the District of Algoma to the Missinabie River, a distance of about 100 miles. To be explored 50 miles on each side of the base line.

No. 3. From Lake Temagami northward to Montreal River, and upon the east and west sides of the Algoma-Nipissing line.

No. 4. The Kabinakagami River, 20 to 25 miles upon both sides, going in from Grassett Station on the Canadian Pacific Railway; also a section of the Missinabie River above its junction with the Moose River, not included in the area of No. 2, and 20 miles on the west side of the Moose below the mouth of the Missinabie.

No. 5. From Jackfish Station on Lake Superior to Long Lake; thence down Kenogami River, the outlet of Long Lake, to Albany River; thence down the Albany to Hudson Bay. English River to be explored for 20 miles on each side, and Albany River by way of any large streams flowing into it from the south.

No. 6. From Ombabika Bay at the northeast of Lake Nepigon, by Ombabika River, and canoe routes to Albany River, and down the Albany to the mouth of the Kenogami. Ombabika River and canoe routes to be explored 20 miles on each side, and the Albany by streams flowing into it from the south.

No. 7. From Wabinosh Bay at the northwest of Lake Nepigon, up Wabinosh River and across the height of land to Albany River, and sections of the Albany and Savant Rivers.

No. 8. West of Lake Nepigon and Nepigon River to Dog Lake, up Gull River from Lake Nepigon to the north of Dog Lake, and the country around Black Sturgeon Lake.

No. 9. From Dinorwic on the Canadian Pacific Railway north to Lake Minnietakie, Lac Seul and Lake St. Joseph, and along the Root and English Rivers and Sturgeon Lake.

No. 10. From the Canadian Pacific Railway between Wabigoon and the western boundary of the Province, north to Lac Seul and English River.

General Duties of an Exploring Party.

Each party will comprise a land surveyor, a land and timber estimator and a geologist, together with guides, canoemen, packers and other helpers sufficient for all the purposes of exploration.

The land surveyor will control and direct the work, which he is required to execute under special instructions, and every member of the party will take orders from him, to the end that a plan of exploration may be conducted in the most efficient manner and with the best economy of time and labor. Neglect or refusal on the part of any one to carry out instructions will subject the offender to suspension or dismissal from service, or to disallowance of wage earnings for a specified number of days, at the discretion of the surveyor; but the surveyor must report the particulars of every such case to the Department at the earliest convenient opportunity.

In the case of No. 3 district, which is believed to be more valuable than the others for land and timber, the exploration work will be in charge of the land and timber estimator, but necessary assistance will be given by members of the party to the land surveyor in the conduct of his special work.

Every person who engages to serve with an exploration party is expected to perform his full share of the general work, in camp, in the field, or *en route*; as well as to report to the land surveyor any interesting facts or observations which may come under his notice in any part of the district not covered by the operations of the surveyor, the land and timber estimator or the geologist of the party.

Care of Instruments and Property.

Travelling Expenses and Wages.

Actual travelling expenses to and from the field of exploration usual and essential to the ordinary comfort of travellers will be allowed, which will embrace the following items of expenditure only, viz: Cost actually paid for fares upon railroads, stages, steamboats or other usual modes of conveyance, together with street-car, omnibus or coach fare from depots and hotels.

When delays at hotels are incident to and necessary for the performance of the duties for which the travel is ordered, charges for actual hotel expenses will be allowed not exceeding $2 per day, and bills should always be obtained and filed as sub-vouchers to the account, giving the dates of arrival and departure and the rate per day. No charges will be allowed for hotel bills when the detention is unnecessary for the execution of the orders under which the journey is performed.

Meals furnished on steamers or other means of conveyance which are included in the charge for fare will not, of course, be made an extra charge.

Whenever special expenditures are made for meals they will be allowed, but for no other items of refreshment than the ordinary food provided for travellers.

When employed on exploration work the members of each party will be furnished with adequate supplies of food by the land surveyor in charge, and also with lodging in tents in addition to their regular pay, but men will provide their own blankets and bags in which to carry their outfit.

The allowance to the surveyor will be at the rate of seventy cents per day for each member of the party, which will cover the whole of the expense on account of rations and lodging—as far as the Department is concerned.

The rate of pay will be as agreed upon per calendar month, and the time will be reckoned from the date of leaving home until the date of return.

In the case of the land surveyor, the land and timber estimator and the geologist, a reasonable additional time will be allowed after return from the field for the purpose of enabling them to complete their respective reports.

All bills and vouchers must be rendered to the Department in triplicate, properly certified by the land surveyer or other officer in charge of a party.

Duties of the Surveyor in Charge.

The land surveyor will be expected to provide himself with the following instruments, viz.: A theodolite, a pocket sextant with artificial horizon, a steel tape and pins, a compass with Jacob's staff, a micrometer, and the usual plotting instruments.

The following articles are furnished by the Department: A Kay Taffrail Log complete, and ruled field-books in which the details of the work are to be entered each evening to a scale; an aneroid barometer; a camera with films wherewith to take views to illustrate the nature of the country, or of waterfalls, or any natural object which may give an idea of the resources of the country; a tin box to carry field-books and drawing materials; a tin case for preserving pressed specimens of the flora of the country; besides tents, canoes and packstraps. These instruments and properties will be returned or accounted for at the close of the season's operations.

The information to be acquired by the surveyor will embrace: (1) The nature of the soil. (2) The various kinds of forest trees and the extent, size and quality of the timber, given in order of its relative abundance. (3) The fixed rocks met with, and economic minerals if any. (4) The flora, collecting and pressing specimens for subsequent identification, taking care to mark in the field-book the localities where such flora are noticed. (5) The fauna, particularly deer, moose, elk, fur-bearing animals and birds, or information concerning any of these which can be procured from the Indians; also the several species of fish in the lakes and rivers. (6) The general features of the country procured from a track survey of the water communications, and also from the explorations made on each side of the rivers and lakes. (7) Valuable water powers, giving as regards each the flow and volume of water, estimating the fall and describing the nature of the banks on ei[illegible] side of the stream.

In using the Kay Taffrail Log the accompanying printed directions should be fc[illegible]ras- ed. In descending or ascending rivers allowance should be made for the current, v[illegible]iver may be determined by floats for a measured distance or by comparison between a measured distance along the river and the distance read off from the log, the surveyor keeping accurate watch at all times on the log so that it does not measure more than the real distance traversed. Records and calculations should be entered in the field-book for subsequent verification.

Numerous observations for latitude should be taken, so as to correct any variation which may arise in the track-survey. Observations to determine the magnetic variation should also be taken at regular intervals.

If any large body of pine or spruce or any considerable tract of good land be met with, the land surveyor in charge will assist the timber and land estimator in arriving at a correct idea of its extent, nature, etc., and if it should be necessary for this purpose to remain longer in any locality than would otherwise be necessary, he will order the movements of the party accordingly; and so also in case a portion of country appears to be of more than usual promise mineralogically. Should he find that any other members of the party can do good work in estimating timber and land, or in collecting rock or mineral specimens, or in map work, he will make use of their services on such occasions, or whenever required.

The surveyor will instruct the men of his party in the use of the pocket compass, with which each one will be supplied. He will also require each man to carry a small axe and a box of matches.

Before each party begins its operations in the field the land surveyor or other officer in charge will read to its members the provisions of the *Act to Preserve Forests from Destruction by Fire*, as required by section 8 of the said Act, and will repeat such caution as often as once a week while the party remains in the field.

Extreme caution is required to be exercised in canoeing upon swift-flowing streams, or upon lakes in rough weather; and when a party is divided to expedite the work of exploration, it should not under any circumstances consist of less than two persons.

Duties of the Land and Timber Estimator.

The land and timber estimator appointed to accompany an exploration party will examine the country traversed by the party for the purpose of ascertaining as accurately as possible the nature and extent of its resources in timber and land. He will place himself under the direction and general control of the land surveyor and will be subject to his orders in all matters connected with the movement, transportation and general conduct of the expedition.

He will be expected to note carefully the kinds of timber encountered *en route*, the size and thriftiness of the trees, the comparative prevalence of the several varieties, and generally all information concerning the forest growth and conditions which may be useful or valuable, and his report will show where there is no timber as well as where it occurs. Should any large or important body of any particular variety of timber be met with, he will make an estimate of its extent and the quantity in feet board measure, or in cords, as the case may be, which such block contains; and in addition he will note carefully its location, so that it can be laid down on the map, and its situation in respect of the nearest or most available river or lake down which the timber may be floated.

He will be expected to take particular cognizance of areas of red or white pine that may be seen, and to obtain the foregoing particulars in respect thereof; and as it is possible that the district to be explored by the party may contain a considerable quantity of spruce, his report will be expected to say where and to what extent this kind of timber is found, together with full particulars as to its size, quality and variety, its fitness for pulpwood or sawlogs and its situation with respect to floatable streams. He will not, however, by any means neglect the other kinds of trees met with, but will make careful observations and preserve full notes as to jackpine, poplar, balsam, cedar, hemlock, hardwoods of all kinds, and in fact every variety of timber capable of present or prospective use.

In addition to reporting on the timber, he will report on the land. He will observe with care the character of the surface travelled over, and mark its nature, whether rock, swamp, water or dry land. Where the surface covering is soil he will note its approximate area or extent, its kind (*i. e.* whether sandy, gravelly, clayey, etc.) and its quality, and whether in his opinion it is capable of producing grain, hay or any other kind of crop. He will carefully note any large areas of cultivable land suitable for settlement, and make

an estimate of their probable extent. Wherever it is found that land remote from known settlements has been occupied and cultivated, as for instance at posts of the Hudson's Bay Company, he will make mention of the circumstance, and give such particulars as he can obtain of the success or failure of such attempts at cultivation. Where swamp or muskeg is found he will note the character of the prevailing vegetation, and whether or not it is a peat deposit. In the latter case, if the bog is a large one, say upwards of 100 acres in extent, he will endeavor by the use of a pole, or otherwise, to obtain an idea of the depth of the deposit.

The object of the investigations of the land and timber estimator is to assist in ascertaining the value of the district to be traversed by the party for farming, lumbering and mining purposes, and any other useful information bearing upon land and timber not covered by the foregoing will be obtained and included in his report to the land surveyor, to be made after his return from the field.

The exploring party will proceed on a route to be indicated by the Department beforehand in general terms, but the land and timber estimator is not to confine his investigations to this route or its immediate neighborhood. He will make excursions into the country to a distance, if possible, of twenty or twenty-five miles on each side of the base of exploration (lake, river or line as may be), and at intervals of about ten miles, so as to get an idea of the general character of the district traversed.

As it is most important that this information should be definite, he will keep a diary in which will be entered his work from day to day, which direction he travelled, whether by land or in canoe, if the latter by what streams, their size, and all particulars thereof, entering full observations as they are made, and under the proper date, and he will report to the land surveyor in charge each evening, or as often as both officers are in camp together, such information as he may have acquired, so that the land surveyor may make a note of it in his field book. It is essential that he should keep in constant touch with the land surveyor, so that he may be in a position to accurately describe the location of either timber or land noted in relation to the physical features of the country.

These instructions as to the land and timber estimator are necessarily general in their character, as it is impracticable to give directions which shall meet every contingency or set of conditions that may arise, and much must necessarily be left to the good judgment of the land surveyor and the man specially charged with this branch of the work. Both officers will keep steadily in mind the object of the expedition, and aim to bring back all information which would be of interest or value concerning the district to be explored, even if not specially referred to or required in these instructions.

Duties of the Geologist.

The geologist accompanying each party is required to note carefully the general topography of the district to be explored,—its plains and valleys, its hills and mountains and their direction, its streams and rivers with their courses, its lakes and ponds with their outlets, with estimated areas, elevations, etc., and sketches of map-making. He will note the rock formations of the district, whether they are Laurentian, Huronian or of later age, and whether eruptive or sedimentary, marking down in his note book the strike of rocks and lines of contact between formations, with estimates of breadth and length of formations where such can be made. Where outcroppings of rocks occur along lakes and streams, or in ranges of hills and mountains, he will make careful examination so as to be able to describe and locate them. Samples of interesting rocks should be taken for subsequent study in the laboratory.

He will note every appearance of ore or mineral. If occurring in a deposit or body, such as iron, copper or nickel ore, he will examine the country rock enclosing the ore and make an estimate of its extent. If it is in a vein or lode, such as silver or gold bearing ore, he will observe in addition whether it is a fissure or bedded vein, and ascertain the width, length and course of it, and its dip if exposed in a sidehill or escarpment. Mineral veins may often be found outcropping on the face of bluffs, on the banks of rivers, on the shores of lakes, and on hills which have been swept by fire, and every such feature should be well scanned. Samples of all promising ores or minerals should be taken for subsequent examination or analysis.

He will note the timber and soil of the country and its suitableness for settlement, and if rapids or falls occur on the rivers and streams they should be studied in order that an opinion may be formed of their value for water power. Indications of coal or lignite should be looked for, and careful notes made if any are observed. If peat bogs are met with it is advisable to ascertain their probable area and their depth, and to examine the quality of the peat as material for fuel and moss litter.

The fauna and flora of the district should be observed, and Indian occupation if any; and generally all data of practical as well as scientific interest should be collected for use in the preparation of a report.

The geologist is expected to examine every object for himself and to describe it in his note book while he is on the ground, or while every observation and incident is fresh in his mind. He should not take the word or report of another for what he may see with his own eyes.

It will probably be found most convenient for the geologist to explore in company with the land and timber estimator of the party, going out from and returning to camp every day if convenient; and as often as practicable a summary of his operations should be handed in to the land surveyor of the party.

The geologist will keep a diary of his movements in addition to notes and sketches of scientific value, so that when he returns at the close of the season's exploration he may prepare a full account of his operations for the surveyor.

Requirements in Common.

It will be a requirement in common of the land surveyor, the land and timber estimator and the geologist in their several reports to describe the localities of promising areas of agricultural land, of timber, or of minerals in each district, to show the advantages afforded by nature for their economic development, to estimate their distance from the nearest towns or settlements and railway or water communications, and to indicate the most feasible ways in which they may be opened up for occupation by roads, railways or waterways.

Copies of notes and sketches made and diaries kept in the field by the land surveyorl the land and timber estimator and the geologist are required to be returned with the fina reports to the Department of Crown Lands, where they may be filed for verification and future reference.

Department of Crown Lands,
Toronto, 31st May, 1900.

E. J. DAVIS,
Commissioner.

SURVEY PARTY No. 1.

Instructions to Ontario Land Surveyor Thomas B. Speight, to survey a Base Line in the District of Nipissing.

Department of Crown Lands, Surveys Branch,
Toronto, June 12th, 1900.

Sir,—By direction of the Honourable the Commissioner of Crown Lands I have to instruct you to proceed to the above service without delay.

You will commence your survey at a point on Niven's line between the Districts of Algoma and Nipissing marked by the 198th mile from the north-east angle of the township of Lumsden, your point of commencement being an iron post with a tamarac post planted alongside, planted by Mr. Niven in 1898 at the 198th mile, where there is a tamarac bearing tree 14 inches in diameter, bearing north 5 degrees west, 31 links from said posts.

Having taken the necessary astronomical observations on your starting point, you will run due east on this parallel of latitude, taking frequent astronomical observations so as to keep on a due east course until you reach a point 72 miles from Niven's line which will probably be in the neighbourhood of the interprovincial boundary and therefore it will be unnecessary to run this line further, as, when the interprovincial boundary comes to be run, it can be connected with your line.

The projected plan herewith, drawn on a scale of 4 miles to an inch, shows the above line in red and also shows the rivers so far as they are known which will be met with in the course of your survey.

You will plant posts of the most durable wood you can find at every mile along this line, marking them with the number of the miles, on the west side of the post, which they are distant east of the meridian line between the Districts of Nipissing and Algoma. At every three and six mile point you will drive in alongside the wooden post an iron tube or post three feet long and one and one-quarter inches in diameter, pointed at the end and forged at the top so as to shed the water and painted red, with a corresponding number to that on the wooden post, and marked with a cold chisel previous to being driven in.

You will take bearing trees on the course and distance to some fixed or remarkable object from the post so planted; and wherever it is possible place a cairn of stone round the six-mile or township corner posts. You will blaze the line in the usual way, that is, with three blazes, one on the east, one on the west, and one on that side of the tree on on which your line passes.

You will take with you two iron posts ($1\frac{7}{8}$ inches) similar to the others except as to diameter, and will plant one at a convenient point where it will not be disturbed near the Hudson's Bay Company's post on Abitibi Lake, driving it in solid and taking bearings to prominent points by which it may be easily found hereafter. You will carefully determine the latitude of this iron post. You will plant another on a prominent point, wherever it can be driven down firmly, say between Upper and Lower Abitibi Lake; and in making your track-survey of the Abitibi Lake connect these two points carefully, and also the track survey down the Abitibi River so as to connect with Niven's meridian line between Nipissing and Algoma, and in this way, on the longitude of Niven's line being determined, the position of the Hudson's Bay Company's post at Abitibi Lake can be determined by calculation.

You will take with you an aneroid for the purpose of ascertaining the height of the hills and the country passed through by your survey; and with a view of procuring data for the Provincial Observatory you will note in your field-book or diary the date and the hour of the day on which the observations are made, and enter them in a small note book ruled in columns, thus:—Date, hour, temperature, inches, direction of wind, light, remarks You will take observations three times a day, the morning, towards noon and in the afternoon, noting the temperature with the thermometer attached to the aneroid. You will be careful in noting these observations of the barometric readings as they will be compared with those of the Provincial Observatory on your return. You will take all reasonable care of this aneroid and return it to this Department when you have finished your survey.

You will be able to get in to your starting point by going in from North Temiscaming on Lake Temiscaming and following the canoe route via the Quinze River to Abitibi Lake, thence down the Abitibi River to Niven's line, which is crossed by the Abitibi River a little south of the 180th mile from the north-east angle of the township of Lumsden.

A track survey of the Abitibi River was made by Mr. Wm. Ogilvie in 1890. A printed copy of this report is herewith handed to you, as a large amount of it will be very useful during your survey. Abitibi Lakes, Upper and Lower, are described by him as being very irregular shaped, having many deep bays and numerous islands. These lakes have never been traversed and a micrometer survey of these combined with the use of Kay Taffrail Log will enable you to make such a survey as will show the size and dimensions of the above lakes. Part of the Abitibi Lake is situate probably within the boundaries of the Province of Quebec, but, as the interprovincial boundary is not known, it would be well to make a survey of the whole Upper and Lower Abitibi Lakes. A large river is said to run into Lower Abitibi Lake and it is quite possible that this may be utilized by you in getting provisions in to your line as you get east of Niven's line. Little Abitibi River and Lake will also be crossed by your base line, but nothing is known of the position of Little Abitibi Lake although I am told that there is a portage into it from the Abitibi River, starting in from the head of Long Sault

Rapids. This also will be very useful to you for getting in your provisions to your line instead of taking them all the way down the Abitibi River and then taking them over to Niven's line or up Niven's line, as the case may be.

You will make a track survey of the various canoe routes you use, so as to be able to lay down the lakes and rivers on your plan. You will determine trigonometrically the widths of the lakes and rivers cut by your line and enter the calculation on the page on which it occurs in your field-book.

You will keep your field notes in the usual style and on the usual field-note paper.

You will be particular in choosing your chainmen, in order that the work may be carefully performed, and will, during the progress of the survey, test your chain occasionally by your standard measure.

There will be two exploring parties, of three men in each, attached to your party and under your authority and control. You will see that all of these men provide themselves with a small compass and an axe. The head explorer will be responsible for reporting on the timber and soil met with, all of which he will report to you as often as possible, in order that you can enter the same in your field-book in the proper place. Both these timber explorers, one of whom will work north of your line and the other south of your line, will be provided with a canoe each and a couple of men, and you will supply them with provisions as they may require them, as they will probably be sometimes several days absent from your line ascending and descending rivers which you cross, using these rivers as their base, and penetrating into the country on either side of said rivers, so as to make as complete a report of all the country on each side of your line for a considerable distance each side as possible. Of course they can utilize your line as a base for exploring the land on each side of your line to a considerable distance, and there will be no danger, by using your line on the ground, of getting themselves lost, as they can keep behind your line instead of going ahead of it. One of the above parties exploring south of your line will be able to utilize the Abitibi River and any rivers running into it from the north as a means of getting into the interior portion until your line gets sufficiently far advanced east to enable them to work southerly from your line. The party to the north will explore from your line up to the Great Muskeg which is represented on Niven's survey of the line between the Districts of Algoma and Nipissing as occurring about the 230th mile from the north-east angle of the township of Lumsden. There will be no use penetrating this muskeg at the present season of the year, but there will be ample scope for this exploration party to explore the country between this muskeg and your line, a distance probably of 30 miles.

As there will probably be many rivers and lakes in the region to be surveyed by you, (in addition I mean to the Abitibi lakes), it would be well for you to have an assistant on your party who will be able to make a track-survey of these rivers and any lakes which may be met, with the Kay Taffrail log and compass, in order that you may produce a map showing all the information possible of this territory of 100 square miles more or less.

As your point of commencement is situate a great distance from the railway and as it will be difficult to transport supplies over the line, you will be allowed, by direction of the Honourable the Commissioner of Crown Lands, $45 00 per mile to cover all expenses connected therewith, including returns; and you will, I feel confident, perform the whole work in good workmanlike style.

You will be allowed at the rate of 70 cents per day each for rations for supplying the above exploring parties as they may require the same.

As this region is to a large extent unknown, you will use your best endeavours to acquire as much information as possible as to the topography and the nature of the timber and soil. You will mention in your report the timber capabilities of the country passed through, as it is probable that valuable spruce and tamarac forests may be found in the course of your survey, and you will instruct your men in their canoe work to take notice of the kind of timber passed through by them.

You are herewith loaned for this service the following:—(*a*) A Kay Taffrail log complete. (*b*) An aneroid and several field books which have been ruled carefully, and in which you can enter the details of your work each evening to a scale. (*c*) A despatch box to keep field book, stationery, etc. (*d*) A tin box for preserving pressed botanical specimens which can afterwards be named by the botanist of the Geological Survey, taking care, of course, to mark in field book the spots where such flora are noticed. (*e*) Two Peterboro canoes. (*f*) Two tents. (*g*) Two tarpaulins. (*h*) Pair of climbers.

The above articles you will return to this Department at the close of the survey, or store the canoes as you may be instructed.

You will, of course, in using the Kay Taffrail log first accustom yourself to its use. The accompanying printed directions give you instructions how to operate it. In ascending or descending rivers allowance must be made for the current, and this you will carefully determine, which you can either do by floats for a measured distance, finding out exactly what the current is and then make the proper allowance, showing your calculation in your field book, so that it may be checked hereafter, or by comparison between a measured distance along the river and the distance read off from your log, keeping an accurate watch at all times on your log, so that it does not measure more than the real distance traversed.

As you have a camera to work with films, it would be wise to take it with you, and where you can get an opportunity of taking a view which will give an idea of the nature of the country, or the nature of any particular object which will be useful for economic purposes, or of a fall on the river, it would be well to use said camera, in order that your report may be illustrated in some similar way as the Geological Survey reports are and have been for some years.

Any valuable water powers you will carefully note, giving the estimated fall and describing the nature of the banks on either side thereof; also any information you can procure from the Indians regarding the fauna and fur-bearing animals and fish in the various waters met with in the course of your survey.

It would be well to take along with the party a pair of climbers such as used by linemen of telegraph companies, by which means a member of your party can ascend any tall tree where an extensive view can be procured of the surrounding country, and make notes of the topography.

The information sought to be acquired on this survey, in addition to the laying down and marking your line with iron and wooden posts, will be: (*a*) The nature of the soil; (*b*) the nature, extent, size and quality of the timber, giving it in the order of its relative abundance; (*c*) the fixed rocks met with and economic minerals, if any; (*d*) the flora, collecting and pressing specimens, which can afterwards be named by the botanist of the Geological Survey, taking care, of course, to mark in field book the spots where such flora are noticed; (*e*) the fauna, particularly deer, moose, elk, and information regarding the fur-bearing animals which can be procured from the Indians, and also the several kinds of fish in the lakes and rivers met with in the course of your survey; (*f*) the general features of the country procured from a track survey, of the water communications, and also from the explorations made on each side of the rivers; (*g*) any valuable water power giving the estimated fall and describing the nature of the banks on either side thereof.

When you have completed the 72 miles of base line, you can discharge what men you will not require and then make a track survey and micrometer survey accurately, of Abitibi Lakes and Abitibi River down to its junction with the Frederick House River, and thence down to Mr. Niven's base line, which he will run from the 198th mile due west. You will also run, from a point on the Abitibi River at or near where it begins to run north-westerly, a compass line due west to Niven's meridian line between the Districts of Algoma and Nipissing, and chain the same carefully, planting wooden posts at each mile, in order to get a tie line between the Abitibi waters and Niven's meridian line, and making a connection with the nearest post north or south of Niven's meridian line. Probably the best place would be to commence at or near Iroquois Falls on the Abitibi River. If there is a canoe route leading from Abitibi Lake to Little Abitibi Lake it would be well to instruct your explorers to make as accurate a sketch of the same as possible.

Where two rivers, as for instance the Abitibi River and the Frederick House River, run almost parallel to one another, the explorers can penetrate half way towards each river, thus covering the whole extent of the country between these rivers, and similarly in any other rivers which may be met with in the course of your survey. The exploration party on the north of your line will not penetrate further north than the southern boundary of the Great Muskeg, which extends around the shores of James Bay, and is shown on Niven's line between Algoma and Nipissing about the 230th mile, as it is not the intention to explore this portion of the territory this season; but there may be, and probably are, some rivers running down towards James Bay, and which will cross your line, which can be utilized as a base line from which to penetrate the country east and west thereof in a similar way as that desired to be performed on your base line, thus getting an insight into the territory between your base line and the southern boundary of the great muskeg.

It is difficult to lay down an exact instruction how the timber explorers and geologists should proceed, but in a general way I may say the timber explorer and geologist should work together, and, assuming there are no rivers crossed by your line, they should penetrate the country north and south at a distance of 12 miles apart, that is to say, if the explorers on the north side went in at 6 miles, the one going in from the south side would go in at 12 miles, and then the explorers on the north side would go in at the 18th mile post and the explorers on the south side would go in at the 24th mile post, and so on, thus leaving about 12 miles apart between each of the main exploring lines. As, however, the rivers can be utilized instead of walking so many miles due north and south, you will instruct them to utilize these rivers wherever practicable as their base, penetrating east and west of these rivers similarly at 10 or 12 miles apart so as to enable them to get a good insight into the whole country. The explorer on the south side will, of course, be able to have a double front so to speak for his work, your base line on the north, and Abitibi Lakes, Upper and Lower, and Abitibi River on the south, by which means he will be able to cover all the country between the Abitibi waters and your base line.

You are herewith handed a Letter of Credit with the Hudson's Bay Company for $100.00 in case you find yourself short of goods. Mr. Chipman, the Commissioner, asks that the surveyor be informed that the Company does not keep large supplies on hand at their inland posts. Mr. Chipman also requests that the Letter of Credit be handed in to the officer in charge of the last post visited by you. You are supplied, for the use of the exploration parties, two canoes, two tents, two tarpaulins, six pocket compasses, tin box for holding field-book, etc., four field-books ruled, three sheets of aluminum for mending canoes if necessary, package of tacks, two small pots of white lead, six pack-straps, book of blotting paper for preserving specimens of the flora, small case of medicinal requirements with directions for use by Dr. Bryce, copy of the Act *To Preserve the Forests from Destruction by Fire*, pay lists in triplicate, and a quantity of vouchers for the travelling expenses of the exploring parties, two blue-print plans of the whole of the northern region embraced by your survey, so far as known, on a scale of eight miles to an inch. I may say that this plan has been compiled from all known sources, from the Geological Survey reports and other available data. You are sent four additional copies of the projected plan herewith sent you, one of which can be given to the timber estimator and geologist attached to each party, also three leather field-book bags, also pair of climbers, also triplicate form of account.

You are also sent a copy of the General Instructions which have been drawn up in pamphlet form, giving generally the duties of the surveyor in charge, the duties of the land and timber estimator, and the duties of the geologist who I may say will form one of each of the other parties mentioned above.

The scale of pay allowed by the Department for the exploration parties is as follows:

Timber Estimator.................. $90.00 per calendar month;
Geologist......................... 60.00 per calendar month;
Canoemen.......................... 35.00 per calendar month.

You will notice that these exploring parties are to be made up of three men each, and you will see that the pay list for these parties is made up by you in the usual way, as you will be responsible therefor.

You will understand that the whole party is under your control and direction in order that the exploration may be conducted in the most efficient manner.

As soon as possible after your return you will furnish this Department with a plan on a scale of two miles to an inch, exhibiting the natural features of the country passed through by you, also field notes and a full report of the whole work duly sworn to, and making the report in every department thereof as ample as possible, in order that the fullest information possible may be had of the region in which your survey operations lie.

You will be allowed the same rate as the other surveyors per month, namely, $210.00

per calendar month for the work that you will do for the Department after your contract of the seventy-two miles has been performed as mentioned in the foregoing instructions. You will re-engage of course as few of your men as you may require for this service and enter their names on the pay list in the same way as those of the two exploration parties, and you will be allowed for this service 70 cents per day for rations, each man.

Having furnished approved security for the proper performance of this survey, you are allowed an advance of $1,500.00 on account of this service for which you have already given a receipt.

I have the honour to be,
Sir,
Your obedient servant,
(Sgd.) E. J. DAVIS,
Commissioner of Crown Lands.

GEO. B. KIRKPATRICK,
Director of Surveys.

THOMAS B. SPEIGHT, ESQ.,
Ontario Land Surveyor,
Yonge street Arcade,
Toronto.

Enclosures.

THE REPORTS.

The results of these extensive explorations, as detailed in the elaborate reports sent in by the surveyors, the land and timber estimators and the geologists, have fully justified the most sanguine expectations in regard to the natural wealth and fertility of Northern Ontario and demonstrated the wisdom of the action taken, whereby some accurate knowledge of the character and extent of its enormous undeveloped resources has been acquired. It has been established beyond controversy that in the eastern part of the territory north of the height of land there is an immense area of excellent agricultural land, apparently equal in fertility to any in older Ontario, with an equable and temperate climate and an abundance of wood and water, which render the inducements it presents to those in search of homesteads as good as those offered anywhere else on the continent. The apprehension entertained by some that our forest resources were very limited has been contradicted by the exploration and estimation of extensive pine areas on the southern slope, as well as the location of great forests of spruce and other varieties of pulpwood north of the height of land, which will enable this Province to take a leading position in the commercial world as regards the growing and remunerative pulp and paper-making industry. While the geological examinations have not resulted in any new discoveries of economic minerals (and it was scarcely expected they would), they have been of material service in identifying and establishing the character of the different rock formations and locating promising indications as a guide to closer investigations in the future. Analyses of the peat taken from the extensive deposits in Nipissing have conclusively shown its high qualities and economic utility, and established the value of this great natural store of fuel, which will probably make it useful in the industrial development of the country.

AGRICULTURAL LAND.

The great clay belt running from the Quebec boundary west through Nipissing and Algoma Districts and into the District of Thunder Bay comprises an area of at least 24,500 square miles, or 15,680,000 acres, nearly all of which is well adapted for cultivation. This almost unbroken stretch of good farming land is nearly three-quarters as great in extent as the whole settled portion of the Province south of Lake Nipissing and the French and Mattawa rivers. It is larger than the States of Massachusetts, Connecticut, Rhode Island, New Jersey and Delaware combined, and one-half the size of the State of New York. The region is watered by the Moose River, flowing into James Bay, and its tributaries, the Abitibi, Mattagami and Missinabie, and by the Albany and its tributaries, the Kenogami and Ogoke. Each of these rivers is over 300 miles in length, and they range in width from 300 or 400 yards to a mile. They are fed by numerous smaller streams, and these in turn drain numberless lakes of larger or smaller size, so that the whole country is one network of waterways, affording easy means of communications with long stretches fit for navigation. The great area of water surface also assures the country against the protracted droughts so often experienced in other countries. The Southern boundary of this great tract of fertile land is less than 40 miles from Missinabie station on the Canadian Pacific Railway; and the country north of the height of land being one immense level plateau sloping off towards James Bay, the construction of railways and wagon roads through every part of it would be a comparatively easy matter.

In the small part of the District of Rainy River which was explored the proportion of good land is not so great, but the clay land in the townships around Dryden was found to extend north in the valley of the Wabigoon River, with an area of about 600 square miles, or 384,000 acres. There are also smaller cultivable areas at various other points.

The Climate.

Another important fact established by the explorations is that the climate in this northern district presents no obstacle to successful agricultural settlement. The information obtained completely dispels the erroneous impression that its winters are of Arctic severity and its summers too short to enable crops to mature. The absence of summer frosts noted by the explorers and the growth of all the common vegetables at the Hudson Bay posts must disabuse the public mind of this erroneous impression. The 50th parallel of latitude passes through the centre of the agricultural belt, and the climate is not much different from that of the Province of Manitoba, lying along the same parallel, with this exception, of course, that the winter is tempered by the great spruce forests and the presence of so large a proportion of water surface. The country, too, has an abundance of wood for fuel, building and commercial purposes, and plenty of pure water everywhere.

The Timber.

Another point equalled only in importance by the existence of a vast area of agricultural land in this country and its moderate climate is the fact that it is largely covered with extensive forests of spruce, jackpine and poplar. The value of this class of timber, as everybody knows, is increasing every day and the market for it is widening; and rich, indeed, is the country which has boundless resources in these varieties of woods. In the District of Nipissing, north of the C.P.R. line, there is estimated to be at least 20,000,000 cords of pulpwood; in the District of Algoma, 100,000,000 cords; in the District of Thunder Bay, 150,000,000 cords; and in the District of Rainy River, 18,000,000 cords; a grand total of 288,000,000 cords. The pine region does not seem to extend much beyond the height of land, but on this side, in the country around Lakes Temagaming and Lady Evelyn, and to the north, an area of red and white pine of fine quality was explored and estimated to contain about three billions of feet B.M.

Water Powers.

A feature of this region, which it is well to note from an industrial point of view, is the existence of many falls on the rivers and streams. These will no doubt be utilized with advantage in the creation of economical power when the country comes to be opened up.

Conclusion.

It was not expected, of course, that the parties would be able to make a thorough and exhaustive exploration of all the territory assigned to them, and the estimates here given of what has been reported are very conservative. Totalling up the figures here quoted, however, we have over 25,000 square miles of good fertile land, or over 16,000,000 acres, and 288,000,000 cords of spruce or other pulpwood. There are also numerous smaller areas, both of timber and land, which are not included in these figures, but which will all be available when the development of the country takes place.

Since the return of the exploring parties two of the surveyors have died of typhoid fever. Mr. Joseph M. Tiernan, who was in charge of party No. 6, died on 13th December last, before he had finished his report of the expedition, so that the information given as regards his district is confined to the reports of the land and timber estimator and the geologist. The decease of Mr. Walter S. Davidson, in charge of party No. 5, took place on the 20th January, and owing to the long illness which preceded his death he was unable to complete his report, which appears in a brief and imperfect form.

SUMMARY.

The following is a summary of the more important features of each of the exploration districts, as indicated by the reports received by the respective parties :

District No. 1.

This district comprised an area embracing 50 miles on each side of a base line run due east from the 198th mile post of the Algoma-Nipissing line across the district of Nipissing to the eastern limit of the Province. The exploration has shown that a large portion of this region is of the same general character and equally well suited for agricultural settlement as the townships around the head of Lake Temiscaming. The section traversed by the base line is clay and clay loam, and the same soil characteristics prevail over the greater part of the territory examined. In general, the land back from the rivers is low lying and marshy, and the impervious nature of the soil prevents filtration and promotes the growth of moss, with which much of the land is covered. The effect of this is to absorb moisture and retard evaporation and also to preserve the winter ice through the summer season, giving the country the appearance of unproductiveness. Nevertheless, the soil is rich and capable of cultivation with proper drainage. The land, which is or could be made suitable for farming coming within the area explored was estimated at one million acres, in addition to which the clay land along the Blanche River above Lake Temiscaming is stated to extend up to the country covered by the expedition. South of Lake Abitibi is a fine rolling area of clay loam. The land on the immediate shore of the lake is sandy, but a short distance inland the soil changes to a clay loam, with merely enough sand to render it light and workable. In the eastern portion of the district are extensive deposits of moss peat, the bogs reaching a depth of ten feet in many places. The peat taken from these bogs on analysis shows a high percentage of volatile combustible matter and fixed carbon, no sulphur, and only a trace of phosphorus with a low percentage of moisture and ash, which render it a valuable fuel. Midway between the base line and Abitibi Lake lies a ridge of rocky hills, three hundred to four hundred feet in height at some points, forming a watershed between the lake and James Bay. The rivers on the south side of this watershed are short and the means of travel correspondingly limited. There are two water powers of considerable magnitude on Abitibi River below the lake, Couchiching Falls, with about 6,000 horse power, and Iroquois Falls, with about half that amount. Upper Abitibi Lake covers an area of 190 square miles of which about 55 square miles lie in Quebec. Lower Abitibi Lake has a surface of 145 square miles. A comparatively small expenditure in lowering the brink of Couchiching Falls would reduce these lakes to about one-half their present area and improve the drainage of an immense tract of surrounding territory, though at the expense of a loss in head and storage for the water power at the falls.

There is little white pine timber north of the height of land, the trees being scattered and inferior in quality. Some small areas of red pine and some jack pine were met with, nearly all of these varieties found being south of Lake Abitibi. The best areas for pulpwood are on Low Bush and Circle Rivers, with their tributaries, where it is estimated that an area of 180 square miles will yield an average of seven cords to the acre or about 800,000 cords. Along Little Abitibi River between Harris Lake and the boundary the pulpwood is estimated at 750,000 cords. A belt reaching from Lower Abitibi Lake along the Abitibi River to Long Sault, eighty miles in length will average seven cords to the acre. There are also considerable pulpwood areas to the west and north of Lower Abitibi Lake.

The rocks to the north of the base line are of the Laurentian formation and to the south either of the Laurentian or the Huronian order. Numerous quartz veins were discovered near the contact of the two systems, but the appearance of the quartz was unpromising and analysis proved it to be of little value as regards gold or silver. The only deposits of a commercial value are the large peat beds.

DISTRICT NO. 2.

This district embraced fifty miles on each side of a base line run west from the 198th mile post on the boundary line between Nipissing and Algoma Districts to the Missinabie River, about 100 miles, and also the tract lying southerly along the Missinabie River up to near Missinabie Lake. Of an area of 7,800 square miles explored, about 1,800 square miles was water and swamp, and of the remaining 6,000 miles, 75 per cent. was found to be choice farming land, the surface in places rolling and the soil a rich friable clay and clay loam. The good land alternated with muskeg, not more than four feet deep, with a clay bottom. If the country were cleaned up a large proportion of the low, wet land could be made productive as pasturage. Of the territory explored 60 per cent, will yield on an averge 5 cords of spruce wood to the acre, in addition to other timber. The prevailing timber is spruce and poplar, there being no pine or hard wood. The spruce, especially along the river banks, attains a size which renders it valuable for square timber, and the poplar is large and abundant, particularly on the Mattagami River. Special acres examined would yield 20 cords of spruce, other acres would cut 15 cords of spruce and ten of poplar. Some of these, if all the timber growing on them were made into cordwood, would show 60 to 70 cords to the acre. Much of the tamarac seen was dead as this tree appears frequently to die after having attained a growth of about 20 inches, and owing to the slight hold of its roots on the clay soil, it is liable to be blown down.

The district is generally flat, with a gradual slope towards the north. Rock exposures are few and of limited area, the prevailing formation being the Laurentian, but isolated outcrops of the Huronian formation were discovered on the Mattagami and Opazatika Rivers. The district appears to be unfavorable to the production of economic minerals, with the exception of some localities where iron pyrites was found, which may be utilized in the manufacture of chemical pulp. The country presents excellent facilities for railway construction. No rock cutting would be necessary after the height of land had been passed, very little cutting and filling would be required, and owing to the level nature of most of the country, the gradient would be easy. Tamarac for ties and sand for ballasting purposes are to be had in abundance. The rivers and streams, more especially the Mattagami and Kapuskasing Rivers, furnish numerous valuable water powers with descents of from ten to twenty-five feet which can be utilized in the development of mechanical industry. The Missinabie Falls could be utilized to furnish the power for an electric railway, and were locks constructed at that point a water route north from Missinabie Station could be secured.

Generally speaking, the climate is similar to that of Manitoba, the weather in mid-summer being equally hot. No destructive frosts were experienced until September 27th, and rains were frequent but not excessive. The fur-bearing animals and larger game have considerably decreased in number of late years, and the fish in the rivers flowing northward are not so plentiful as on the southern slope.

DISTRICT NO. 3

District No. 3 included the territory from Lake Temagami northward to the Montreal River and upon the east and west sides of the Algoma Nipissing boundary line. Owing to the even distribution of waterways, this area is well drained and the prosecution of lumbering greatly facilitated. South of the height of land the country is largely unfit for agriculture excepting in the river basins. The territory lying between the surveyed townships of the Temiscaming region and the Montreal River and a belt on the west side of the river, comprising 500 square miles in all, is arable soil of fine quality. The Obabika and Sturgeon River valleys contain from 200 to 250 square miles of good agricultural land. North of the height of land two-thirds of the territory explored is fertile, a level tract of clay soil extending northwards along the Mattagami River.

The country north of the height of land has been largely ravaged by fire but some fine belts of timber have survived. On the southern slope there are extensive pine areas The territory east of Lake Temagami and extending south to the surveyed townships is all a pine forest with the exception of brulés. The country west of Lake Temagami as far as Sturgeon River is fairly well timbered with pine and the area between the Sturgeon and Wahnapitae Rivers and northwards to Smoothwater Lake is also largely a pine district.

The total area of pine-bearing land is estimated at 1,650 square miles with an average yield per mile of 1,750,000 feet B. M. giving a total of 2,887,500,000 feet, B.M.

The spruce is a good deal more scattered, the best timber being on the upper waters of the Sturgeon River covering an area of 125 square miles, where it grows largely intermixed with balm of gilead and poplar. The region north of the height of land is to a considerable extent a spruce and balm of gilead district, many of the trees being large Jack or banksian pine is found on both sides of the height of land. The pulpwood territory explored comprises some 4,500 square miles, the yield from which would average two cords an acre, making a total of 5,436,000 cords.

The rocks of the district belong to the Archean age, including both the Laurentian and Huronian formations. The former prevails in the western part of the district, the Huronian system extending from Peter Long's Lake eastward and southward throughout the basins of the Sturgeon and Montreal Rivers. Economic minerals were found in nine different places indicating the presence of gold, silver, copper, lead and iron distributed through the region explored as far north as Fort Matachewan. A number of valuable water powers were noted, more especially on the Sturgeon and Montreal Rivers.

District No 4.

No. 4. District No. 4 comprises both sides of the Kabinakagami River for 20 or 25 miles northward from Grasett Station on the Canadian Pacific Railway, a section of the Missinabie River above its junction with the Moose River not included in District No. 2, and 20 miles on the west side of the Moose River below the mouth of the Missinabie River. The land taken as a whole is level, rising slightly along the watercourses, where it is rolling. The soil is clay and sandy loam covered in the lower levels with boggy peat and moss varying from two to four feet in depth. The country can be easily cleared and for farming purposes the soil will be equal to the best in the older portions of the Province. Much of the area that is at present swampy will secure natural drainage when the country is cleared, owing to the incline of the land. The mixture of the clay forming the prevailing subsoil, with the surface soil, will prove rich and productive. The district is well watered by numerous rivers and streams some of which are well stocked with fish.

The district is heavily timbered with spruce and tamarac interspersed with other varieties. Owing to their density of growth the spruce and tamarac are for the most part too small for any other commercial use than pulpwood, their diameter not being proportioned to the height they frequently attain. In some places, however, they are of larger dimensions. The quantity could not be estimated. The spruce will yield in some localities 40, 50 and in one instance 60 cords to the acre, being especially fine in the country along the Kabinakagami River. There is also a heavy growth of spruce along the Mattawishguani River which will produce from 20 to 35 cords to the acre. The dense spruce and tamarac forests of the Moose River basin are of great value and cover an immense area. In the southern portion of the district there are some areas of red pine, but no large pine is found north of Lake Kabinakagami, the main portion of the territory explored being north of the pine limit.

The rock formations of the district are for the most part Laurentian, with so Huronian exposures, especially on the Kabinakagami River. Near the Missinabie Ri were found boulders of fine grey slate which cleaved readily, and which is also known exist in other localities. The peat found in the lower levels below the moss on the s face is inferior as fuel owing to the shallowness of the beds and the amount of mois it contains. Similiar soil to these peaty tracts, at Brunswick House on Missinabie I has been found capable of raising good crops.

DISTRICT No 5.

This district comprised the territory from Jackfish Station on the Canadian Pacific Railway to Long Lake, thence down the Kenogami River to Albany River, and down the Albany to Hudson's Bay. The party was not able, however, owing to the limited time at its disposal, to descend the Kenogami River further than Pembina Island. The territory explored extended from Pembina Island to Howard's Falls on the Kawakaska River, being 120 miles in length by about 25 miles in width, making an area of 3,000 square miles About half of this is good arable soil, which is not found in large continuous areas, but principally in the neighborhood of the streams. In the northern part of the district there is a good deal of muskeg and the flatness of the surface will be an obstacle in the way of drainage. Much of the district has been burned over so that the timber is not generally of a large size. About one-third of the total area is timbered, making 640,000 acres, half of which will, it is estimated, yield good pulpwood or timber. The trees growing along the river banks have usually attained a fair size. Inland the timber is generally small and scrubby. The best timber district is between the Kawakaska River and Lake Eskeganaga where extensive groves of spruce and tamarac up to 36 inches in diameter are found. The poplar which grows everywhere along the river is singularly free from "black heart" which renders it of value for pulpwood.

The rocks of this district belong to the Niagara series, the Laurentian formation obtaining to the north of Long Lake and northwest to the lower Kawakaska River. Huronian rocks appear on the northern portion of that stream, on the Little Long Lake River, in the Pic River country and elsewhere. Silurian strata and sedimentary rocks are found in the northern part of the district. No mineral deposits of economic importance were found. Iron pyrites occurs in considerable quantities on Pine Lake, but it carries only small traces of gold, nickel and copper. The most promising region is the country on the Kawakaska River below Wawong Portage. Samples from quartz veins here showed traces of gold, which further prospecting may disclose in paying quantities.

The climate is similar to that of parts of the North-west lying in the same latitude. Frost is unusual during the summer season and all the ordinary garden vegetables are raised without difficulty. Barley and oats can also be matured successfully.

DISTRICT No. 6.

District No. 6 extended from Ombabika Bay at the north east of Lake Nepigon, in the district of Thunder Bay, by Ombabika River to the Albany River, including the country tributary to that river, to the mouth of the Kenogami River. Although some portions of the territory explored are unfit for agriculture on account of their rocky or sandy character, there are considerable areas of fertile land. The valley of the Ogoke River is a wide, level tract of good clay soil but interspersed with smaller areas of sand. The upper portion of this valley is the most extensive and promising stretch of agricultural land met with. The lower section down to the Albany River is wet and contains numerous peat bogs, but as the land lies considerably higher than the river bed it could be easily drained and rendered suitable for cultivation. The total area of arable land 10 miles inland on each side of the Ogoke River for a distance of 140 miles is estimated at 1,500,000 acres. There is a comparatively small tract of black alluvial soil along the Ombabika River and some areas of good clay soil down the Kapikotongwa River. There are great quantities of excellent pulpwood throughout the district, the principal varieties being spruce and jack pine From the mouth of the Ombabika River to the Albany River the land, exclusive of brulé, will yield 38 cords to the acre, or a total of 56,346,400 cords The Ogoke River country will average 44 cords to the acre, making a total estimated output of 78,846,000 cords, being 135,194,400 cords in all from the territory tributary to these two watercourses.

The geological characteristics of the territory are mainly Laurentian, but the Ombabika route crosses a belt of the Huronian formation 10 or 12 miles in width. Silurian limestone, overlaid with beds of drift, prevails near the Albany River between the Ogoke and Kenogami Rivers. Traces of gold were found in the quartz veins in the Huronian rocks about Cross and Summit Lakes, the samples taken yielding sufficient gold to encourage further prospecting. Extensive water powers exist on the leading rivers, and the streams and lakes abound with fish. Pike, pickerel and whitefish are generally distributed, speckled trout are plentiful and sturgeon were caught in the Ogoko and Albany Rivers. The climate is much the same as that of the Temiscaming townships. No frosts were experienced until the 25th of September, and throughout October the weather continued fine and warm. All kinds of vegetables produced in temperate climates flourish at the Hudson Bay posts.

District No. 7.

District No. 7 lies immediately west of No. 6 and comprehends the northwestern section of Thunder Bay. The course of exploration was from Wabinosh Bay at the north-west of Lake Nepigon up Wabinosh River and across the height of land to Albany River, including sections of the Albany and Savant Rivers. The district is generally rocky and barren, and not adapted for agricultural settlement. Small tracts of arable land exist at the head of Lower Wabinosh Lake and along Highland Lake and River. Along Little Mud River the soil is fitted for root crops and fodder, but not for general agriculture. In other parts potatoes and hay can be grown. On Little Mud River, near Lake Nepigon, there are deposits of a fibrous, peaty character, but of no great depth or area. There is little timber of commercial value, and such tracts as are intrinsically valuable are practically unavailable on account of location, and the limited quantity of the different areas, except along Mud River, where large spruce and tamarac were found. The prevalent rock formation is Laurentian granite, but the Huronian order occurs at the northern end of Lake Nepigon, and there are indications of the same formations elsewhere, the contact between the two systems being irregular and difficult to follow. The only mineral of value discovered was at Poplar Lodge on the east shore of Lake Nepigon, where red jasper was found mixed with a silicious iron ore. Small game is fairly plentiful but the larger animals are scarce, owing to the barrenness of the country. There are some good water powers on the principal rivers.

District No. 8.

District No. 8 includes the region to the west of Lake Nepigon and the Nepigon River to Dog Lake, up Gull River to the north of Dog Lake and the land around Black Sturgeon Lake. While the country is not generally adapted for agricultural settlement, its principal characteristics being stone, rock and swamp, it comprises considerable areas of sandy loam which would make good farming land. These fertile tracts are, however, isolated by intervening stretches of rough and barren country. The district is largely timbered with spruce and tamarac but in many parts jackpine is predominant. In most sections where timber exists the yield is estimated at between 15 and 30 cords, taking all kinds. There are indications that the district is capable of development as a mining region. The two principal geological formations are the Keeweenawan series of rocks and the Laurentian gneiss, some forms of the former containing a considerable amount of magnetic iron and iron pyrites. The Huronian areas are small and occur in strips. There is an extensive iron ore deposit on the east side of Black Sturgeon Lake which has not yet been thoroughly examined and there are known to be other iron deposits elsewhere. A number of brine springs were found which formerly furnished the Indians with their supplies of salt. The rivers are all rapid with numerous falls available for water power. The large game and fur-bearing animals are diminishing in numbers, with the exception of moose and red deer which have only made their appearance in the district of late years and are not plentiful.

District No. 9.

District No. 9 covered the north east angle of Rainy River District, extending from Dinerwic on the Canadian Pacific Railway north to Lake Minnietakie, Lac Seul and Lake St. Joseph and along Rat, Root and English Rivers and Sturgeon Lake. There are some areas of good land but they are small and scattered, and most of the district is unsuited for successful agriculture. There is a great deal of rocky broken country and the soil in some places is scanty, the underlying rock being covered with sand, sandy loam or clay, the surface soil sufficing for the growth of timber. There is no pine excepting in isolated clumps. Spruce timber is thickly scattered throughout the territory but much of it is too small to be marketable, though on the higher land it reaches a good size. Jackpine prevails towards the south and poplar in the northern portion of the district but in low-lying areas the average size is small. If they were accessible these tracts would furnish a large supply of pulpwood and timber. The rock formations vary greatly, the lower part of the district being almost entirely Laurentian but changing to Keewatin near Sturgeon Lake, which lies in a narrow belt of these rocks. A number of quartz veins occur on and near Sturgeon Lake, some of which carry free gold, and a mine located on King's Bay is being actively developed. Gold-bearing veins, the assays of which give encouraging results, have also been discovered at Abram Lake, where a good deal of prospecting has been done. Galena is found in the Minnietakie Lake region where an attempt to develop it is being made.

District No. 10.

District No. 10 embraces the north-western portion of the Rainy River District, including the territory between Lake Wabigoon and the western part of the Province north of the Canadian Pacific Railway, to Lac Seul and English River. Owing to the rocky character of the greater portion of the area embraced in this district the opportunities for agricultural settlement are limited. There are large tracts of clay soil on the Eagle and Wabigoon Rivers, and on the English River above the Mattawa River, which, judging by the natural vegetation they support, will prove available for cultivation. The timber on the English River and its tributaries is exceedingly valuable, especially the spruce and poplar, which are sufficiently large in diameter to yield many million feet of lumber, in addition to a vast quantity of smaller timber suitable for pulpwood. The logs can be driven to the mouth of the Wabigoon River for manufacture. Large quantities of spruce and poplar are available in the Wabigoon River region, where there is about 3,500,000 feet of red and white pine which could be taken to the mouth of Canyon River. The timber on the banks of the Winnipeg River and its tributaries, including the waters of the Black Sturgeon, Swan and Sand Lake regions, comprises a large quantity of good poplar and spruce, and some red pine. The geological characteristics of the district are mainly Laurentian, and as is usually the case with that formation, the rocks of that order include no minerals of value. In the Huronian areas on Linklater Lake and lying between Boulder and Lacourse Lakes the outlook is more encouraging. Quartz veins are frequent in these localities, and may be found on close examination to contain gold in paying quantities. Game is plentiful throughout the district, including moose, caribou and red deer, the latter being a recent arrival, but protection against indiscriminate slaughter is urgently required to prevent their extermination. The numerous falls and rapids on the rivers provide valuable water powers.

SURVEYOR'S REPORT

OF

EXPLORATION SURVEY PARTY No. 1.

TORONTO, November 16th, 1900.

SIR,—I have the honor to submit the following report on the surveys performed by me under instructions from your department, dated 12th June, 1900 :—

The object of the work entrusted to me, as set forth in those instructions, was the exploration of the comparatively unknown part of the District of Nipissing, bounded on the north by the Great Muskeg, adjoining the southern shore of James' Bay, and on the south by the southern water-shed of Abitibi Lake. In this exploration was contemplated the acquiring of all obtainable information regarding the topography, the nature and extent of timber lands, the soil and its capabilities, the minerals, water-powers, and water-ways, the flora and fauna of the country, in brief, all the natural and possible resources of the territory included, were to be noted. A further object was the acquirement of necessary geographical information.

For the purpose of systematic exploration, the instructions required the establishing of a permanently marked base-line formed by six mile chords of a parallel of latitude passing through the one hundred and ninety-eighth mile post, north from the north east angle of the Township of Lumsden on the west boundary of the District of Nipissing, and extending across that district to the approximate inter-provincial boundary, that boundary not being, as yet, defined upon the ground.

Two exploring parties, each composed of a timber and land estimator, a geologist and a canoeman, were attached to the usual survey party, the duties of these exploring parties at this stage of the work being to examine the country on either side of the base line, as far north and south therefrom as was compatible with the keeping up of communication with the main party during the progress of the line survey, all available water-ways being utilized for the purpose of penetrating the interior. Messrs. P. F. Graham Bell of Toronto, and Thomas G. Taylor of Gravenhurst, were appointed as timber and land explorers, and Messrs. R. W. Coulthard of Toronto, and M. B. Baker of Stratford, attended to the geological department.

Upon the completion of the base line and examination of the adjacent country, the Department directed a micrometer survey to be made of the shores of Upper and Lower Abitibi Lakes, and of Abitibi River from the outlet of the latter lake down the stream to some suitable point in the vicinity of Iroquois Falls, whence a compass "tie line" was to be run west to connect the river and lake survey with the west boundary of the District of Nipissing. These surveys together with available water-ways were to be utilized for the exploration and examination of the surrounding country by the exploring parties.

Agreeably to instructions, the party was organized as speedily as possible, and on the 18th of June I left Toronto, accompanied by the explorers and assistants from this part of Ontario, and on arriving at Mattawa, several additional men were engaged, the remainder of the party being secured at North Temiscaming.

By prior arrangement, the necessary supplies had been shipped in advance from Mattawa, and preceded us to Quinze Lake. A brief description of the route may be of interest. Our journey from Mattawa was by Canadian Pacific Railway to Temiscaming Station at the foot of Lake Temiscaming—a distance of thirty-nine miles—and thence by steamer to North Temiscamingue at the head of the lake of that name. About seventy-five miles from this point, which is near the inter-provincial boundary, the canoes were taken by way of Quinze River to Klock's Depot, those of the party not required for this service, going over the Portage Road to the Depot, about sixteen miles of lumbermen's wagon road connecting these points. The supplies were taken by wagon over this road.

From Klock's Depot about ten miles of canoeing, easterly and northerly, brought us to the mouth of Riviere Barrier, up which stream at a distance of nearly three miles, a portage on the east side of fifteen chains in length is necessary to pass rapids. At the north end of this portage Lac Barrier is met with. An ancient Indian practice of constructing dams at the outlet of this lake, for the purpose of facilitating their fishing operations, is said to have been the origin of its name. The length of Lac Barrier is about fifteen miles to the point where a small river, called Lonely River, enters it. Following the winding course of this stream for a distance of eight miles we entered an arm of Long Lake. Passing through this arm longitudinally, through the main body of the lake, and keeping to the eastern arm of its upper end, in all about twenty-five miles, we reached the mouth of a creek. As this creek is too shallow for canoe navigation, a portage of about fifteen chains is necessary to reach a lake about one mile in length, at the farther end of which begins the "Height of Land Portage." Crossing this portage, which leads half a mile in a north-westerly direction, we came to a small lake forming the head waters of the James' Bay water-shed, though in high water, a part of its contents flows towards the Ottawa.

Beyond this lake, which is about one and one-quarter miles long, we entered a small winding creek, and followed its meanderings for about one mile to its entrance to Island Lake, so called from the number and variety of its islands. After a delightful trip of about ten miles on this lake and its northerly arm, we arrived at its outlet, where rapids and a fall necessitate the making of three portages within a mile, known as "The Three Carrying Places." Seven and a half miles farther down the river we came to a short portage to pass a fall, after which navigation to Upper Lake was uninterrupted, the total distance between Island and Upper Lake being estimated at twelve miles. Crossing Upper Lake north-westerly and north easterly for about six miles, we reached its outlet, and continued down the river for two and a half miles, whence a short portage to pass a fall of about twelve feet was made. Five miles farther down the stream we arrived at Abitibi Lake. and crossing to a peninsula two and three-quarter miles from the mouth of the river, reached the Hudson's Bay Company's Post, which was made the base of supplies for our season's operations.

A whole week had been consumed in transporting our five tons of supplies, camp equipage, etc., over the ninety miles lying between Klock's Depot and Abitibi Post through the Province of Quebec, with a force of twenty-five men, all told. The road traversed had been in almost constant use by the employees of the Hudson's Bay Company during the summer months since the advent of steamers on the Temiscaming and railway connection thence with Mattawa, and the portages have been much improved, good landing docks having been constructed where necessary.

Abitibi Post was established in 1755 by the Hudson's Bay Company, and has been continuously occupied since that date. It is situated on a picturesque point which extends into the lake from the eastern end. Up to the time that steamers began to ply on Lake Temiscaming the Post was supplied from Moose Factory, on James' Bay, whence the goods were laboriously conveyed up the river after the arrival of the annual vessel from England. The population at last census (1898), including Indians belonging to the Post, was 450 souls.

Having stored the bulk of the supplies at this point, we proceeded westward about thirty miles to the "Narrows" between Upper and Lower Abitibi Lakes, where a trader named W. F. Biederman has a store for trading with the Indians.

Passing through the "Narrows" about four miles in a northwesterly direction, we entered Lower Abitibi Lake, and continued in a general northwesterly direction about sixteen miles to the mouth of a river called Tapa saqua (Low Bush), which enters the most northwesterly bay of Lower Abitibi Lake.

Proceeding up this stream for three days, during which ten portages were rendered necessary by rapids and falls, we reached the height of land portage, about thirty-five miles from the lake. We next crossed this portage, which is about two and a half miles in length, and leads northwesterly to a small lake, named by us "Welcome Lake," about half a mile in length, and arrived at another portage about thirty chains in length, leading to a small lake about one and one-half miles long, which we named "Michel Lake." Having crossed this lake, we reached another portage of two and a half miles, which leads over a rise of more than three hundred feet. This brought us to another small lake, which we named "La France." Crossing La France Lake, about one and a half miles in extent, we entered Little Abitibi River, which is here a shallow, weedy stream, with moderate current and increasing depth of waterway. We followed down it and its expansions, without further portages, about ten miles to Little Abitibi Lake. We crossed this lake on a course slightly west of north six miles to its outlet. Half a mile down the stream a portage was made to pass rapids and a fall about eight feet in height. Another half mile brought us to a lake one and a half miles long by one mile broad, which we named "Williston Lake." Continuing down the stream for a mile and a quarter, we reached a lake five and a half miles long by three miles broad, which we named "Pierre Lake." This lake contracts at its lower end, and is divided by only a short "narrows" from Montreuil Lake, an irregular body of water about four miles in length on a winding course Another "narrows" at the lower end brought us to Harris Lake, which is about two miles long by two miles broad. Having decided that we were then about opposite the initial point of the base line to be run, we cached a part of the supplies, and turning westerly followed a small tributary stream, and thence by means of a series of small lakes and intervening portages, we reached the west boundary of the District of Nipissing, two and a half miles south of the initial point of the base line, on the 10th of July. On the following day, the course having been determined by astronomical observation, the work of running the base line was begun at the 198th mile post, it being marked by an iron post with a tamarac post standing beside it. From this point I ran due east astronomically on chords of a parallel latitude for seventy and three-eighths miles, the line being deflected six minutes north at every sixth mile post. Except where such point occurred in a lake or river, a wooden post was planted at the end of every mile, and an iron post three feet long and one and one-quarter inches in diameter at the end of every third mile, the number of the mile being marked on the west side of the post or posts. The wooden posts were made of the most durable timber to be found in the vicinity, and wherever practicable a mound of stones was erected about the posts, and bearing trees marked and noted in the usual manner. Where a mile terminated in a lake the post was planted on the line on the nearest land and marked with the number of the mile, plus or minus the number of chains or links. Astronomical observations for the purpose of verifying the course of the line were taken at short intervals. The magnetic declination was noted throughout, and found to be generally uniform at about eleven degrees west.

From the time of leaving Lower Abitibi, the exploring parties were active in examining the country traversed by our route so far as opportunity permitted, and from Harris Lake Messrs. Taylor and Baker proceeded down the Little Abitibi river to explore the country to the north of the proposed base line, Messrs. Bell and Coulthard having remained behind at Lower Abitibi lake to examine that region. Both exploring parties again connected with the surveying party at the thirteenth mile on the base line, and examined the district on their respective sides of the base line as far as the forty-fifth mile, when, finding from the unchanging character of the district traversed, it was useless to pursue their quest in that direction, they returned to Lake Abitibi, examining the geological features along its shores and from time to time penetrating the interior by means of the rivers met with while the survey of the base line was being completed.

The country crossed by the base line may for the purposes of description be divided at about the 25th mile post. The land for the first 25 miles is chiefly good clay, but the frequent occurrence of small lakes, muskegs and marshes detracts materially from its value for agricultural purposes. The surface is almost level and well watered by small streams of excellent water, the beds being sufficiently below the general surface to afford drainage when improved. Spruce and tamarac of a maximum diameter of 16 inches, but average of about 8 inches, are found in fair sized groves along the banks of streams and extending inland therefrom about half a mile. Beyond this distance the growth is stunted. Other timbers, including poplar, balm of Gilead, white birch and cedar are less abundant, and cannot be counted as of sufficient value for more than passing mention.

Along the remaining forty-five miles of the base line the surface of the country is similar in appearance to that of the first twenty-five miles, but becomes lower and more even towards the east; swamps and muskegs become more numerous and extensive, until, at the eastern boundry of the district, it terminates in a vast muskeg, where no timber, except an occasional scrub spruce or tamarac, is seen. Great difficulty was experienced here in transporting the necessary camp outfit, owing to the generally swampy and wet character of the country. In many cases it was necessary to build a flooring of poles, covered with boughs, to make the tents habitable. The soil continues a heavy clay throughout, but covered to a considerable depth by moss and vegetable matter. Soundings in many places showed a depth of ten feet of overlying vegetable matter. With the exception of the immediate vicinity of the streams and lakes, where belts of about a quarter of a mile in breadth of fair timber are found, the easterly forty-five miles of the line may be said to be devoid of timber of any commercial value, the growth diminishing until the confines of the "Great Muskeg" are reached.

The effect of at least seventy per cent. of the territory tributary to the base line being covered by a thick coating of moss has been to prevent drainage, and as a result of the protection from the sun's rays, the winter's ice is retained all summer. This serves to account for the retarded growth of timber remote from streams, and for the icy temperature of the water in the muskegs. In many places the accumulation of ages of moss growth has produced beds of peat. On the line between the thirtieth and fortieth mile posts, large areas of that fuel-of-the-future were found, and our geologists assure us that its quality is unsurpassed.

About midway between the baseline and Abitibi Lake, lies a ridge of rocky hills, in many places three hundred feet in height, forming the water-shed between the waters of that lake and of James' Bay proper. This ridge is broken by sandy plains, muskegs, and numerous small lakes. The rivers on either side of the water-shed are necessarily short, and the means of penetrating the country to any considerable distance correspondingly limited.

Having completed the base line, we returned as far as Burnt Bush River, near the centre of the fifty-first mile, whence I made a track survey down that stream to its junction with Hannah Bay River, and thence up the latter stream to the height of land. whence by portage of nine miles I reached White Fish River, of which I made a track survey to the mouth, about five miles north-east from Abitibi Post.

Burnt Bush River, where it crosses the base line, is about two and a half chains in width and runs in a southerly direction far about twenty miles, to the point where it is joined by Mud River, running north-easterly, from which junction it runs easterly and north-easterly about twelve miles to where it empties into Hannah Bay River running northerly. About fifteen miles further up the latter stream, the inter-provincial boundary is crossed.

Along the banks of all three of these rivers, poplar, spruce, balm of Gilead, tamarac, balsam, white birch and banksian pine to a maximum diameter of twenty inches, and an average of about eight inches, grow in abundance; at distances varying from five chains to half a mile from the streams the timber becomes stunted, and beyond that limit diminishes.

The soil is generally clay and clay loam with occasional sandy ridges.

The river banks vary from four to ten feet in height, while the general surface inland is lower than the river banks. The spring freshets carry down great quantities of ice to be stranded on the banks as the water recedes. As the ice melts under the action of the sun, deposits of mud, small timber, etc., are released, and the banks are thus slowly but surely built up. This, in a measure, accounts for the lack of good drainage from the surrounding swamps.

The same is true also of nearly all the rivers and larger streams flowing through flat country in the Abitibi region.

Having reached Abitibi Post, at the end of the track survey above described, I discharged eight men not required for the latter work, made the necessary arrangements for supplies, and proceeded to rejoin the remainder of the party at Biederman's.

Leaving Biederman's on the 1st of September, we journeyed westward to the outlet of Lower Abitibi Lake, and thence down the Abitibi River to the Iroquois Falls, one of the exploring party branching off at Black River, and the other continuing down the Abitibi to explore the territory between Iroquois Falls and the district boundary. At a point about a mile down the stream from Iroquois Falls, I ran by compass a tie line due west, a distance of eighteen and a half miles to connect the river at this point with the district boundary. In the first four miles of this line the soil was a fine clay loam with rolling surface. The timber included spruce, poplar, tamarac and balm of Gilead, with a few balsams, birch and banksian pine. The fifth and sixth miles were broken by rocky and sandy ridges. Banksian pine, birch, poplar and occasional balsam and balm of Gilead comprised the timber met with in this stage. The seventh, eighth and ninth miles passed through continuous swamp and muskeg, with none but scrub timber. Early in the tenth mile a rocky ridge occurred and extended for half a mile. The remainder of the line passed through flat clay country with occasional sandy ridges, spruce, tamarac, poplar, white birch and occasional balsam comprised the chief timber. Some of the poplar and balm of Gilead here attain a diameter of sixteen to twenty inches.

Arriving at the boundary between the Districts of Nipissing and Algoma, which we intersected in its 143rd mile, we retraced our steps to Abitibi River, up which we made a traverse with compass and micrometer to Lower Abitibi Lake. Next taking the north shore, I made a traverse of Lower Abitibi Lake with transit and micrometer. By this time the season was so far advanced that to have remained to make a traverse survey of the shore of Upper Abitibi Lake would have detained us beyond the closing of navigation. We therefore contented ourselves with running a traverse line from island to island, on as direct a course as possible, to connect the lower lake survey from Biederman's at the "Narrows" with the Hudson's Bay Company's Post at the eastern end of Upper Abitibi Lake, planting in suitable places at each end of such line an iron post one and seven-eighth inches in diameter and three feet in length and thus concluding the survey for the season.

By survey of tie line and calculations by latitudes and departures the following distances are obtained:—

Difference in longitude between west boundary of Nipissing and east end of tie line	18m.	36c.	38lks.
Difference in longitude between west end of tie line and outlet of lake	23m.	41c.	10lks.
Outlet of lake to Biederman's	11m.	66c.	44lks.
Biederman's to Abitibi Post	26m.	62c.	89lks.
Total	89m.	46c.	81lks.
West boundary Nipissing to east bounday of Ontario	71m.	12c.	....
Difference equals distance of post east of east boundary of Ontario	9m.	34c.	81lks.

On each of the three occasions on which I visited Abitibi Post the weather was unfavorable for astronomical observations and I therefore quote the latitude from observations taken by Mr. William Ogilvie ten years before, viz.:

Latitude	48 39' 32"
By oaculation the latitude at Biederman's is found to be	48 43' 48"

During the whole course of the tie line and traverse surveys the explorers were engaged in examining the country to the west, south and north of Lower and Upper Abitibi Lakes, where they report large areas of good clay land and a fair amout of pulpwood timber. As in the cases already mentioned the timber was found to be much better along the banks of streams than in the interior.

A brief summary of the results of the season's work may be given as follows:—

TIMBER.

White pine to the north of the height of land is found in only a few places, and where seen, consisted of scattered trees of inferior value. Here and there small areas of red pine occur but are not of considerable importance, the largest being on the south shore of the outlet of lower Abitibi Lake, where the quantity is estimated at 100,000 feet board measure. On Lake Montreuil another small tract of 50,000 feet occurs. Long Point in Lower Abitibi is estimated to have 60,000 feet of red pine, together with about 800,000 feet of Banksian pine, running about forty feet board measure per log.

Pulp-wood timber of fair quality, though small in size, occurs in many places, the most valuable tract being on Low Bush River and Circle River, with their tributaries Mr. Graham Bell estimates this at 806,400 cords. Along the Little Abitibi River between Harris Lake and the District Boundary, Mr. Taylor reports 750,000 cords of pulp-wood. He also reports considerable areas of pulp-wood to the west and south of

Lower Abitibi Lake. For details of timber found by the explorers, reference may be had to the reports of Messrs. Graham Bell and Taylor, accompanying this report. As to the present value of the pulp-wood of the territory little can be said. When, in the future, railway communication is opened some point on the Moose River may be utilized for the collection of the raw material, or, the water powers at Couchiching and Iroquois Falls will furnish energy for manufacturing pulp and paper. As a large percentage of the pulp timber consists of poplar and balm of Gilead, the difficulty of floating for long distances will have to be considered.

Soil.

As already stated, the section traversed by the base line is clay and clay loam. The same is true of the greater part of the whole territory examined. In general the land, beginning at a distance varying from a few chains to two miles back from the rivers is low-lying and marshy, the impervious nature of the clay preventing filtration, and thus promoting the growth of moss. This moss absorbs moisture in such large quantities that evaporation is retarded, and the winter's ice is preserved throughout the summer. The effect of these conditions is that an immense territory with, generally speaking, good rich soil, is rendered comparatively unproductive in a climate which, so far as our observation and information extend, offers no serious obstacle to as successful farming as Manitoba is capable of. We experienced but two frosts during the entire summer. At Abitibi Post potatoes were planted on the 29th of May of this year, and taken up on 27th of September. The yield was satisfactory and quality good. Oats ripened, and a good harvest was reaped. Timothy hay is grown and thrives.

Within the scope of our cursory examination, during the season, the land which is, or, with proper drainage, could be made, suitable for farming, amounts to at least one million acres. In addition to this, we are informed by residents in the vicinity that the well known clay soil found on the Blanche River above Lake Temiscamingue, extends all the way up to the land above described.

From this it will be seen that the agricultural resources of Ontario are capable of expansion to an extent hitherto little dreamed of.

Minerals.

The geologists accompanying the expedition report no minerals of commercial value Gold-bearing quartz in small veins was seen, but the highest assays from picked samples gave only $2.20 per ton.

Kaolin also was noted in various places, but not in large quantities. The extensive deposits of moss peat, before referred to, would seem to be the most valuable resources discovered.

The Huronian formation extends from the southern limit of the territory examined to a considerable distance north of the Abitibi River and Lower Abitibi Lake; and again appears near the intersection of Little Abitibi River by the District Boundary, from which point it stretches to the north as far as the season's explorations extended.

From the north shore of Upper Abitibi Lake, and following the northern limit of the Huronian tract first referred to, the Laurentian formation includes all the remainder of the territory examined. The exact lines of contact are, however, difficult to follow, owing to the rarity of rock exposures.

Water Power.

The only water powers of considerable magnitude met with are the two on Abitibi River, below Abitibi Lake. The first of these, at Couchiching Falls, is roughly estimated at about five thousand horse-power, and the second, at Iroquois Falls, at rather morethan half that amount. Both are included within surveyed locations, indicating that application has already been made to your department for the right to develop them.

Abitibi Lakes and River.

These lakes, locally known as Upper and Lower Abitibi Lakes, are connected by "Narrows" about two miles in length, with a minimum breadth of about two hundred yards. The area covered by the Upper Lake is approximately one hundred and ninety square miles, of which about fifty-five square miles lie within the Province of Quebec. The length from the narrows to the eastern extremity is about thirty-one miles, and the breadth from north to south shores varies from three to eighteen miles. For a lake of its dimensions, it is remarkably shallow, from four to ten feet being the prevailing depth throughout. Owing to its shallow nature, ordinary winds make canoe navigation difficult and but for the shelter afforded by the islands travel would be extremely dangerous. The water being impregnated with clay, the winter's ice reaches the bottom of the lake for a considerable distance from the shore in the shallower parts. It is said to be customary for the Indians at the Post to be obliged to travel a distance of five to six miles in order to reach water of sufficient depth for fishing in mid winter. Notwithstanding this fact fish are abundant. This lake has a very irregular shore line of about one hundred and fifty miles, roughly estimated. Nearly fifty per cent of the shore line is rocky, another twenty-five per cent. boulder-strewn, and the remainder about equally divided between sand and clay beach, the general height of the banks being from four to ten feet. Islands innumerable, of all shapes, and varying in size from a few square feet to about six square miles, are scattered all over the lake, giving it a natural beauty not excelled by the far-famed St. Lawrence. A range of hills three hundred to four hundred feet in height, beginning two to three miles from the southern shore, affords a grand view of the lake. From one of these hills two hundred islands were counted. On a few of the larger islands much of the original forest has been swept away by fire and succeeded by second-growth timber. Where this has not occurred, the usual mixed timber prevails, and in some instances small areas of good pulp timber are found. Considerable areas of fair agricultural land were seen on several of the larger islands.

Lower Abitibi Lake covers an area of one hundred and forty-five square miles, and has a shore line one hundred and fifty miles in length. Its greatest breadth from east to west is seventeen miles, and from north to south nineteen miles. The general depth slightly exceeds that of the Upper Lake. From the centre of the south shore a sandy peninsula, named Long Point, extends about seven miles into the lake, and includes an area of probably twenty square miles. The proportion of rock boulders and beach on the whole shore line is about the same as estimated for the Upper Lake. Islands are less numerous, the more important being in the northern part. As in the other lake, the larger islands contain much agricultural land of fairly good quality.

The official meteorological records kept at the Post show the following dates of the opening and closing of navigation on these lakes :—

Opened.	Closed.
1897 (not on record).	8th November.
1898, 11th April.	27th October.
1899, 24th April.	11th November.
1900, 21st April.	(Not yet received).

Abitibi River, which discharges the waters of the lakes above described, leaves the Lower Lake from a point on its southwestern shore and flows westerly for a distance of about twenty miles to the junction of Misto-ago River from the north. It then takes a northwesterly course for about nine miles, where it is met by the Black River, from the south. From this point it runs in a general northwesterly direction, crossing the district boundary near the one hundred and eightieth mile post.

The general breadth is from two hundred and fifty to three hundred feet, with depth varying from four to ten feet, where not interrupted by rapids or falls. At the time we visited it, the water was about four to eight feet below the general level of the banks, which were composed of clay. The most important falls are Couchiching, which occurs about four miles below the river's mouth, and Iroquois, about thirty miles farther down. In the former the main chute is about thirty feet in height, but supplemented by about ten feet more in the rapids below that cataract. Iroquois Falls has a descent of about fourteen feet. Rapids and other falls of minor importance occur frequently throughout the whole course of this river, and at a point about eight miles above the district boundary Long Sault Rapids is said to begin and to extend a distance of five miles. Not having been able to reach Long Sault, I cannot speak from my own knowledge, but the total descent is reported to be about seventy feet in that distance.

With regard to the feasibility of a plan for reclaiming a part of the lands covered by the waters of Abitibi Lakes, I may say that, in my judgment, a comparatively small expenditure in the lowering of the brink of Couchiching Falls would reduce those lakes to about one-half their present area. Whether the land so reclaimed would be found to be valuable is an open question. The loss in head and storage for the valuable water power at the falls should also be seriously considered. On the other hand, the improved drainage facilities for an immense area of land surrounding the lake to be afforded by such a work should not be lost sight of.

Game and Fish.

The fur-bearing animals of this territory include moose, cariboo, red deer, bear, wolf, lynx, fox, beaver, otter, marten, fisher, rabbit, mink and muskrat. Of these, wolf, mink, rabbit and fisher are scarce.

The feathered tribe includes duck (chiefly black duck and redhead), loon, crane, partridge, hawk, owl, and many small birds. Fish were found in abundance, among the varieties being pike, pickerel, white fish, tulabie, white and red sucker, and, below the falls, sturgeon. Botanical specimens were taken at various points throughout the course of the work, and are returned herewith together with notes as to their location.

.I regret that the brevity of the season precluded our covering as much ground within the allotted territory as the instructions contemplated, but a complete examination was out of the question.

The uniform courtesy and assistance accorded by the officers of the Hudson's Bay Company at all points touched did much to further the success of the expedition, and should not be passed over without special mention.

Accompanying this report I beg to submit

(*a*) A general map on scale two chains.
(*b*) Field notes of the base line.
(*c*) Field notes of the tie line.
(*d*) A traverse plot on tracing cloth, showing traverse work.
(*e*) Special report on land and timber, by Messrs. Graham Bell, and Taylor.
(*f*) Special reports on geological features, by Messrs. Coulthard and Baker, with maps.
(*g*) Photographs and negatives, etc.
(*h*) Accounts and vouchers in triplicate.

I have the honor to be, Sir,
Your obedient servant,
(Sgd) T. B. SPEIGHT,
Ontario Land Surveyor.

The Hon. E. J. Davis, Commissioner of Crown Lands, Toronto.

LAND AND TIMBER ESTIMATORS' REPORTS ON EXPLORATION SURVEY PARTY NO. 1.

Toronto, November 30th, 1900.

Sir,—In connection with Exploration Survey Party No. 1, under your charge, I have the honor to report to you, that according to instructions received from the Crown Lands Department on May 31st, 1900, I attached myself to your party on June 18th, 1900, when we left Toronto for Lake Abitibi and that portion of Northern Nipissing that we had to explore.

After a very pleasant journey we reached the Hudson's Bay Post on Lake Abitibi on the morning of June 28th, and were most kindly received by the officers of the Company. We left the post at noon and started down the lake. After going about fourteen miles, a heavy sea got up so we had to go to one of the islands for shelter. Here we were wind-bound for three days. The following day we made down to the foot of Upper Lake Abitibi, and down the river and to the far end of Lower Abitibi Lake, and about eight miles up a river that flows into Lower Lake Abitibi at the north-west end. This river i known as Tapa-saqua-sibi (Low Bush River) and is about four chains wide and quite deep. The water is cooler than the lake water, and not quite so muddy. The river flows from the north-west and is about thirty-eight miles long, with about twelve small portages in it, just over small rapids which would not interfere with driving timber. I went inland in different places and found the timber chiefly spruce, poplar, balm of Gilead, balsam, and tamarac. There are a few white spruce trees on the banks of the river. Most of the spruce, however, is of the black variety and scrubby in growth. Towards the head of the river we ran into jack pine, very small and not large enough for logs. The poplar is the largest timber, some few trees running up to sixteen inches in diameter, but these, as a rule are faulty, being full of punk knots and covered with fungus, the larger portion of the poplar however, is about seven inches in diameter and fairly sound. The tamarac all through this country is dead. I learn from the Hudson Bay Company's officers that about eight years ago worms got into the timber and killed all the trees between here and Three Rivers, Quebec. Certainly, in all the country I have been over this summer, there are very few green tops, and these have been small trees. The balsam attains a maximum of eight inches diameter, and will average four inches diameter. There is no great quantity of balm of Gilead and what there is is faulty. The white birch is small and very crooked, being only fit for firewood.

I made a careful estimate of the timber along this river, and would judge that for the first twenty-five miles of the river from the lake and for a distance of half a mile on each side of the river, one could cut three hundred and fifty-two thousand cords of pulp wood, poplar and spruce, five inches in diameter and larger and for a very large area behind this, the timber will run four cords of five-inch pulp wood, chiefly spruce, to the acre. Above the twenty-five miles to the head of the river there is no timber to speak of. Here we came on heavy clay rolling land covered with very open bush of small jack pine and spruce. This had evidently been an old brule. This brule runs back sixty chains from the river and falls away into a spruce swamp which is very wet. The average size of the trees in the swamp was four inches, diameter. All the tract along the river for a distance back of say one mile, would make fine farming land. As a rule, it is perfectly level, the soil being heavy clay, and could be easily drained into the river by the numerous small creeks running inland. Flowing into this river at about one mile from the mouth, is Circle River which runs from the north-east and is about two and one-half chains wide. I went up this river for about twenty miles when I got stopped by rapids and low water. I think that for a distance of about seventeen miles up this river and a distance of a quarter of a mile in from its banks, one should get one hundred and nineteen thousand, six hundred and eighty cords of pulp wood (spruce and poplar). Beyond this, the timber will only cut four cords to the acre, all spruce. For the last three miles of the river, the banks are much higher and covered with jack pine. At the rapids I went inland on a jack pine ridge and found the timber from five inches to seven inches diameter and limbed to the ground. All up this river is fine farming land, being heavy clay soil which can be easily drained. There is very little current in this river, and the banks are low till you ascend the stream for fourteen miles. There are two smaller streams entering this river which I went up for a short distance, and found the water in them very low. The timber is just the same as on Circle River. that is to say, fair timber for a few chains in from the bank, and after that swampy land with clay bottom and much smaller timber. I would judge that there might be eight hundred and sixty-four thousand cords of pulp wood (four inches in diameter and over) brought down to the mouth of Low Bush River with a draw of not over two miles. About one and one quarter miles south-west of Low Bush River there is another river entering the lake about two chains wide, with banks very low and a sluggish current. For the first two miles up this river, the timber will cut twenty cords of pulp wood per acre for twenty chains on each side, and for the next eight miles the only timber is poplar, which grows just on the river banks, not running back more than two and one-half chains, and will cut about ten cords to the acre. Behind this is all an open spruce and tamarac swamp with dead trees. There is about sixteen thousand cords of pulp wood, six inches in diameter, on this river. The land is level and the soil heavy clay, with only about four or six inches of moss on it. There is a large area here that would make first-class farming land. This river we called

Dokis River, on account of an Indian by that name who hunts on it. After we ascended it for ten miles we were stopped by fallen timber, too much to cut our way through.

The next largest river after Low Bush River, flowing into Abitibi Lake on the north shore in the Province of Ontario, is Rotten Jack Pine River, or, as it it is sometimes called, Moose Factory Road River. This flows into the deep bay on the north side of the Upper Lake Abitibi, and is about seventeen miles long and five chains wide, with only one small rapid on it, up which we had no difficulty in paddling (although we were told before we started that we would need poles to go up it). The timber up this river is almost altogether second-growth poplar, spruce and jack pine, most of it running from one inch to four inches in diameter, though on a few points or bends in the river good-sized poplar, balm of Gilead and a few white spruce are met with, but not in quantities to be considered of any commercial value. Near the forks of the river a jack pine brule is met which extends three or four miles on each side of the river. I went inland at the forks for about two miles north and found only jack pine, very scrubby, and not over eight inches in diameter, and limbed to the ground, and small spruce of no value. The land all the way up this river is good and rolling and would make splendid farms; the soil is heavy clay. I learn from an Indian who hunts here that the land is similar for six or seven miles back, when small lakes are met with, and after that low swampy land with sandy bottom. This is just before coming to the mountains, where the timber gets smaller. As you go back on the other side of the mountains there is no timber to speak of, and the land is mostly muskeg. There are several other rivers flowing into the lake on the north side, which we explored and went inland from them from points as far up as we could get. Most of them are about two chains wide at the mouth and have very little current. After ascending them seven or eight miles they simply dwindle away to nothing. The timber on the banks for from two to five chains is fair, mostly averaging eight to ten cords per acre, but beyond this the land falls away to flats and swamps, with much smaller timber which would only average about three or four cords of four-inch pulp wood per acre. The timber on the north shore of Lake Abitibi is very small and of no commercial value, chiefly small poplar, spruce, white birch and balsam, and a few balm of Gilead and some scrubby cedar and jack pine. A few small ash and elm trees were met with on two or three points, but none of any size. The timber back from the lake is chiefly small spruce, from three inches to seven inches in diameter, and quite one quarter of it dead. All the tamarac is dead, which gives a most dismal appearance to the whole country. The land is mostly flat, with clay bottom. I think there will be sufficient area of good clay land fit for farming to make several townships six miles square on the north shore of the lake.

From the headwaters of Low Bush River to the first lake met with over the portage, the land is rolling and the soil clay and timbered with small jack pine and spruce from four to eight inches in diameter. This runs for about one mile. There is about one mile of swamp timbered with spruce four or five inches in diameter, but very scattered and scrubby. For the last half mile the land is sandy and timbered with spruce and jack pine that will cut fifteen cords of pulp-wood to the acre. Between the first and second lakes, there is a portage of half a mile. The timber around the shore is spruce and poplar with a few jack pine and will run seven cords per acre. Here we found some soft maple underbrush. The water in these lakes is clear. From the second to the third lake there is a portage of two and one half miles, chiefly over rocky hills, timbered with small jack pine and scrubby spruce. The highest point on this ridge met with was three hundred and five feet above the third lake, which we called Lac la France. For forty chains south of this lake there is good farming land though rather hilly. The soil is clay, timbered with good poplar, spruce and balsam that will cut twenty-six cords of pulp-wood per acre. Along this valley, that runs for about three miles the timber is all the same. I went back over the second rocky bill, which is covered with small scrubby jack pine and spruce, and came on some good rolling clay land with fair spruce and poplar, which would average fifteen cords of pulp-wood per acre. This only runs about half a mile. Two and a half miles further back, that is to say to the west of the portage track, the land is very rocky with pockets of from twenty to thirty acres of good land between the hills. In these pockets the land is mostly sandy with a clay sub-soil; the timber on them is chiefly poplar and spruce which grows very tall. On the west shore of Lac la France the chief timber is poplar, spruce and balsam and some balm of Gilead along the lake shore. For one and one half miles from the shore the poplar and spruce will cut twenty-six cords of pulp-wood per acre. After this you come on a swamp with scrubby spruce, cedar and black alder bushes, which runs for about forty chains, when we arrive at a rocky ridge which rises three hundred feet. The only timber on the ridge is scrubby jack pine and spruce. We got a good view of the country from this hill, and could see another range of hills away to the north. The land between here and Little Abitibi Lake is very flat and timbered with spruce. We could see several small lakes but no great body of water. After we walked for a mile and a half over this ridge we came to low land, soil clay, timbered with small spruce and poplar, averaging four inches in diameter or seven inches maximum, which will cut about four cords of pulp-wood per acre. These trees are not over twenty-five feet high.

From Lac la France to Little Abitibi Lake, the country is very flat, the timber being small and of no commercial value. At the south-west end of Little Abitibi Lake I went into the bush due east and found it a nice gentle rolling clay land with a covering of moss about eight inches deep with a clay sub-soil. The timber here is small, the maximum diameter twelve inches, spruce, balsam and birch averaging six inches, and will cut about ten cords of pulp-wood per acre. The walking is very bad as there is so much fallen timber. Along the west shore of Little Abitibi Lake, the land is low and timbered with spruce, too small for pulp-wood. In the west bay of the lake we found a river (Louis River) flowing in from the south-west. It is about three chains wide at the mouth and twelve feet deep with very little current. Three miles up it is only two chains wide and ten feet deep. Four miles up the river branches, one branch one chain wide, going south-west, the other, one chain wide, going south-east. We only got up these branches for a little over a mile when we were obstructed by logs. There is very little timber up this river, and most of the country is covered with a second growth of poplar from one to three inches in diameter. The land is heavy rolling clay and could be easily drained. Up this river we saw a great number of moose and bear tracks. At the far end of the west bay, the country is all swamp, covered with dead tamarac. Going north along the west shore for a distance of one hundred yards from shore the timber will cut twenty cords of pulp-wood per acre, and behind this will cut five to eight cords for half a mile. There are some white spruce trees along the shore of this lake, running up to twenty-eight inches diameter on the stump, but they are so few and scattered that I did not think it worth while going round to them all. I would judge there were about thirty trees that would run on an average twenty inches diameter. The land close to the shore is a little sandy, but back a few yards is heavy clay. The timber on the river running north out of this lake is good and will cut thirteen cords per acre, chiefly spruce and poplar, though the poplar is faulty. The timber round Pierre Lake is very small, scrubby popular and spruce not fit for pulp-wood. The land back from the lake is very flat and wet with clay soil.

On Montreuil Lake, spruce, balm of Gilead, poplar, balsam and white birch are the only trees, and these as a rule are under five inches in diameter. At the north end of Montreuil Lake we found two hundred and fifty red pine trees that would average two logs to a tree, fourteen inches in diameter, sixteen feet in length, containing 50,000 feet B.M. From Montreuil Lake to where we found your base line, the timber is all small, and along the line from the twelfth mile post to the twenty-fourth is similiar, and will only average three cords of six inch pulp wood to the acre. I went inland to the south of the line in several places between the twelfth and twenty-fourth mile posts, and found the land all the same, with the exception of an area of about five hundred acres that had no timber on it whatever, covered with short grass and very wet. Here I could put a pole down nine feet before striking clay. Several other stretches of over a mile wide were met with timbered with very low open scrubby spruce, not over four inches in diameter and the highest of them about fourteen feet in height. All through the country here, there are occasional white birch, poplar, balm of Gilead, and balsam trees. From the twenty-fourth mile post to the thirty-second I was following up a waterway. To the south of a line along this waterway and up from it to the thirty second mile post the timber is too small for pulp-wood, it being chiefly small spruce, poplar, balsam and a few balm of Gilead and white birch. The land is very low and wet with clay soil. It would be very hard to drain as the banks of the lakes and rivers are so low. From the thirty-second mile post to the forty-fifth mile post the land as a rule is very flat and wet, though there are a few sandy ridges of from ten to twenty chains wide, running north and south with spruce on them that would cut from seven to ten cords per acre, but the average on the whole is about three cords per acre where there is timber. Your plan will show how much of the land is muskeg, and to the south of the line I found a good deal more swamp and muskeg than on the line. At this point we decided to return to Lake Abitibi. As the land all round here was only muskeg and swamp we thought we could employ our time to better advantage and get more useful information for the department by continuing along the line on Abitibi River and south-west bay of Lake Abitibi.

Abitibi River flows out of the lake at the south west end, and is about one hundred yards wide at the mouth with a strong current. At the mouth of the river on the north side the timber is chiefly poplar, spruce, balsam and a few white and red pine containing about five thousand feet B.M. and some white birch. At a distance of one and one half miles in a north-westerly direction a large sandy and gravelly ridge was met with, which had been timbered with jack pine, but has been burnt over about three years ago. I walked for about one mile across this to a point where the land runs down into a spruce swamp, On the skirts of the brule there are about ten thousand feet B.M. of jack pine from nine inches to fourteen inches in diameter. After this you get into spruce swamp, the trees being from four inches to seven inches in diameter, and would cut five cords of pulp-wood per acre. This swamp has a clay bottom, and although it was very wet when I went through it, it could be easily drained into Abitibi River. The walking on the ridges and also in the swamps is very bad, there being so much fallen timber, and black alder underbrush, not to speak of the thick moss and water.

I have made all the estimates of the distances I walked and found that I was going at the rate of one mile an hour. About five miles down the river we came to the Couchiching Falls, which with the rapids above has a fall of forty feet and would make a splendid water-power. The land here is chiefly rocky and sandy for a distance of twenty-five chains from the falls, timbered with balsam, white birch and small spruce. After going back twenty-five chains we came on a very wet spruce swamp, where the average size of trees is five inches diameter. From the Mis-to-a-go River to the Couchiching Falls on Abitibi River the timber on the north bank will cut seven cords per acre of pulpwood six inches in diameter for a distance of three chains from the river bank After this the land falls into swamp and the timber is much smaller. All the tamarac and about thirty-five per cent. of the largest of the spruce is dead. In the swamp the sub-soil is clay. From the Mis-to-a-go River down to the Iroquois Falls the chief timber is spruce and poplar, and it will cut ten cords per acre of pulp-wood, six inches in diameter, for a distance of three chains from the bank. After this comes spruce swamp with trees four inches to six inches in diameter and very short. The land is clay and can be easily drained. Mis-to-a-go River flows from the north into Abitibi River and is about two chains wide at the mouth. One mile up stream we came on a rapid and fall of about fifteen feet. We went up this river for about thirteen miles, when we were stopped by low water and logs across the river; here it was about thirty feet wide and very shallow. Two miles above the falls the timber is chiefly poplar and spruce and will cut six cords of pulp-wood per acre. Above this and for the next three miles is brule grown up with second growth poplar from one to three inches in diameter. For the next seven miles the timber is chiefly poplar about eight inches in diameter and will cut about six cords per acre for forty chains on each side of the river; beyond this is a spruce and dead tamarac swamp. I walked north west from camp for three and a half miles, forty chains from the river, along a clay ridge covered with poplar and spruce which will cut six cords to an acre. Then came five chains of spruce four inches to six inches in diameter succeeded by one mile of spruce swamp with no timber of any value, chiefly windfalls and black alder bushes. This led up to a small sandy ridge with white birch, poplar and spruce about twenty chains wide which falls away into very open bush with small spruce and some dead tamarac. Around the Iroquois Falls on west bank the timber is the best I came on. From about four miles below the falls to half a mile above them and for a distance of sixty chains from the river the land is rolling and timbered with poplar and spruce, balsam and a few white birch and on some of the rocky ridges jack pine. One can cut fourteen cords of pulp-wood per acre, chiefly poplar, which will average twelve inches in diameter. The bush is open and the walking very bad, there

Priest's house, Baie des Peres, Lake Temiscaming.
Party No. 1.

Rapids on Quinze River. Party No. 1.

Sunday afternoon tea. Height of Land. Party No 1.

H. B. Co.'s canoe on Summit Lake. Party No 1.

Height of land near Summit Lake Party No. 1.

Fort Abitibi, looking west. Party No. 1.

Indians on wharf. H. B. Co.'s post, Upper Abitibi Lake. Party No. 1.

Fort Abitibi, looking west. Party No. 1.

Factor Robert Skean and Family, Fort Abitibi. Party No. 1

Factor's house at Narrows, Abitibi Lake. Party No. 1.

Biederman's house at Narrows, Abitibi Lake. Party No. 1.

Indians and winter supplies on Lower Twin Portage,
Abitibi Lake. Party No. 1.

being so much down timber. Beyond the ridges near the river we came on good leve' land for two or three miles, clay soil bordered by another ridge of jack pine; this would all make first class farming land. The timber on the east shore is smaller than that on the west, averaging only ten cords per acre. From four miles below Iroquois Falls to Duck Deer Rapids the timber is much smaller and only runs back about two chains (about nine trees per acre five inches in diameter that would make pulp-wood). At Duck Deer Rapids there is a brulé grown up with poplar two inches to five inches in diameter. A few miles below Duck Deer Rapids I went into the bush on the west side for about four miles and found no timber after I left the banks of the river over four inches in diameter. Here the land is rolling and the soil clay and would make first class farming land. This is one of the few places where we were not travelling through swamps. There is only about four to six inches of dry moss on the top of the clay. The trees grow very close together and will never come to anything until they are thinned out as there is no room for them to grow.

On the east bank we found an Indian by the name of Ez-haw, who has a house and is considered by the Hudson Bay officers to be the best hunter up here. He told us that the country we were over yesterday runs just the same until you come to the district boundary line run by Mr. Niven in 1898. There the timber gets larger but very little of it is over seven inches in diameter, though some odd poplar is fourteen to fifteen inches in diameter. About eight miles below this Indian's house and about one and one-half miles above the Long Sault I went inland for a day's walk on the east side, slept out, and returned next day. I found for the first ten chains the usual river timber, averaging about seven cords of pulp-wood per acre. After this, for the next forty chains is spruce swamp, trees three to five inches in diameter, clay bottom, then a dead spruce and tamarac swamp is met with, white sand bottom with moss (peat) on it two feet, six inches deep. This runs from north to south; we crossed it from west to east, walked on it for about three miles, and found ice in a number of places under the moss. Then we came on a small clay ridge with some good balsam, a few balm of Gilead and some spruce. This ridge is only about twenty chains wide. After that we got into another spruce swamp with black alder bushes very thick in it, and about one-half the spruce trees dead, and many of them fallen, making the walking very bad. This extended for about two miles, the maximum size of spruce in this area being seven inches in diameter. Then we struck another ridge of heavy clay with poplar, white birch, balm of Gilead and a few big spruce, but no quantity. This ridge is only about fifteen chains wide with a small creek on the east side, running north-west. After this, there was another large muskeg where the timber was very small and open. I would judge that we only came in from the river ten or eleven miles. On the following morning we struck west by north and walked on muskeg to within three-quarters of a mile of camp. From an Indian I learned that there is a portage and a chain of lakes running from this end of the Long Sault to Pierre Lake, which one can travel over in two and a half days by going light in summer. The land, he informs me, is all swamp and muskeg. The timber he describes as "just what a man can see over." The lakes are clear water with sandy shores. We returned up the river and on the 20th of September met you.

Along the south-west bay of Abitibi Lake, and west of the river, the timber is fair, chiefly poplar, spruce, balsam, and a few red and white pine, about five thousand feet B.M. Along this bay for a distance of one mile back from the lake, spruce and popular will cut twelve cords of pulp-wood per acre. Along the westerly shore and for a distance of five chains from the foot of the bay, the timber is chiefly poplar, white birch, balsam and spruce. Very little of this runs over eight inches in diameter, although a few odd trees run up to sixteen inches and will cut seven cords of six-inch pulpwood per acre. At the foot of the bay we ran into a spruce swamp for two or three chains, where the timber is small, and beyond this into a very open spruce swamp with no timber to speak of. What there is, is mostly dead. Along the north shore of the Lower Lake Abitibi the timber as seen from the lake is chiefly white birch, seven inches to twelve inches in diameter, and very crooked and only fit for firewood, poplar, six to seven inches diameter and very small spruce. The shore is rocky and sandy. This seems to run back to a large spruce and tamarac swamp as we can see dead tops in the distance at the foot of every bay. We went up a river seventy-five feet wide where it entered the bay, and called it "Mountain River." We expected to be able to reach by this route a range of mountains which we could see from the lake. We found that the river only wer up a mile and a half, when it was blocked with logs; then we walked inland north eas for two and a half miles. We fonnd the land undulating with a clay soil covered wi spruce and jack pine, too small to be available for pulp wood. Then we came to an ope spruce swamp, with scrubby timber and peat deposit, three feet six inches deep with clay bottom. We walked on this for one mile, when we got into spruce swamp with clay bottom that would yield four cords of pulp-wood per acre. Most of the large trees are dead. This tract extended for two miles and was succeeded by a sandy stretch of two and a half miles with white birch, balsam and a few large spruce.

When the ascent of the mountain was reached we found the timber to consist of small scrubby jack pine and spruce. The mountains are almost three hundred and forty feet high and all rock. As far as we could see from the top of this elevation, looking north, the country seems to be flat and timbered with spruce. Here and there, there was a line of poplar, evidently along the bank of some creek. Working still along the north shore of the lake, the timber is small, chiefly white birch and small spruce, a few white poplar, scrubby cedar, and small jack-pine, five inches in diameter and under. The shores of the lake here are rocky. At the far end of the bay there is a large spruce and tamarac swamp with a lot of dead timber. This swamp has a clay bottom. We entered Fork River which is two chains wide at the entrance and went up it for six miles when we were stopped by low water and logs. Going inland on the west side, we found small spruce and poplar timber for three chains, and then came to spruce swamp, containing very few trees over six inches in diameter. The poplar was very faulty and full of punk knots. On the east side of the river the timber for the first thirty chains is small jack-pine and spruce, five inches in diameter; the land rolling with clay soil. Beyond this is a tract of very low wet land. About one mile from this river we came on another river about three chains wide with very little current. We ascended it for about four miles until we came to rapids, and there was too little water to go further. No doubt one can go a long way up this river in the spring, or when the water is high. The timber along this river is chiefly poplar of an average diameter of nine inches, which would cut fifteen cords of pulpwood per acre. The other timber is balsam, spruce, jack-pine and a few balm of Gilead. The land is rolling, with clay soil, and well adapted for agriculture.

A short distance along the shore, west of the river, and about one mile inland, there is a hill one hundred and ninety feet high. This I went up, and found the land sandy with here and there a ridge of rock, the chief timber being white birch, small spruce and jack-pine. From the top of this hill we got a good view of the country and could see a long way to the north, north-west and south. The land between this hill and the range of mountains to the north-east is rolling and covered with spruce with quite a lot of poplar along the river and in some places there are large open spruce swamps. I would judge that all the land is clay, as we found it so on the banks of the different rivers we went up. Further west along the north shore the land is very rough and rocky for a distance of forty chains back from the lake, when it falls away into spruce swamp. On this ridge I found sixty white pine trees, containing about seventeen thousand two hundred and eighty feet B.M., averaging sixteen inches in diameter; each tree will make two logs. Behind this ridge is a large spruce swamp with trees from three to seven inches in diameter, though most of them run from 3 to 5 inches in diameter. At the far end of the west bay in Lower Abitibi Lake, there is a large brule extending to the narrows near Biederman's Trading Post. This is covered with second growth of poplar, three inches to five inches in diameter, and growing very close together. The soil is clay and the surface of the land undulating.

This tract of fertile agricultural land extends behind the hill on the north-west side of the narrows. Here we left you and, as it was getting late in the season and the weather very uncertain, we decided to meet at the Hudson's Bay Post on the following Thursday, so that we could only give the Upper Lake Abitibi a very hurried run over, and had only time to go up the chief rivers. For the first four miles along the north shore there is considerable rock, and the land is mostly stony and sandy until we get inland five chains, when we come on spruce swamp with clay bottom. The chief timber along the shore is poplar and spruce, very small. After this, appear high clay banks timbered with second-growth poplar extending back one mile and a half to a creek called Snake Creek on account of its tortuous course. One can paddle for three hours up this creek and you will not have gone more than three miles in a direct line from the lake which can be plainly seen. The timber here is small spruce and poplar, not fit for pulp-wood. In the next bay, the shores of which are covered with rocky boulders, the timber is chiefly white birch, poplar, spruce and balsam. At the end of the bay there is a spruce and tamarac swamp with a creek flowing through it one chain wide. Entering another bay which is about ten miles deep, we went up one creek for a short distance, when we were stopped by logs. This creek is known as Gooseberry River, and is about one and one-half chains wide at the mouth. There is very little current in it. The timber comprises small spruce, and a few poplar, very few of the trees being over five inches diameter. The shores were clay land.

From here we went up Rotten Pitch Pine River, which I have described before, and thence to the Hudson Bay Post on Lake Abitibi, where I met you the following day, when we started for Toronto, arriving here on October 12

SUMMARY.

The following is a brief summary of the result of my work :—

TIMBER.—Scarcely any pine was seen north of Abitibi Lake. In general, the pulp-wood is stunted, but along the margins of streams it is larger and of fair quality, the best areas being found along Low Bush and Circle Rivers. In the vicinity of these rivers for an area of about one hundred and eighty square miles, I estimate the average at about seven cords per acre of pulpwood (exclusive of faulty timber) down to four inches in diameter. This includes spruce and poplar. Immediately along the banks of Abitibi River, and extending for only a short distance therefrom, is a belt reaching from Lower Abitibi Lake to the Long Sault, a distance of over 80 miles. This I estimate at seven cords of pulpwood, down to four inches in diameter, to the acre. In rear of this belt, so far as I have knowledge of the country, the timber is of little value, not averaging more than three cords per acre. Along the north shore of Abitibi Lake the timber is generally small, but along some of the tributary streams there are occasional small areas of fair pulpwood.

LAND.—With the exception of occasional sandy and rocky ridges and the water-shed, the soil throughout is clay of uniformly good quality. Moss to a depth varying from two inches to ten feet (inclusive of resultant moss peat), covers almost the entire country. Between Abitibi Lake and River on the south, and the water-shed on the north, about fifty per cent of the land is, in my opinion, good farm land. Drainage would considerably increase this proportion. To the south of the base line, and east of its twenty-fifth mile, a large proportion of the country is broken by swamp and muskeg, the intervening parts being chiefly clay and sandy ridges.

Yours truly,

P. F. GRAHAM BELL,

Land and Timber Estimator, Exploration Survey Party, No. 1.

T. B. SPEIGHT, Esq., O. L. S.

I, P. F. Graham Bell, of the City of Toronto, County of York, having been appointed Land and Timber Estimator by direction of the Commissioner of Crown Lands, do solemnly swear that the enclosed statement is a true account, to the best of my belief, of my work.

P. F. GRAHAM BELL.

Sworn before me, at Toronto, in the County of York, this, the 30th day of November, 1900.

GEO. B. KIRKPATRICK,

A Commissioner.

LAND AND TIMBER ESTIMATOR'S REPORT OF EXPLORATION SURVEY PARTY NO. 1.

Gravenhurst, Nov. 8th, 1900.

T. B. SPEIGHT, Esq., O. L. S.
Toronto, Ont.

Sir,—Having had instructions from the Honorable the Commissioner of Crown Lands to proceed with my duties as land and timber estimator upon receiving instructions from you, I accordingly joined your party at Gravenhurst station on June 18th, and after a pleasant trip via Mattawa and Lake Temiscaming, we arrived at the Hudson Bay Company's post on Lake Abitibi on June 28th.

After leaving part of our effects and supplies not needed just then, we left for your line, but as Lake Temiscaming is large and shallow and a heavy sea was rolling, we could not go more than fourteen miles, and had to camp on an island which is quite large and excellent soil; here we were compelled to remain until July 2nd.

We left for Low Bush River and got about seven miles up from the mouth and camped after making three short portages commencing about a mile below camp. Along this stream are some very fair spruce and poplar, but the timber does not go back any distance before it gets smaller and shorter. Mr. Graham Bell and I walked back about half a mile from the banks of the stream. Spruce and popar will run from the bank back about ten cords per acre of pulpwood, diameter six inches and over. The chief timber along the banks are spruce, poplar, balm of Gilead, white birch and, scrub jack pine, but the last three are of no value except for firewood.

There is only an occasional tree suitable for saw-logs, and these will not average more than eleven inches diameter, as all large poplar is faulty and covered with fungus. We continued up to the last portage. This stream gets very narrow and crooked, and so rapid that we could not paddle, but had to pole up.

From the lake up the spruce and poplar will not exceed fifteen cords per acre along the banks, and inland not more than seven cords per acre. All the soil is fine clay loam cut in places by wet swamps, but has splendid drainage by small creeks running to the river.

From here we portaged to Lake Welcome about three miles, where we found no timber of value. For the first mile and a half is good clay soil, to a swamp for one mile and white sand to the shore of the lake. The timber around the shores of Lake Welcome is spruce and jack-pine. The spruce will run seven or eight cords per acre, but the jack-pine is valueless. About one mile from here we came to a half mile portage, to another small lake about one and one-quarter miles; the spruce here will average about seven cords per acre. After crossing this lake we came to our last long portage. The soil is sand on the south side to a high rock which extends almost to the north end about three miles. The centre of this portage is quite high, and having been burned over affords a good view for a considerable distance. The rock ends in clay soil near the north end of the portage. Here we came to a chain of small lakes running into Little Abitibi Lake; around these the timber is all very small and scrubby. There were only thirty-seven trees that I could see that would make pulp-wood or small saw-logs.

We crossed little Abitibi Lake; on the north side are some fair sized poplar, but the largest are very faulty. The other timber is white birch, spruce, balm of Gilead, and tamarac, but these are of no value, being small and scrubby. All the tamaracs are dead, killed by worms.

On continuing down the Little Abitibi River, I saw some good poplar but no quantity. There is no other timber of value in sight, the trees being all scrub white birch, balm of Gilead and spruce. The soil along the banks is white sand, and back only a short distance is swamp, low and wet. Descending this river, we came to Lake Pierre, around the shores of which there is no timber of any value. The banks are all low. We continued down to Lake Montreuil where you left us (Mr. Baker, Thos. Pierre and myself) and you proceeded to your base line. On the south-west end of this lake I found about two hundred and fifty red pine trees that are small and will not average more than one hundred and thirty feet per tree, board measure, growing on a narrow ridge from fifty to two hundred feet wide, between this lake and a lake about two and one-half miles long and from three-eights to one half a mile wide. This lake empties into Pierre Lake, and runs from north-east to south-west from the west. I walked about three miles and turned north one mile and back to the canoe. Spruce and poplar would run about eight cords per acre. The soil is all fine clay loam, covered in some places by a spongy moss from four to six inches deep. From the west shore walking north-west, the timber is taller and better, and will average about twelve cords per acre, both spruce and poplar, six inches in diameter and over. The soil is all good clay loam, cut by large swamps, which were full of water. I would judge from the high water-mark that this whole country must be inundated, as it is quite flat and has very low banks. On the east side of this lake the timber will not exceed eight cords per acre. We came to Harris Lake, which is almost round. Beside the water from Lake Montreuil, there are two streams entering this lake, one on the north and the one you went west on. Little Abitibi River runs out on the west side of this lake. It is about twenty chains wide at the outlet of the lake, but gradually narrows to about two chains at the first rapids.

Here there is a ten-chain portage on the north side, about one mile below the lake. There are some fair-sized spruce and poplar on this stream, and around the lake, but they do not extend back any distance. After passing this portage one mile, the stream running in a north-west direction, we camė to an old brulé, which extends for about six miles, grown up with very small white birch and poplar, which stands very thick. The banks are low in places. The swamps come to the river, but the stream is quite swift and very shallow, being filled with large boulders in places. It averages to the second portage about two chains wide, the portage on the south side being about ten chains long. We saw some good spruce and poplar after leaving the brulé, also some tamarac, but mostly dead. We reached Niven's north and south line or district boundary which this stream crosses about thirty chains north of the two hundred and eleventh mile post. I followed the line south about one and one-half miles. After leaving the river bank the timber gets very scrubby, and in places I found open swamp and no timber except short trees three to seven or eight feet high. We turned east and walked two miles through an almost continuous swamp, over two ridges with some poplar on them, and then north to the river bank. The timber will not average more than ten cords per acre. We came back to the second portage and camped. There is a fall here of fourteen feet over rocks. but the banks are so low, above, that I am of the opinion that, were a dam put in to raise a head, the water would find an outlet around. We again walked south from this point for two and one-half hours, and found after leaving the camp about one-quarter of a mile, a ridge thirty feet high, on which are some good spruce and poplar, but immediately on going down the other side we came to a swamp covered with balsam and spruce with some balm of Gilead, but nothing over twenty to twenty-two feet high. The swamp extended for three-quarters of a mile. Then we ascended another ridge about three-eighths of a mile wide, principally poplar, and a few good sized spruce. Turning south-east for one and one-half miles we crossed four small creeks running to the river out of the swamps, none of which are over six to eight inches deep. The soil is all clay, low and wet, and the only good timber is on the ridges and on river banks.

We came back to the river through three large swamps, covered with small balsam and spruce. Some odd spruce and poplar near the banks will run eighteen and twenty inches in diameter. After walking north, we found the country somewhat higher and the timber better. The spruce and poplar will average seven inches in diameter for four foot cut. There is also some tamarac, but it is nearly all dead, as is also fully one-third of the large spruce. In the tract extending from the lake to Niven's Algoma and Nipissing boundary, about twenty-four miles, taking one and a half miles on each side of the stream, about seven hundred and fifty cords of poplar and spruce pulp wood could be got, six inches and over in diameter, in the proportion of seventy-five per cent. spruce, and twenty five per cent poplar. We went north on a stream running in from the north; the banks are very low and swampy. We canoed about eight miles, but could get no farther as the stream is shallow and full of driftwood, not in any place over three feet deep, and in many not more than four inches, over which we would have to lift our canoe, and ending in a swamp. I walked two miles east through swamp which is full of ice-water, ending in muskeg. There is no timber of value on this stream, only an odd poplar of any size, and small spruce and balm of Gilead.

North-east of this lake I walked for three and one-half hours and east for another three hours, being north of the fifteenth and sixteenth mile posts. The soil is good clay, which I believe can be drained to Montreuil and Harris Lakes. There are some very heavy windfalls through this territory. The spruce will only average ten cords per acre. From the sixteenth to the twenty-second mile post the timber is about the same, averaging about ten cords per acre. The land is very wet and swampy with a clay bottom. From the twenty-third to twenty eighth mile post the land is mostly all muskeg, or spongy moss covered with open swamps, in which we could sink to our ankles. This we concluded were good peat beds. The timber will not average more than four or five cords per acre for at least five miles north of these posts. From th twenty-ninth to the thirty fourth mile post, five miles north, the country is valueless. There is no timber, all being open spruce and tamarac swamps. The only good timber is right on the line, but it is no depth.

Baker and I walked east three quarters of a mile on a large sand ridge covered with poplar and white birch ; the large poplar is very faulty, and the birch only good for firewood. We turned north and followed the ridge one mile to where it ends in a swamp, in which the timber is short, small spruce of no value. After passing through we came to a large open muskeg running from south-east to north-west. We could see no end, and as it was very wet didn't follow ; it would be more than one mile wide. The poplar ridge runs from east of the thirty-fourth mile post almost to the thirty-sixth mile post, but north of the line it gets very narrow. Due north of the thirty-fifth mile post is a string of lakes, so cutting up this territory that there will not be more than four hundred and fifty to five hundred acres covered with poplar and spruce, which will run about eighteen to twenty cords per acre. The white birch is only fit for firewood. From the thirty-sixth to the fortieth mile post for five miles north, there are only small patches of timber that would make pulpwood, but not in quantity. There is no soil suitable for agriculture. From the fortieth mile post to the forty-eighth mile post, we walked six miles north and came through one open swamp after another, passing several small lakes running south. Mr. Baker has taken samples of peat from one open swamp north of the forty-fifth mile post. We arrived back at your camp on the line August 8th, and as there had been no change in the country since leaving the twelfth mile post, you considered we could work to better advantage from Lake Abitibi, Baker and I taking the south shore and as far back as we could go.

We accordingly started back and arrived at the mouth of Tapa-Saqua River on August 15th but our canoe being too small for three men we did not arrive at the narrows for supplies for three days owing to very high winds. Our canoes with three men, our dunnage and two weeks' provisions are, in smooth water, not more than three inches out of water, and if any sea is rolling we could not help taking water and are not safe ; they would be all right with two men and the load.

We commenced work on Big Point on Lower Lake Abitibi. I walked from one side to the other ; the point will average over a mile in width. There are some good spruce and poplar, also balsam from four to ten inches diameter, and averaging seven inches, from forty to fifty feet high, also white birch, but only good for firewood.

Part of the point has been burnt over and other points are covered with windfalls. Some spruce and poplar will run twenty-two inches diameter on the stump, but only an odd tree, and they are very scattered south of the portage, which is about seven miles from the north end of the point. I walked to the south-west shore at the foot of the bay, and then east, passing five small ponds in about one and one-quarter miles. Large ridges between these ponds are covered with very fine jack-pine ; about eight hundred or nine hundred feet of logs could be got off this point averaging forty feet per log About two hundred thousand feet of red and white pine, only fourteen trees white, and from sixty thousand to seventy thousand cords of spruce and poplar pulp-wood are also obtainable from this area. If balsam were taken from fifteen to twenty thousand cords more could be got. Roads could be made anywhere as the point is nearly level. The soil is all white sand mixed with gravel, until near the mainland, when I came to sandy clay. The whole point is covered with rolling stones. After leaving the point we went to the south shore on the west side of point, and found no timber ; for the first mile and a-half spruce and tamarac has been growing close to the shore, but is now all dead. Behind this is all open swamp, covered with a thick moss from six to seven inches deep, only scrub trees standing here and there. The swamp is about two miles wide by three miles long and for half a mile from shore is mostly sand bottom, then clay.

We continued along the shore to a creek running into the lake from the south, about forty feet wide at the mouth, up which we paddled about three and a-half miles. The land is only low, wet swamp for one mile, then spruce and poplar ridges on each side of the creek, which on the east side only extends about one-quarter of a mile back, behind is all open swamp. The west side is also narrow, being about one-half a mile wide to a recent brulé. All the soil, including the brulé, is good clay loam.

On these ridges poplar and spruce will run about fifteen cords per acre. Owing to rocks and driftwood we could go no farther so came back and continued along the shore towards the outlet. When within about three-quarters of a mile we walked over a sandy clay ridge covered with jackpine. These trees are fairly tall but are not large, being from 7 to 12 inches in diameter, about 50,000 feet in all. Leaving the ridge we came to a small sand and gravel strip covered with balsam, white birch, poplar and spruce, the yield would be 35,000 feet poplar and 30,000 feet spruce logs to average 64 feet board measure per log. The balsam and white birch is of no value as it is small and scrubby.

After walking southeast for one and a half miles I came to a large brulé which the Indians informed me was burned three years ago by lightning. I could not see the end of it. The last mile we travelled was also a brulé with clay soil. We came back northwest through a tract of red and a few white pine to the shore of the lake at the outlet. There will be about 100,000 feet board measure in this bunch. A few red pine are 24 inches on the stump and from 80 to 90 feet high. From the east side of the big point to the south side of the big shore of the next bay, along the shore there is no good timber. One mile south of the second point I came to a brulé of considerable size, of a recent date. There is a creek running into the lake from the west near the foot of the bay up which we could not canoe more than half a mile. At this point I left the canoe and walked due west one and a half miles to a small lake, the source of the creek. Here the brulé ends in this direction, but continues south a long distance. All the soil is fine clay with a splendid chance to drain. When we came to the end of the bay Mr. Baker and I canoed up a stream coming in from the south for about three miles. It is all marsh on either bank with small feeders running in at short intervals. On the west side the brule extends, on the east there is no timber except dead tamarac and spruce.

From this bay to the east side of the next point around by the narrows the timber will not exceed six cords per acre, of six inches diameter and over, for pulpwood. There is a lot of scrub cedar, white birch and balm of Gilead, but nothing of value. The soil is all clay loam. The width of the point is from 1¼ to 1½ miles and the land is in some places wet, but can be drained. The shores are in places white sand and elsewhere rock. We then met you at the narrows and left in your company to go down the Abitibi river and up Black river. About four miles below the lake we came to Couchiching Falls, with a descent over rocks of more than 40 feet, furnishing a splendid water power. From this point we canoed down and over two portages to Black River, you going on down the river and our party up Black River. This river is almost four chains wide at the mouth, and for the first eight miles is timbered with balm of Gilead, spruce, poplar and white birch. The spruce and poplar near the shore is fairly good and thick. The spruce will average 14 inches, and the poplar 11 or 12 inches, some trees being 80 to 90 feet high. The heavy timber does not extend back any distance. There is only an occasional good balm of Gilead, some of which will measure 20 inches on the stump.

About six miles from the mouth of the river a stream, about 40 feet wide at the mouth, runs in, up which we canoed one and a half miles, but owing to the banks washing down and filling up the stream and a quantity of large boulders and driftwood we could go no farther. This stream the Indians use in high water to a small lake, from which they have a portage over to Albitibi River above the falls, so as to escape the swift current in that river. Both north and south of this stream the soil is fine clay. There is no quantity of timber of value along this stream. We continued on up Black River about two miles and came to a recent brulé which extends for at least seven miles, to within a short distance of the falls, on each side; the brulé extends back a long distance. The soil throughout is an excellent clay loam with odd wet spots, from which small creeks run to the river. About three miles below the falls a small creek runs in from the west through the brulé but is so filled with driftwood and fallen trees that we could not canoe up it any distance. We walked back onto a ridge from which we could see for a long distance. A large rolling area of agricultural lands with odd standing dead trees. We walked west 3½ miles, for the first mile and a half of which I saw no timber of value except for fuel purposes. Then followed two miles of rolling clay land, rather lower than the rest but not wet, with a growth of small spruce and poplar. Then I came to a sand plain covered with small scrub jackpine from three to eight feet high. I turned back here and came to a small lake, from which the Indians have a trail to the River Blanche waters from below the falls on the river on which I returned. It is a very poor trail not cut out and passed another small lake with marsh all around it. The only spruce I have seen since leaving the riverbank is right around it, and will run seven or eight cords per acre. There is also about 300 jackpine trees around this lake that will average 200 feet per tree board measure, but there is no timber nearer the river, as the brulé cuts in, I next canoed up the river about two and one-half miles, and from there walked east over a clay ridge covered with spruce and poplar. Then through a rolling area of good clay soil, with somewhat smaller timber. After walking about two and a half miles, I came to a low, wet swamp, very open, the only trees being short spruce, three to eight feet high, whence I came back to river and continued on up stream. Before reaching the next portage we came to a creek running in from the east about thirty feet wide at the mouth. Up this we could not go more than one and one half miles, on account of shallowness and floodwood jams; good clay soil was seen from the banks, but very little timber. Returning to the river we went on up to the falls in the neighborhood of which are a number of high hills. I walked west over one, and immediately on going going down the other side came to a large wet muskeg or open swamp, which I did not follow. On the hills are some fair-sized poplar and spruce, but not any quantity. I walked east and after walking about one mile came to a very large open swamp, very low and wet with short spruce from three to eight feet high on it. I came back to the falls, and as the river appears to be one continuous rapid, I concluded not to follow it farther.

An Indian family by name of MacDougall are camped on the west side of the river below the first falls. They planted some few potatoes about June 10th that were very good looking but had never been attended to. Some were seven inches high and of a good size. We then returned to Lake Abitibi, from the Falls on Black River to the mouth, say twenty miles. Seven miles of this are brulé, leaving thirteen miles under timber. Taking one-half mile on each side of the stream this would cut five hundred to six hundred thousand feet board measure of spruce and poplar saw logs per square mile, in the proportion of seventy-five per cent. spruce, and twenty-five per cent. poplar. The same territory will cut two thousand cords pulpwood per square mile, six inches and over. I have not counted anything for logs under ten inches, as they would be too rough; back behind the one-half mile of each side will not cut more than six thousand cords per mile and will yield no saw-logs. On coming up Abitibi River,, I twice walked south a distance of three and four miles over three ridges and through fine rolling clay land. On these ridges are some fine clay soil covered with some good poplar and spruce, but all from the lake to Black River will not average more than seven or eight cords per acre for pulpwood. I turned back in both cases on coming to low, wet swamp. All the large poplar is very faulty. There are three creeks on the east side of the river between the first fall and the lake, the first and third we could not go up any distance, the second one we called Dokis Creek on account of an Indian of that name living on it. Up this stream about one-quarter of a mile we came to a fork, the east branch of which we could not follow, as it was filled with trees and driftwood. On the shore is spruce and poplar, scrub ash, balm of gilead and dead tamarac, but nothing of value. The main stream we followed for about four miles and found no timber to speak of—such as there is will not cut more than three cords per acre of poplar and spruce pulpwood. We here left the canoe and walked over a sand ridge with clay on both sides and continued for over one mile through open spruce bush, too small for pulpwood. I turned here and came to the canoe and back to the forks. I then walked north about three-quarters of a mile, and came to one stream with very low and wet swampy banks. I came back and continued to the lake and on to the narrows.

We now paddled down the east shore of Biederman's Point to a stream about eight miles, up which we canoed about fifteen miles, mostly between marsh banks. About seven and a half miles up we came to a fork. On following the main stream we arrived at a mountain about three hundred and fifty feet high, consisting of clay soil, but so thickly covered with small poplar, spruce, white birch, balm of Gilead and jackpine, all scrub, that it was impossible for us to get a view from it. On coming back we took the branch and followed it for about three miles, finding the country almost one continuous windfall, the higher ground being good clay soil. We walked east for two hours and north for two hours, and came back to camp, having seen no timber that would exceed four cords per acre. Returning to the lake we saw no timber of value from the shore for the next seven miles. We then came to Lightning River, about three chains wid at the mouth. We canoed up this river about six miles. The banks are all marsh, but gradually rising to higher land, where the river divides into a series of small streams, one after another, so that we had to turn back When within about two and one-half miles of the mouth, we left our canoe and climbed a bare rock about two hundred feet high, from which we got a grand view of the surrounding country, for fully ten or more miles, showing an almost complete circle of mountains ranging from one and one-quarter miles to fifteen miles from the lake shore and from two hundred feet to four hundred feet high.

Some of these mountains have been burnt over, affording good views, others are so thickly covered with scrub jack pine, poplar, spruce and balm of Gilead, that a view can not be had from them. All the soil in the valley is fine clay loam. We climbed one mountain north of the burned over one, but could get no satisfactory view from it. We came back to the mouth and on to the foot of the next bay, and again walked in about one and one half miles, to a bald mountain about four hundred feet high, affording a magnificent view for miles of both land and lake, showing fifteen mountains in a circle. All timber in the valleys will not exceed six cords per acre of six inches diameter and over. Spruce is not plentiful, and all poplar over ten inches is very faulty, covered with fungus. We followed along the shore south of Dominion Island. Here the mountains reach the shore, but are densely covered with scrub poplar, balm of Gilead, white birch and balsam, so that we could not get a view from it except at long range. We here came to a small stream running in from the south, but could not follow it any more than three-quarters of a mile. The soil is all clay but grows no timber to speak of. We came back and continued on down the lake to the next large point, and here came to another stream, which we could not follow owing to its being very shallow and full of driftwood. The timber is about the same average, and the shore becomes more rocky. Some of the islands have a few scrubby red and white pine on them, but nothing of value. We were then about seven miles west of the Hudson's Bay Company's post, and from your instructions I concluded we were past where the line will come, so I came on to the post and started for home. Before concluding I wish to express to you my thanks for your kindness and help with my work.

Respectfully yours,

THOMAS G. TAYLOR,
Land and Timber Estimator, Exploration
Survey Party No. 1.

I, Thomas G. Taylor, make oath that the foregoing Report is correct and true to the best of my knowledge and belief.

THOS. G. TAYLOR.

Sworn before me at Webbwood, this twenty-seventh day of February, 1901.

E. GARROW, J.P.

GEOLOGIST'S REPORT OF EXPLORATION SURVEY PARTY, No. 1.

Kingston, Ontario, December 24th, 1900

MR. T. B. SPEIGHT, O.L.S.,

Surveyor in Charge Exploration Survey Party, No. 1, Toronto.

SIR,—I have the honor to make the following report of my work as geologist on your exploration and survey party during the past summer.

Acting under instructions from Mr. A. Blue, then Director of the Bureau of Mines, I joined party No. 1 in Toronto on June 18th, 1900, and proceeded with it to our district. At Gravenhurst I met Mr. Thomas Taylor, the land and timber estimator, with whom I worked throughout the summer. We travelled with the main party, via Lake Temiscaming, Lake Abitibi and the Little Abitibi River route as far as Pierre Lake, as shown on our new sketch for map. I have not described the area through which we travelled, as it was agreed that we should confine our exploration and examination to that part of the district north of the base li e run this summer; while Mr. Coulthard should report on the area south of the base line.

Having arrived at Pierre Lake, however, and believing that the line would cross somewhere about here, we began our examination from this lake, while the remainder of the party proceeded to the Niven boundary line between Algoma and Nipissing to begin the running of the base line.

Pierre Lake.

Pierre Lake is about six and a half miles long and varies in width from one hundred and fifty yards at the "narrows" to about two miles in the lower stretch of the lake. It lies with its longer axis about north and south. The Little Abitibi River enters Pierre Lake near its south-eastern corner and leaves it again at the north end of the lake. The river then flows to the west, or a little south-west, to Montreuil Lake about one mile distant, and enters it at the eastern side of an expansion of the lake; crossing this in a west south-westerly direction we enter Montreuil Lake proper.

About three-quarters of a mile due west of the north end of the narrows of Pierre Lake is a narrow gap, which leads to a long bay running due north and south. This bay is about three miles long and about two hundred yards average width. Going through this way affords a much shorter route to Montreuil Lake, but requires a short portage or lift of twenty yards over a low sand ridge. This route has some advantage as it avoids the current in the river from Pierre Lake to Montreuil Lake, and this in high water must be considerable, as it was quite strong even in July.

Montreuil Lake.

This portage meets the same expansion in Montreuil Lake as does the river mentioned above. Crossing this water then we pass through a narrow place into Montreuil Lake proper, which is about two miles long and three-quarters to one mile wide. Entering near the south end of the lake we travel north-west and leave by the Little Abitibi River or really a long narrows connecting Montreuil Lake with Harris Lake. These narrows are about three quarters of a mile in length, and are cut by the base line at 12 miles forty hains, at a point about a quarter of a mile from the north end of the narrows.

Harris Lake.

Harris Lake is nearly round and is about two miles in diameter. In the centre of it is an island of sand and gravel, which we called Gull Island from the fact that it is dotted with gulls' nests, and is the home of all the gulls in that district.

Harris Lake has four water-ways to and from it. We have first the one by which we have just entered, flowing into Harris Lake at its south-east corner. At the north-easterly point of the lake we have North River entering, which will be described later. Again at the south-westerly side of the lake and about half a mile from the south end of the lake we have a third stream entering. This stream with a series of small lakes makes a water route to the Niven boundary line between Algoma and Nipissing, and meets it near the one hundred and ninety-seventh mile post, but as I did not travel up this stream, I have no report to make of it.

Little Abitibi River.

The fourth stream, the Little Abitibi River, flows out of Harris Lake at the north-west corner of the lake, and after twenty miles or so in a north-westerly direction it crosses the Niven line fifty chains north of the two hundred and eleventh mile post, and flows on to join the Frederick House River.

To give a more detailed description of this river, it will be seen by the map that there are eight rapids proper, and two falls, with many other places of strong current almost rapids. The total fall in the twenty miles of river, i.e., to the Algoma line, is about forty feet I should judge, making an average of two feet to the mile. This gives quite a strong current, probably equal to a mile per hour. Most of the rapids can be run in any cañoe provided a careful watch be kept for stones, as the river is quite shallow in most places. There are three portages, however, which must be made. The first is at a rough rapids about two miles down stream from Harris Lake. This rapid is about four hundred yards long, with a total fall of about five feet. The portage is on the right or eastern bank of the river and is two hundred and fifty yards long, the rest of the rapid is run safely either up or down. Having passed this rapid we can canoe for about fourteen miles without any obstruction, but here we come to a fall of twelve feet with a rapid at its foot of two feet fall. The whole is passed on the left or west bank of the river by a good portage two hundred and fifty yards long.

About one mile farther down stream is a second fall of about four feet with a long rapids at the base having a total fall of three feet. There is a portage on the right or eastern bank used in high water, but in low water a short lift of thirty yards over the rock, passes the falls, and the rapids, some four hundred yards in length, are run as usual. Going on down stream about three miles below this fall we cross the Niven boundary line between Algoma and Nipissing. The river cuts the line at fifty chains north of the wo hundred and eleventh mile post, and is here very shallow, not more than twelve nches of water were flowing over it at the time of visiting it.

North River.

Coming back to Harris Lake we next took a trip up North River, which I mentioned above as entering Harris Lake at its north-eastern corner. This river at its mouth is about thirty yards wide and has the appearance of a fine stream and one likely to continue some distance. Upon examination, however, we found it only extended about eight miles nearly due north when the water became so shallow and the stream so narrow that we could not canoe further. Walking on north north-east we found the stream to be of swamp origin, having many small feeders forming the drainage of this area, which is for a great part wet land. There is very little current in the stream as a whole, but in seven places the current is quite strong; three of these places being old beaver dams. They are hardly worthy of being called rapids. The country through which this river runs is a rolling clay district as a whole, the valleys between the ridges are as a rule spruce swamps forming the sources of water for this stream. The ridges are clothed with poplar, balsam and spruce and would make a fine agricultural area when cleared.

District Drained.

Having now described in a general way all the water-ways met by us in our work about the line, let us go into a more detailed description of the area drained. Starting again then at Pierre Lake we find the land to be very sandy all around the lake, and this is also true of Montreuil, Harris and all the lakes of this district. Upon going back, however, from the lake shore from one-half to a mile, we find the sand to run out more or less and clay areas to come in, with several sandy clay ridges separating the clay valleys. Some of the highest of these ridges are composed of very coarse sand and are mixed all through with stones varying in size from a hazel nut to boulders of granite and gneiss four feet in diameter. Along with these Archeon rocks are innumerable fossils of several Devonian corals. Among these I found many examples of Zaphrentis, Bothryophyllum, Heliophyllum, Favosites, Halysites, Michelinea and others. These fossils of course prove the existence of Devonian strata farther north, as they have been carried south by the ice of the glacial epoch. Strange to say I saw none of these, nor any limestone about Abitibi Lake which, is some forty-five miles farther south, although the rounded boulders of gneisses, granites, diorites, gabbros, etc., were there plentifully. This whole north country has been heavily glaciated and the underlying rocks have been deeply buried under clay, sand and gravel, so much so that it is very difficult to find any outcrops of the underlying rock in place; the only ones seen all summer north of the Base line were on the Little Abitibi River, and these will be described a little later.

This whole country is a rolling clay area and is well suited for agricultural purposes as far east on the base line as the twenty-second mile-post, and for a distance of twenty-five miles north of the line, after which we pass into the low wet muskeg belt, which, though underlaid by clay, could never be cleared and worked as agricultural soil.

The timber of the district, apart from about two hundred and fifty red pine trees at the south-east corner of Montreuil lake, is of no value commercially. It would at best be only suitable for pulpwood, and there is only sufficient of it for settlers' firewood and building purposes. It consists entirely of spruce, poplar, balm of Gilead, birch and balsam.

Laurentian Granite.

As mentioned above the only rocks found in place in this whole district were on the Little Abitibi River, and even these only occur at the falls and are immediately lost under the heavy clay capping upon reaching the shore. About five miles down stream from Harris lake, and on the east shore we have an outcrop of gray granite. This low, flat rock is only a few square feet in area. It is a typical gray granite of Laurentian age. It shows a marked flow structure in one part of it and also shows glacial striæ running due north and south.

About eleven miles farther down stream, at the first fall, we find the underlying rock again exposed. Here we find the main mass of rock to be a very coarse-grained gray granite, almost a gray pegmatite, which has caught up into it pieces of a very micaceous gneiss, and thus given the whole mass a more or less inter-layered appearance. The softer and more altered gneiss has weathered more rapidly than the granite and thus left the net-work of granite ridges protruding from one to two and a half inches like a series of little dikes.

About one mile further down stream we have the second fall over this same gray coarse-grained granite, but here we find no inclusions of the gneiss, but the granite has a very marked jointing, so much so that the whole rock is split in slabs from eight inches to two feet in width, all standing on edge with the joint planes nearly vertical, and the strike of the upturned edges is north fifteen degrees east.

Strange to say, about one hundred yards farther down stream another outcrop of the gray granite like that found at the first fall above shows the flow structure and the inclusions of gneiss as before, but here these inclusions are rounded or in knots, instead of layers as they were in the first fall. The difference is probably due to a subsequent shearing action having drawn out the inclusions at the first fall, while here they have not been so squeezed or drawn out.

These, then, are the only outcrops of the underlying rocks seen in the whole district north of the base line, and even they could not be examined very well as they were buried beneath a heavy capping of clay loam upon reaching the bank of the river. These are sufficient, however, to show that a Laurentian belt runs through this country in a more or less easterly and westerly direction, for I believe Mr. Coulthard reports Huronian rocks a little further south at Little Lake Abitibi.

Land Along the Line.

I mentioned above that the good agricultural rolling land ran out about the twenty-second mile-post on the base line. From this point east the land begins to drop away to low spruce swamps and wet land generally, there being very little high land till we reach the thirty-fourth and a half mile post, when we met a series of gravel ridges, carrying many large rounded boulders of granites, syenites, gneisses and some gabbros, etc. In the valleys between these ridges are lakes or spruce swamps, and the ridges themselves soon dip down in the north to low spruce swamps and finally to peat bogs. The ridges strike, as a rule, about north ten degrees east, and are timbered with poplar, spruce, birch and would average about five cords of pulpwood to the acre, but it would not pay to to cut and take out at that rate.

Peat Bogs.

After passing the thirty-sixth mile post we find the land again drops away to low spruce swamps and peat bogs as it was on the western side of the ridges. This low land continues past the forty-second mile post when the land rises again for about one and a half miles, then drops away again to low wet swamps and peat bogs, which continue without much change to the eastern end of the line at the Quebec border.

These peat bogs lie more or less north and south or a little north west, they are as a rule from one half to one mile in width, and are no doubt the outstretched fingers, as it were, of the great muskeg belt of the north. This belt, I think, must strike a little south-east for at the western end of the base line it is twenty-five or thirty miles north of the line and so we have a few,if any,peat bogs below that,while just as we go east they creep farther and farther south, till at the eastern end of the base line, the land is almost continuous bog.

The peat-bogs are composed of the compressed moss, ferns, etc., of ages ; each year's growth burying the previous year's growth beneath it, and more or less under swamp water, which prevents the decay of organic matter, due, it is claimed, to the presence of apocrenic acid in swamp water. This moss keeps accumulating for years till the hollows are filled up to the draining level, when the water will not lie any longer, and the preservation of the moss will cease largely. These peat-bogs that are omnipresent in this northern district and the supply of peat is practically unlimited ;—as a pole can be shoved down ten or twelve feet in almost any of them and not reach the solid clay then in many places. Beneath the peat in several of these bogs is a layer of ice water about one foot thick, then we come to the hard clay of old lake bottoms, or low swamps.

Analysis of Peat.

A sample of this peat from a bog at the forty-second mile post on the base line,—and this sample by the way is typical of the whole area, as they are all exactly alike,—was sent to Mr. J. Walter Wells, B. Sc., Analyst at Belleville, who reports it as follows :—
Air-dried it analysed thus :

Moisture	No. 1.	9.68	No. 2.	9 28
Volatile combustible matter	"	66.24	"	66 26
Fixed carbon	"	20.78	"	20 74
Ash—lime, silica, etc	"	3.30	"	3.32
		100.00		100.00
Sulphur	"	none.		none.
Phosphorus	"	traces.	"	traces.

Now it will be noticed at once how very low this peat is in moisture and in ash, which qualities speak for themselves no mattter to what purpose it is put. Notice also how very high this peat is in volatile combustible matter and in fixed carbon, which qualities, with no sulphur and only a trace of phosphorus, make an ideal peat for fuel. I can find no other peats whose analysis compare with these on a whole for a peat fuel. They are usually higher in ash and lower in volatile combustible matter and carbon. So while Ontario is barren so far as coal is concerned we have here an ample supply of fuel for the future.

Most peat-bogs being formed in hollows between ridges have much silt and sediment washed into them from the surrounding hills, and this accounts for this high percentage of ash, but these bogs of northern Ontario are formed in a flat low country and hence we see the reason for this low percentage of ash.

Their rapid formation has been largely due to the immense growth of moss annually, brought on largely by the very warm days, cold nights, and excessive rain-fall ; keeping the whole land damp, and the flat country retaining what moisture it receives.

In addition to the large deposits of peat-fuel we have here endless quantities of the looser top sphagnum mosses which are lately being proved so well adapted for moss litter, deodorizing, absorbing, manurial and sanitary purposes. The higher portions of this wet area are, as stated, spruce swamps. These are underlaid by good clay loam, and might possibly be cleared up and drained for agricultural purposes ; but I doubt very much whether it will ever be profitable to work any great area east of the twenty fifth mile post, as it is all quite low and wet. The timber in these swamps is all spruce and would run six inches to ten inches or about four cords of pulpwood to the acre, but it would never pay to take it out at that rate.

The flatness of this area is also well shown by the great number of small lakes scat tered indiscriminately all over it. An idea of the general topography may be had by a glance at the sketch of the line area from the thirtieth to the forty-fifth mile posts. In this short distance we find ten or eleven lakes ; and all within about eight miles of the line, and no doubt there are many more a little farther north, which it was impracticable to visit as this whole area had to be examined by walking, there being no waterways except the one described above at the western end of the line. This stretch, then, of fifteen miles showing the lay of the ridges, lakes, peat bogs and spruce swamps, is characteristic of the whole area east of the twenty-fifth mile post.

Flora.

In describing the flora of this district we can say it is that usually found in wet land, and includes such plants as the iris, yellow water lily, but no white ones, strange to say, horse-tail, pitcher plants and innumerable swamp grasses and reeds. No marsh hay is to be found in the whole country and this accounts for the absence of the red deer to a large extent. The great amount of moss and water lilies afford good pasture for the cariboo and moose, however. Among the berries usually found in bush lands we saw only blueberries, and even they were in scattered patches. We found few, if any, raspberries, blackberries or thimbleberries, or cranberries in this whole northern area ; we did, however, find all these farther south, where the land was dryer. There was a great deal of alder bush along the small streams and about the small lakes.

Fauna.

Regarding the fauna of this district we can say that moose and cariboo are quite plentiful, and bears have been numerous, judging from the number of skulls found around old Indian camps, but they are not very plentiful now. There are many marten and some beaver, lynx and otter, and occasionally a gray fox is seen. Among the birds of the district I saw the following : partridge, gray and spruce, northern shrikes, ruby crowned kinglets, golden crowned kinglet, chick-a-dee, Canada bird, ducks—black, wood, buff-buff-head and shelldrakes, gulls of three varieties, loons, Wilson's thrush, cedar bird or wax-wings, fly-catchers, ravens, robins, fish-hawks, night-hawks, cat-owls and wood-peckers of many varieties.

The chief fish in the waters of this northern country are pike, pickerel and a flat variety of white-fish ; while in the cool streams we found plenty of speckled trout. The Indians live almost entirely upon fish during the summer months, and even dry many for their winter use. They use chiefly white-fish, of which there seems to be plenty.

Indian Occupation.

In connection with the Indian occupation of the country, we can say that it is occupied by the Cree tribe—a very friendly and peaceful tribe. These Indians are in the habit of leaving the Hudson's Bay Company's posts about the beginning of September, to go to their different hunting grounds for the winter months. There they hunt all winter and return to the Hudson's Bay posts about the beginning of June with their furs, to pay for the provisions, etc., supplied them by the Hudson's Bay Company. They then idle about the posts all summer and set out in the fall again. Many of them are employed in the summer months by the Company to transport provisions from Baie des Peres, on Lake Temiscaming, to Abitibi Lake, using large birch bark canoes twenty-four and thirty feet in length.

A few families of the Abitibi Indians go up to this district about the base line, but only along the water-way, that is, along Harris, Montreuil and Pierre Lakes, etc. We saw several of their old camps, but, of course, saw none of the Indians, as they were all at the posts at that time of the year.

Lake Abitibi District.

Upon finishing the work upon the north side of the line, Mr. Taylor and I, acting upon instructions, returned to the Lake Abitibi district. It was decided that wes hould here confine our examination to the south side of the Abitibi Lake and River so as to let Messrs. Coulthard and Bell report on one large belt south of the base line and north of the lake.

Accordingly, I will begin my report at the Black River, for the district west of this river has been reported upon by Mr. Arthur Parks in company with Mr. Alexander Niven's Survey Party, 1898-99.

The Black River enters the Abitibi River at a point where the latter turns sharply to the north-west and about twenty-two miles from the outlet at Lake Abitibi. The Black River is so named from the dark colored water flowing through it. This water, however, is clean and good, especially so compared with the dirty water of the Abitibi, which is never clean. The Black River at its mouth is about seventy yards in width and quite deep, with no noticeable current. It is, in fact, a succession of deep pools and shallow rapids and falls, with little or no water flowing over them. In high water, however, that is, in the spring and fall, the rush of water in this river must be very great, for I could see everywhere the marks that drifting ice and logs, etc, had made upon the trees along the shore, barking them and breaking limbs from them fully ten feet above the present water level.

Huronian Rocks on Black River.

Upon ascending the Black River about two miles we came to the first outcrop of rock in place. Here occurs the dark green pyritores diorite, which is the most abundant rock in this whole country. This green diorite will be described in more detail later. This outcrop was very small and immediately covered on reaching the shore.

About three miles further up stream, and on the same shore, that is the west shore, we come to a second showing of the rock "*in situ*" Here we have what appears to be a more silicious schist cut by bands of the green diorite, also schistose here, and carrying pyrite. In several places this diorite shows an amygdaloidal surface on weathering. Both these schists are dipping here about vertically, and they are always at right angles to each other.

We have no more outcrops for the next eleven miles, when we come also to the first check to navigation. Here we have a fall of ten or twelve feet over a ridge of massive diorite, green as before, and having many little veinlets and splashes of quartz scattered through it. These veinlets are not over one to two inches wide and one or two feet in length, so they are of no economic value. The ridge of diorite strikes north-west and the fall takes place through two gaps, like the prongs of a horse-shoe, the centre being solid rock. The western gap is about seven feet wide, and the water now flowing through it would be five deep. The eastern gap is ten feet wide with four feet of water flowing through it. These falls could be easily converted to a water-power, and would make a splendid one for smaller mills. The fall is passed on the east side by a good portage of about sixty or seventy yards.

About half a mile, and again one mile above this fall we have two small barriers of this massive diorite as before. At the upper one is a small rapid of six inches drop but it required no portage. In the next six hundred yards we have several outcrops of the green diorite, but at the end of that distance we have the diorite becoming schistose again, with the little veinlets of quartz, usually found wherever this green diorite becomes schistose. The schistosity here seems to have been caused by the intrusion of a hornblende granite. We have here a very rough rocky rapid, and at the lower end on both shores we have the green diorite schist, but about the middle of the rapid and on the east shore is a dyke—like intrusion of hornblende granite, which has baked the schist for fifteen and twenty feet around it.

Some of the schist near to the granite is baked so as to be almost flinty in appearance and hardness, and I could trace the schist less and less baked right to the ordinary green diorite of the whole country. Above the hornblende granite the schist comes in again and continues on up stream as we found it below. The ridge runs up on either shore to a height of fifty feet, and runs back from the river for a mile or so, but it is covered all the while by a clay loam; it then dips down and becomes the general rolling land as usual. The rapids are passed on the east side by a portage of one hundred and fifty yards.

In the next four miles we pass three small rapids, easily run and need no portage in ascending. Six miles above this rough rapid, however, is another fall of about nine feet with a short rapid at the foot. At the lower end of this rapid we have the green diorite schist in place, but the fall proper is over a ridge of a green diorite porphyrite. The porphyritic constituent appears to be felspar which has altered to kaolin, the rock shows in addition phenocysts of hornblende apparently altering to serpentine or chlorite. The fine-grained ground mass is felspar, and the whole rock is a pretty example of a porphyrite.

A good portage of one hundred yards affords a passage for the worst part of the rapids, and the rest are easily passed in canoes. Glacial striae was very plain on the porphyrite running north ten degrees west.

About two miles above this fall is a long rapid, or really a series of small falls with a total drop of ten or twelve feet. For the lower one hundred yards we have the fine-grained pyritous diorite schist, with many veinlets of quartz as usual, but for the uppermost fifty yards we have a coarser grained diorite, not schistose and little or no quartz in it. I could find no reason for the lower part of the rapid being over schistose rocks, unless it be that a fault occurs at the fall, and that metamorphism caused by the shearing, etc., produced the schistosity along cracks, and formed lines of deposition for the quartz veinlets.

The stream is only about fifteen wide here, and the walls of rock run up twenty feet on either side, making a rough, narrow gorge. The whole is passed on the west bank by a portage of one hundred and fifty yards.

Above this fall and rapid, we have the green diorite and the green schists coming in again as below. We found the river to be almost continous rapids now, and so shallow at this time of year that we did not consider it advantageous to go farther.

Area Drained.

Having dealt with the rock exposures along this river let us now examine the nature of the country drained. It will be seen from the character of the rocks described above that this area is a Huronian one, but like the Laurentian area north of the base line, it has been deeply buried beneath clay, sand, gravel and boulders carried by the glaciers. One very noticeable thing about this glacial drift, is the absence of any limestone, either as small boulders or as fossils, like those mentioned as being so numerous about the lakes farther north.

The soil all along this river and its tributaries is a fine sandy clay loam, with an occasional ridge of gravel; but forming on the whole an area admirably adapted to agriculture, easily cleared, and naturally drained, as the whole country through here is a rolling one.

The land is timbered with spruce chiefly, but has also a fairly good supply of poplar and balm of Gilead. This timbering extends for eight miles up the stream when we come to a large brulé, which is cut by the river for the next seven miles, or really to the first fall mentioned above as being about sixteen miles from the mouth of the river. This brulé was burned three years ago this summer and the Indians say it was started by lightning. It extends in a northeast and southwest direction for miles on each side of the river, and apart from a few blackened poles, it has only berry-bushes growing upon it now. So that this large area is practically cleared already for farming, and the soil is splendid. In fact a family of Indians named Miss-shib-a-gee have planted some corn, onions and potatoes on a small piece of this brulé and they were doing well. Miss-shib-a-gee and his family live at the first fall some sixteen miles up stream and he has built there two or three very neat huts, and they were remarkably well kept and clean. He has also a horse and some cattle, which he brought in on the ice from lake Temiscaming via the Riviere Blanche, then by portage route to this fall on Black River. He gets his hay for winter from a marsh around a small lake on the portage, and about one and a half miles from the Black River end of the portage. This portage route is used only by this family, who hunt on the Riviere Blanche. Other Indians going from Black River to Riviere Blanche take either the portage route from the head of Black River, or canoe up Driftwood Creek, which enters Black River near its junction with the Abitibi River.

Portion of Speight's party on shore of Lower Abitibi Lake.
Party No. 1.

West shore of Lake Abitibi. Party No. 1.

Forks of stream flowing into Lake Abitibi. Party No. 1.

On Lake Abitibi. Party No. 1.

At 3rd mile on O. L. S. Speight's base line. Party No. 1.

Poling up stream near beginning of O. L. S. Speight's base line. Party No. 1.

Pulpwood forest at 10 miles on O L.S. Speight's base line. Party No. 1.

Camp at 5 miles on Speight's base line. Party No. 1

Pulpwood forest at 10 miles on O.L.S. Speight's base line. Party No. 1.

On 38th mile O.L.S. Speight's base inc.

Looking nor h along the 35th mile on O.L.S. Speight's base line.

At the 51st mile on O.L.S. Speight's base line.

TRIBUTARIES OF BLACK RIVER.

While speaking of the tributaries of Black River, I might add here, that there is only one other river that is navigable for any distance, and even it is only so in high water, being nearly dry when we visited it.

This shallow river enters the Black River about six and a half miles from the latter's mouth, and comes in a south-westerly direction from a point nearly south of the first fall on the Abitibi River. A portage from the head of this river to the fall forms a route by which the Indians often travel in high water, when ascending the Abitibi River, as by so doing they escape the strong currents and rough rapids and falls of the Abitibi River. We did not canoe very far up stream before it became so shallow as to stop any further progress. The soil was still the fine sandy clay loam as before.

There are two other streams of considerable size entering Black River and although they are of no importance as water-ways yet they deserve some mention. About 13 miles up stream is a river entering from the south west. At its mouth this stream is ten yards wide, and about 6 feet deep. The current is strong and the water clear and cold, but like all the streams of this country it is soon blocked up by driftwood, which the Indians never seem to clear away once it lodges.

About twenty-two miles from the mouth of the Black River we have another small stream entering from the north east. This stream is about ten yards wide and about three feet deep, but is not navigable owing to driftwood.

We thus see the Black River flows in a more or less north westerly direction for twenty-six miles through a country timbered with spruce, poplar and balm-of-Gilead. The soil of this area is exceptionally suited to agriculture being a clay soil carrying just enough sand to make it easily workable, and not heavy, hungry clay. The area too is rolling, being thus easily drained and easily cleared.

ABITIBI RIVER.

Leaving the Black River we now go up the Abitibi River towards Lake Abititi. In the first four miles we have only one outcrop, and even it is very small and at once covered up on reaching the shore; in fact it is likely covered in high water altogether. It occurs at one and a half miles above Black River and is an outcrop of green diorite carrying pyrite. At the end of the four miles however we have a large exposure.

THE TWO PORTAGES.

Here we have two falls, the first or lower fall has about nine feet of drop, over a ridge of the same green pyritous diorite, more or less schistose, according as we examine nearer or farther away from cracks or fractures. In these cracks are little veinlets of quartz as usual about one to two inches wide and two or three feet long. The ridge strikes about north-west, and the fall takes place through two gaps, each of which would be thirty feet wide and would afford good water powers. The ridges rise on either side to a height of forty feet but drop away inside of half a mile to rolling land as usual. Glacial striae were seen here running north fifteen degrees west.

The green diorite continues all along the eastern shore to the second or upper fall, two hundred yards above the first. This second fall might also be called a rough rapid of eighty yards, with a total fall of six feet. The diorite here is a little more schistose than that at the lower fall, but differs from it in no other way. Glacial striae was very plain here running north six degrees west. These falls are called the two portaging places as they are both passed on the south east side by portages of one hundred yards while the small stretch of river between them is crossed by the canoes.

In the next four miles we pass three outcrops of green diorite, not schistose and lost under the clay, etc., upon reaching the shore. At the end of the four miles we come to a stiff rapid, which however is navigable, here the rapid is over the same diorite but it is schistose, and has the small quartz veinlets mentioned above. We usually find these veinlets occurring at rapids, falls, etc, or places where fracture, etc., has caused lines of weakness, along which the quartz vein could form. Glacial striae were observed here running north fifteen degrees west.

After two miles of quiet water, we reach another rapid which is the first of a series, which extend more or less regularly for one and a half miles, all over the same green diorite schist.

At the head of these rough waters we have a mile of smooth quiet water, then a very rough rapid which can be run in descending but must be portaged in ascending. The stream is divided into two channels by an island, over which is the portage of one hundred yards. The fall is over green diorite schist, filled with small veinlets of quartz, none however are of economic value. The ridge strikes north ten degrees west, and is covered upon reaching the shores by the clay loam of the district.

About one and a half miles above this rapid is a stream twenty-five feet wide and two and a half feet deep entering from the south-west. We canoed up this stream for about a mile, when flood wood stopped us as usual. Leaving our canoe we walked on and came to a low, wet marsh of about fifty acres, in the middle of which was a small lake, from which the stream flowed.

About three miles farther up we have another small stream entering from the south, south-east, but it is not navigable for over three-quarters of a mile.

About half a mile above this stream we have two bluffs of green massive diorite on the south side of the river, while on the opposite shore are low, flat plains. These bluffs rise twenty feet from the water's edge, and the bluffs are about three hundred yards apart; they strike off south east, and reappear several times for a mile or so and then are lost in the rolling area.

During the next four miles up stream we pass several small outcrops of green schist carrying the veinlets of quartz, and all show the marked effects of weathering. At the end of this four miles we come to a high fall with a total drop of forty-five feet, including the short rapid at the foot. The fall takes place over a ridge of quartzose diorite showing in several places somewhat of a schistose structure. The strike of the ridge is north thirty degrees east and glacial striae were well shown running north thirty-five degrees west.

There is a great flow of water in the Abitibi River and the water power at this fall is enormous. There was no way of actually measuring the quantity of water passing; but when we remember that all the water of the Abitibi River, with its strong current, passes over this fall, we can see the power is very great. The fall takes place through two gaps, with high rocks on either shore and the central mass of rock, around which the water flows is also considerably higher than the level of the water in the river, so we see that this fall is admirably suited to become a good water power.

Dokis River.

About one and a half miles above this fall is a river of considerable size entering from the south. We called the river Dokis River from the name of a hunter who has built two huts on it and who lives here. The river is about fifty feet wide and four feet deep, but forks about one quarter of a mile from the mouth. The main stream above the forks comes from the south, and is very cold and clear. We canoed up it two and a half miles when we had to leave our canoe and walked on south for three miles but did not see where it came from; Dokis however afterwards informed us that it flowed out of a chain of small lakes not far from where we turned back. While walking up this stream I crossed two ridges of white quartz sand but the general country was fine sandy clay loam of good quality.

The small feeder, which enters Dokis River about a quarter of a mile from its outlet into the Abitibi River, is navigable for about half a mile; we followed it on for another half a mile when we could step across it. The water in this feeder is dark, blackish swamp water and the stream as a whole a stagnant one almost.

Coming back now to the Abitibi River we go up about three quarters of a mile when we see what appears to be another fine stream entering from the south east. On entering it however we find its size is due to the backed up water of the Abitibi River. We canoed up this stream about one and a quarter miles when it became very small and unnavigable. In that one and a quarter miles we have three large creeks entering the main stream and hence we see the reason for its falling away so soon to a small creek itself.

Area Drained on South Side.

In describing the soil of this area we can safely say that the whole area from Black River to within a mile of Lower Lake Abitibi is admirably adapted to agriculture, being a fine rolling area of sandy clay loam, just sufficient sand to make it easily workable. The ridges are rarely over twenty feet high, but are generally only slight undulations. The timber of this whole district is chiefly spruce and poplar with some balsam and balm-of-Gilead. The spruce and poplar would average in pulp-wood about four cords to the acre, that is, of trees six inches and upwards in diameter. Some of the higher ridges carry a few jackpine, but there is only sufficient of all timber to provide for settlers' buildings and firewood.

Lower Lake Abitibi.

Having now reached Lower Lake Abitibi we will examine the south shore of the lake as far as the Quebec boundary line. Starting east then we pass about two miles of high sand and gravel shore, which extends back from the shore for a mile or more in some places then shades off to sandy clay loam. These hills are timbered with spruce and poplar as usual. But we have also a good percentage of red pine, jackpine and a few white pine on the hill nearest the outlet of the Abitibi River, we have therefore called that whole point Red-pine Point. After passing this higher land we have lower sandy and gravelly shore, which continues all along the south shore of the lake, including the long point six miles east of the Abitibi River, and farther along the south shore to include the next shorter point, as shown in the map of the district and labelled "sand and gravel."

Huronian Diorites.

The shore as a rule is lined with irregular angular boulders of granite, syenite, etc., but we have only three outcrops of the underlying Huronian rocks in place, between the outlet of the Abitibi River and the long point on Lower Lake Abitibi. These are all green diabases carrying pyrite and appear to have been part of a big trap-flow, for in several places the columnar jointing so characteristic of traps and basalts is very plainly shown, the columns being five, six, etc., sided columns. A piece of one of these, upon examination under the microscope, shows it to be a dark trap containing basic plagioclase most of which is still fresh but some pieces are altered to sausswrite. Through the rock are many distinct crystals of augite and some of these are rapidly altering to chlorite. The section also shows considerable magnetite and some pyrite, as well as several small needle-like crystals of apatite. We thus see that the rock is a basic diabase showing very marked weathering.

This green diabase outcrops in two places about a quarter of a mile apart on the south shore and about two miles south east of the outlet of the lake. Its rapid weathering has deposited much magnetite in the sand along the beach near it, so that the sand is quite magnetic, and by drawing a magnetic through it will collect quite a load of magnetite.

About fifteen feet from where I took the sample described above, I took another which appeared to be more weathered and was also nearer to a vein or series of little veinlets. An examination of it shewed it to be a basic rock as before, containing much felspar, some of which was altered to scapolite, but most of it gone on to sausswrite. This rock showed very much magnetite both as allotriomorphic grains, and as little groups of octahedra. The sections showed very much hornblende which no doubt gives the rock its greenish cast. The sections also showed several crystals of augite, or rather parts of them, as they are rapidly altering to chlorite, which also accounts to some extent for the green color of the rock as a whole.

Nearer to the veins or cracks, the diorite has quite a schistose structure; and nearer still looks quite like a chlorite schist; and nearer still, and closer to the water, where it is subjected to constant wave action it is quite soft and putty-like and appears quite like a talc-schist. Now these green diorites carrying magnetite and pyrite are the general rock of the whole country; we find that everywhere one large Huronian are of diorite and green schists with occasional bands of porphyrites as mentioned on Black River and some others which will be mentioned later, also some diabases, etc., which will also be mentioned later; so that when this diorite or green diorite is referred to all through this report, it is the rocks as described above, that I refer to.

Low River.

Passing around the small point about two and a half miles east of the outlet of the lake, or about half a mile east of the outcrops described above, we come to a round bay at the south-west side of which, we have a stream entering from the south. This stream is about forty feet wide at its mouth and fully six feet deep, but little or no current flowing out of it. It soon falls away to a small stream, as it is fed about every quarter or half mile by a side stream of very good size.

The first mile of this stream is through a low, wet sand flat, covered by moss and low brushwood; we then came to the western side of the flat and met a series of sandy and gravelly ridges clothed with spruce, poplar and a few pine, these ridges extend in a south-west by westerly direction to the Abitibi River.

Our river now follows the eastern edge of these ridges for about a mile, then flows between them for another one and a half miles, and then through lower rolling clay loamy land, and bends off to the south-west, then west and then north coming from a small lake or rather two lakes, connected by a small creek as shown in the sketch for map.

At the north end of this lake is a brulé which extends for two or three miles towards the Abitibi River and Lake. The Indians of the district go here to pick berries for the summer. The brulé also extends in a south- south-easterly direction for many miles.

Rocks on Low River.

There are several outcrops of rocks on this river. In the first three-quarters of a mile we met three outcrops of the same diorite as we found at the lake, but here we pass into a very acidic rock, a quartzite belt which strikes east-south-east. I afterwards found this same belt outcropping on the east shore of the long point on Lower Lake Abitibi; but this will be referred to in its place. In some places the quartzite was white, its natural color as a rule, but in some places it was reddish very much like a jasper, and had a baked or burnt appearance, probably caused by the green diorite referred to above. About two miles farther up stream we came to a small rapid with a portage on the east shore of twenty yards. The rapid is also over a quartzite which may or may not be this same belt, as the whole is covered up by clay between this outcrop and the last one two miles below, but is probably the same.

Above this rapid the river was too narrow and shallow to canoe far, but leaving our canoe, we found the land to drop away to fine clear loam, timbered chiefly with spruce, running five cords of pulpwood to the acre.

Long Point.

Coming back to Lake Abitibi we go east, then north, and after one mile we come to another narrow point vith green diorite in place; going about half a mile farther north we come to a portage of five hundred yards across the narrowest part of the long point in Lower Abitibi Lake. This portage is on the route from the Abitibi River to the narrows of Abitibi Lake and besides saving all the rough water of Lower Lake Abitibi, also cuts off several miles of the trip. In the middle of this portage is a small clear lake having neither inlet nor outlet. The long point is one ridge of sand, gravel and boulders, and would average about one mile in width and about six or seven miles in length. The whole point is timbered with spruce and poplar, but at the north-west corner we have several small ridges timbered with red pine and a few white pine. The west side of the point is much indented, while the east side is one long curve to the north-east from the portage.

Starting again then at the portage, and on the east side of the point we travel south to a round bay. The sand and gravel of the point continues past the foot of this bay, but soon begins to die out. As we go inland we meet a series of sandy, clay ridges running north-west and south-east. These ridges are close together and the valleys between them are very narrow. We passed eight small lakes on our walk inland at this point. They-were about fifty yards in width and varied in length from seventy-five yards to a quarter of a mile. The ridges all carry some jack pine, red pine, and a few white pine as well as spruce and poplar; beyond these ridges the land becomes rolling clay loam.

About one and a half miles south of the portage and about half a mile from the foot of the bay is an outcrop of the green diorite, in contact with a band of quartzite, the same band as was described above as occurring on the small river on the west side of the point. This quartzite strikes east-south-east. The quartzite is stained quite reddish like jasper near the contact, but farther away it becomes typical white quartzite. The contact cannot be examined as the outcrop is very small and is at once covered by the sand and gravel of the point.

Laurentian Granite.

Going in a north-easterly direction from this bay, we pass around a small point and enter a long bay extending fully five miles south. The west shore runs about north and south. After passing about three miles of this shore, which is still sand and gravel, we come to an outcrop of Laurentian age, namely, a horn-blende granite. This ridge strikes north twenty-five degrees east, crosses the bay and appears on the opposite shore and forms the whole north end of Biederman's point. It then crosses the narrows and appears again on the opposite main shore, for its continuance after that, see Mr. Coulthard's report.

Glacial striæ were observed on it running north twenty degrees west. The ridge is not very wide here, and is at once covered, above the water line, by the glacial drift. This band of hornblende granite is not more than a half a mile wide here, for within half a mile from meeting it, I find the green pyritous diorite to come in again. This hornblende granite is quite coarse grained, very much coarser than that found on the Black River, and in some places the collection of the hornblende, etc., into a more or less layered set of bands, gives the rock a gneissoid appearance. I do not think, however, that it can properly be called a gneiss, for by far the greater bulk of it has no gneissoid structure whatever. On the weathered surface of this granite the felspar has gone to kaolin and the distinct pseudomorphs of kaolin after felspar are very plain. The felspur has so altered in some places as to give the whole rock a whitish appearance and from a distance makes it look like a quartzite or if a narrow strip shows, it looks like a quartz vein. A petrographical examination of this rock shows it to be a coarse grained hornblende granite, with the felspar rapidly altering to kaolin, as is also shown in the hand specimens. The felspar too is full of inclusions which, however, are unrecognisable on account of the dull and opaque kaolin. The rock also contains considerable quartz, which when near the felspar surrounds it, or at least includes it sufficiently to show that it crystallized since the felspar did.

The section shows a great deal of hornblende which is as a rule considerably altered, some of it being quite fibrous; and from many different colored pieces, as well as different amounts of pleochroism visible, we can see that we have several stages in the alteration present. The cleavage of the hornblende at sixty-two degrees and one hundred and twenty-four degrees is very well shown in most of the hornblende in this granite. Several of these pieces of the mineral look like biotite, but in the examination of my sections I can find none in reality. Some of the hornblende too looks quite like an orthorhombic pyroxene from its greenish yellow color and strong double refractive colors, but I believe it is simply a different stage of alteration in the hornblende. Moreover, the rock is too acidic to expect to find pyroxene present.

After leaving the hornblende granite we again meet the green diorite as mentioned above. Here also is the beginning of a recent brule, burned probably in the fall of 1899, as the trees are all dead and no green shrubs or plants of more than one year's growth are to be seen on it. The soil is a very fine clay loam.

Coming on to the south end of the long bay mentioned above, towards the east side, is a river twenty-five feet wide and fully six feet deep entering from the south. The water is dark brown and there is little or no current in the stream up which we canoed for three miles when it became not more than five feet wide. The whole basin is a low wet marsh, grown up with marsh hay, willow and alder brush, bordered by a dead tamarac swamp. In the three miles of river travelled we passed eighteen feeders, some larger and some smaller, but this will show why the river falls away to so small a stream in so short a distance. All these feeders are but the drainage of this large marsh, and the dark, dead water is almost stagnant, from the low flat country, giving it no current to speak of.

BIEDERMAN'S POST.

At the east side of this bay is another long point which forms the west shore of the "Narrows" connecting Upper and Lower Lake Abitibi. The point is about one mile average width and three miles in length from the foot of the bay on its western side.

About one and a half miles from the foot of the bay, and on the west side of the point, is an outcrop of the green diorite which shades off gradually into a diabase, which continues for a quarter of a mile when we pass into the Laurentian hornblende granite again which, as I mentioned above, constitutes all the north end of this point. After crossing the bay it continues in an easterly direction, and crosses the "Narrows" and reappears again on the main shore. For its description from here on, see Mr. Coulthard's report.

To return to the diabase just mentioned. The passage from the green diorite to the diabase is quite gradual but coming up nearer to the granite we find a trap dike separating the hornblende granite on the north side, from the diabase on the south side. This dike is a very fine grained one and is about two and a half feet wide, striking north thirty degrees west and dipping towards the north at an angle of sixty degrees. Farther from the dike the diabase becomes an olivine gabbro, very magnetic and rapidly weathering, forming a magnetic sand which has deposited all along the beach.

The olivine diabase however has a distinct ophitic structure, and small lath shaped pieces of plagioclase penetrate both the augite and olivine crystals present in the section, showing that the usual order of crystallization has not been followed, that is, the felspar crystallized before the pyroxenes instead of after them, as we would naturally expect. The plagioclase felspar it very basic, and is much twinned, some pieces showing both the "albite and pericline" lines clearly. In addition to augite, olivine, and plagioclase we have considerable magnetite in allotriomorphic grains, biotite, mica, and many minute needle-like crystals of apatite. We thus see how very much this olivine diabase resembles the olivine diabase from about Sudbury, Ont. They are alike in almost every way.

This whole point is covered with beautiful clay loam and timbered with spruce and poplar of very fair quality. I saw one poplar tree here fully thirty-six inches on the butt.

On this point. and at the south end of the "narrows" is a trading post belonging to W. F. Biederman, who has been among the Indians there for four years. Like the Hudson's Bay Company, he provides them with provisions, etc., for the hunting season, and buys their furs in the spring. He has a house and a two storey store, also a garden in which he has grown successfully potatoes, onions, carrots, corn and turnips; he has never tried to grow cereals yet. The soil then on this point is splendid and continues so as we go south to the lower side of upper Lake Abitibi.

DIORITE PORPHYRITE.

Coming south then from Biederman's post we have the clay soil continuing to the water's edge for four miles, at the end of that distance, however, we came to an outcrop of a green diorite porphyrite, quite like that mentioned on the Black River. The groundmass of this rock is a very fine grained felspar having many large and distinct crystals or phenocrysts of felspar all through it. This felspar has nearly all altered to kaolin, so that we have distinct pseudomorphs of kaolin after felspar. In addition to these phenocrysts of felspar we have many others of hornblende, which is rapidly becoming fibrous and some appears to be even gone to serpentine or chlorite. Through this rock too are many scattered grains of magnetite. The rock then has a distinct porphyritic appearance, and from its constituents is properly a diorite porphyrite. The rock only shows for a very short distance from the shore when it is covered by clay loam. Along the water's edge it soon shades off into the usual green diorite carrying pyrite. These rocks continue to the foot of the bay south of Biederman's point and then they occur in a continuous belt all along the south shore to and into Quebec. This south shore is rocky from end to end, and is all the same green diorite; but a few feet from shore they are covered by clay.

Interesting Glacial Striae.

A very interesting set of glacial striae are found on the rocks of this south shore. At the south west corner of the lake we find glacial striae running north twenty-five degrees east. About three miles east of this we find the striae running north eighteen degrees east. Then about one mile farther east they run due north and south. One mile farther east they are seen running north ten degrees west, and about one and a half miles farther east than these we find them running north thirty degrees west. They now continue thus till we pass "Cauliflower Point" where we find striae north thirteen degrees west. We thus see that within a radius of six and a half miles we have sets of glacial markings, which vary in direction fifty degrees, and this I think is rather unusual.

Some of the variations may have been caused by a series of hills which occur here, varying from one half to twelve miles from the shore, and ranging in height from two hundred to four hundred and fifty feet. These hills are all green diorite and no doubt it is one of them that Mr. Parks reports as occuring a few miles east from the head waters of the Black River. I counted fifteen of these hills, most of them being covered by clay and timbered with spruce, poplar and birch about their bases, and jackpine on their tops. A few however are bare, having been burned off by lightning. On one of these bare hills two and a quarter miles from shore I found glacial striae running north ten degrees west, while at the shore of the lake two and a quarter miles due north I find them running north thirty degrees west.

Ghost River.

Going back now to the southwest corner of the lake, we have a river thirty yards wide and quite deep entering from the south. This river is called Ghost River by the Indians, from some thrilling experiences through which some of them claim to have passed. At the mouth of this river are three log houses, well-built and neatly kept. About each is a small garden of potatoes, onions, etc. The soil is a fine sandy loam, as we find all through this area, and well-adapted to agriculture. We canoed up this river about eleven miles, and found it still to be flowing through fine arable land.

About five miles from the mouth, the river branches to the west, we followed this branch west, then northwest for three and a half miles, when it again forked and the branches were too small to allow canoeing farther. The land back of this branch is higher than that immediately along the main stream. The whole country through here is timbered with spruce, poplar and balsam, with a few balm of Gilead; but would only suffice for settlers' purposes.

Coming back now to Upper Lake Abitibi, we travel along the south shore in an almost due easterly direction. The shore is very much indented, the points being the ends of innumerable small more or less parallel ridges of the green diorite, all striking about north ten degrees east.

Lightning River.

The south shore then is unbroken except for these little points, for seven miles east of Ghost River. Here, however, we find another river entering from the south. This river is about forty feet wide and very deep with little or no current. It has been called Lightning River by the Indians from the fact that in some very severe thunder storms, a couple of the above mentioned hills were burned over by the lightning. The river passes between three or four of the hills and some of the Indians were caught in the storms of this river, and were very much frightened, so they called it Lightning River.

We canoed up stream six miles when the river forked and became too small to allow us to canoe farther. About 2¼ miles from the mouth and on the west bank of the river is a diorite hill 200 feet high. This hill comes right to the river's edge, and being one of the hills burned clear it affords a grand view of the surrounding country. I could count three other hills from this one, burned clear, while the remaining 10 or 12 were densely wooded with spruce and poplar, like the valleys between them. The summits of all these hills are covered with scrubby jackpine, but their sides are timbered as described above. We climbed four other hills and found them all alike. The soil is the usual sandy clay loam, and on the hillsides are many rounded boulders of granite, gneisses, etc., which have been dropped by the glacial drift and exposed by the later washing down of the soil by water, rain, etc. This hill is the only rock exposure seen on the river, the other hills being covered by soil and timber as described.

Returning to the lake we go east about one mile, when we come to another long point about three miles in length, and running about north and south. It is in shape so much like a cauliflower that I have called it by that name in order that there may be no confusion of these long points. Landing at the foot of this point and at the west side of it we walked southeast about one mile to a high, burned-off hill of diorite. This hill is 450 feet in height, I should judge, and affords a most magnificent view of the whole country for 20 miles around. Lake Abitibi can be distinctly seen from here, and an exact idea of its shape and appearance is obtained. The lake is seen to be dotted with islands and I think we would be quite safe in saying there are 500 of them in the lake. Some of these islands are quite large, especially four of them lying a little northeast of the head of Cauliflower point. They are, like the main shore, green diorite and diorite schist, covered by clay loam and timbered with the usual spruce, poplar, balsam and some balm of Gilead. One of these large islands we called Dominion island, as we were stormbound on it for three days, one of which was July 1st, Dominion Day.

About the middle of Cauliflower point is a dead tamarac swamp, but this is only about ten acres in area, while the rest of the point is fine soil and timbered as usual. At the foot of the point and on the east side of it is a deep bay with a sandy beach, but soon gives way to the usual much-indented rocky shore, which continues in an east and westerly cirection for about three miles. The shore then turns off to the southeast, and after two miles in this direction we come to a small river about eight yards wide and very deep, but on entering it we find its size is only due to the backed-up water of the lake, for inside of half a mile it becomes unnavigable. The soil through which it flows and that behind it is still fine sandy clay loam.

Coming back to Lake Abitibi we travel southeast about 2½ miles to another deep bay, the south shore of which is clay banks, but towards its eastern side it soon passes into the usual rocky shore, which runs out due north like a long point, but is not such, for at its north end it strikes east and passes into Quebec in about one mile, still continuing its easterly direction.

General Character of Soil.

As regards the soil of this whole area south of Lake Abitibi and extending in a southwesterly direction to the Black River, we can say that it is excellently suited to agriculture, being a fine sandy clay loam, just sandy enough to be easily workable. The immediate shore of the lake, and more especially Lower Lake Abitibi, is sandy but this soon gives way to good clay loam back of it. The cause of the sandy beaches, I think, is that the clay has been washed out and carried away for, as I have stated before, Lake Abitibi and River Abitibi are always muddy, and in fact the Indians tell me that Abitibi means "the water that shines white," referring to the turbid clayey colour of the water. The lake is very shallow. Mr. Biederman told me it would not average eight feet, so that it is always stirred up. The absence of much sand in Upper Lake Abitibi is likely because of the rocky shore in contrast to the clayey ones of the Lower Lake Abitibi.

There is then fine clay loam in general all over this area as indicated on the map, and is covered by a rather dense growth of spruce, poplar, balsam and balm of Gilead, but not in sufficient size or quantity to be of commercial value as timber nor even as pulpwood. The trees are, as a rule, small, and the land would be easily cleared and being of a rolling aspect would be naturally drained.

The great difficulty at present of settling that country is transportation. Of course the present canoe route from Baie-des-Pere's on Lake Temiscaming to Hudson Bay Post on Lake Abitibi is out of the question for settlers. Once a railroad runs through or near that district there is nothing that I can see to hinder its rapid settlement.

Indian Occupation.

The Indians have lived there for many years, longer in fact than most people have any idea of, I found a very interesting proof of this, for while camping on Dominion island in Upper Lake Abitibi I was examining the shore when I found a flint arrowhead well made. Those we know were the weapons of warfare, etc., used by our earliest Indians. I consider that this is a very interesting fact in connection with that northern Ontario so new to us now.

The Indians now living there are members of the "Cree" tribe, but are by no means that noble red-man described by Fenimore Cooper. They are an idle, careless and indifferent people, earning a hard and scanty living by hunting in the winter months and fishing in the summer. They live in tents the most of the year, a few of them build teepes or wigwams in the winter, but most of them live in tents even in the winter.

The half breeds and there are many of them among these Indians, usually have huts or log-houses and show more thrift than the Indians, for they usually have a little garden of potatoes, onions, etc. They as well as the Indians get a livelihood by hunting furs during the winter months for the Hudson Bay Company. They leave the post each September with their winter's provisions, etc., and go to their several hunting grounds, living there in tents and huts till June, when they return to the posts, give their furs for the provisions of the previous fall, then lie around the posts all summer doing nothing. Among the fur-bearing animals hunted by the Indians I would name the following in order of importance, numerically, marten, muskrat, moose, bear, mink, beaver, caribou, lynx, otter, fisher, gray-fox and an occasional silver-gray fox. There are no red deer in this district, partly because there is no marsh hay for pasture, and partly on account of the wolves, and the Indians say that the caribou are being driven out by the wolves also.

General.

In the way of general information regarding this district, I learned from Mr. Biederman and others, that the lowest temperature of winter is thirty-five degrees below zero, but as a rule the lowest is twenty degrees below zero, while the general run is not over ten degrees below zero. Snow usually flies early in October or even in September, but winter does not set in properly before the first week in December, and usually breaks about the last week of April to the second week in May. The snow is seldom over two feet deep on the level, as winter rains keep it down. The travelling is done on the crust and on the lakes with toboggans and dog-trains.

The lake freezes about the first week in December and usually breaks up about the second week in May, after which there are no frosts till the following September.

M. B. Baker,
Geologist, Exploration Survey Party No. 1.

I, M. B. Baker, of the City of Kingston, County of Frontenac, geologist, make oath and say that the above statements in this report are correct and are the results of my explorations in Exploratory Survey Party No. 1. M. B. Baker.

Sworn before me at the City of Toronto, County of York, this eighth day of January, 1901. (Sd.) J. A. G. Crozier,
A Commissioner.

GEOLOGIST'S REPORT OF EXPLORATION SURVEY PARTY NO 1.

Toronto, Jan. 15, 1901.

T. B. Speight, Esq., O.L.S.
In charge of Exploration Survey Party No. 1.

Sir,—In compliance with instructions received from the Director of the Bureau of Mines on the 12th day of June, 1900, to the effect that I act as one of the Geologists on Exploration Party No.1, accordingly I put myself in communication with you so as to be in readiness for the day of departure.

The duty of Party No 1 was to run a base line due east from the 198th mile post of the Algoma-Nipissing line across the District of Nipissing to the boundary between Ontario and Quebec, a distance of 70 or 75 miles, to be explored 50 miles on each side of the base line.

Our party consisted of yourself as Surveyor in charge, Messrs. P. F. Graham Bell and Thos. G. Taylor, land and timber estimators; Mr. M. B. Baker and myself, geologists; together with the surveyor's assistant, chainmen, axemen, packers, guides, etc. making in all a party whose number varied between 23 and 26 men.

My exploring excursions during the summer were made in company with Mr. P. F. Graham Bell; also we had as helper Louis Montreuil of Mattawa, who proved himself, entirely satisfactory and agreeable on all occasions.

The part of the country assigned to me to be explored geologically, is that portion lying south of the base line which Mr. Speight ran during the season, and north of the Abitibi River and Lakes.

Because of the large tract of land which this involved, and because of the limited time at our disposal, it must be at once apparent that a great part of the district was left untraversed by us. Nevertheless I feel satisfied that sufficient country was covered to give a fairly definite idea of the whole—geologically and otherwise.

With regard to the specific rocks on the north shore of both Abitibi Lakes, I was unable to examine very many of the islands, but gave my time mostly to the shore work.

During the course of the summer I took samples of some quartz veins which I encountered, for the purpose of having them assayed for their gold content. On my return following the instructions of the Bureau of Mines, I sent these to Mr. J. Walter Wells, Assayer for the Bureau of Mines at Belleville. The results as obtained from him I have included in this report.

Leaving Toronto on the 18th day of June our route lay via North Bay and Mattawa to the northern terminus of the Temiscaming Branch of the Canadian Pacific Railway at the foot of Lake Temiscaming. Here we took steamer to the head of the Lake. From this point some of us took canoes and ascending the Ottawa River reached Lac des Quinze, passing on the way fifteen rapids or chutes where portaging was required. Here we joined the remainder of our party who had walked over from the Head of the Lake.

Our course to reach Abitibi Lake now lay by way of Barrier Lake, Long Lake, over the Height of Land, Island Lake, Upper Lake, down the Abitibi River to Abitibi Lake. Eight portages were necessary between Des Quinze and Abitibi, only one of them being of any length, that at the Height of Land, which is over half a mile.

This waterway has been described by Mr. Wm Ogilvie* and by Mr. Walter McOuat† so that redundancy on my part is not necessary.

Owing to the weather and to certain other circumstances, we did not reach the Abitibi Lakes until the 28th of June.

After spending a few hours at the post of the Hudson Bay Company, of which Mr. Robert Skene is the Abitibi factor, we proceeded west along Upper Lake Abitibi. Wind and rain bound us for three days on this Lake, and it was July 2nd before we arrived at the extreme north end of the Lower Lake, entered the Tabasaqua (Low Bush) River, and began its ascent for the purpose of reaching the starting point of the Base Line. By reference to the accompanying map it will be seen that by following up this river to its head, crossing over a height of land, and passing down a series of lakes which form the headquarters of Little Abitibi River as far as Harris Lake, then turning westward through another series of small lakes, the beginning of the line can be easily reached.

Low Bush River.

Low Bush River is the largest one flowing into either of the Abitibi Lakes from the north. It is four or five chains wide at its mouth. We found it possible to ascend by canoe in a north-west direction about 36 miles, at which point the width was reduced to an average of about six feet.

The current is not noticeable until about three or four miles from the mouth. From here up, however, it increases gradually as the width diminishes. There is a fairly good volume of water.

All along evidences of higher levels were seen on the banks. This is due to the tremendous increase in volume toward the latter end of spring. Near the mouth it was

*See "Report of Exploratory Survey to Hudson's Bay" by Wm. Ogilvie, D. L. S., 1891—Dom. Gov.
†Report of Progress, Dom. Geological Survey—1872-3.

seen that the water rose about 2½ feet higher, while about 25 miles up a height of over 12 feet above the present level was obtained. During a dry spell it is impossible to accomplish 38 miles on the river.

It was necessary during our ascent of a considerable portion of the upper and narrower portion to chop a passage because of the numerous fallen trees and snags which barred the way.

In all there are about 10 or 12 rapids or small chutes met with. None are over a few chains long, nor does the fall in any one case amount to over 4 or 5 feet. About 9 portages are necessary. I should imagine that some of these rapids could, because of the high banks, be utilized for motive power; but not to any considerable extent.

There are only 3 outcrops of rock on the river. These are at the first, sixth and eighth portages from the lake At each the chutes, which are very short, flow over bed rock, causing a difference in level of no more than 5 feet in any one case.

The rock at the first portage is schistose, but my sample went astray, so that I am unable to state definitely its composition. It had a strike almost east and west and dipped slightly to the south. My recollection is that it is Huronian and serpentine in character.

At the other two portages, which are 15 and 18 miles from the mouth respectively, the rock is pink granitoid gneiss with an indeterminate strike. The last portage, about 20 miles up, is necessary because of a landslide which carried down trees, etc.

The rapids at the other carrying places run over Laurentian boulders.

The banks of this river are typical of the rivers running into the lakes from the north. They rise gradually from the water to an elevation varying from 5 feet to 25 or more. This continues for a few chains back when the country falls into swamp, draining being prevented by the banks which represent the sedimentary deposit of centuries during the spring seasons, the river in the meantime gradually wearing out a lower bed.

The soil is all good clay with a covering of humus and moss, the latter being the natural result of the wet ground.

As the upper part of the river is approached the banks are more abrupt, also more broken. The soil becomes more loamy in character near the river, the banks showing gravel and sand at the water's edge. As mentioned before this is the typical river of this region.

Good spruce and poplar, also balm of Gilead, small white birch and balsam, soft maple saplings, etc., are found along the banks for the first ten or twelve miles up, but run thus for only 5 chains from the river, where it falls into scrub. Farther up the timber becomes smaller and small Banksian pine makes its appearance.

The upper part of the stream, which is brownish in color, abounds in speckled trout, some of which are of good size. But two tributaries of importance flow into Low Bush River, and both from a north north-east direction. The first called Circle River, is 3 chains wide at the forks, which are about three quarters of a mile from the mouth of Low Bush River. The other, named White Water River, is met about 14 miles further up stream, where its width is equal to the other branch, i.e. 1½ chains.

Circle River showed very little volume at the time we ascended. 18 or 20 miles up the first rapid is encountered and flows over boulders; it is about two chains in length, with a fall of say 5 feet. The volume is evidently much greater in the spring. The soil through which it runs is all good clay, splendid elevated stretches of which extend for some distance near the rapids and lower. Beyond, the country is rolling, with swamp in lower levels.

Some very good patches of poplar and spruce grow along the banks but continue for the usual distance away from the river. Small Banksian (commonly known as pitch or jack) pine is the prevailing timber on the high level stretches spoken of. It is of small consequence, as is also the case with the balm of Gilead, balsam, white birch, etc.

I met with 4 outcroppings of rock on this river. The first one near the mouth being a dark green, fine-grained chlorite schist, with a strike east north-east, and dip slightly inclined to north. Half a mile further is a fine-grained Huronian gneiss, which in thin section under the microscope showed a somewhat similar structure and composition to Lawson's Couchiching series. The next outcrop, over a mile above this one, is a dark compact altered diabase. The last one seen was just a few chains below the rapid, it being a medium-grained grey granite, with a suggestion of gneissic structure.

The feeders which run into Circle River could be ascended for but short distances by canoes. White Water River is so called from the color caused by the clay which it carrries in suspension. It is not of great length. At the head of Low Bush River a portage 2½ miles long leads in a west north-west direction over a height of land, which extends east by south and west by north and passing into the Province of Quebec. The waters to the north of this height flows directly into James Bay, while those to the south find their way to the Abitibi Lakes, reaching James Bay eventually by the Abitibi River.

The long portage just mentioned runs for the first mile or so over a series of clay hills; then through some swampy land, when it rises, the first half mile being over a level sand plain, which is open except for some scattered scrubby Banksian pine. This brings us to a small clear water lake, called Welcome Lake. Crossing to the north end of this lake another portage of half a mile brings us to Lake Michel, also a small lake. This trail passes in a north north-west direction over sandy soil, and is comparatively smooth, but drops quite abruptly at the north end.

The timber observed after leaving the Tabasaqua is small spruce, poplar, Banksian pine, birch, balsam; also mountain ash and soft maple saplings. Two outcrops of coarse pink gneiss on the S. W. shore of Lake Michel were noticed. At the north end of Michel Lake another long portage to the north-west connects with a series of three small lakes (Lacs La France) from which James Bay can be reached by water all the way. This portage, which is 2½ miles long, begins in low wet ground but soon rises and for over 1½ miles, runs along the west side of a range of granite hills where it again falls until Lacs La France are encountered.

Passing now northward through the three small lakes we descend about 5 miles of Little Abitibi River, of which these lakes are the source. It is about two chains in width and carries a good volume of water, most of which is received from another river emptying into the most northerly of the lakes just referred to, at but a few chains eastward from where Little Abitibi River runs out. This river receives its water in turn partly from a small lake a few chains up, and partly from a tributary coming in from the east and skirting the lower end of the lake.

At the end of the 5 mile descent on Little Abitibi River we emerge into another lake which has an area of 10 or 12 square miles, and is called Little Abitibi Lake. The Indian knows it by the name Abitibi-shi.

The water in all these lakes and rivers is of a dark brown color, which is due to the non-cultivation of soil and consequent accumulation of decayed vegetable matter or humus through which all the water must originally drain.

Lakes Welcome, Michel and La France are situated in the valley between two ranges of granite hills. The most easterly of these ranges has been already mentioned as being the one along the side of which the last long portage follows. It runs for perhaps six miles in length in a north-west and south-east direction when it falls away at both ends. At the north end it begins to fall away considerably before reaching a point opposite Lacs La France, dropping then into a rolling clay and swampy country. Two of the high hills at the north end of the range rise about 315 feet (aneroid) above the level of the surrounding lakes. From these hill tops an excellent view of the surrounding country is obtained. On both the east and west sides the range drops away abruptly.

Another lower ridge of continuous hills can be seen about 10 miles to the eastward, extending northerly and southerly for as far as the eye can reach. From their appearance this elevation is composed of soil and not hard rock. Its southern extremity seems to terminate in two high rocky hills at which point the Height of Land, which was crossed at the head of Low Bush river, is apparently encountered at right angles. The country between appears to be rolling, alternating between clay upland and swamp. Occasional glimpses of small lakes or ponds are obtained and Little Abitibi Lake can be distinctly seen to the northwest, with gradually falling country intervening.

The range on the west side of the valley extends further north before dropping away than does the eastern range. It extends almost north northwest and south southeast. Three quarters of a mile from the small lake at the north end of the last long portage spoken of, we come to the foot of one of the highest hills on this side Its height above the surrounding water level is, by aneroid, 275 feet. This is the most northerly high hill of this range, for beyond its drop is abrupt, the land rolling between it and Little Abitibi Lake. On the western and northwestern slope, however, the range breaks into a series of irregular barren granite hills until the valley is reached. About one mile distant to the northwest is a fairly high rounded hill which goes down steeply to the west and north. Four miles to the west of this again is seen still another range of hills running apparently parallel with the one just described and probably composed of granite.

The valley between contains two or three small clay ridges coursing irregularly north and south. Much of the country is swampy.

It will thus be seen that the land on both sides is of suitable clay, as far as we examined it, and if proper drainage could be effected it would be quite suitable for agriculture.

Both lines of hills upon which I have just reported are composed of granite of different textures, but chiefly of the very coarse variety. The gneissoid structure also was observed in many places, while strewn all over are coarse irregular patches of orthoclase felspar.

The outcrops on Michel lake are from the same formation, they being at the base of the hills. A dike of dark green weathered diabase strikes north and south directly over the top of the second hill (going south) on the range along which the portage follows. It is 1½ chains wide and stands up from the surrounding granite, especially so at the west wall which the trail follows for a short distance. In some places near the edges I found what were seemingly felspathic inclusions in the diabase.

A similar dike, 2½ chains wide, with the western wall also standing up prominently nd striking in the same direction comes up on the top of the hill on the range just west of the small lakes. In places this one contains considerable evidences of magnetite. Both dikes have a perpendicular dip. The best patch of timber seen was one of good poplar on a clay ridge rising up from the west shore of Lac La France. This ridge falls into swamp before the granite range is reached. Some fair spruce was also observed. In the valleys the higher ground has larger and better spruce and poplar than in the lower and swampy parts where it is chiefly small spruce.

Little Abitibi Lake is shallow and contains four or five small islands along the east shore. It is drained by a continuation of the river of the same name, which runs out in a most unlooked-for place near the northwest corner of the lake on the north shore.

Driftwood River.

Besides the river already mentioned there are two other important ones emptying into the lake. The larger one called Driftwood or Floodwood River comes in at the extreme north east end of the lake, where it is three chains wide. It has its origin in a series of lakes which lie east-north-east from the lake and south from between the 28th and 33rd mile of the base line. I might here mention that these lakes lie at the high land which was seen from the top of the granite hills south of Little Abitibi Lake, and which might therefore be called a subsidiary height of land running irregularly south from the 35th mile post of the line to the already mentioned height. The waters are thus divided east and west.

The distance from the mouth of Driftwood River to its source is over 30 miles by water, but the first lakes are met with about 18 miles up. The course is very irregular and after ascending about 7 miles we come upon the first rapids. For the next 10 miles a considerable number of small rapids are encountered, and all running over boulders. These can be poled up 13 or 14 miles up stream, a portage of half a mile saves over two miles of swift water. The river just beyond the beginning of this portage takes a quick bend to the south south east. This bend is the most northerly point on the river, it being only about 2 or 2½ miles south from the 27th mile. Eighteen miles up the volume is 50,000 gallons per minute. In general the upper portion of the river has more continuous, higher and steeper banks than at the lower part. Also the surrounding country is higher, although the usual swamp is encountered a short distance away from the banks. The soil is all clay with a topping of moss and humus.

Another smaller river emptying into Little Abitibi Lake comes in at the south shore about 2 miles west from where Little Abitibi River runs into the lake. This is called Louis River. Ascending this about three miles through low, swampy or marshy country, the land becomes higher. Up 4 miles it is divided into two branches because of an elevated stretch of good clay land. The branches soon become narrow and frequent fallen trees and snags bar the way. The volume of water is comparatively small.

West of Little Abitibi Lake the land appears to extend, for as far as could be seen from an elevated position, as flat or slightly undulating swampy ground.

On the east side it rises to a low clay ridge, which runs parallel with the shore at a short distance from the lake. The east side of this elevation slopes down into swamp, but in general the country is rolling.

The depressions are all swampy, while the higher portions, although covered with moss, show good clay soil.

Spruce and poplar is the most prevalent timber around the lake and the surrounding district, but the only places where it is of useful size, are in the immediate vicinity of the water, or on the ridges or higher ground. In the swamps small spruce or dead tamarac prevail and there are also found balm-of-Gilead, white cedar, white birch, balsam, dogwood, etc, but aggregating only small quantities.

At the south west part of Little Abitibi Lake are seen a couple of exposures of coarse garnetiferous, white pegmatite ; the felspar is chiefly orthoclose although plagioclose is also shown. On the west shore about 1¼ miles south of the outlet are two outcrops of similar pegmatite, containing sheared and drawn out inclusions of gneiss. Stringers of quartz stand out from the face of the rock.

The outlet from the lake cuts at right angles through a band of micaceous schist for about 250 yards, when it turns abruptly to the west along the strike of the rock which at this point may be considered a very schistose micaceous gneiss. In fact the whole band may be called this. After running along the strike of gneiss for 100 yards the river again turns northward. The walls of the rock by which the water flows with rapid current are not continuous but occur at intervals throughout this distance. Dikes of the coarse, white garnetiferous pegmatite come up through the folds of the schist, some of them containing large crystals of muscovite mica. The strike of the schist or gneiss, and consequently the dikes, is east, seven degrees north. The dip is almost vertical with slight inclination towards the north.

Following the shore along from the outlet towards Driftwood River, I came upon an outcrop of the same pegmatite dike containing inclusions of the above referred to gneiss. But deeper in the bay at this end of the lake several exposures of very coarse pink pegmatite granite occur on the shore and inland at this point. A couple of them were binary in character, and all were extremely coarse. None of the schist was observed at the extreme end of the bay.

Going down the outlet from this lake for one half mile we come to the head of a rapid about 30 chains long and running over boulders. An expert canoeist can run it in a lightened canoe. A few chains farther and the river emerges into Gillestan Lake, which is about 1½ miles in length and one mile in width. Before reaching the rapid two or three outcrops of coarse gneissoid granite are passed.

At about the middle of the east shore of Gillestan Lake is shown some pegmatite which in one place has the peculiar graphic structure containing twisted and contorted inclusions of biotitic gneiss.

Running out from an arm of the lake at the northeast corner, the river continues for 1¼ miles in a north-west course and comes out into Pierre Lake. The current on this part of the river is swift in two or three places and quite noticeable throughout.

Pierre Lake is about 6 miles long in a north north-west and south south-east direction and 2½ miles at its greatest width. The north end is broken up into two irregular arms, from the eastern one of which the river continues in a north-west course for only two or so miles before entering another very irregular lake (Montreuil). From the extreme end of the western arm Montreuil Lake may be reached by a portage of about 50 yards over a small clay elevation. The latter route is shorter, and in ascending overcomes the current which is met in the river.

At its northern end Montreuil Lake narrows down and remains thus for a short distance when it again opens out into another small lake called Harris. It is over this narrows that our base line crosses at about the 12th mile post. Near the north-east end of Pierre Lake there are three or four outcrops of granite jutting up in the lake.

Other than the rocks already described I saw no more *in situ* in this part.

Neither throughout that portion of the base line, nor in the contiguous lakes or country lying between the 12th and 45th mile, did I see any exposures whatever. As regards the rest of the line, although I did not get an opportunity to examine for myself, yet no bed rock was reported by any of the party.

The timber from Gillestan Lake to Montreuil Lake along the water route consists of the usual small trees with occasional patches of fair poplar and spruce.

At the south-east end of Montreuil Lake, however, are about 300 sticks of red pine, being the only ones, with a few individual exceptions, seen by us in the whole district lying north of the Abitibi Lakes.

The soil of that country, on both sides, between Gillestan and Harris Lakes is all clay, but covering much of this are areas of swamp on which stand scrub spruce and dead tamarac. The eastern shores of this waterway seem to rise up higher, showing better land than those on the west The shores of the lakes are for the most part sandy in the bays and bouldery on the exposed points.

Eastward from the Narrows between Montreuil and Harris Lakes the line runs for a long stretch, through country with a gradual upward grade and consisting of moss-covered clay soil and swamp. With the exception of here and there a depression where a creek wends its way or a stretch of barren meadow intervenes, the surface of the land is but slightly undulating.

At about the 31st mile is encountered flat, spongy muskeg, which, however, is of sufficient age to render walking possible. This continues as far as the 34th mile alternating with swamp on which is found scrubby spruce, dead tamarac (small), also small poplar, balsam and white birch. The muskegs grow only very small spruce and even these are sparsely scattered.

A trail made by us leads from the north-west part of Baker Lake, which is near the head waters of Driftwood River, and leading north for 3½ miles, touches the line a few chains east from the 32nd mile post. For 20 chains back from Baker Lake the ground rises to a small sand ridge coursing east and west approximately. To the east of the trail on the north side of this rise is a small lake of 20 chains diameter. The trail then leads along another sandy ridge for about 25 chains.

The next three quarters of a mile runs over moss-covered clay soil, with occasional wet parts. This brings us into a muskeg area similar to that described above and which I will more fully describe later This continues to the line.

A short distance beyond the 34th mile a series of sand ridges, running generally north and south, are crossed. These continue until near the 36th mile. In the depressions between the ridges are enclosed several small lakes, which mostly find their way easterly or south-easterly from this water-shed before turning northerly towards James Bay. The elevation of the sandy ridges reaches 45 feet above these lakes.

Beyond and going eastward from the lakes, along the line, the land drops again into swamp and muskeg, continuing thus as far as I traversed the line (45th mile) or south of it. From the reports of the survey division of our party, the same character of country is met to the end of the line, except that the last few miles appear to touch the southern limit of the Great Muskeg, which is much more spongy than that already mentioned. These muskeg areas to the south of the line are composed of deposits of sphagnum moss, the samples of which taken by me at depths of only 2 to 2½ feet, burned quite readily when dried. The deeper down the deposit the darker was the color of the peat owing to maturer decomposition. The actual average depth was very hard to ascertain as poles 8 or 10 feet long failed to reach the clay bottom in most places.

The area of the deposits was equally indeterminate, but much of the region between the Base Line and the Abitibi Lakes, especially the eastern portion, consists of them.

The beds of peat are covered with a coating of about 6 inches of green moss. Immediately below this it is yellowish green. which gradually fades off into light brown, brown, dark brown and black brown, in order of descending depth. At 2 feet, as a rule, it is dark brown in color. Every year, no doubt, new moss germinates and thrives while a proportionate amount drops away from the roots and decays.

The only timber growing on these peat deposits is spruce, averaging 2 to 3 inches of diameter, and growing very sparsely.

ABITIBI RIVER.

Our examination of the Abitibi River and adjacent country went only as far as the head of the Long Sault Rapids, which are about 70 miles down river.

The general characteristics of this river have been described by Mr. Wm. Ogilvie D.L.S.,* and that part from the two portages down has been observed and reported upon by Dr. Parks.† These reports can be recommended as far as we followed the river, but I may mention that in one or two places we found swift water which had not been mentioned. This, however, is probably due to the local changeable condition of high or low water level, and not to non-observance on the part of the previous explorers.

At the very head of the Long Sault is a clay island some four or five chains in length. A quarter of a mile down the rapid is another small island of grey gneiss. Here also occurs a drop in the river, which runs over a very micaceous gneissoid schist.

On the east side of the river, 25 or 30 chains above the Sault, an Indian trapper's winter trail leads to Pierre Lake in a general north by east but very irregular course. As will be remembered Pierre Lake forms part of the upper waters of the Little Abitibi river. The information regarding this trail was given us by an Indian, who said that it crossed or passed 16 or 17 small ponds or lakes. Also from him we learned that the country was swampy, with occasional clay upland. An excursion of about eight miles inland from a point one and a half miles above the Long Sault found this information correct.

The timber after leaving the river bank is all small, consisting of spruce and dead tamarac in the swamps, while the clay uplands, rising but little above the swamp levels, support in addition small poplar, balsam and some white birch.

Nearly seven miles up stream from the Long Sault, is a small rapid with a fall in level of about two and a half feet. The water runs over Laurentian boulders. Above this the river widens out, and one or one and a half miles brings us to a quickening of the current, caused by the shallowness of the river and by the presence of many boulders in the river bed. From here till we reach Duck Deer Rapids, a distance of 17 miles, no rapids are encountered.

The surrounding country is slightly undulating for some distance inland, similar to that already described. The timber along the river banks consists of fair spruce and poplar, together with some white cedar, balm of Gilead, white birch, dead tamarac, dogwood, alder, etc.

A good deal of the spruce is blighted. At a very few chains from the river the trees are all small and practically useless.

The banks themselves are of clay (alluvial) and fall into low country. This prevails up to the Duck Deer Rapids, at which point the west side appears to run into more rolling clay land. The first rock seen in place by me ascending from the Sault occurs one or one and a half miles below Duck Deer Rapids. This is a small island near the west bank composed of pink hornblendic gneiss. Several similar outcrops come out as far as the head of the rapids which flow over it.

These rapids (Duck Deer) are 35 chains in length and can be run in a large canoe if care is taken to avoid the rocks, which are sometimes hard to discern, owing to the very muddy water. In ascending they can be poled up along the west bank. Two miles above the rapids is a constriction in the river, followed by a second one about 25 chains beyond; the water in both cases running between walls of granitoid hornblende gneiss, which are the only exposures seen on this part of the river.

We now ascend for about five and a half miles before another outcrop is passed, it being the similar gneissic rock. Within the next 30 chains several of these exposures are met with. This brings us one and a quarter miles below Iroquois Falls.

The strike of the gneissoid structure observed on coming up the river gradually changed from an east-south-east to a north-east direction. The next rock in place occurred three-quarters of a mile below the Falls, being a dark compact eruptive diabase with inclusions of gneiss. Since all the rock from here to the head of the river is Huronian in character, I place the contact between it and the Laurentian at or near this

* *Vide* "Report of Exploratory Survey to Hudson's Bay by Wm. Ogilvie, D.L.S., 1891.
† Bureau of Mines Report for 1899, 2nd part, pp. 181-2.

Natural log jam Speight's base line at 70th mile
Party No. 1.

Open muskeg (spruce), Speight's base line at 70 mile.
Party No. 1.

The 70th mile post on O.L.S. Speight's base line. Party No. 1.

Burnt Bush River, north of Lake Abitibi.
Speight's Party No. 1

Hannah Bay River at its junction with Burnt Bush River.
Party No. 1.

Upper Twin Falls, Hannah Bay River. Party No. 1.

Island Falls, Hannah Bay River Party No 1.

Hannah Bay River Party No. 1.

Abitibi River. Party No. 1.

On O.L.S. Speight's base line, showing forest. Party No. 1.

point. The strike of the gneiss also provides a clue as to the probable course which the line of contact follows.

Iroquois Falls tumble over a coarse weathered diabase, the difference in level caused thereby being between 12 and 14 feet (aneroid). Two islands which crop out at the brink cause them to be divided laterally into 3 chief cascades. The river at the foot widens out considerably but closes in again about three hundred yards below. The portage is on the south side, and because of a log jam which closes up the passage at the head of the long island dividing the fall, a second portage of one chain over this island is necessary.

One-half mile below the falls a tie line of over 18 miles was run from the south bank of the river due west to meet Mr. Niven's Meridian line. I was able to examine only the first few miles of this. It runs at first over a series of clay ridges which are cut up by draws or gullies by means of which the water finds its way to the main river.

The poplar along these clay elevations although sparse, is of good size. The spruce also would be valuable as pulpwood were it not that much of it has been killed by worms. Balsam, white birch, etc., also grow here.

Sixty-three chains from the beginning of the line a small diabase hill rises up, but falls away on all sides. This hill affords a very good view to the west, where in the distance is seen a long ridge extending north-east and south-west. Higher land can also be seen to the south. Proceeding up Abitibi River from Iroquois Falls low-lying land is found on both sides until Black River is reached (five miles). The timber is similar to that already seen on the banks. Only one rock outcrop was noticed in this stretch at about three-quarters of a mile above the fall. It was a dark fine-grained basic Huronian probably diabase.

Four miles above Black River Junction we come to the lower of the two portages which are necessary because of the two cascades, separated about 8 or 9 chains from each other. The upper one changes the level 8 feet, while the lower one has a fall of 5 feet (aneroid). The portage in each case is on the south side, and about four or five chains in length. One half mile below these cascades I found a light green schistose rock which on examination proved to be a mixture of serpentine and dolomite, the latter very impure. This is most likely the result of considerable leaching and weathering of some basic rock, which occurring as it does, was of Huronian age probably. Farther up I encountered more of this rock, while the cascades themselves poured over a dark green altered chlorite schist.

Ascending about six miles farther, and passing two small rapids on our way, we reached the mouth of the Mis-to-ogo River, which is the largest one flowing into Abitibi River from the north above the Long Sault. The two small rapids run over a similar altered chlorite eruptive schist, of which several outcrops are observed on the way.

Mis-to-o-go River.

The Mis-to-o-go River is 2 chains wide at its mouth. An Indian informed me that it was called Sand River, but whether or no this is the meaning of the original name I did not find out. Its general course is from the north-north-west first, then it turns and runs towards the south-south-west. It carries a good volume of water, and a chute of 2 chains' length 1 mile from its mouth would make a very fair water power. The drop is 18 feet (aneroid). The portage is on the west bank. No other portage was necessary until 12 miles up where a log jam blocks the stream. The banks assume higher proportions as the river is ascended.

A tramp inland in a west-north-west direction from a point 12 miles up, took us for the first five-sixths of a mile over some clay elevations with steep gullies cutting them. After this the land is fairly level and high but very swampy. The popular and spruce on the ridges is good but otherwise the timber is small. The chute aforementioned has cut its way through a chlorite schist, which is, pyritous in places, at right angles to its strike, the latter bearing east and west. The dip inclines slightly to the north. This rock contains some bedded and lenticular veins of quartz, not very continuous or very wide, nor did the quartz have a very promising appearance, a picked sample assaying in gold only $2.20 per ton. A fine-grained hornblende schist outcrops 30 chains farther up the river. Another half mile and the altered chlorite schist is again met. A very impure dolomitic rock containing considerable plagioclase is found 1 mile above this last. It is schistose with a strike of south 45 degrees east and dipping slightly to the south.

Eight miles up river an outcrop of diabase comes out on the east side. Patches of popular and spruce, which would cut into pulpwood. grow along the banks of Sand River, but beyond the banks it is as elsewhere, i.e., small or scrubby. On Abitibi River, half a mile above where Mis-to-o-go River enters, several outcrops, including a couple of small islands of the altered schist, occur—strike east 25 degrees south, dip vertical; 30 chains beyond this the strike is east and west, dip 10 degrees north. Two small rapids are then ascended, a small island of dark green diabasic rock lying at the foot of the lower one which runs over boulders, while the upper one passes over the altered chloritic schist.

A mile above this again a portage of about 8 or 9 chains brings us to the head of another rapid, which can be easily run by keeping close to the north side of the island. It can also be poled up. A light green, hard schistose rock runs across the island at its highest point, and the rapid passes over the same rock.

The next 10 miles of the river is quiet and the banks show a better quality of poplar and spruce. Four miles up this stretch is seen the altered eruptive schist, several exposures of which are shown during the next three miles, the general strike being east, 12 degrees south. At the head of this quiet portion a rapid runs over a ledge of similar rock with a portage of one chain on the south bank. The strike of the schist is east, 10 degrees north, with vertical dip. The river then widens out to 30 or more chains width for 2 miles when it again contracts at the foot of a very rough rapid. Half way along this expansion on the north shore I found an outcrop of diorite.

The rapids just referred to are at the very base of Couchiching Falls, the first encountered in descending the river.

The falls, as viewed from a rocky bluff 200 yards below them on the south bank, present a spectacle of impressive grandeur. The river, taking a sharp turn to the north with its great volume of water, rushes over two quickly succeeding cascades. These latter are divided laterally by an out-jutting of rock at its centre. Near the north bank an island of rock again divides the stream, but the north-easterly cascade, which is the smallest, is not seen from below, it being hidden by the island. The drop or fall is about 25 feet. The water then very swift and turbulent rushes onward for about 12 chains, when it turns abruptly to the westward. The total length of the rapid is over half a mile. The portage on the south side is about 30 chains long. The difference in level between the top and the bottom portage landings is 35 feet (aneroid). The rock exposed on both sides of the falls and rapids is altered diabase, containing pyrite in places.

From here to Lower Abitibi Lake the distance is 4½ miles, and the only rock seen in situ is about 20 chains from the lake, where a badly-weathered, coarse-grained and probably altered diorite outcrops on the north banks.

Between Mis-to-o-go River and Abitibi Lake the banks of the river are all clay. In general, the south side rises to higher ground than that on the north. The timber also is good in poplar and spruce, but only in places.

THE ABITIBI LAKES.

The Lower and Upper Abitibi Lakes form the valley between two ranges of Huronian hills, one to the north and one to the south, both running in a general easterly and westerly direction. That to the south is higher and more continuous than the one to the north, which is connected with a continuous line of sand hills between the higher ones of eruptive rock. The height of the highest hill to the north is probably 350 feet above the lake level (aneroid).

Both of the lakes present very ragged and deeply indented shores, which show considerable rock *in situ*. They are dotted with numerous islands, most of them composed of bed rock. The Lower lake lies to the north-west of the Upper, a narrows about three miles long connecting them. Both are very shallow, it being said that 10 feet is the average depth. One of the deepest parts is at the south end of the narrows, where, we are told, it went down to 13 or 14 fathoms.

Generally the timber along the north shore of the lakes is small, although occasionally good stretches of pulpwood are observed. The river banks on higher land to the north grow the largest and soundest timber. No red or white pine are seen on this side. Mr. Bell and myself ascended all the navigable rivers coming in on this shore (the north). We found them all very similar in character, flowing between banks of alluvial clay, with little volume or current as the general rule. With the exception of Tabasaqua and Okikodosik Rivers (tho latter name means Rotten Pitchpine) none of these rivers could be ascended by canoe, except for a few miles, because of the very small volume of water carried at this season (Sept. and Oct.)

Okikodosik River flows from the north-east into the north-east end of the deep bay on the north shore of Upper Lake Abitibi. It is almost wholly in Quebec Province. We ascended it for 16 miles, when it branches into two small streams. About a quarter of a mile up the easterly one of these branches a trail leaves the shore, running north for some distance over clay ridges, and finally connecting with the main winter trail to Moose Factory. The latter starts at the north-west end of the deep bay before mentioned.

About 1 or 1½ miles south west from this stream the extreme south west part of the lower lake, after passing over a succession of low sand, sandy loam, and clay ridges, we come to the apparently highest part of this locality. It consists of a sand ridge with burnt Banksian pine, running east and west. At the north east, and south east sides of this brulè the land falls abruptly into swamp. To the north and west the drop is more gradual. The brulè is probably 60 feet above lake level. The timber in the swampy portions is as elsewhere shown, while the higher portions carry besides poplar and small Banksian pine.

Passing up the west shore 3 miles north of the outlet, is a small island showing considerable serpentine. Another 2½ miles is a wooded island composed of dark green coarse diorite. A mile farther and we enter a small bay, containing several small islands of altered chloride schist, showing in one place a band of gneiss only a few feet wide and not continuous. This might indeed be an inclusion.

Up to this point the shore is sandy and bouldery with no rock in place. Beyond, however, as far as Dokis River which flows into the lake a short distance east from Low Bush River, the shore exposes the fine-grained chlorite and hornblende schists, striking generally east 5 degrees north and dipping slightly to the south. Near Dokis river a harder and more compact variety is shown.

Some lenticular quartz beds are found in the folds of the schist, but do not appear to be very continuous. From one vein 4¼ miles south from Low Bush River, I took picked samples. These assayed $1.20 per ton in gold. This vein was 4 feet at its widest, and contained considerable pyrite, as also did the walls which were chlorite schist. The quartz was glassy.

Leaving the Tabasaqua River and following down the east shore we go about six miles before touching any bed rock. Beyond this point, however, is exposed much rock, the bulk of it being altered chlorite and hornblende schists with also some bands of slaty looking felsites (phyllites). The strike of these schists is most irregular and variable, comforming in general with the direction of the shore line at the particular spot where it occurs. One thing noticeable is that at the northern half of this eastern shore, the strike is generally to the north of east, while on the southern half the rule is to the south of east.

The dip of the schists on the northern shore of the first deep bay met on the way south, varies from 70 to 85 degrees south. South from this point to the Narrows, however, it oscillates unexpectedly from one inclination to the opposite, but not very far from the vertical. This shore, especially the southern half, is quite hilly with rock which runs up from the water's edge to eminences reaching as high as 200 feet above the lake level (aneroid). The same rock which composes these elevations runs down to the lake where it is exposed on the shore. It is all Huronian formation.

About 4 or 5 miles inland north from the bay mentioned above, is one of the hills which form part of a range on this side of the lakes. On ascending it I found the following formations : At the southern base was chlorite schist dipping south. A little above this I passed over a peculiar hornblende biotite Hurorian schist or gneiss, in which the crystals of hornblende showed a typical ophitic structure. Above this again (in elevation, I mean) is a band of coarse hornblende schist, while at the top a dike of porphyritic granite runs along parallel with the strike of the schists which is east and west. Just over the top of the north slope I found a band of the gneissic rock already found on Circle River, and described as being similar to Lawson's Couchiching gneisses.

At the south west side of the head of land which divides the two bays at the north east part of this shore, I passed a band of Laurentian gneiss (coarse) about three-quarters of a mile wide. The eastern contact showed basic dikes. Immediatedly east from this band I found a distinctly gneissic rock similar to the Couchiching series. In this, thin layers of biotite alternate with the thicker ones of the fine grained matrix. Just beyond this and seen on the shore coming up through hornblende schist is a dike of quartzless porphyry. An island lying off this point is composed of diorite with a band of the above Huronian gneiss interposed. Other rocks met with on the way down to the Narrows were diorite schist, also non-schistose diorite, gabbro and diabase with occasional dikes of eruptive granite showing. Much of the rock has been altered.

On the shores of the bay to the east of the Narrows on the Lower Lake, bands of Laurentian gneiss, granite or syenite, alternate with those of the more basic Huronian. This whole eastern shore is evidently near the line of contact between the Laurentian and Huronian.

There were numerous bedded veins of quartz seen while coasting down this part of the lake. I took samples from three of them. Two of them which were average samples, showed only a trace of gold on analysis. The third from a true vein running across the strike of the rock, gave not even a trace, although a picked sample was taken. Some of the deposits were quite wide in places, and, did their contents warrant it, would make extensive beds for quartz mining. The walls of the veins sampled are composed of the altered chloritic schists.

Proceeding north of the Narrows, we come upon outcrops of a coarse brownish, syenitic rock, which has an obscure gneissic structure. On the east side, over half way up, a coarse, eruptive olivine diabase is seen.

After leaving the Narrows and following east along the north shore of Upper Lake Abitibi, we encounter for the first few miles or so, a coarse-grained Laurentian rock, which might be classed as a syenite verging into granite or because of the considerable amount of plagioclase contained it may also be called a very acid diorite. More quartz is contained in some portions than in others. Some basic eruptive dikes are forced up through it. This rock then gives place to pink gneiss and granite, which is the only kind found along this shore, or the numerous islands adjacent, until we reach the Quebec side of the inter-provincial boundary. Similar Laurentian rock is seen 15 miles up the Okikodosik River.

The east side of the very deep bay, which runs for 9 or 10 miles into the north shore, exposes some eruptive dikes. At the south end of this side of the bay a prominent bluff of altered chlorite and hornblende schist rises up from the water, having at its base a snow-white glassy bedded vein of quartz, 12 feet wide at one place, and extending in a tortuous manner between the folds of the schist. The strike of the schist follows the shore line. A picked sample from one of the many little veinlets of quartz on this bluff, assayed $1.40 per ton in gold. A few chains east a very coarse hornblendic rock occurs, and for a few miles the Huronian rock now predominates along this shore, but in the deep bay at the extreme eastern part of the lake the Laurentian gneisses again appear. At one or two places on the east shore of the deep bay at the north of the lake I found bands of the gneiss resembling the Couchiching variety.

The strike of the gneiss (Laurentian) between the Narrows and the deep bay is south of east. This assisted me in determining the line of contact between the two formations at this part.

The region to the north of the upper lake consists of rolling clay land with considerable swamp. As the hills which are seen from the lake to the north are approached the soil becomes sandy. There is considerable small Banksian pine found in this district.

Glacial Striæ.

I took numerous readings of direction assumed by the glacial striæ at various parts of the country examined, with the following results:

Tabasaqua R., 8th portage	N. 22° W.
L. Michel	N. 15° W.
Little Abitibi L. (N. end)	N. 15° W.
" " (N.E. end)	N. 15° W.
Circle R. (1 mile up)	N. 35° W.
" (1¼ ")	N. 30° W.

Abitibi R.—
Iroquois Falls (several at foot) .... N. 15° W.
½ mi. below Two Portages .... N. 17° W.
Foot of Couchiching rapids .... N. 50° W.
Chute at Mis-to-o-go R .... N. 20° W.
2 mi. up " 2 sets shown crossing one another ...... N. 45° W. } N. 25° W.
Further up Mis-to-o go R., also 2 sets .... N. 45° W. } N. 28° W.
11 mi. " " .... N. 18° W.
West shore of Lower L. Abitibi, average of 3 readings, varying from N. 31° W. to N. 15° W .... N. 23° W.
N.-E. and E. shore of Lower L. Abitibi from Low Bush R. to Narrows, average of 16 readings varying from N. 10° W. to N. 23° W .... N. 16° W.
N. shore of Upper L Abitibi as far east as Deep Bay, average of 7 readings, varying from N. 27 ° W. to N. 35° W.... N. 32° W.
E. shore of Deep Bay on N. shore (near end of bay) average... N. 15° W.

SUMMARY.

Line of Contact Between the Laurentian and Huronian.

All the rock *in situ* encountered by me was either of Laurentian or Huronian age and I have placed the line of contact as nearly as I could from the few outcrops exposed inland, also using as a clue certain general strikes and dips. The points on the south shore of the Abitibi Lakes where the line touches I am unable to locate, but Mr. Baker, who examined this part, has filled in this in the accompanying map.

The line as fixed by me is as follows :—Starting at a point three-fourths of a mile below Iroquois Falls it runs north, north-east, cutting Low Bush River (say) 10 miles up and Circle River about the same distance form its mouth. Diverting its course now eastward for a few miles I now make it turn sharply toward the south-west until it comes out at the north-east part of the Lower Abitibi Lake. I place the high hills to the north of this point as the Huronian side of the line. About three-fourths of a mile to the east of where it enters the lake it goes inland again, circles around the remainder of the shore southward, and comes out near the lower end of the Narrows between the two lakes. Again altering its course sharply to the south and cutting off a part of the long point between the two lakes to the west of the Narrows, it apparently curves gradually, running up through the Upper Lake until it again enters the land. The place where it enters is on the north shore of the lake at the eastern side of the mouth of the deep bay running into this shore. Cutting off about 10 miles of shore line, it now courses south for a few miles, and leaving the south-east part of the lake turns eastward.

Natural Resources, Etc.

Mineral.—Except the large area of peat beds to which ample reference has been already made, I encountered no deposits of mineral which could have any immediate commercial value.

Although quartz veins bearing pyrite of the bedded or segregated type were numerous at points near the contact of Laurentian and Huronian, yet the analysis of samples picked and assayed, as performed by Mr. J. Walter Wells, has proved them worthless as regards their gold or silver content. The appearance of the quartz in these veins was altogether unpromising.

Timber and Vegetation.—Spruce is the most common, and poplar next, of the prevailing timber. The only places where these were of any value as pulpwood were along the shores of small lakes, the banks of rivers, or the clay and sand ridges—but not, I should judge, to any startling amount.

The other timber growing in the region examined by Mr. Bell and myself, is balm of Gilead, white birch, white cedar, tamarac (mostly dead), balsam and Banksian pine.

Most of these are of little use. A few red pine and elm were observed in one or two isolated places. One also sees small dogwood, maple (soft) and cherry saplings, alder bush, etc., scattered throughout.

Of the ground bushes I might mention that most of the ordinary berry-producing varieties are found, such as the black and red raspberry, the strawberry, the blueberry, etc , etc.

SOIL.—The whole region may be included in the great clay belt which extends from the west. All the soil is covered with moss to a greater or less extent and with windfall, which latter is considerable. The moss tends to make the ground swampy, especially in the depressions. Nevertheless innumerable tracts of good farm land can be found along some of the rivers, which after clearing would prove themselves valuable agriculturally. The sun being excluded ice is frequently found at the bottom of the water-holes contained in the deep moss.

DRAINAGE.—The swamps into which the banks of the rivers slope down, are of higher elevation than the rivers themselves, and since their drainage is prevented in most cases by the banks or from innumerable small moss water-reservoirs which are strewn over their whole area, it is quite reasonable to suppose that these swamps could be turned to good account when once the land is cleared and intersected with drains. These latter could eventually be led toward the rivers.

The question as to the best manner of drying the swamps surrounding the two Abitibi Lakes requires more careful study, however. Were the level of the lakes lowered this might be done with comparative ease, but perhaps the only manner in which this accomplishment could be performed would be by lowering the Couchiching Falls on Abitibi River.

Even then the efficiency of the project is but conjectural, the result depending on the depth of the river as compared with that of the lake. If the lake levels were reduced much land might also be reclaimed, because of the shallowness of the lakes. The value of the land thus obtained, however, is a matter of speculation only.

In objection to this scheme may be urged that the loss incurred by the lessening of the splendid water-power on the Abitibi River is not recompensed by the gain obtained in an agricultural way. Again, whereas boats constructed for shallow waters can at present navigate these lakes, yet the depth being decreased this might become impossible.

Therefore, as previously stated, before the matter can be undertaken, serious consideration is necessary.

CLIMATE.—Although the lakes freeze up at any time after the 20th of October and do not open up until some time in May or the first part of June, yet there is no reason why this region should not be as well adapted for agricultural pursuits as is Manitoba. The Hudson Bay Company's factor at Abitibi grows potatoes and timothy with success, and I also noticed that he had a small private root garden which seemed to thrive.

FLORA.—During the summer, specimens of the different varieties of the flora were collected and pressed between blotters for subsequent examination by the Department.

LARGE GAME AND FUR BEARING ANIMALS.—I have enumerated the large game and the fur-bearing animals, stating their abundance or scarcity as given us by Mr. Skene, the Hudson Bay Company's factor at Abitibi, or as judged from our own observations as follows :—

Moose—not so abundant as farther south.
Caribou—not abundant.
Red Deer—very scarce, occasionally a strayed one.
Bear—not uncommon.
Wolf (gray)—very few.
Fox (red)—scarce.
Fox (black)—have been seen.
Lynx—scarce.
Muskrat—very common.
Marten—common.
Weasel—common.
Otter—not uncommon.
Beaver—some colonies found in isolated lakes and small streams.
Mink—not common.

Ermine—not common.
Rabbit—once plentiful, but not so at present.
Fisher—some.
Skunk—few.
Porcupine—scarce.

BIRDS.—The various birds seen by us most commonly during our exploration were:—
Game Birds,—Duck (chiefly black and redhead), Canada grouse.
Water Fowl,—Loon and hell-diver.

Other Birds,—Bald-headed eagle, white-headed eagle (neither of these common, however), hen-hawk, night-hawk, bittern, hoot owl, raven, gull (three varieties), sand piper, wheatbird (whiskey jack), highholder, woodpecker, kingfisher, blackbird (two varieties), red robin, grosbeak, fly-catchers, yellow canary, several varieties of warblers and thrushes, besides many others of the ordinary Canadian birds.

It is noticeable that birds are not so numerous in this north country as they are farther south.

FISH.—We saw and caught the following varieties of fish:—
Speckled trout—In some of the rivers.
Toulibi, a flat white fish, } In the larger lakes.
White fish (proper), }
Pike, }
Pickerel, } In both lakes and rivers.
Suckers, }
Perch, }
Sturgeon,—In Abitibi River below Iroquois Falls.

INDIAN OCCUPATION, ETC.

From a census taken a short time since by the Roman Catholic missionaries, it was shown that there were 450 Indian souls dependent upon the Hudson Bay Company for supplies (at Abitibi).

These Indians, who are of the Algonquin race, gain their livelihood by hunting and fishing.

Late in the spring or early in the summer, they bring their winter's capture to the post, where all trading is performed. Abitibi Post was established some time between 1755 and 1760.

They remain about the post or the large lakes, where most of them live chiefly on the fish from their nets until the latter part of September or the first of October. Then they return to their hunting grounds with supplies in proportion to their capacity as fur-getters.

It will thus be seen that they escape the mosquito and fly season in the bush.

The Indian is well educated up to the highest value which his fur will bring, and, as is the case with the white man, some are industrious and some are not, the former making good livings by their dealings.

The missionaries abide a short time among them each summer, services being held in a little but neat frame church situated on a hill at the back of the post.

Unlike the Iroquois, by whom their fathers were driven to these northern wilds, the Algonquin, even as in Champlain's time, is decidedly not of an agricultural turn, preferring rather the roving, though sometimes precarious, existence of the trapper and hunter.

R. W. COULTHARD.
Geologist Exploration Survey Party No. 1.

Toronto, January 15, 1901.

I, Robert Wilson Coulthard, of the City of Toronto, in the County of York, Geologist, make oath and say, that the above report is correct and true to the best of my knowledge and belief, so help me God.

R. W. COULTHARD.

Sworn before me at the City of Toronto, in the County of York, this 15th day of January, 1901.

GEO. B. KIRKPATRICK,
Commissioner, etc.

SURVEYOR'S REPORT OF EXPLORATION SURVEY PARTY NO. 2.

Haliburton, Ontario.
November 24th, 1900.

Sir :—I have the honor to submit the following report on surveys made by me during the past summer, under instructions from your Department, dated 8th June, 1900.

My party was known as Exploration Party No. 2, of the general exploration of Northern Ontario during the present year.

The work assigned to the party was the running of a base west from the 198th mile post on the Nipissing and Algoma Boundary to the Missinaibi River, about 100 miles, and the exploration of the country for a distance of fifty miles on each side of this line, and thence southerly up the Missinaibi River to near Missinaibi Lake; also the continuation of my base line of 1899 from the 120th mile point, a distance of 24 miles west, and thence due south 6½ miles to Dog Lake, and connecting with O. L. S. Stewart's survey of 1893 along the Canadian Pacific Railway.

The object of the exploration was the obtaining as much information as possible regarding the topography of the country, the timber, the soil, the minerals, water-ways, and water powers, the flora and fauna, and, in short, everything that would give a proper idea of the country and its resources.

Two exploring parties were attached to the survey party, each consisting of a land and timber estimator, a geologist and a canoeman, and their duties were to examine and report upon each side of the line as far north and as far south as possible, to ascend and descend the rivers crossed by the line, with a view of obtaining all possible information during the time at their disposal. Messrs. J. L. Bremner, of Admaston, and J. M. Milne, of Queensville were appointed to accompany the party as land and timber estimators, and Messrs. A. G. Burrows B. A. of Queen's University, and E. L. Fraleck, of the Kingston School of Mines, as Geologists.

Leaving Toronto on the 11th of June, accompanied by the two geologists, the timber and land estimators joining me at Mattawa, I proceeded via North Bay and Mattawa to Temiscaming Station, the Northern terminus of the Canadian Pacific Railway, at the head of the Long Sault Rapids, on the Ottawa River, and the south end of Temiscaming Lake; thence by steamer to the head of the lake, seventy miles; thence by the usual canoe route through Quebec, which has been fully described by Mr. Ogilvie, over the height of land, and down stream to Abitibi Lake; thence by the Abitibi River to a tributary know as Jawbone Creek, where the exploring parties commenced their work; thence easterly to the District boundary, and north fourteen miles along said boundary to the initial point at the 198th mile post; being about two hundred and ten miles north of the Canadian Pacific Railway at Sudbury, and in latitude 49°-35′-30″.

The time occupied for the journey was twenty days. After obtaining the necessary astronomical observations, I commenced the survey of the base line on the 2nd day of July, running west astronomically on six-mile chords to the Missinaibi River, which I crossed on the 100th mile. The line was continued to the end of the 102nd mile to complete a six mile chord, and after putting up a conspicuous cross on the east side of the river, and ascertaining that no line had been run from the west by exploration party No. 3, within twelve miles to the north, and tying my line to an island, I proceeded up the Missinaibi River to Missinaibi Lake. From the end of my base line of 1899, which was run west from the 120th mile post on the Nipissing and Algoma boundary, I continued west twenty-four miles to the end of the 144th mile, from which point I ran south astronomically through Wabatongashene Lake to the Canadian Pacific Railway and Dog Lake. Here I connected my line with O. L. S. Stewart's survey of Township boundaries along the Canadian Pacific Railway and returned to Toronto at 10.20 p.m. on the 13th of of October. The instructions called for a traverse of Missinaibi Lake, but as the Kay Taffrail log, with which I was supplied did not work satisfactorily, and the season was well advanced, I thought it advisable to leave the traverse to a more suitable time.

The line crosses the Abitibi River on the 14th mile and just north of Island Portage the Mattagami River on the 33rd mile, and immediately north of where Poplar Rapids River comes into it, the Ground Hog River on the 40th mile, the Kapuskasing River at the 55th mile, and the Opazatika River on the 79th mile. The lines were well cut out and blazed and carefully measured. Wooden posts, nearly all tamarac, were planted at every mile, and iron posts every 3 miles, marked with a cold chisel on the east side "III M.." "VI. M.," "IX. M.," &c., up to "CII. M." Mounds of stones, where stones could be found, were built around the posts. Bearing trees were also taken, marked "B. T.," and their size, course and distance from the posts noted. Where the end of a mile came in a lake or river the posts were planted on the line on the nearest land, and the distance noted. In these cases the iron posts were marked with a plus or minus sign, as the case might be. No mounds were built on the 102nd mile line as no stones could be found, but on the 1899 extension line it was different, and nearly all posts planted were surrounded by stone mounds.

Astronomical observations were taken frequently, the details of which will be found in the field notes. The magnetic variation of the needle ran between 5 and 12, averaging about 7 west on the 102nd mile line and about 4 west on the line of 1899.

General Description.

The initial point of the line, the 198th mile post on the district boundary, is in a level clay country, timbered with spruce and tamarac, being in the Abitibi valley. The country along the whole length of the line is almost level. There are a few very slight elevations and the depressions are only at the rivers and creeks crossed by the lines, and as soon as it crosses these it comes up to the general plane. The soil is generally clay—not a hard, white clay, but open, and resembling grains of wheat. There are no stones to be seen, as a rule, excepting in the river-bed, and no exposures of rock to speak of. A large part of the country is covered with a heavy coating ot moss which, in many places, retains the winter's frost until late in the summer and retards the growth of the timber. The land east of the Abitibi River—about 14 miles and for about 9 miles west, or to about the 23rd mile—is generally wet and swampy, but the clay is ever present as a foundation, even in the muskegs, of which there are many.

The timber on these 23 miles will run from 4 to 12 inches in diameter, probably 7 inches average, being, of course, smaller in the muskegs. From about the 23rd mile to the Mattagami River on the 33rd mile the land is not so wet, and the clay comes to the surface in many places. The 29th, 30th, 31st and 32nd miles are about parallel to the river, running through splendid land and timber, large poplar, white spruce and balm of Gilead. From the Mattagami River to the Groundhog River—33 to 40 miles—the country is tolerably dry, being drained by the two rivers, and there is a change also in the country from the Groundhog River to the Kapuskasing River—40 to 45 miles—muskegs occurring less frequently than further east. From the Kapuskasing River to the Opazatika River, at the end of the 78th mile, the country is dry but generally covered with a heavy coat of moss, in some places 12 to 24 inches in depth. West of the Opazatika River and on the 87th mile I made the following note in my field book: "The country for a number of miles back is about two-thirds clay, flat, covered with heavy moss, and about one-third clay and sometimes sandy loam ridges with very little moss, and frequently clay to the surface, with poplar and spruce timber." From this mile to the Missinaibi River on the 100th mile, the land is generally dry, in places a little rolling and very little muskeg, while from she Missinaibi River west to the 102nd mile, I may quote from my notes: "Level clay land, spruce, poplar, and a few tamarac of fair size, with balm of Gilead."

Summing up the whole line, it may be said that from start to finish it runs through as fine a tract of farming land as can be found in Ontario, Where else in Ontario can a tract of land 100 miles square be found, all alike level and good? Muskegs there are in it, of course, but 75 per cent. of the whole country could be cultivated as soon as cleared and the moss burned off it, and of the 25 per cent. remaining, a considerable portion could be drained and cultivated.

Timber.

Spruce, both white and black, is the principal timber along the whole line. Tamarac comes next in order, while poplar mixed with spruce is the principal timber on the dry level or sometimes undulating land. The general average of trees is from 5 to 10 inches in diameter, but along rivers and creeks where the land is drained the trees are much larger, often 18 to 20 inches in diameter and frequently two feet.

I would say that the best timber is to be found between the Mattagami and Opazatika Rivers, or from about the 25th to the 75th miles.

Balm of Gilead is found in large quantities among the poplar and spruce on the dry land along the rivers and creeks. There is no great quantity of cedar in the country, and only a fringe along the river banks. There is no white or red pine along the line, and pitch or Banksian pine only occurs in two or three places and in small areas. Other woods are balsam and white birch.

Water.

The line crosses all the rivers in the country flowing north to James' Bay, and a few small lakes, but there are no lakes of any importance along the line.

The Abitibi River, where the line crosses on the 14th mile, is 8 chains wide. It is one of the principal tributaries of the Moose River. It drains Abitibi Lake, and varies in width from 3 chains near its source to three-quarters of a mile near its junction with the Moose. It has numerous falls and rapids in its course, and is capable of developing a large amount of electrical power, if required. It was speciallly described by Mr. Ogilvie in 1890. The water is muddy from the clay through which it runs. The banks are from 80 to 100 feet high at Island Portage and Falls where we crossed it. The rock formation at that point is the Laurentian.

The Mattagami River where we crossed it (on the 33rd mile) is over a quarter of a mile in width, with banks from 20 to 50 feet high and its current runs three miles an hour. The water is bright, its source being in the high rocky lands of the south. It is larger in volume than the Abitibi River. It received many other rivers north of the line before entering the Moose, and is the principal tributary of that river. It has many fine waterfalls and numerous rapids, and almost unlimited power could be obtained for industries of every kind that may hereafter be required. Poplar Rapids, river 2½ chains wide, enters it just south of the line. The Groundhog River, on the 40th mile, has a width of 9 chains where the line crosses it, with fast current and banks 40 feet high. It is a fine stream, with numerous islands and a heavy flow of clear water. The explorers reported it to be one of the finest rivers in the country. It falls into the Mattagami River about 9 miles north of the line. The Kapuskasing River at the 55th mile, 6½ chains wide, and flowing between banks from 20 to 40 feet high is next in order. It rises over 100 miles to the south of the line, and the explorers report it as having many beautiful waterfalls and an interesting river to canoe upon. It falls into the Mattagami River about seventeen miles north of the line.

After leaving the Kapuskasing we cross the height of land between the Mattagami waters and those of Missinaibi River, and at the beginning of the 79th mile cross the Opazatika (meaning poplar) river, the first stream of any importance that falls into the Missinaibi. It is 4 chains in width and 9 feet deep where the line crosses it. The banks are low, and during high water the shores are flooded. The water is dark and the current slow, and along the shores the high bush cranberries are abundant. It falls into the Missinaibi River about 50 miles to the north of the line. This river is much travelled by Indians coming from Moose Factory to Missinaibi, being an easier river to paddle than the Missinaibi River, which goes to show that the water stretches of the Opazitika are longer and the falls higher than on the Missinaibi River.

Last, but not least, is the Missinaibi River (Swift Water) on the 100th mile, being part of the great highway between Lake Superior and Hudson's Bay. It is twenty chains wide where the line crosses, and at a rapid which I supposed was the KaKagee (Crow) Rapids. I have since been told by Mr. Baird that his Indians, who knew the river well, called it the Black-feather rapids. I am not now sure which it may be, but the point could be easily settled by any one knowing the locality, as the line is plainly visible from the river. I thought the water was low when my party ascended the river (in September) to Missinaibi Lake, but the Bishop of Moosonee, whom I met on his way to Moose Factory, assured me that he had seen it very much lower.

It was my intention to have defined some points along the river by measurement and observation as I came up, as required by the instructions, but I found it impracticable to do this, with such a large party as mine, and moreover it was cloudy and raining nearly the whole time I was on the river. I managed to get an observation for latitude on Sunday the 9th of September, on an island in a bend of the river, some distance south of Conkling's River, making it 48°,,42′ North.

After seven days of hard paddling and poling up rapids, I arrived with my party at Missinaibi Lake, on the 11th of September, and after getting to the end of my base line of 1899,—the 120th mile post—and getting in supplies, commenced the prolongation of this line on the 15th of September. I continued west astronomically to the 144th mile, which I reached on the fourth of October, and then turning south through Wabatougashene Lake, ran down to the Canadian Pacific Railway as before described.

These twenty-four miles of base line were through an exceedingly rough and difficult country, much cut up by lakes and high, rocky hills. There is practically no land fit for farming purposes along the whole line west of Missinaibi Lake, and the timber, often for miles together, was nearly all blown down; the progress of the line was therefore necessarily slow, and the moving of the camp and supplies along the line a very difficult matter The soil is sandy, and the whole country rocky and broken, and covered with boulders. The timber is of all kinds,—pitch pine, birch, poplar, and spruce, and in many places of large size, especially the pitch pine and spruce. On some of the high hills the timber is nearly all white birch of large growth with sandy loam soil.

Much of the country along the line has been burnt and is practically useless. Away to the west of Wabatongashene Lake, the country is apparently all burnt to the Canadian Pacific Railway, and the white, rocky hills can be seen for miles. The geological formation along this line is the Laurentian.

EXPLORATION.

As the rivers crossing the 102nd mile line all run north to James' Bay, it was not practicable for the explorers to be much on the line, and as a rule they only came to it where the rivers crossed it. Considerable intervals of time, therefore, elapsed without my seeing the explorers and geologists; but so far as I know, they performed their duties faithfully, and their reports are submitted herewith. I have read the joint report of the land and timber estimators, and may say that I agree with them in what they say regarding the land and timber thereon, judging from what I saw of the country along the line and from my previous knowledge of it. It was, of course, impossible during the months of July, August and September to explore the whole territory assigned to No. 2 Party, but I am satisfied that the whole James' Bay slope is very much alike, and that what is recorded of the part seen is a fair sample of the whole. Much of the information on the place has been gathered from the geologists regarding the rivers, falls, lakes and portages. The portages were often difficult to find; and if the explorers were sometimes useful to my canoemen, the canoemen were often useful to the explorers.

The matter of getting in supplies was a very important one. Four of my men devoted their whole time to it, and I seldom saw them. Two of them I did not see from the time I left them at Abitibi River until they came to me at Missinaibi River, and I had frequently to send men off the line to meet them and get supplies to keep the line going. Routes across the country from one river to another were difficult to find, and long trips had often to be made through territory unknown to the white man and seldom travelled by Indians.

The first frost was on the 6th September. Wild fruits were abundant,—strawberries, raspberries, and red and black currants, cherries, etc. Signs of moose and caribou and bear were often seen, and beaver, otter, marten, rabbit, mink, and muskrat are the principal fur bearing animals of the country. Partridge were very plentiful, and the rivers contain fish of the usual kind, viz.: pike, pickerel, white fish and sturgeon (below the falls).

Specimens of the flora of the country were taken at various parts of the territory by Mr. Burrows, and are sent herewith. I regret to say that my aneroid got out of order immediately after leaving Toronto, and although sent back to be repaired, it never reached me, and consequently I have no barometric observations to submit.

Accompanying this report I beg to submit,

(a) A general map of the base line of 1900.

(b) A map of the base line of 1899 additional and meridian line to Dog Lake.

(c) Field notes of base line of 1900 and 1899 additional.

(*d*) Account for base line of 1900 in triplicate.
(*e*) Pay list exploration party, with statements of expenses in triplicate.
(*f*) Account for base line of 1899 additional in triplicate.

I have the honor to be, sir,
Your obedient servant,
A. NIVEN,
Ontario Land Surveyor.
Exploration Survey Party No. 2.

HON. E J. DAVIS,
Commissioner of Crown Lands,
Toronto.

NOTE.—There are no barometric observations returned, for the reason that the aneroid given me went out of order immediately after leaving Toronto. Upon getting to North Bay, I returned it to be put in order, with instructions to give it to Mr. Speight, who was following me a week later. Mr. Speight, however, had no opportunity of sending it to me, and consequently I was without it all the season. A. Niven, Ontario Land Surveyor.

LAND AND TIMBER ESTIMATOR'S REPORT OF EXPLORATION SURVEY PARTY No. 2.

TORONTO, Oct. 29th, 1900.

SIR,—Acting under instructions from your department, dated June 5th, 1900, we left home on the 11th and proceeded to join Mr. Niven's survey party at Mattawa, whence we travelled by rail and boat to the head of Lake Temiscaming; thence by the usual canoe route over the height of land and down stream to Lake Abitibi; then down the Abitibi River to Jaw Bone Creek, where we were ordered by Mr. Niven to begin our explorations, and where we arrived on the 28th of June. Here we separated from Mr. Niven and party, they going east to meridian line running north and thence to the one hundred and ninety-eighth mile post, from whence they were to run a line west to the Missinaibi River.

We spent a few days in the vicinity of Jaw Bone Creek, and found abundance of timber, spruce and poplar being predominant, also tamarac on the lower lands, a very considerable portion of the latter being dead or in a state of decay from some unknown cause. We also found balsam on the higher ground nnd cedar along the water's edge, but in no place did it extend back any distance, being only a fringe.

The banks of the Abitibi River rise in places to a height of from twenty to fifty feet usually with a gradual slope, and after leaving the water and getting back one-half mile, or sometimes more, the country becomes generally flat in appearance, but in no place does it drop behind the same as it rises in front, which would indicate easy drainage, the plane seemingly going off from the top of the banks without any apparent dip.

The timber along the banks is large spruce and poplar from six to twenty inches in diameter, and as thick generally as it can grow. However, after getting back from the water from one-half mile to one mile on the flat country, there is usually no poplar except on patches of rising ground, and the timber on the flat land is almost exclusively spruce and tamarac, and is smaller in size than what is on the banks, the spruce being from four to eleven inches in diameter, and the tamarac from six to ten inches in diameter, and as before mentioned, a reat deal of the latter is dead.

There are also considerable areas of what might be called muskeg. Although no naturally wet or boggy, it is low and flat and covered with a scrub growth of spruce and tamarac from two to five inches in diameter, very short and of no commercial value, but which where large enough would do for fuel. There are also small areas of this so-called muskeg with practically no timber on it at all; a heavy moss surface, with an occasional scrub spruce from two to six feet in height is all that can be seen.

The soil in this section is clay. Even in the muskeg when we cut down through this covering of moss, usually from one to four feet in thickness, we invariably found clay bottom. After spending a few days on the Abitibi River, between Jaw Bone Creek and the mouth of the Frederick House River, Mr. Bremner went up the latter river, and Mr. Milne descended the Abitibi River to Lobstick Portage. We ascended the Frederick House River for a distance of about ten miles, and would have gone further, only there were so many rapids, we might say miles of them, that we considered it a loss of time trying to get up. The water at this time being high and the current strong we had to pull our canoe up with a line, there being portages only in the worst places, and we could only make three or four miles in a day. We travelled east and west of this river and found the timber to be much the same as on the Abitibi River, and the ground seemed to be much higher and more rolling back inland, therefore there was more poplar in this section, there being quite large tracts of windfalls and all over a great deal of lying timber. Still there was a fair amount of good spruce and poplar standing, which, owing to their great height would cut a fair average of cords per acre. In this section we saw some magnificent poplar, twelve or thirteen inches in diameter, with fifty feet of clear trunk without limbs. The country west for two miles might be called slightly rolling, then for three miles further, the usual flat land, almost muskeg, in places appeared, which, although almost barren of timber in spots, was quite dry to walk through and under the moss, good clay soil, the same as on the higher ground, sand showing up near the rapids, but only in spots. West of this river five miles above its junction with the Abitibi River, where a creek of a chain in width enters from the west, we also found good timber and fine rolling land, the spruce being very thick on the flat land, although somewhat smaller than on the banks. Ten miles farther down on the west side of Abitibi River, is Driftwood Creek, south-west from the mouth of which we explored, and found here quite an extent of brulé, extending along Abitibi River for about three miles, and inland from one half to one mile.

This brulé is beautiful rolling clay land, the banks along the creek being fifty feet high in places, and altogether a fine farming country. This place was burned over probably five or six years ago, and is now growing up with young poplar, cherry, etc., the greater portion of the burnt timber having fallen. After getting inland through the brulé for a distance of three miles, there is abundance of good spruce, also good poplar in places.

Eight miles further down, and west of the rapid known as Three Carrying Places, we travelled inland three miles, and for one mile back found rolling clay land with good spruce and poplar on it, and for the other two miles the country seems to alternate between rolling land and muskeg with a scrub growth of spruce, poplar and white birch, although there is little of it of any size or value; the land, however, is good and quite suitable for farming purposes.

The timber along the Abitibi River, from Three Carrying Places down to Red Sucker Creek, is practically the same as what we saw further up, good for a distance of about one-half mile from the banks, then muskeg with scrub alternating with belts of fair timber as one goes farther inland, and we noticed the farther north we got the poplar and white birch seemed to become more plentiful. However, this might possibly be the case for a short distance only, and had we gone farther north the spruce might have been more in evidence again.

Returning now to Driftwood Creek, we ascended it on our way across country to Mattagami River, and found good spruce and poplar along the banks, also some cedar along the water, but no quantity of the latter. About fifteen miles up this stream we travelled inland about three miles on the north and south sides, and as usual found large good timber for about one mile back, then on the flat lands the growth was small and only in a few places was the spruce large enough for pulp-wood.

On the north side we found a large sized wild meadow with a most luxuriant growth of grass on it, and fine locations for farms, the land being rolling, with good clay soil. Ascending this creek some fifteen or twenty miles farther, there is good spruce, poplar and balm of Gilead, with the usual fringe of cedar on the banks, and close to the water's edge.

Turning westward from this point, which is about thirty-five miles from where this creek enters the Abitibi River, we made a nine mile portage, one mile of which is water, to Mattagami River. After leaving Driftwood Creek, there are some fair spruce and tamarac for one-half mile inland, but for the rest of the distance across this portage, there is nothing but scrub timber with only an occasional clump, large enough for pulp-wood, say from five to eight inches in diameter, and a considerable muskeg is to be seen with only a scanty growth of small spruce thereon, this muskeg being close to a small lake and being rather wet.

We now struck Mattagami River at Loon Portage, when we travelled west again and found beautiful rolling land with good clay soil and fair timber on the greater part of it, some places having only a scrub growth of timber and other parts being what might be called dry muskeg, with practically nothing but a growth of alder, etc., thereon.

Here we saw the first pine tree since leaving Lake Abitibi, there being a small red pine growing on the banks of Loon Portage.

We next proceeded down the Mattagami River to Fish rapids and found good timber along the banks all the way down. We travelled west from these rapids five miles and saw a good deal of dry muskeg with nothing but alder bushes, etc., growing on it, also some rolling land covered with poplar and white birch, and in other parts a growth of small spruce and tamarac from four to seven inches in diameter, but no great quantity of it.

However, if there is not much timber in this section it is splendid farming land, being high and slightly rolling with good clay soil and interspersed with some fine wild meadows, altogether a most desirable place to begin farming in.

On the east side, contrary to the general rule, the timber was small along the banks, but we got into some good patches after we got back from the river. Also saw some wild meadow on this side and slightly rolling land with clay soil. Going down stream again to Poplar Rapids River, we saw good timber on the banks all the way down, and after ascending this latter river for the first five miles, the banks were twenty-five feet in height, well timbered for a considerable distance inland; after that the banks got low and were not well timbered, only a small growth of spruce and tamarac on them. At about sixteen miles up, we decided to return as we could not get our canoe further, the water being so shallow. On our return we entered a small creek coming in from the west, which took us to two small lakes, around which there was abundance of spruce. Here, also, we saw a ridge three-quarters of a mile in length, by two chains in width, covered with red pine twelve inches up to eighteen inches in diameter, and forty to fifty feet in height. This ridge was sandy soil on top with clay bottom and along the river the soil was clay, as a rule, with signs of sand wash in places. We proceeded down stream again to within three miles of Cypress Chute, where we examined a creek, coming in on the east side, for three and a half miles up, where we found abundance of good spruce and tamarac. Here, also, we saw the first jack pine we met on our journey, there being some sand hills along the creek, the banks being more uneven, we might say hilly than in most places in this country.

We travelled ten miles across this section to a point opposite the mouth of the Kakozhisk River, and it gave us a good opportunity of seeing the country inland, and in those ten miles we crossed three small muskegs, none of them of any extent, and we saw plenty of good timber from 5 to 12 inches in diameter. When we mention the word timber we mean spruce, poplar, and tamarac of commercial value.

West of Cypress Chute, for three miles inland, we also found good timber nearly all the way through, excepting two small muskegs which we crossed. On the higher grounds we found outcrops of sand, but it is clay soil as a rule, the same as in the other sections. On down stream again to Devil's Portage, the most northerly point reached by us in our explorations. The timber from Cypress Chute to this point is spruce and poplar, alternating, and generally of good size, although the poplar in some places is scrubby. Opposite the mouth of the Kapuskasing River, on the east bank of the Mattagami River, there is a stretch of burnt country with a frontage on the river of about two miles and extending back inland from one to four miles, nearly all of which has been fairly well timbered. The fire has run here last spring, early in the season, and it gives the country a different appearance altogether, inasmuch as the ground seems higher and has a more rolling appearance than where the forest is all green.

The soil is clay, and more suitable or beautiful land for farming purposes we have never seen than presents itself in this particular section. Adjoining this and north of it there has been a fire some twelve or fifteen years ago, and the land is now covered with a growth of young poplar, presenting the same fine agricultural appearance as the other, this old brulé seeming to be of considerable extent.

South-west from Devil's Portage, we examined the country for two miles inland. Along the banks for a quarter of a mile back the timber is large and long, and here we saw a poplar thirty six inches in diameter, but after we got inland the timber was much smaller, five to nine inches in diameter, but long and clean and good. Here also we found the first exposure of rock inland and Mr. Burrows made a thorough examination of it.

We also noticed that the further north we got the outcrops of sand along the river became more frequent. We now returned to the Kakozhisk River, which we ascended for a distance of about fifty miles, and sixteen miles from the mouth, on the east side, there is a burned district of probably twenty-five squrae miles in extent. The timber along this river is no larger than what we had previously seen, but it is good inland as far as we travelled, and the soil is clay as in other places. We do not think it necessary to particularize any portion of this stream more than another, as it is much the same throughout its length.

We now returned to the Kapuskasing River and ascended it, expecting when we got about seventy miles up to cross the country by portage to Lake Opazatika. The Kapuskasing river is a beautiful stream, but not so large as the others we have been on. Thirteen miles from the mouth, on the west side, we travelled inland for three miles. On the first half-mile we saw fine large timber eight to twenty inches in diameter, and for the other two and a half miles the timber is smaller, although quite large enough for pulp-wood, and there is abundance of it. The soil is clay and would make good farming land, although inclining to muskeg in some places. We might mention that here on the bank of the river we saw the first ash tree we met with on our trip, but only one as far as we could see. We went on up stream again and crossed Mr. Niven's line at probably seventeen miles from the mouth of the river and we found here that some Indian had a patch of potatoes which he was cultivating, fine healthy looking potatoes they were, with vines almost three feet long and what is wonderful, no potato beetles on them.

Travelling west on Niven's line for two and a half miles we found plenty of good timber all the way in, also some jack pine on a knoll which we passed. This section is quite rolling with clay soil and would make good farming land.

Proceeding up stream again when about thirty-three miles from its mouth, we travelled on the east side inland for about three miles. High rolling and low ground seemed to alternate with good timber on the higher land, and fair-sized spruce and tamarac on the low land. Here we saw a fine wild meadow of probably five acres in extent, covered with a rich growth of grass, and good farming land all around, clay soil as a rule, but showing a gravel wash around this meadow and in places along the river bank. Up stream again for the next ten miles the timber seen was generally good, although in places nothing but scrub poplar appears on the banks from three to seven inches in diameter. The soil is clay with occasional outcrops of sand near the water.

There was such a sameness of appearance in this country and along those rivers, that we found it impossible to describe one place differently from another; it is merely the same thing over and over again. After travelling up stream again for ten or twelve miles we found the banks and timber as usual with a few small black ash and cedar along the water's edge. Then we got into brulé on both sides which extends for a distance of probably ten miles up along the river, this fire seeming to have taken place ten or twelve years ago, all the other timber except small patches being dead. We turned west here up a large creek and explored it for some fifteen miles up, passing through a small lake and were still in burnt country with some odd patches of green timber of fair size. Over the burnt land is a growth of young poplar and this section of country seems to be in general much higher and more rolling than that we had passed through, and there were outcroppings of rock to be found over a great part of it, in some instances rising to a height of forty feet, but the soil is good clay and workable right to the base of those outcroppings of rock.

Returning again to Kapuskasing River and for six miles up it, we found good timber, but here at this point for about one mile we found on examination that the land back from the river is low and level and marshy with nothing but scrub timber on it, but again for three miles up, the timber is good. We now turned westward, up a creek which we found to be the portage route to Lake Opazatika. Before doing so however, we explored the Kapuskasing River for about ten miles above this creek and found the banks as usual with abundance of good timber on them and good clay soil.

On this creek leading westward to Lake Opazatika, the timber is good in places, and scrubby in other places, and we passed through three small lakes with scrub timber on the shore but fair further inland. From the third lake, say eight miles from Kapuskasing River, we followed the old H. B. Co's portage to Lake Opazatika, a distance of about nine miles, six of which is land portage. All across this route the timber is splendid large spruce, poplar, tamarac and white birch, also good cedar on the swampy ground. This is the first place that we saw what might be called a cedar swamp, or any quantity of cedar back from the river's bank, and there seems to be quite an extent of it in this section.

Here also, some rolling stone is seen in places, but the soil is still clay as a rule, although sand appears on the higher ground where the stone is. The lake shores are quite sandy and rocky in places, the bluffs being twenty feet high, but this occurs only close to the water, the land in the interior being undulating and the soil clay.

We now reach Lake Opazatika, a beautiful sheet of water, full of long points and islands, the latter being chiefly timbered with small poplar and the mainland with good spruce, poplar and white birch, and inland the spruce and tamarac are splendid. Towards the north end of the lake, on the east side, we got into a brulé which extends for a distance of about fifteen miles along the lake and river. We suppose that this is the same brulé which we saw on Kapuskasing River, and if so its extent would probably be two hundred square miles. The timber down the Opazatika River for a distance of about thirty miles is mostly poplar of fair size along the banks, but inland spruce of good size seems to predominate. After getting down the river about fifteen miles further to where Mr. Niven's line crosses, the shores are low aud marshy, with some places on which scrub timber is growing, but inland there is some spruce and tamarac of fair size, the soil all over being good clay.

Returning now to the south end of Lake Opazatika, we found a creek, marked on projected map, Beaver creek, entering from the south, and about a chain in width, which we explored for a distance of about eight miles up, the first four of which were marshy on both sides with scrub timber to be seen. Inland and for the next four miles the timber is splendid, with a considerable quantity of cedar along the water's edge, also some small black ash.

Inland from the banks the land is rolling, with clay soil, and the same may be said of all the southern portion of the lake shore. The banks are rocky and sandy, but inland the soil and timber are good. In this section, from here to Kapuskasing River, we found clay with sand sub-soil, also sand with clay sub-soil. Here we saw on a point five white pine trees, twenty-four to thirty inches in diameter and of good length. These are the first white pine we have seen since leaving Lake Abitibi.

Leaving Lake Opazatika, we crossed south-westerly through a portage of some nine miles of land and water to Missinaibi River, and along this portage found good timber throughout as well as good land, but more sandy loam than usual, also more rolling stone than met with elsewhere.

We now went down the Missinaibi River about fifty miles to meet Mr. Niven at the line, and on our way we passed some brulé on both sides of the river at a distance of some eight or ten miles below the portage, but from appearance this seemed to be of only small extent, and to have been burnt years ago. The land and timber along the banks of the river are just the same as on the others, good spruce and poplar being seen nearly all along. Even where there appeared to be little else than scrub, on examination further inland we found equally as good timber as where it made a better showing on the banks, and after arriving at Niven's line we travelled north-west on the west side of Missinaibi River for three miles inland and found good timber all the way.

Near the river it would be eight to sixteen inches in diameter, and inland five to eleven inches in diameter, long and clear, and growing thickly on the ground. The soil is a good clay and the general character of the country the same as on the other rivers and all good farming land. We now started on our return trip up the Missinaibi River with Mr. Niven's party, who were going out to the Canadian Pacific Railway to finish

Dinner Time on Quinze Lake. Party No.

Quinze Lake. Party No. 2.

Quinze Lake. Party No. 2.

Quinze Lake. Party No. 2.

Pointer Boat on Quinze Lake. Party No 2.

Where the line crossed the Missanaibi River. Party No. 2.

On the Missanaibi River. Party No. 2.

On the Missanaibi River. Party No. 2.

some surveying. We remained to make some further explorations on Missinaibi River and Brunswick Lake and River, and also on the Pozushkootai River, situate about ten miles above Brunswick House, and to meet Mr. Niven at Missinaibi Station on the Canadian Pacific Railway about the 10th of October.

We separated from the survey party at Albany Rapids and proceeded to examine the branch of that name. Coming in from the west we ascended it about nine miles and found the water so low that we decided to return.

The timber is good everywhere we travelled. After leaving the Missinaibi River a few miles we found the banks sloping and the soil a rich sandy loam, but going inland a mile or so clay land is found again, the whole being slightly rolling and magnificent farming land. We next came to the mouth of Brunswick River, whence we travelled east of the Missinaibi River three miles inland and saw good timber all the way. Here we also found some very large white birch twelve to eighteen inches in diameter—splendid timber and splendid land as far as could be seen on every side.

Along Brunswick River and Lake we found the timber, as usual, good, the poplar predominating, near the banks on the north end of the lake; inland from the southwest portion of the lake, the timber is fine, there being less muskeg and less scrub as we got further south. This is a beautiful lake about fifteen miles long and from one to five miles wide. About three miles from where the river flows out was Old Brunswick Post of the Hudson Bay Company, which was abandoned by the Hudson Bay Company about twenty one years ago and the post removed south to Missinaibi Lake. At one time, some five or six acres of ground had been under cultivation here, and on examination we found that apples, plums and red currants had been grown in the gardens in bygone days, also garden celery which is yet to be found growing in a wild state, and from which we gathered some seed. Timothy grass and red clover, some of the latter in bloom when we were there on the 12th of September, were also to be seen in abundance, as well as a most luxuriant growth in places of the farmer's pest (couch grass) or as it is called by some, quack grass. The very thing that is giving farmers in the older parts of Ontario to day a great deal of trouble in trying to get rid of it flourished here in the back woods, probably a hundred years ago, and was probably imported from Scotland in grass seed when this post was first established.

We now left Brunswick Lake, and portaged across about a mile from the south end into Missinaibi River. Again through this portage there had been fine spruce and tamarac but a fire last spring has burned a small area here between the lake and rivers, this dividing ridge being rocky most of the way through.

From here we went up the Missinaibi River three miles and ascended the Pazushkootai River which enters it from the west, for a distance of about thirty-five miles, when we found the water so low and the stream so blocked with driftwood, that we decided to return and get out to the Canadian Pacific Railway as soon as possible, as our provisons were getting low. Along this stream the timber is generally good and the banks high in places, with the soil usually sandy loam, but on going inland from one-quarter to one mile the clay lands are again met and good timber to be seen on all sides with fine rolling land and a beautiful farming country.

Having reached Missinaibi River again we started up stream on our way to the Canadian Pacific Railway. The stream here presents the same general appearance as farther down; there was good timber to be seen nearly all along until we reached Missinaibi Lake, with jack pine in several places since we left Pazushkootai River, with some cedar and small black ash by the water's edge.

Around Missinaibi Lake and close to the water, small poplar and white birch predominated, but on examination farther back the timber is good. At the south end of the lake we began to see an odd white pine now and again, and they continue until we reached the Canadian Pacific Railway at Missinaibi station.

The general features of the country over which we have travelled are almost the same, rough swift streams and comparatively level flat ground after leaving the river bank, with usually a smaller growth of timber than that which is near the water.

A great deal of the tamarac is dead, or half dead, and what is living and thrifty is what might be called a size too small for railway ties, although in time it will certainly be used for that purpose. There is very little dead spruce except what is on the ground. After it attains a growth of twenty inches and upwards it seems to die and fall to the ground, as the woods in many places are strewn with lying timber. However a great deal of this is blown down as the trees are so long, and in this clay ground the roots have no hold and are practically on top of the soil and the least wind blows them over. This is the only way we can account for so much lying timber in a virgin forest. There is very little balsam, it being only on the high dry ground and usually small in size and not worth mentioning as a wood product. The same may be said of the cedar; although good in many places there is no extent or quantity of it, it being confined almost altogether to the river banks, we having only seen it once in any quantity away from the water and that was on Opazatika Portage.

There is a considerable quantity of poplar of large size and also some large balm of Gilead and white birch, but the latter is generally scrubby. Spruce is the staple wood and there is an almost endless supply of it, if fire can only be kept out of the country. As for the land it is practically all good and quite suitable for farming purposes if the climatic conditions are favorable. During our stay in this Northern country this season, we could see no difference in the climate from that of eastern Ontario. The first frost we had was on the night of the 6th of September, and only on two other occasions was there a frost from that date until we arrived at Missanaibi Station on the 4th of October. All the wild fruits seem to flourish in this section. We saw strawberries, raspberries, red and black currants, red and black cherries and June plums, all these fruits were luscious and good and fully matured. We did not happen to see any gooseberries or wild plums, but would not say that they do not grow there.

In conversation with Hudson Bay Company's employées, we understand that all garden vegetables are grown successfully at their different posts, including Moose Factory, and even as far north as Rupert's House all vegetables are grown except tomatoes, cucumbers and melons. Oats also mature fully at Moose Factory and barley occasionally. We were also informed by a missionary from there that all garden fruits, such as strawberries, raspberries and all kinds of currants and gooseberries are successfully grown in the Bishop's garden at Moose Factory.

SUMMARY OF REPORT.

The district included in our investigations comprises an area of about seven thousand eight hundred square miles, and leaving off one thousand eight hundred square miles for water, etc., we have left six thousand square miles, seventy-five per cent. of which is choice farming land, and in the event of the country being cleared up, a good portion of the remaining twenty-five per cent., and what we now condemn as being too low or wet, could be brought under cultivation as grass land. And of this six thousand square miles of territory, sixty per cent. of it will cut five cords of spruce wood per acre, taking it down to five inches in diameter, besides tamarac, poplar, balsam, balm of Gilead and white birch. We might add that we have examined special acres that would cut twenty cords of good spruce, also some acres that would cut fifteen cords of spruce and ten of popular, good clean trunk wood without going into the limbs, and some of those same acres, if all the timber on them was made into ordinary fire-wood, would cut sixty to seventy cords, and those were by no means the best lots that we saw either. We cannot see that there is much more to be said. We have tried to describe the country with the land and timber as nearly as possible from our point of view, and we refrain from giving the topography of either the rivers or land as that part of the work was done by the geologists.

Hoping this report may be satisfactory to you,

We remain, your obedient servants,

J. L. BREMNER.

J. M. MILNE.

Land and Timber Estimators Exploration Survey Party No. 2

The Assistant Commissioner of Crown Lands, Toronto.

I, J. M Milne, of the Village of Queensville, make oath and say, that the contents of this report are correct and true to the best of my knowledge and belief.

JAMES M. MILNE.

Sworn before me at Queensville,
this 25th day of January, 1901.

GEORGE WRIGHT,
A Commissioner.

I, John L. Bremner, of Admaston, Land and Timber Estimator, Exploration Survey Party No. 2, make oath and say that the contents of this report are correct and true to the best of my knowledge and belief.

J. L. BREMNER.

Sworn before me at Biscotasing,
this 29th day of October, 1900.

WILLIAM ROBINSON,
A Commissioner.

GEOLOGIST'S REPORT OF EXPLORATION SURVEY PARTY No. 2.

NAPANEE, Ontario, Dec. 17th, 1900.

To A. NIVEN, Esq., O.L.S.,
Surveyor in Charge Exploration Survey Party No. 2.

SIR,—I have the honor to present the following report of the work done by me, as one of the geologists, associated with your survey in Algoma. The work was undertaken in connection with that of the land and timber estimator, Mr. Bremner, of Admaston, and we were given the assistance of Mr. H. G. O'Leary as canoeman, who rendered good service. The explorations were carried on chiefly along the waterways, from which numerous trips were made into the country. The geological character of the country was studied almost wholly from those rocks found outcropping along the streams and lakes, since, as a rule, no rocks were found outcropping in the country, owing to the deposits of clay and occasionally sand, which appear to form a continuous layer over that portion of new Ontario, explored by our party.

The party left Toronto on June 11th via North Bay and Mattawa for the north country. The journey by rail ceased at the foot of Lake Temiscaming, after which the steamer "Meteor" conveyed the party to the head of the lake. From here a twelve mile waggon road leads to Quinze Lake. The rest of the journey was continued in canoes through a system of lakes and rivers to the height of land separating the waters of the Hudson Bay slope from those of the St. Lawrence. Beyond this the journey was continued by a water-way to Lake Abitibi, thence down the Abitibi River to Jaw Bone Creek, where the main survey party separated from the exploring parties.

FREDERICK HOUSE RIVER.

Where the Frederick House River joins the Abitibi River it is about four chains wide. The clear water of this river forms a marked contrast to the turbid waters of the Abitibi, which latter is due to the clay shores around Lake Abitibi, which is also very shallow. The Frederick House River was found to be well wooded along the shores with large spruce, poplar and cedar. Four and a half miles from the month, a small rapid breaks the splendid canoeing. This is caused by an exposure of a light colored gneiss. A portage of one chain over an island passes this. One half mile below here, an exploration trip was made to the west. A shallow rapid creek, a chain in width, enters the river, which was followed for a mile. The creek is clothed with 12 inch spruce and poplar. West from here is a wet country for a mile but well wooded with spruce and tamarac. Small ridges of poplar and balm of Gilead were observed with good clay soil. This stretch of country is considerably higher than the river, and if properly drained should make good agricultural land. Above the rapid, smooth water continues for four miles to a hard rapid nearly a mile long, the upper portion of which is provided with a portage on the east bank.

GNEISS. Beyond this is a chute of an eighth of a mile, over mica schist and gneiss, striking north, seventy-eight degrees east and dipping thirty degrees south. An exploration trip was made southwest from the chute for three and a half miles, which revealed rolling clay land for two and a half miles clothed with 15 inch to 30 inch poplar, birch and spruce. This passed into a muskeg, which was followed for a mile; a digression east showed muskeg passing into rolling land, with good timber, which extended to the river. Numerous signs of moose, cariboo, and bear were observed on this trip. Clay banks were noted along the river, one place eight miles up showing four feet of banded stratified clay.

DRIFTWOOD CREEK.

The description now turns to Driftwood Creek, which flows into the Abitibi River, five miles below Frederick House. This creek is used as a highway between the Abitibi and Mattagami Rivers, by the Indians of this region. About one mile up, a trip was made to the west. A steep hill, covered with burnt timber, was ascended for fifty feet, beyond which is a small beaver lake fifteen by twenty chains, which is connected with Driftwood Creek by a small creek. Half a mile further the brulé passed into the green bush. Spruce and balm of Gilead twenty inches in diameter were observed on the ridges. Level clay land followed for a mile, passing into a semi-muskeg, with three inch to six inch spruce for half a mile. This was succeeded by three-quarters of a mile of good land with poplar, spruce, and tamarac of good size, passing beyond into a muskeg with scrub timber. A series of ridges of dry land followed alternating with wet land for a mile. The return trip to the Abitibi River showed low land for two miles with fair-sized spruce. From here to the river, a distance of three miles, showed a series of ridges with sandy soil, which for two miles had been burned over. Near the Abitibi River is another small lake, half a mile long with brulé to the water's edge. The general direction of Driftwood Creek is somewhat south of south-west while its lower portion is so tortuous as to render any detailed work difficult. The first twenty miles of creek is characterized by hills, reaching in places an altitude of a hundred feet, with numerous steep rapids and falls, whereas the upper portion is comparatively flat with shallow rapids. The hills are for the most part clay with accumulations of sand in places. The creek is well wooded with poplar, spruce and tamarac, which are very vigorous. At all the rapids were noted exposures of Laurentian gneiss. Fifteen miles up stream, just above a small fall, an exploration trip was made inland to the northwest. The trip revealed a steep bank, fifty feet high with spruce and poplar, passing to a slightly lower tract with spruce, beyond which is a dry muskeg, which alternated with ridges of clay land to a large creek two miles back. This creek drains an extensive meadow which at one time had abounded in beaver, but now is a grazing place for moose, numerous signs of which were seen. The country revealed by this trip is high above the creek, and if drained, should be adapted for farming.

The topography of the first twenty-five miles of this creek was given by Mr. Parks in the Bureau of Mines Report of 1899. Above this a shallow rapid extended for a half mile, followed by good water, for three-quarters of a mile to a small rapid which was tracked. Just beyond is a bad rapid over hornblende gneiss, which requires a portage of thirty-seven chains, on the west shore. Good water extended for a mile and a half to a flat rapid which was poled. Above this, the creek flows through a flat country, with small spruce and tamarac for over a mile to a rapid provided with a twelve chain portage on the east bank. One-half a mile beyond is another rapid requiring a portage of five chains on the west bank. Above this good canoeing extended for five miles to a small rapid. One and a half miles below this, a twenty-five foot creek comes in from the east, flowing over Laurentian boulders. Around this creek were seen Indian winter encampments. The Indians evidently trap beaver here, for fresh work was observed around the creek.

TRAIL TO MATTAGAMI RIVER. Half a mile beyond the rapid is the trail to the Mattagami River, just opposite an Indian encampment. Here the Indians had left their cooking utensils, guns, traps, etc., which were as safe as though under lock and key. The portage route from here to the Mattagami is about nine miles, in a westerly direction.

The river bank extended for a quarter of a mile passing into a swampy tract with clay soil, and six inch to ten inch spruce for a mile and a quarter. Beyond this is a muskeg (very wet) for a half a mile. This muskeg approaches a peat bog, with several feet of vegetable accumulation. This is followed by swampy ground with six inch spruce to a belt of dead timber which extends to a swampy lake, a mile long by nearly half a mile wide. This lake is nearly filled with silt and vegetable matter having only a depth of from two feet to three feet. Beyond the lake the trail extends over three quarters of a mile of good clay land with 6 inch to 10 inch spruce, to a creek flowing from the lake, but which was too shallow for navigation. Two and a half miles of nearly level spruce-tamarac area brings us to another creek which is navigable with difficulty for a half mile. The trail from here to the Mattagami River continues over swampy ground for a half mile, followed by two miles of good clay land with small poplar and spruce to the river.

MATTAGAMI RIVER.

HURONIAN SCHIST. The portage from Driftwood Creek comes out on the Mattagami River just below Loon Portage, which is around a fall of 12 feet divided into three parts by two rocky islands. This fall is occasioned by an outcrop of greenish hornblende schist which strikes about north sixty degrees west. At the head of the fall the rock is finer grained and stained with iron. Here was also observed several jack pine, the first seen on the trip. Below Loon Portage is half a mile of swift water, followed by Davis' Rapids, a portion ofwhich is passed on the east shore by a fifteen-chain portage.

LAURENTIAN GNEISS. At the foot of these rapids is an exposure of hornblende gneiss which marks the beginning of the Laurentian rocks, which were observed continually as far as we explored on the Mattagami River. Below the rapids the river expands to twenty chains around some islands. Three miles further brings us to Yellow Falls, which is a vertical drop of fifteen feet. The water plunges over in four lateral cascades and presents a grand view. This fall is over a hard dark rock showing weathered felspar.

GABBRO. A thin section of this rock shows plagioclase felspar, much kaolinized, augite, which in places is altered to uralite, also a little orthoclase, aud magnetite, thus making the rock a gabbro. This fall is provided by a five chain portage over a hill on the west bank. Fine large white spruce and poplar were observed along this section of the river. Two miles of swift water brings us to Island portage, where there is a fall of ten feet over hornblende gneiss. The portage is over the island and is five chains long. Just below is a very dangerous eddy, which renders navigation difficult for loaded canoes. Beyond is a heavy rapid for half a mile, which can be run if canoes are not too heavily loaded. Below this, good canoeing extended to Smooth Rock Portage, a distance of ten miles, four miles above which a large river, one and a half chains broad, comes in from the west. On this river near its mouth spruce, poplar and balm of Gilead are very vigorous. At "Jump Over Place," or Smooth Rock Portage, there is a fall of fifteen feet in four chains, the gorge contracting to a chain and making good facilities for water power. The fall is over hornblende granite gneiss, which shows much evidence of contortion. Several trap dikes cut the strike in the gneiss at this point, one of which, six feet wide, strikes north-northeast and stands out in marked oontrast to the gneiss. Along the walls epidote has been developed and the gneiss is much stained with iron oxide. The portage known as "Smooth Rock" is six chains long on the east side.

Three and a half miles below the fall, clay bluffs, from twenty to fifty feet high were observed, with large ten to twelve-inch spruce on the banks. Beyond this for five miles the river flows between sparsely wooded banks, the timber being chiefly scrub poplar. Nine and a half miles below the fall, an exploration trip to the west revealed a river bank, forty feet high, beyond which is a ravine followed by a comparatively level country with small scrubby timber. Several beaver meadows were observed, but no fresh cutting was noted.

Below here the river flows in a direction west by north for eight and a half miles; after which it swings to the west for three miles to Poplar Rapids River, a tributary from the west. The banks of the river along this section are clothed mainly with poplar, passing into spruce further back. Olay soil was noted repeatedly along the river and sand occasionally presents itself, though generally below the high water line. The shores are much cleaner than those of the Abitibi River, there being fine sand and gravelly beaches in places, in contrast to the commonly occurring mud banks of the Abitibi, with scrub alder and willow to the water's edge. At the junction with Poplar Rapids Mr. Niven's line crosses at the thirty-third mile. Below this the river is swift and shallow for three miles. In places the canoe has to be dropped down, while one spot requires a portage round a drop of two feet.

GARNETIFEROUS GNEISS. At the foot of the rapids is an exposure of white garnetiferous mica gneiss, with strike north, eighty degrees west. From here to Cypress Portage, ten miles distant, the river has a northwesterly course and the navigation is good though the current is strong. Five miles above the portage are high sand banks clothed with poplar. Sandy shores are becoming prominent and sandy soil is observed inland. The timber along this section of river is passing into large poplar with spruce less plentiful along the shore. Two and a half miles above Cypress Portage a creek of some size comes in from the west. This was followed in a northeasterly direction for three and a half miles when it forked. The right branch was followed for a half a mile, to a shallow rapid over boulders. Considerable Devonian drift was seen along the creek, in which were observed zaphrentis, favosites, etc. Numerous 8 inch to 10 inch Jack pine, with large spruce are to be seen along the creek, and sand is very much in evidence and forms high banks in places overlying the clay. The left branch was followed in a north west direction for a mile, finding it blocked with driftwood. High sand hills are frequent, sometimes reaching an altitude of a hundred feet. On these ridges 6 inch to 10 inch jack pine flourish. A trip further up the creek revealed jack pine all along, also large spruce. In the valleys clay soil was noted. The growth along the creek is very luxuriant and should denote a good country for agriculture. A trip was made from here to the mouth of the Ground hog River, a distance of nearly ten miles. The creek was followed for two miles showing sand ridges with large jack pine and spruce. From here a digression was made to the Mattagami River. After half a mile, the river bank passed into a muskeg for three-quarters of a mile, followed by low land with fine spruce, alternating with strips of muskeg for four miles. After this we passed several ridges covered with spruce and poplar, alternating with low land to the river bank. This country was found to be well timbered, and to have sandy clay soil. Cypress Portage, on the west bank, has a length of twenty chains over Laurentian gneiss. The fall here is in two parts with a drop of eight feet in ten chains. A trip west from the portage, shewed a low spruce area for three miles. The spruce would not average more than eight inches but would be suitable for pulpwood.

Between here and the Kakozhisk River, a distance of three miles, are two small rapids which may be run in ordinary canoes. Several small well-wooded islands are passed just below the falls. On the east shore above the Kakozhisk River are high banks of clay and sand, in one place showing five feet of a stratified bluish clay capped with twenty feet of fine white sand. From here to the Kapuskasing River is about six and a half miles, in a direction a little west of north, broken mid-way by two stiff rapids over mica gneiss. The first of these is passed by a portage over the island on the west side and the second requires a lift out. At the former was observed a vein of quartz-stained with iron oxide. The quartz is impregnated with pyrite and garnets. This vein is thirty feet wide and showed for forty feet. A sample of vein material, nearly all iron pyrites, showed on assay $1.40 a ton of gold. Below the Kapuskasing the river expands to ten chains and is very rapid for a mile when it broadens to twenty chains around several islands for three miles. Four miles below here is Devil's Portage. Here the rock exposed is a mica gneiss striking north, eighty degrees west and dipping 45 degrees west. In high water several islands exist on the west side, but in low water, as we found it, numerous blind channels were observed. The rapids at this place are a quarter of a mile long with a pitch at both ends. A trip was made from here to the west showing 6 inch to 15 inch poplar, and spruce for two miles. The soil is a light sandy loam and not very deep, as gneiss was observed in several places two miles back which gave a rolling appearance to the country. The return trip southeast was over a nearly level country timbered with spruce from six to ten inches in diameter. Near the river several poplars of at least 30 inches were seen. Seven miles below the Kapuskasing River a trip was made to the east. The first mile is timbered with 12 inch spruce and poplar and a few birch. Very much windfall was encountered which rendered progress difficult. Beyond is a brulé at least fifteen years old, overgrown with small poplar. On this brulé the soil is a clay loam, sandy in places. The burnt area afforded an excellent opportunity of seeing the country which presented a rolling appearance. An exposure of a light mica gneiss is seen three miles inland. Just below the mouth of the Kapuskasing River is a brulé which extends three miles down stream. The burn appears to be this spring's, and was found to extend at least four miles inland. The burnt area comprises a valuable farming locality of slightly undulating clay land. The bluffs opposite the Kapuskasing River are deposits of clay, overlaying sand and gravel and reach an altitude of forty feet. A mile below this, the bank shows fifteen feet of clay over ten feet of sand and gravel.

Kapuskasing River.

The Kapuskasing River is about five chains wide at its mouth and enters the Mattagami from the west. For twenty miles this river was followed in a general direction south west, after which it became south-south-west. From its mouth to Sturgeon Falls, a distance of thirty-five miles, it is broad and shallow, with its channel full of boulders. Above this it is narrower and deeper. In the former distance are several rapids, which require to be tracked, while above are numerous falls. Nearly twenty miles up a trip was taken to the west. The river bank, with large spruce, poplar, and balm of Gilead, passed into an area of tamarac and spruce, followed by three-quarters of a mile of low land, with fair-sized spruce to a high ridge with large spruce, birch and poplar. The soil where observed is clay, and the timber was uniformly large. Four miles above, Mr. Niven's line crosses the river at the fifty-fifth mile. Both above and below the line the gneiss is cut by trap dikes. Three miles above the line the dike is stained red by specular iron ore disseminated through it. A thin section shows the rock much decomposed with kaolinized felspar, hornblende and hematite. The first real impediment to navigation is a chute, known as "Sturgeon Falls," or the Indian name, "Ne ma-boi-tik." Here there is a fall of twelve feet in ten chains. A portage of six chains on the east over the rock passes this. This chute is so named because the sturgeon do not get up the river beyond this. Between here and the next fall, a distance of two and a half miles, are two rapids, the first is tracked and the second passed by a portage of two chains on the west side. Three-quarters of a mile beyond the river takes a wide bend to the south-west around some islands to a rapid requiring a short portage over the rock. A quarter of a mile beyond the river bends to the east for half a mile, to a fall coming in at right angles from the south. The fall has the Indian name of "O-Ke-Kes-boitkik," or "White Spruce Rapid." The descent here is fifteen feet in three plunges. A portage of eight chains over the point avoids this. The rock is hornblendic gneiss, which is much contorted, showing a fine ribboned structure in places. The gneiss strikes south, eighty degrees west and dips forty-five degrees north. Above this good canoeing extends for nine and a half miles. The river banks of this part are clothed with large spruce and poplar. Three and a half miles above this fall, a trip was taken to the east. The river bank of clay loam, timbered with good spruce and poplar, passed into a level spruce area, with clay soil Three-quarters of a mile in is an old beaver meadow, beyond which is a spruce area, passing into a swampy flat with eight-inch spruce and tamarac, and much windfall. At one and a half miles the country is higher, with a series of ridges, with clay soil, and is well timbered. Beyond this is a swampy tract with eight-inch spruce and tamarac to a level spruce area, with alternating poplar ridges—to a large creek—flowing to the Kapuskasing River evidently from beaver lakes further north.

The next obstruction to navigation is "Big Beaver Fall," or "Me-som ko-boitek." At this place there is a descent of thirty feet in ten chains, which should furnish an excellent water power. The portage avoiding this is nine chains long on the right side. The country in the vicinity of the fall is rougher and more rugged than any before seen. Ten chains above here there is a drop of four feet, passed by a short portage on the north shore, and a quarter of a mile further is a long rapid, provided with a twelve chain portage, also on the north side. Two chains beyond, the river takes an abrupt bend to the north-west for a mile and then resumes its south-westerly direction.

A Brulè.—Below this bend a brule comes in which extends for miles up the river. At the bend is a bad rapid of a quarter of a mile requiring a portage over the rock and beyond, the river expands to fifteen chains to a fall of five feet, broken into three parts by two islands. A mile and a quarter of good paddling is followed by swift water and rapids for a mile and a half. At the foot of the rapids the river flows round an island. Here the hornblende gneiss strikes a little south of west and dips 60° north. It is cut by a fine grained hard trap dike striking south-southeast. The numerous exposures of gneiss along here make many bendings in the river, with some magnificent falls. At the head of the rapids a lift out is required on the east side. A mile and a half further a beautiful fall comes in from the east, broken in two parts by a large island. The southerly portion has a drop of twenty feet in three chains, while the upper one is a long steep rapid. A portage of ten chains over a hill is provided, which shows signs of much usage in the past. In the brulé the soil, where observed, proved to be a clay loam. A fair growth of poplar covers the brulé, which is at least ten years old. In the next three and a half miles are two small rapids which can be run, while above them is a large fall known as "Old Woman Rapid," or "Min de-moi-ye." This fall, which has a descent of twenty feet, is portaged through the brulé on the east side. One half mile beyond this, green bush comes in again and continues as far as we journeyed up this river. A mile and a half of good canoeing brings us to "Ka-Kij-uan Falls" which consists of a fall of five feet, below which is a heavy rapid. The rapid is tracked and the fall is provided with a good portage on the east bank. A half mile above this fall a large creek comes in from the east. This creek, we were informed, forms part of a portage route to the Ground hog River. Good canoeing extends beyond as far as the river was travelled. Four miles up a large creek, coming in from the west, was explored. This creek, a chain in width at its mouth, has a general direction of north west, though it is very tortuous. It was followed for seven miles through a low tract of country to a lake over a mile long, having the Indian name "Ba-ke-dan-e Ka." It is very shallow and marshy and abounds in black ducks. Its shores are low with brulé on all sides, which forms part of the extensive burn seen on the Kapuskasing River.

Gneiss—Augite Gabbro.—Laurentian gneiss outcrops on the east shore while beyond was observed a weathered gabbro. Under the microscope this rock shows a basic plagioclase-felspar, and augite, with much magnetite in grains disseminated through it. Due north from the outlet a creek flows in which was also followed in a north west course. For two miles the canoeing continued good but above this for a mile the creek is rapid and jammed with driftwood, while beyond good canoeing extended for three miles, when the creek again narrowed and was found blocked with driftwood, making further navigation impossible. The whole area traversed has been burned over, with a few patches of green bush. The country is rolling clay land, which should be suitable for agriculture. Along this creek, recent beaver work, which is of rare occurrence, was seen, and the creek showed signs of much use by Indian trappers.

On the Kapuskasing River above here the surrounding country is very low, with scrub spruce, tamarac and cedar. The river banks are low, being but a few feet above water. Several long marshy bays extend in to the country on either side of the river. Eight miles further up stream, a large creek from the west, was also explored. This was followed in a north-west course for two miles to "Kich-awa tofe" Lake nearly two miles long. The shores of the lake are very precipitous, the rock being Laurentian gneiss. At its north end a creek flows in, which was followed for two miles in a north-west direction thro' two ridges of gneiss to a second lake about a mile long.

Route to Opazatika Lake.—This lake has high rocky shores along its south side, at its west end a narrow marshy creek, a quarter of a mile long, flows into it from a third lake nearly a mile long. Beyond this is a fourth lake, and these three together are called "Mang-Ki-We-yawis" or Big Herring Lake. These lakes abound in pike and pickerel of large size and no difficulty was experienced in catching a good mess. At the west side of the third lake the portage route to Opazatika Lake begins.

First Three Mile Portage.—The first portage is three miles long and nearly due west. This showed signs of much usage in the past but is now blocked with fallen timber in many places. The low lake shore passed to a ridge with much slash to swampy land for half a mile followed by a high ridge with fine spruce and poplar for a mile to a small creek flowing north. Swampy land, alternated with ridges for another mile, passing into a muskeg for a quarter of a mile, to higher land which continued to a lake. Clay land and good timber was noted along this portage. The portage comes out in a bay of the lake, which has a sandy beach, whence the name "Wan na-ta-wany-za" mean ing sandy beach. This lake is a mile and a half long and over a mile wide at the portage. The only rock found outcropping on the lake is a light colored mica gneiss.

SECOND THREE MILE PORTAGE.—Nearly west across the lake from the first portage the second begins, which is also three miles long in a western direction. The landing is in a low marshy place, which passes beyond into a cedar swamp to a ridge with fine timber, and rolling-clay land for one and a half miles. Then there is a very wet muskeg and low spruce tract for half a mile, followed by swampy land to an old brulé. Beyond this is low land to another brulé, followed by a cedar swamp, and high land to a small lake about a third of a mile in diameter. From the west side of this lake a creek flows to Opazatika Lake, about two miles distant. The creek is followed for a quarter of a mile when a portage of thirty chains is required to avoid an obstruction, after which good canoeing extends to the lake.

OPAZATIKA LAKE.—A half mile west of the mouth of the creek is a poplar point from which the lake derives its name, "Opazatika" meaning poplar. Opazatika Lake is made up of a number of smaller lakes which are connected by narrow channels. Numerous islands make the lake very picturesque. At the poplar point before mentioned is a channel a chain wide and a quarter of a mile long, connecting two portions of the lake. At its middle, a large creek flows in from the south. This creek was followed for nine miles, at which distance it was found to be blocked with driftwood. It flows through a marshy country with scrub spruce for five miles, but above this the shores are higher and are well wooded with large spruce and poplar.

GABBRO.—The only rock seen outcropping along the creek is a massive gabbro outcropping four miles up. This rock has a coarsely crystalline structure with weathered felspar. Under the microscope it shows twinned plagioclase, altering to kaolin, augite, secondary hornblende and magnetite. This creek is called Hay River because of the marshes at the mouth, and we are informed that a route exists by this way to Missinaibi Lake.

A trip to the southern end of the lake showed numerous exposures of Laurentian gneiss, cut through in places by narrow basic dikes. The islands are generally wooded with spruce and poplar, but some of them have been burned over and are clothed with a growth of small poplar. Several large white pine were noted at the southern end, and were the only ones seen on the trip. Beyond the narrow channel mentioned above, the route to Missinaibi Lake is followed to the west for a mile and a quarter, when the lake narrows between some islands and the course turns to the northwest for a mile. At this point a dike of gabbro sixty feet wide and striking north, 20 degrees east, cuts through the gneiss. A thin section of this rock taken from the middle of the dike shows partly kaolinized plagioclase felspar, augite, secondary hornblende (uralite) and a little biotite, magnetite, pyrite and quartz as accessory minerals.

MICA GNEISS.—The gneiss observed along the lake is chiefly micaceous, while in the vicinity of the dike it has been rendered coarsely crystalline, showing in places well developed crystals of quartz, felspar and mica.

EPIDOTE.—Here also epidote is observed as a contact metamorphic mineral. Beyond this the course turns again southwesterly for three miles, passing numerous islands to the south, many of which are clothed with numerous jackpine. At this point the lake narrows to a narrow channel half a chain wide for a quarter of a mile, when it broadens for half a mile. At this channel are high bluffs of mica gneiss to the east. The lake again narrows and continues so for two and a half miles to an expansion over half a mile wide and nearly two miles long. Beyond this the lake becomes marshy with a winding channel for two miles to its southern end. Here a small rapid creek flows in over mica gneiss cut through by a six foot trap dike striking south, ten degrees west.

PORTAGE ROUTE TO MISSINAIBI RIVER.

At the foot of this rapid is the first of the portages to Missinaibi River. It is three-quarters of a mile long over a hill in a southwest direction. The soil where observed on the portage is sandy, but is covered with large spruce and poplar. This leads to a small lake (fifteen by twenty five chains) which is shallow and very marshy around its margins. The Indians have a long name for it which means "The lake we walk along shore," because they sometimes portage around it. This lake is crossed in a south southwest direction to the second portage which is nearly a mile long, also south west, over rolling land, showing clay soil in the bottoms and sandy soil on the ridges. The country is timbered with large spruce, poplar and tamarac. This portage leads to a lake nearly a mile long by half a mile wide, which is crossed in a southwest course to a creek flowing from it. This creek was found to be very shallow and the canoes had simply to be pushed through the mud and the alders which overgrow the banks. For probably a mile this creek was followed in a circuitous direction through a low country which is well timbered with spruce and tamarac. Here it was found incapable of further navigation, and a land portage is used to the Missinaibi River. At this point the Indians have constructed a small dam to make the upper portion of the creek navigable and from this the creek gets its name Ka-o-Kne-e-me-kad-da, meaning Beaver Dam River. The portage from here is nearly west for fifty chains. At thirty chains this creek is crossed, and it flows into the Missinaibi River a quarter of a mile below the portage. Fine large spruce and poplar, also tamarac, were observed along the portage, where the soil is a sandy loam.

Missinaibi River.

The description now turns to the Missinaibi River, where Mr. Niven's line crosses it below Kakagee Rapids. A trip was made in the country from the hundredth mile to the northwest. As usual this showed the river bank to be heavily timbered with large spruce and poplar, and to have good clay soil. This passed further back into a comparatively level spruce area with a slight covering of moss. Beneath the moss good clay land was exposed. This moss "sphagnum" is of very common occurrence back from the rivers, being in some places, but a few inches in depth, while in others several feet. The river at the line is very shallow and rapid. An eighth of a mile below is an outcrop of mica gneiss flanked to the south by a grey mica schist, striking south 80 degrees west and dipping 80 degrees north. Above the line are the Kakagee or Crow Rapids which are over a fine-grained mica schist, which passes further south into quartz schists and mica schists cut through in places by trap dikes and diabase. Two miles above the line a mica schist striking north-east and south-west forms a barrier for a rapid, while beyond is a quartzitic schist, flecked with mica. At Sandy Bay portage is a light hard schist, showing quartz and felspar, probably a quartzose schist, with strike north eighty degrees west, while to the north is a dark grey mica schist, showing in thin section quartz which is very much fractured, biotite, and a little plagioclase. At Sharp Rock portage a few miles beyond, is a fall of four feet in three parts over a hard grey mica schist. The schists exposed below Beaver Portage resemble those at Sharp Rock so that the belt of mica schist likely extends to here. At Beaver Portage the river expands to twenty chains while there is a fall of twenty feet. At the lower end of the portage is a quartzite which is light colored and shows in thin section quartz grains, a little muscovite, felspar, hornblende, calcite and magnetite. To the north it becomes schistose, and is flecked with mica. This rock is cut by a vein of white quartz, two feet wide and striking westerly. On analysis a sample showed a trace of gold. At the upper end of the portage an eruptive fine-grained rock cuts through and shows in thin section plagioclase felspar with a ground mass of pyroxene so the rock is probably a fine-grained diabase. Above here dark schists cut with trap dikes extend to near the Albany Rapids, where the rock is a mica gneiss having a granite structure, also it is cut through by narrow trap dikes. From the line to this point the banks are high and uniformly clothed with spruce and poplar. High clay banks were noted near the line and clay soil all along the river.

A half mile further the Albany Branch or Mata-wa-ge-wan River comes in from the west. For nearly a mile the river is shallow and rapid over mica gneiss and boulders. A quarter of a mile beyond this is another rapid over gneiss, striking south southwest. Above here good canoeing extended for over two miles to an island where a lift out is required. Beyond this for seven miles the river is shallow with several flat rapids. The channel is filled with large boulders, which made navigation difficult. We were unable to navigate above this, for lack of water. The river is from three to four chains wide as far as travelled. Clay banks were observed in many places, capped with fine white sand. In the shallow stretches the channel is filled with large gravel and sand. A trip to the south, nine miles up showed rolling land with sandy soil. The rapids for the most part are occasioned by barriers of fine grained mica gneiss. At eight miles is an exposure of a green chlorite schist, with pyrite striking south 80 degrees west and dipping 70 degrees north and is stained red by the oxidation of the pyrite. The country on either side of the river is rolling and quite open in places. This river is used as a route to the Kabinakagami River.

On the Missinaibi River above here the river is rapid and filled with boulders for a half mile, while beyond good canoeing extends to Wilson's Bend. There are few exposures of rock in this stretch. Near the bend is observed a fine-grained granite mica gneiss, a diabase dike with strike north thirty degrees west cuts the gneiss. It is in fine-grained and in thin section shows plagioclase, augite, secondary hornblende and magnetite. Above the bend the country becomes higher with bluffs of Laurentian gneiss. An eruptive mass of hard diabase, showing phenocrysts of weathered plagioclase, outcrops two miles above here. In this section it shows the felspar, nearly all changed to saussurite, set in the augite. At the first of Devil's Rapids, is a green hornblende schist, striking north of west and dipping 70 degrees north. An eruptive mass of pegmatitic granite is also seen here. At the second rapid is a fine-grained hard schist with scales of mica. The third is caused by an outcrop of micaceous schist, as is also the fourth where there is also an eruptive diorite. The rock at the fifth rapid has less mica and approaches a siliceous schist. The sixth, seventh and eighth rapids are nearly continuous and are passed by a forty-five chain portage on the east bank. The schists here strike south 85 degrees west with nearly vertical dip. A quarter of a mile beyond is the last of Devils Rapids known as Sugar Loaf Portage. Here there is a fall of about six feet, nearly vertical, the rock exposed here being a light grey quartzose schist. There is a three chain portage on the west. At Aschuter portage just beyond, the rapid is occasioned by an outcrop of hornblende schist which is also seen at Double Portage two miles beyond. At the lower end of the portage the schist is coarse-grained, becoming fine-grained further north and these rocks are cut by several dikes of hornblende porphyrite showing in thin section phenocrysts of hornblende in a ground mass of felspar. There is a fall of four feet with rapids below, which are passed by a quarter mile portage on the east. One and a quarter miles above here fine-grained mica gneiss is observed which marks the beginning of the Laurentian. The Brunswick or Ka-ba-sching River which enters from the west two miles above, flows in a northerly direction from the lake of the same name. A detailed description of the navigation of the river shows one and a half miles shallow water with boulders and gravel, one quarter mile good water, chute 5 feet, half mile good water, fall 3 feet, quarter mile rapid, fall 2 feet, one and half miles good water, one mile broken water with three rapids, three miles good water through a low marshy stretch to the lake. Brunswick Lake for five and a half miles is very narrow when it broadens, forming three deep bays.

On the west shore of this lake was formerly a flourishing Hudson Bay Company post but it has since fallen into decay. The old buildings have been burned and on the grounds were found apple trees and red currants which had formerly been cultivated. Clover and carrots were growing wild and luxuriant This fort had a splendid situation, being on a clay bluff, overlooking the lake. At the southern end of the narrow portion the shores are low and swampy with scrub willow. Beyond this the lake expands forming the deep bays. The western arm is at least four miles long and on its west shore are high, rocky bluffs. The central bay extends two miles south, while the eastern branch, which is long and narrow, has a length of nearly four miles. From the southern end is the portage to Missinaibi River. The lake end is low and marshy which becomes higher passing a rocky ridge running south from the lake. The portage is three quarters of a mile long, coming out on the Missinaibi River five miles above the Opazatika portage.

Geology. Huronian rocks were observed on the Abitibi River from its source to near the head of the Long Sault, where the contact with the Laurentian to the north occurs. Below this on the Abitibi River to Three Carrying Places and on the Frederick House River, gneiss and mica schist were the only rocks seen outcropping. These rocks were also noted all along Driftwood Creek, but where the portage from this creek to the Mattagami River came out, namely at Loon Portage, Huronian schists outcrop. At the foot of the portage a coarse hornblende schist outcrops which becomes finer-grained to the north. Hornblende gneiss outcrops at Davis Rapids just below, so this marks the contact of the two series of rocks. At Yellow Falls, gabbro cuts the gneiss, while below here numerous narrow trap dikes were observed at intervals. This river was explored north as far as Devil's Portage, below Kapuskasing River, and Laurentian gneiss was the only rock outcropping beyond these. Along the Kapuskasing River are exposures of mica and hornblende gneisses, which form the barriers for all the falls, and the portage route from this river to the Misssinaibi River reveals only gneiss, with eruptive masses of gabbro.

On the Missinaibi River below the portage, Laurentian gneiss extends to near Double Portage when Huronian schists consisting of hornblende, quartzose and mica schists form a belt reaching nearly to Wilson's Bend. Between here and Albany Rapids, a distance of nine miles, a fine-grained mice granite gneiss outcrops. Below this another belt of Huronian schists, consisting of mica and quartzose schists and quartzite, cut by diabase. extends to within a few miles of Kakagee Rapids, near Mr. Niven's line. These belts of Huronian schists were found to have a general strike of south, 80 degrees west, with nearly vertical dip.

The geological map of northern Ontario compiled from Dr. Bell's reports depicts the Huronian rock, observed at Loon Portage as stretching across country to the Missinaibi River, completely enclosing Opazatika Lake. The Ground Hog River was explored for fifty miles by Mr. Fraleck, who reports no other rock exposures beyond Laurentian gneiss and trap dikes. The writer travelled the Kapuskasing River for seventy miles, during which distance hornblende and mica gneisses, cut by a few trap dikes were observed outcropping continually, also on the portage route to the Missinaibi River. Mr. Fraleck reports a band of Huronian schists on the Opazatika, which belt extends in a westerly di- irection to the Missinaibi crossing the river in two narrow belts. This would make an solated Huronian area crossing these two rivers.

AGRICULTURE.—The soil over most of the region travelled proved to be clay, while occasionally areas of sand, or sandy loam were met with. Especially along the rivers is the soil good if it is to be judged by the luxuriant growth of trees, shrubs and small fruits. Few particular areas can be mentioned as better than others, for the country presented a very uniform appearance. After journeying back from the rivers from a half to three miles, the rolling land of the river bank generally gave place to a comparatively level spruce tract which is covered with a layer of moss; clay soil was observed repeatedly under the moss when it was removed. As one travelled from east to west the country ap parently became higher and drier, and so the muskegs decreased in proportion. The land back from the river is generally higher than the river and if a system of drainage were established a good section of farming land could be obtained. Potatoes were found growing luxuriantly on the Kapuskasing though they were not cultivated. Between the Kapuskasing River and Opazatika Lake, the portage route revealed a good stretch of country, low in places but with good clay land predominating.

CLIMATE. The climate of this district is moderate during the summer months, only a few days of excessive heat being encountered. Frost was not seen from the time we entered the country until Sept. 9th, on the Missinaibi River. The climate is moderated by the large body of water to the north, and the large rivers all flow northward, their warmer waters pushing the cold water of James Bay further out.

TIMBER. The varieties of trees met in this north country are not numerous. The prevailing tree is spruce, while next in importance is the poplar. The white variety of spruce of good size was seen continually along the rivers and on ridges back from them. Black spruce, generally scrubby, clothes the muskegs; poplar and the related tree, balm of Gilead, are very common. In many places poplar is to be seen as second growth on the old brulés. Tamarac is a common tree in low, swampy districts and is of considerable size, but it was noted that many of the trees, especially the largest, were dead, due, no doubt, to some insect which infests them. Birch and balsam are also common on high lands, while only a scrub variety of white ash is to be seen along the Kapuskasing and

Opazatika rivers. Pine trees are of limited extent, being seen but seldom. Pitch pine was observed in some quantity near Cypress portage on the Mattagami river, while red pines dot many of the islands in Opazatika Lake. Here also several large white pine, over two feet in diameter, were noticed.

TIMBER. Of smaller trees and shrubs the most common are alder and several varieties of willow, which are luxuriant along the river banks. Dog maple and thorn are also common. No hardwood of any kind is found in the country.

FRUITS. Of small fruits, raspberries are plentiful on the old brulés, those of the Kapuskasing River being very productive, while blueberries, except along the Kapukasing, are not abundant. Deerberries and June berries are very common along the rivers. Red currants, blackberries and gooseberries grow profusely in the woods. Wild cherries and the common chokecherry are not plentiful.

FAUNA. The stretch of country explored during the summer cannot be said to abound in fur bearing animals. The beaver is fast becoming an extinct animal, very little fresh work being observed. On trips inland many old beaver meadows and dams were noticed, but few were of recent date. Signs of marten, otter, muskrat and mink were seen along the rivers, especially the Kapuskasing. Moose are still plentiful and numerous signs of these, along with some caribou, were noted in the muskegs and along small streams. Bears are plentiful along the upper Kapuskasing and on the Mattagami River. The common partridge and the spruce variety were seen repeatedly and are numerous, while rabits are very scarce. The writer was informed by a Hudson Bay employee that one old hunter who had formerly brought in as many as a thousand rabbit skins could this last year get but twenty. Wild ducks cannot be said to be plentiful, though sometimes observed in large numbers, especially on the portage route from Kapuskasing River to Opazatika River.

FISH. North of the height of land, the waters cannot be said to be very abundantly supplied with fish. The large rivers like the Abitibi and Mattagami are nearly barren of fish. The lakes on the portage route from Kapuskasing River to Opazatika abound in pike and pickerel of large size, one pike of ten pounds furnishing a good repast. Sturgeon are found in the large rivers, but are not plentiful. In Opazatika Lake the Indians catch whitefish in nets. In contrast to the Hudson Bay slope is the country to the south where the lakes and rivers are teeming with pike, pickerel and whitefish.

WATER POWER. An enormous water power could be developed on the rivers of Northern Algoma, where there are numerous falls and rapids. Mention may be made of several rivers which are capable of furnishing good water power. On the Mattagami at "Yellow Falls" with a drop of twelve feet "Island Portage" drop of twelve feet, "Jump Over Place" drop of fifteen feet in two chains, good power can be developed. On the Kapuskasing, White Spruce Rapid with a fall of fifteen feet, and Old Woman Rapid with fall of twenty feet may also be mentioned.

OCCUPATION. With the exception of several hundred Indians scattered over a large area, the country is practically uninhabited. During the summer months scarcely an Indian is to be seen, outside the Hudson Bay posts where they congregate during these months, living chiefly on fish. The male portion is hired by the Hudson Bay, in transferring their stock from one place to another. During the winter they return to the wilds, where they hunt and trap. We were informed that often the Indians have great difficulty in keeping from starvation, due to the lack of game during some winters. The Indians of Northern Algoma belong to the Cree and Ojibbeway tribes. The language of the Abitibi Indians resembles that of the Missinaibi, whereas these have great difficulty in conversing with the Mattagami Indians.

OPENING UP OF THE COUNTRY. The opening up of the country should not present many difficulties. The nearest point of access is Missinaibi Station, from which point there is a water route of about forty miles consisting of Dog, Crooked and Missinaibi Lakes. These lakes are separated by narrow strips of land not more than sixteen chains in width and as mentioned by Mr. Borron, in his report of 1890, on this section of country the navigation of these lakes could be opened by constructing locks at these points. The Missinaibi River with its numerous water falls is too much broken for navigation but its falls would furnish abundant power for an electric road through this section of country. It may also be mentioned that the country below Missinaibi Lake differs materially from that around the lake, being more level and less rocky, which should make the building of a road not very difficult.

I have the honor to be,
Sir,
Your obedient servant.
A. G. BURROWS, M.A.,
Geologist of Exploration
Survey party No. Two.

Napanee, Dec. 17th, 1900.

Dominion of Canada
Province of Ontario
County of Lennox and
Addington.
To Wit:

In the matter of the report of A. G. Burrows, M.A., Geologist of Exploration Survey party No. 2.

I, Alfred Granville Burrows of the Town of Napanee in the County of Lennox and Addington, Master of Arts make oath and say:—

1. That the within 53 pages (of original manuscript) containing report to A. Niven, Esq., of my work as Geologist of Exploration Survey party No. Two to the best of my knowledge is correct and is based upon field notes made by me between June 30th, 1900, and September 10th, 1900, which are correct, and upon my diary for the period beginning June 9th, 1900, and ending September 19th, 1900, which diary and field notes accompany the report.

2. That the signature of "A. G. Burrows, M.A" appearing at the end of said report is in my own proper handwriting.

Sworn before me at the Town of Napanee
in the County of Lennox and Addington
this 1st day of January, 1901.
N. M. WILSON,
A Comm'r &c. in H. C. J.

A. G. BURROWS, M. A.

GEOLOGIST'S REPORT OF EXPLORATION PARTY No. 2.

BELLEVILLE, ONTARIO, December 26th, 1900.

To ALEX. NIVEN, ESQ., O.L S,
Surveyor in charge of
Exploration Survey Party No. 2.

SIR,—In accordance with instructions received from Mr. Blue, then Director of the Bureau of Mines, I joined Mr. Niven's party at Toronto on June 11th and proceeded with them via Mattawa and Temiscaming to the junction of Jaw Bone Creek with the Abitibi River, arriving there on Thursday, June the 28th. Here we entered upon our territory and separated from Mr. Niven and his party.

The country has been divided topographically into three great zones, i e. the height of land plateau, the intermediate plateau, and the low lying coast belt. Our territory lay between the Abitibi and Missinaibi Rivers in the second of these divisions and was wholly within the great clay belt.

INSTRUCTIONS. My instructions repuired me to make a track survey of the lakes and rivers that we traversed and to note all points of an economic or scientific interest. My work was to be done in conjunction with the timber estimator, Mr. James Milne, of Queensville, and we were supplied with a canoe and canoe man. Another party similar to ours divided this territory with us.

ABITIBI RIVER.

After making some explorations in country at this point we proceeded down the Abitibi River. Here the river turns to the west and for four miles we pass with swift current between high banks, thickly wooded mainly with spruce and poplar, when we came to an island and a rapid with a fall of three feet. The portage is on the left side of the island and is four chains long, over a large exposure of mica schist with gneissoid segregations. Below this the river widens, numerous islands are passed, the shore becomes low and swampy and in four miles we come to Kettle Falls which has a five foot drop over gneiss. The portage is on the north side and is four chains long. At the end of the next two miles the mouth of Frederick House River is passed and six miles further brings us to Driftwood Creek, the River meanwhile resuming its north by west direction.

At this point an excursion was made inland to the east. A burnt area extends on both sides of the river for three miles and the next three miles between high even shores with deep gullies brings us to Three Carrying Place. Here are two rapids with a total fall of fifteen feet over a great gneiss outcrop. The portage is on the west side and is fifteen chains long. This should be used, as the two portages on the east side are rather dangerous. For the next four miles the water is very bad, and we arrive at a small rapid with a portage of four chains on the west side and across a pool is Island Portage, three chains in length. For the next eight miles the river has cut through high and almost precipitous banks. Here Red Sucker Creek enters. Seven miles further up we made a trip inland to the west and with the canoe explored several small creeks flowing in from the east. The clay soil characteristic of this tract of country is here covered in places by a gravelly Devonian drift. A mile below the river makes a bend to the west and a large exposure of gneiss strikes north 80 degrees west. This is cut by a small trap dike running south 40 degrees west.

Returning, we canoed up Red Sucker Creek, but found the water very low. At the end of two miles we encountered a long shallow rapid over boulders of drift which had probably come from a Huronian belt to the north of us. From the appearance of the banks a large volume comes down this stream in high water. Continuing our journey up the Abitibi River we examined the high banks and found although there were small, irregular patches of sand near the base they were practically composed of clay. The soil inland was also clay. We passed Mr. Niven's line at the foot of Island Portage and met the other exploring party at Three Carrying Places. Here the timber estimators explored in the country to the west, and on Thursday, July the 12th, both parties returned to take the Driftwood Creek route to the Mattagami River.

Driftwood Creek. We canoed up Driftwood Creek in a general south by west direction for a distance of twenty-nine miles, examining some large clay hills near the mouth and making an excursion inland to the south east fifteen miles up. This stream varies from two to three chains in width throughout the distance that we traversed.

Portage Route. The first two miles of the land route passes through good soil and timber. Then we experienced one-half mile of wet muskeg coming to the shore of a small, shallow lake. Resuming our portage on the other side we cross another creek at the end of a half mile, and two and one-half further placed our canoe in another small creek bounded by muskeg and marsh. After canoeing down this creek one-half mile, a two mile and one-half portage brings us to the bank of the Mattagami River just below Loon Portage. This route runs east and west and with the exception of the two muskegs mentioned traverses a dry and thickly-wooded country; the first two miles and the last two and one-half to a marked degree.

Loon Portage. The timber men now made a trip in the country to the west, while Mr. Barrows and I examined the rock exposure about the falls, which have a descent of fifteen feet with a portage on the west side seven chains long. Up to this time we had been in Laurentian, but here was found a large outcrop of hornblende schist. A quartz vein, two and one-half feet wide, was observed on the east side, and another six feet wide on the opposite bank. These veins strike north-east, and although not mineralized would be considered worthy of investigation, if situated in one of our mining districts The contact between Laurentian and Huronian is situated a short distance below these falls.

Mattagami River. On Friday, July 20th, we commenced our journey down Mattagami River. Not far below we came to a rapids over Laurentian boulders and were forced to take a portage on the east side one quarter of a mile long. Then passing through a low country covered with spruce we came to Yellow Falls over a large dike with portage on the west side five chains in length A few miles below this, one-half mile of rapids brings us to Island Portage four chains long over gneiss. Then after somebad waters and other half mile rapids, we have good canoeing between banks clothed with splendid spruce until we come to Smooth Rock Falls, an almost perpendicular drop of twenty feet over gneiss, with portage on the east side six chains long. Here two trap dikes were observed, the first four feet wide with strike north north-east—the second six feet wide and parallel; and the dark green of the dikes showed clear against the light grey gneiss. As we proceed nine miles below large spruce along the shore gives way to poplar and white birch. At this point we made a trip inland to the east. As we canoe down the stream, running two small rapids in the next four miles, a dense growth of poplar and birch, with spruce and balsam in patches, is replaced by good spruce, and that again by large poplar, with some birch, which is characteristic of the shores at the mouth of the Poplar Rapids River, where Mr. Niven's line crosses.

Poplar Rapids River —Here the exploring parties separated, and on Wednesday, July the 25th, we paddled up Poplar Rapids River, which is three chains wide near its mouth. It was noted that the shores were low, and scrubby poplar and tamarac prevailed, although there were clumps of good spruce in places. Two miles up, a series of flat shallow rapids were ascended. The banks are now getting higher and the timber larger. Seven miles further, we passed the mouth of a creek one and one-half chains wide coming in from the west and the next three miles takes us past a series of marshy ponds. We landed our canoe at the foot of a shallow rapid over three-quarters of a mile long. There was a very small volume of water coming over; it was too shallow to permit of navigation, and no sign of a portage was visible. Returning, we explored the ponds mentioned, finding that one on the west side opened into a lake two miles long. A portion of the soil was composed of white and red sand covered with Norway pine averaging ten to twelve inches in diameter, the remainder of the country being heavily timbered with spruce. The creek above mentioned was also explored in a westerly direction for four miles, where it choked with driftwood. An excursion was also made in the country to the east opposite the mouth of this creek.

Mattagami River. We canoed down the Mattagami River passing the Poplar Rapids, running the canoe most of the way but were forced to portage a few yards here and there, arriving at Cypress Falls, two drops of eight feet each, the portage being on the west side one quarter of a mile long. Then a few miles of bad water delivers us at the mouth of Kakozhisk or Ground Hog River, which we ascended.

Ground Hog River. During the first nine miles we encountered a very stiff current and passed many large islands. The river is twelve chains wide at its mouth, but on account of the islands frequently widens over a quarter of a mile. A small rapid was here met with and after paddling up a ripple one half mile further on, Mr. Niven's line was crossed. Our course so far was south, but now keeps working to the west. Two mile and one half of good canoeing, then a rapid one hundred yards long with a three foot drop necessitates a lift out on the east side where a dike eight feet wide strikes north-east. At the end of the next half mile a flat rapid was paddled up and two miles further the river falls in two cascades with a total drop of fifteen feet. A large dike two chains wide cuts an outcrop of grey and pink gneiss. A portage on the east side one quarter of a mile long passes the falls and the rapids above. After one and three-quarter miles more we ascend a rapid with a short portage on an island, and one half mile further a falls with a five foot drop and a portage on the east side twelve chains in length. We continued our journey, ascending a flat rapid at the end of four miles, and a short distance further on portaging around a seven-foot fall. The portage is six chains long and is on the east side. After passing some large islands during the next two miles a rapid forced us to lift out on the west side. We had now six miles of good canoeing when we came to the foot of a large island four and one-half miles long. We took the west channel, lifting out once at a rapids and passing the mouth of two large creeks. This stretch of country has, years ago, been subjected to fire, the brulé ending when we reached the end of the island, from which point we have good canoeing for six miles, passing the mouth of a river one and one half chains wide, entering from the west. A small rapid necessitates a lift out and a short distance above the river falls eight feet, then separating into three parts has another eight foot drop. The portage is on the east side and is six chains in length. These falls are caused by a large outcrop of pink gneiss. Two miles above the river falls two feet, then separates into two lower falls with five foot drop each. The portage is three chains in

Long Rapids, Missinaibi River. Party No. 2

Towing up Missinaibi River, 2nd day from line. Party No. 2.

Rapids on Missanaibi River.   Party No. 2.

Towing up rapids, Missanabie River.   Party No. 2.

Falls on the Missanaibi River, well up towards the lake.
Party No. 2.

Victoria Mine, Sudbury. Party No. 3. (See page 83.)

Camp at Moose Point, Lake Chimcoochichi. Party No. 3.
(See page 83.)

On Stouffer Lake. Party No. 3. (See page 83.)

Rapids on Sturgeon River below Kettle Falls.
Party No. 3. (See page 83.)

length on the west side. The rock here is a fine-grained grey mica gneiss. Four miles further a half mile rapid was encountered. An excursion was made inland to the east and upon our return the large tributary from the west was explored some distance. Having canoed forty-seven and one-half miles up the Ground Hog River, we arrived back at its junction with the Mattagami, on August 4th. The current on the Ground Hog River is very swift, the lower thirty miles especially so. There are numerous islands in the river, nearly all diamond or delta shaped. The banks are high and clothed with splendid spruce, poplar and balm of Gilead. Dikes may be observed continually but many exposures may be of the same dike as they strike to a large extent in the direction of the rivers.

MATTAGAMI RIVER. We now proceeded down the Mattagami River to the mouth of the Kapuskasing passing over Island Portage on the way. A peculiar vein-like formation in the gneiss was here observed. A band of highly oxodized ferruginous matter and pyritiferous quartz runs with the strike of gneiss south, 20 degrees west. A sample sent to the Provincial Assayer yielded traces of gold.

KAPUSKASING RIVER. Canoeing up the Kapuskasing River for about twenty miles passing many large islands, and ascending many flat rapids, we crossed Mr Nivern's line and joined the other exploring party. After ascending this river a considerable distance the two parties explored the creeks coming in from the west and the country generally for one of the two portages which were said to lead to Opazitika Lake.

PORTAGE TO OPAZATIKA LAKE After investigating without success a series of creeks flowing through a low, flat marshy country, our party also proceeded without success twelve miles further up the river and upon our return found that the portage had been discovered near the head waters of a creek, which empties into the river about forty miles above the line. There were two land portages of two and one half miles each, which had to be made passable by clearing out considerable fallen timber. A creek brings us out into the upper of the lower Opazatika Lakes.

LOWER OPAZATIKA LAKES AND RIVER. On Monday, August the 27th we commenced our journey down the Opazatika River. Canoeing to the foot of a lake one mile and three quarters by one mile and one half, we entered a narrow channel two chains and one half wide and very shallow. At the end of two miles we pass a large marsh and creek coming in from the west and in three quarters of a mile more enter another lake one mile and three quarters long by one mile and one half. These lakes are confined by high bluff banks of gneiss, well wooded with poplar, sprinkled with spruce. For the following three miles the river presents a peculiar seriated appearance, the banks alternately narrowing to three chains and expanding to one quarter of a mile and sometimes one half a mile. Here trips were made inland on each side of the river. The rock at the river and to the west is gneiss, but inland to the east, several low ridges of light grey quartzose schists were crossed, the contact being one quarter of a mile east of the river. We now portaged over boulders on the west side round a rapid five chains long. Two miles below the rapid a small exposure of quartzose schists seamed with small stringers of segregations of quartz outcrops. This is flecked with small particles of iron pyrites. In the next three miles—which are almost continuous rapids, except for one expansion of the river, one quarter of a mile wide—the river nearly makes two great loops flowing toward almost every point of the compass. The river resumes its northerly direction and we have good canoeing for six miles with the exception of a small rapid one hundred feet long, where the river is compressed to a width of twenty feet over boulders. For over a mile we have a series of flat shallow rapids and a short distance further a fall of four feet forces us to portage on the east side a distance of five chains. A brulé with large patches of green poplar which occupies a large territory on both sides of the Kapuskasing River and which runs to within a short distance south of the portage route and to the east of the lower Opazatika Lakes now appears on both sides of this river. After a little way a long rapid with a descent of four feet necessitates a portage on the west side of ten chains, and across ten chains of clear water is another portage on the same side fifteen chains long. Three miles bring us to a steep rapid and falls with a total descent of twenty-five feet and a portage of six chains on the west side passes this. The rock here is grey quartzose schist shading off into mica schist. After more flat rapids, we have clear water for five miles, where we pass a small exposure of a coarse hornblende schist coming to the end of the brulé a couple of miles down stream. Just below this two small outcrops of hornblende schist occur containing fine disseminated particles of iron pyrites. One mile and one-half brings us to a fifteen foot fall over a granite gneiss with portage on the east side six chains long. The contact evidently lies between these last two exposures but is covered. The gneiss is of a grey micaceous character and shows a foliated structure in the mass that is not apparent in hand specimens. From this point to that where Mr. Niven's line crosses there is good canoeing with the exception of one small rapid. No rock was seen and for the lower eight miles the country inland appeared to be a series of great marshy ponds and creeks with a fringe of poplar around the banks.

UPPER OPAZATIKA LAKES. On Monday, the 3rd of September, we returned to take our way through the middle and upper Opazatika Lakes and across the portage route to the Missinaibi River. The route through the lakes is so tortuous that I would not attempt to describe it for any one's guidance. The shores are rocky and of the prevailing gneiss cut by numerous dikes.

PORTAGE ROUTE TO MISSINAIBI RIVER. The first portage, to the shore of a small lake, is in a westerly direction and three quarters of a mile in length. It is somewhat hilly but good. Landing at the south east corner, we portaged for one mile to the shore of another lake one half mile long. This portion of the route runs somewhat south west and is dry and good. Proceeding to the outlet we pushed our way down a very shallow creek overhung with alders. At the end of a mile we came to a small dam one and one half feet high (erected by the Indians) and a land route of one half mile to the west brought us to the Missinaibi River. This portage is also good but somewhat wet in places and is used by the Opazatika and some of the Kapuskasing Indians in obtaining supplies from Brunswick post. Rock exposure is very slight and of a grey gneissoid character.

MISSINAIBI RIVER. We now canoed up the Missinaibi River to the Brunswick Lake portage finding the other exploring party camped at the far end. An excursion was here made inland to the west of the Missinaibi River.

PAZSHUSHKOOTAI RIVER. On Monday, September 17th, we proceeded up the Missinaibi River to the south of the Pazhushkootai River and up that river eight miles. After making a trip inland to the north we explored up this river, a distance of thirty-eight miles. At this point another trip was made inland to the south and we returned. The following is a synopsis of the navigation of the Pazhushkootai River:

1. Rapid 4 feet fall P. on N. side, 2½ chains.
2. Falls 14 feet " P. on N. side, 12 chains.
3. " 4 feet " P. onS. side, 3 chains.
4. Two rapids, paddled and poled.
5. Falls 15 feet fall, P. on N. side, 15 chains.
6. Two rapids, P. on S. side, 16 chains.
7. Falls 20 feet fall, P. on S. side, 4 chains.
8. Falls 6 feet fall, P. on N. side, 5 chains.
9. Rapids, P. on N. side, 15 chains.
10. Two rapids, paddled and tracked.
11. Rapids, P. on N. side, 5 chains.
12. One mile rapid, P. for last on N. side, 15 chains.
13. Steep rapid P. on S. side, 12 chains.
14. Driftwood P. on E. side, 3 chains.
15. Small chute, P. on W. side, 4 chains.
16. Small chute, P. on N. side, 1 chain.
17. Chute 4 feet fall, P. on N. side, 4 chains.
18. One mile rapid, steep, no portage.

Gneiss cut by the usual trap dikes was the only rock met with on this river. At the twenty foot fall, noted above, a large exposure of gray and pink gneiss highly altered and even decomposed in places strikes north 25 degrees west and dips about 50 degrees to the south-east.

On Thursday, September 27th, we proceeded up the Missinaibi River and through Missinaibi Lake, Crooked Lake, and Dog Lake to Missinaibi Station on the main line of the Canadian Pacific Railway.

Geology of the District.

Surface in General. Generally speaking, the district we traversed may be regarded as absolutely flat or more correctly as sloping evenly and gradually to the north. When only one river is considered the descent appears to be in a succession of flat terraces, the steps being situated at the falls, but the latitude of the falls on different rivers does not in any way approximately coincide. The cause, therefore, that gives rise to a fall *i. e.*, an elevation in the underlying gneiss, must be purely, local and not a great ridge running across the country. Great gullies, ramifying through this level country have been cut out by the rivers, the land lying as a general rule anywhere from sixty to a hundred feet above the high water mark, and sloping back gradually to that height from the water's edge. The immediate banks would be covered by high water but are generally six to eight feet above the low water mark, except on the Ground Hog River where they rise to fifteen feet. There do not appear to be any discernible evidences of a river having changed its course in past time, although the Abitibi River carries down a large amount of sediment. These rivers although very violent do not erode their banks in any appreciable degree. Landslips do occur and portions of the bank topple down but this is probably due to the agency of the frost and not to the undermining action of the water. Fluctuations in level are very considerable and sudden, all the rainfall seeming to flow off the surface of the land immediately. Although there had been no excessive rainfall, we found the level of the Abitibi River two feet higher upon our return.

Surface in Detail The great brulé, on the Kapuskasing and Opazatika Rivers, being denuded of its forest growth, afforded a good opportunity of observing the surface in detail. The shores slope up gradually, and then the land stretches away in successions of low rounded hummocks. This appearance is heightened by the numerous ravines through which the creeks and brooks run to the rivers. To the west of the Opazatika Lakes the country is quite hilly and up the Pazhushkootai River several sandy ridges were seen.

Origin of Soil and Drift. The country everywhere is covered by a thick mantle of blue and drab colored clays, the origin of which is doubtful. These clays showed evidence of being deposited very irregularly or having been greatly disturbed after deposition. The high banks, on the Abitibi River, which had been reported sand, were examined by us and were found to be composed of clay with a few small irregular patches of sand and similarly with some large banks up Driftwood Creek. The nearest approach to stratification observed was a layer of sand eight feet thick lying uniformly on the clay on the east bank of the Mattagami River opposite the mouth of the Ground Hog River. The prevailing soil was the blue clay which overlays the drab colored clay down the Opazatika River but sometimes the latter appeared on the surface. Indications were observed of glacial striae having been formed before the superficial deposit was laid down. While making our excursion inland from Driftwood Creek, Mr. Milne and I found Devonian fossils, zaphrentis, heliophyllum, etc, and bits of limestone, three miles inland, lying on the top of the clay under six inches of humus As it was impossible anywhere, to get a cross section of the soil, except at the rivers, and there the original character of the deposit would be obliterated by the actions of the frosts and water, its origin could not be finally ascertained. I am however, inclined to believe, from general impressions and the points above cited, that severe glacial action has been succeeded by subsidence and a long period of quiet and that again by elevation and renewed glaciation, during which the stratified deposits were worked over and considerable additions probably made, upon the recession of the glacier.

Rock Exposures, Laurentian.—Rock exposures were few and of small extent The prevailing rocks were the grey and pink Laurentian gneisses composed of quartz and felspar, sometimes with mica and sometimes with hornblende. At almost every exposure, examples of both may be found, shading off one into the other. Small segregations of mica schist frequently occur. The characteristic formation of epidote, where the ferro magnesian constituent has weathered in contact with the felspar, may be everywhere observed. Alteration to epidote due to contact metamorphism may be seen along the edges of every dike. The weathering, also, of mica schist to chlorite is everywhere in evidence. The gneiss is cut by numerous gabbro and diabase dikes running in a general direction of north east to south-west. Specimens from the large dike on the Ground Hog River when examined under the microscope exhibited the structure of a typical diabase being composed of plagioclase felspar, augite and magnetite.

HURONIAN.—The Huronian formations exist only at Loon Portage on the Mattagami River and on the Opazatika River. It has been thought hitherto, that these were portions of the same belt that ran across country from Lake Abitibi to the Missinaibi River, but I was up the Ground Hog River forty seven miles, and both parties were up the Kapuskasing River, and although rock exposures were few and slight, a Huronian belt could not cross this country without having been observed. From their similarity it is probable they were once connected, but severe erosion has completely isolated them. With the exception of the pyrites mentioned elsewhere, practically the whole of our district is unfavourable to the production of economic minerals.

CLIMATE AND AGRICULTURE.

CLIMATE.—The climate of our district was apparently not very dissimilar from that of Manitoba. The rainfall was frequent but not excessive, and the weather in midsummer seemed to be just as hot. Our party frequently discussed and formed the opinion that there was a frost line along the height of land that did not exist in the low lying country to the north. Dr. Bell says, "The climate in going northward from the "height of land, or towards James Bay, does not appear to get worse but rather better. "This is due to the constant diminution in the elevation more than counterbalancing for the increasing latitude—etc."* Borron, however, considers that the climate of our dis-district is superior to that of the coast. Be that as it may, we experienced no frost until September the 7th and that was extremely light. We were then camped at the foot of the lower Opazatika Lakes and the only destructive frost experienced by us was on September the 27th, when we were up the Pazhushkootai River.

SOIL.—The soil, as I have stated, is a stiff clay with minute patches of sand in places and occasionally covered by a mixture of sand and decayed vegetable matter. On making a trip inland we almost always passed through good land and timber, then across a strip of muskeg into rising ground, this being repeated as we advanced. These strips of muskeg never seem to exceed four feet in thickness, clay soil being attainable at that depth. One large muskeg previously mentioned occurs around a shallow lake on the Driftwood portage. The formation of this has resulted from the filling in of the lake and is now in progress. The shallow muskegs are reclaimable and have originated in an entirely different manner. Where windfalls or beaver dams have obstructed the natural drainage of the land, conditions have been formed facilitating the growth of the sphagnum moss. The land and timber thus lost is never actually reclaimed as the sphagnum moss is destructive to other forms of vegetation. I have pointed out the uneven nature of the country, and we thus have the lower parts, where imperfectly drained, muskeg, and good timber on the higher and well drained parts. That this muskeg would not prove intractable is shown by an examination of the great brulé on no part of which was muskeg to be seen, the thin layer of peat and moss having all been burnt off.

FUTURE PRODUCTION OF CEREALS.—As to whether cereals will ripen. Borron is definitely of that opinion, Bell favors that view and Parks leaves the question open. Their discussion however rests upon obtaining a length of season in the north country equal to that required in southern Ontario. As to whether that would be necessary I append the following extracts from the Director of Experimental Farms reports.

"PRODUCTION OF NEW VARIETIES OF CEREALS.

"The experiments in cross fertilizing have been successfully continued and a large number of new varieties produced particularly of fruits likely to prove hardy on the Northwest plains."—Report for 1897.

"Among the many lines of scientific investigation carried on at the Central Farm none have attracted more general attention than the work done in the production of new

*Geographical Survey Report, 1875-6, page 341.

varieties of cereals by cross fertilizing and hybridizing. Since the Experimental Farms were established more than seven hundred new sorts have been produced. Some of the varieties of grain used as a basis for this work have been brought from the northern parts of Russia, others from the high altitudes in the Himalaya Mountains in India. In these localities some of the earliest ripening varieties of grain are found. These have been crossed with standard sorts of the highest quality and productiveness with the object of producing new varieties combining the high quality and productiveness of the one parent, with the earliness of the other. After a careful test all those of less promise are rejected, but there are still under trial more than two hundred varieties of these hybrid and cross-bred sorts consisting of wheat, barley, oats and peas. Some of these new kinds have produced heavy crops for several years past and seem likely to occupy a prominent place among the best sorts in cultivation. Many new fruits have been similarly produced, especially of hardy varieties, likely to be useful in the Canadian Northwest."

CROSS-BRED WHEATS.—Date of sowing, April 20th to 22nd. Date of ripening, July 26th to August 2nd—Report for 1898, page 27.

Whether the extended duration of sunlight and daylight in the north country would not further conduce to early ripening should also receive consideration.

TIMBER.

This country, although well wooded, contains no white pine and no hardwood. The prevailing timber is spruce and poplar. The former, especially along the river banks, grows to a size that makes it very valuable for square timber. The poplar also is large-sized and uniformly abundant, especially so on the Mattagami River. These are the two most desirable woods in pulp manufacture. At Lower Island portage on the Mattagami River there is every indication of a considerable body of iron pyrites while Borron reports another deposit on one of the tributaries of the Opazatika River. To the north are large formations of limestone and dolomite. If it were desired to purify the pulp in that country the raw material for the manufacture of the sulphite liquor and the elimination of the resins, is there available. There is also considerable white birch and tamarac and to a less extent balsam and cedar. Two groves of Norway pine were found by our party, one nine miles up the Poplar Rapids River, one on the Missinaibi River four miles inland east of Brunswick Lake Portage and others doubtless exist. With the exception of muskegs and marshes the country is heavily wooded and deductions will have to be made for brulés and windfalls. These windfalls pursue a straight narrow path in a general north and south direction, and hardly an excursion was made inland without passing through one or more of them. These deductions, however. would not be very significant when the whole country is taken into consideration.

RAILROAD FACILITIES AND WATER POWER.

After the height of land is well passed and our district is entered, all rock cutting disappears, cuts and fills are slight and gradients easy in every direction. There is an abundance of tamarac, and enough cedar available for ties and sand exists in sufficient quantity for ballast. Owing to the excessive snowfall, an extremely large factor of safety must be allowed all culverts. Falls ranging from ten to twenty-five feet occur at intervals on all rivers and power from them could be readily utilized. Although the cost of installation for an electric road would be comparatively higher it would probably be more economic in the long run.

FUR BEARING ANIMALS.

These comprise beaver, marten, otter, fisher, moose, caribou, bear, rabbit, muskrat, etc., and as far as I could learn are becoming rarer. I was informed by Mr. Spence of Brunswick Post that only half the furs are received there in comparison with those of former years. Pike and pickerel are abundant in the rivers, and up the Pazhushkootai

River speckled trout were obtained by our party. The Indians in the summer subsist mainly on white fish which they obtain by means of small gill nets. Sturgeon also appear to be present in these rivers, as one was caught by our party floundering in a shallow rapid on the Poplar Rapids River.

The Indians inhabiting our district do not at the most number over eight hundred individuals. Each household has its hunting ground allotted according to tribal custom. They are provided only with the old flintlock muskets. It is therefore to the interest of each household not to exterminate the game in its own district. The opening up of the country would in my opinion lead to a rapid depletion of the fur-bearing animals.

I have the honor to be, Sir,
Your obedient servant,
E. L. FRALECK,

Geologist with Exploration Survey Party No. 2.

Dominion of Canada,
Province of Ontario,
County of Hastings,
To Wit :

IN the matter of a certain Geological report of Ernest Leigh Fraleck referring to that part of the Province of Ontario lying north of the height of land and between the Abitibi River and Missinaibi River.

I, Ernest Leigh Fraleck, of the City of Belleville, in the County of Hastings, Gentleman, do solemnly declare that I wrote a certain report in this matter and sent same by express to the Bureau of Mines, Toronto, on the 26th day of December, A. D. 1900, which said report bears my signature at the foot or end thereof.

That the statements made therein by me as to the facts are true and correct in every respect.

And I make this solemn declaration conscientiously believing it to be true and knowing that it is of the same force and effect as if made under oath and by virtue of "The Canada Evidence Act, 1893."

Declared before me at the City of Belleville, in the County of Hastings, this 5th of January, A.D. 1901.

E. L. FRALECK.

J. J. TRILLE,
A Commissioner, etc.

SURVEYOR'S REPORT OF EXPLORATION SURVEY PARTY No. 3.

SUDBURY, Ontario, December 22nd, 1900.

SIR :—We beg to submit the following report of our work, during the season of 1900, in connection with Exploration Party No. 3, sent out under your instructions, in charge of Mr. Geo. R. Gray.

OUTLINE OF WORK.

The party left Sudbury on June 27th, and proceeded via Wahnapitae and Metagamasing Lakes to Dewdney Lake.

Commencing where the north boundary of the Township of Mackelcan crosses Dewdney Lake, a micrometer, log and compass survey was carried forward through Dewdney, Chinicoochichi, Saw Horse, Adelaide, Button, Dougherty, Frederick and Stouffer Lakes, to the Sturgeon River.

The survey was then carried up the river to where Stull's branch comes in from the north-east.

The route through to Shusawagaming or Smoothwater Lake, via this branch and its chain of lakes, was then explored; and the survey carried through, connecting at Shusawagaming Lake with Sinclair's traverse of the Montreal River.

The party then proceeded down the east branch of the Montreal River to its confluence with the west or main branch, and after calling at Fort Matachewan, turned and travelled down the main Montreal River to Bay Lake, returning to Mattawapika Lake, the outlet of Lady Evelyn Lake.

After exploring the vicinity of Mattawapika and Lady Evelyn Lakes, the Lady Evelyn River route was traced through to Apex Lake,just south of Shusawagaming Lake on the route to the Sturgeon River.

A compass and canoe traverse of this route was made on the return trip to Lady Evelyn Lake.

The party then proceeded via Non-wakaming or Diamond Lake to Lake Temagami.

From this lake as a centre, the territory around White Bear, Net, Cedar, Rabbit, Cross and Gull Lakes was examined. Obabika Lake was next visited, with Wakemika and Round Lakes and the Obabika River.

From the mouth of the Obabika River to Sturgeon River was followed down to the portage route leading westward to Maskinonge Lake.

Via this route the party passed through to Metagamasing and Wahnapitae Lakes, and down the Wahnapitae River to Wahnapitae Station on the Canadian Pacific Railway.

Leaving Sudbury again, via the Canadian Pacific Railway to Metagama Station, we crossed over through Lost Channel into the East or Mattagami branch of the Spanish River, and following this River and its lake expansions, crossed over the height of land into the Mattagami River, and on down to Fort Mattagami, making a compass and canoe survey from the railway.

This survey was then continued down the Mattagami River to the mouth of the Kapiskong River.

From here the crossing of Niven's Base line of 1898 with the Mattagami River was located, and a careful micrometer survey carried up the river to the mouth of the Kapiskong River and thence up the Kapiskong River to Peter Long's Lake, where an old survey line was pointed out by Indians, and stated to be Sinclair's line.

From Kapiskong Lake a canoe and log survey was continued along the Kebskwashising or Grassy River, through Loon Wing and Kebskwashising Lakes, into Little Hawk Lake, where it was connected again with Niven's District line.

Crossing over the divide between the Moose Montreal waters, via the "Hawk portages," the canoe and compass survey was again continued down the Montreal River through Opishgoka, Metikemedokagoda, Suwamnakong, Mistinigon and Matachewan Lakes to Fort Matachewan.

From Matachewan the party returned to Wahnapitae Lake, via Lady Evelyn, Temagaming and Obabika Lakes, Sturgeon River and Maskinonge Lake.

After reprovisioning and repainting the canoes, the party returned to Chinicoochichi Lake, and began a series of explorations east and west of this lake and of the lakes to the northward, leading to the Sturgeon River.

A well-travelled route leads from East Bay of Lake Chinicoochichi, via Lakes Laura, Rawson and Halleck to the Sturgeon River.

Another, via Lake Marjorie and Rathwell Lake, south to Wolf Lake.

Another, south from Rawson Lake, via Lake Evelyn, to Metagamasing Lake in the township of Mackelcan.

The routes from the Button Lake to the Sturgeon River, via Parsons' Lake, and from Parsons' Lake south to Halleck Lake, via Lakes Alma and Josephine, are merely trappers' trails, very little used; as are also those on the west side of Lake Chinicoochichi, the lower one of which, leading from West bay, goes through via Sam Martin's Lake in the township of Aylmer, to the Wahnapitae River.

A careful micrometer survey was made of the Sturgeon River from Stouffer Lake portage down to the mouth of the Obabika River connecting there with the traverse made by the Geological Survey.

Bobs' Creek was traversed for a considerable distance, and we were informed, constitutes another route to Smoothwater Lake, via Lake Florence and Lady Evelyn River.

Leaving the Sturgeon River again, the party proceeded to Emerald Lake, via Wawiashkising and Manitoupeepagee Lakes. The more southerly of the two routes shown has been recently cut out, and is by far the preferable.

Another route from the Emerald Lake to the Sturgeon River is via Kibble's Lake and Creek, but is very little used.

Heavy frosts and the formation of ice on the small lakes brought operations to a close at Emerald Lake; and passing to the northward through the south arm of Obabika Lake, we returned via Round Lake and Sturgeon River to Wahnapitae Lake, leaving there for Sudbury on November 15th.

General Features.

The country around Dewdney Lake and the South East and South West bays of Chinicoochichi Lake is very rough, the hills rising to a height of five hundred and sixty feet above the lake in some places. Further north on this lake and around Dougherty, Frederick and Stouffer Lakes the hills become less prominent, and away from the water there are considerable stretches of flat country.

Eastward towards the Sturgeon River the surface is mostly comparatively level, the rock ridges being of no great height. A very high and precipitous range of hills, however, continues right across the Sturgeon valley, about two and a half miles below the Parsons' Lake portage, forming a prominent feature of the landscape.

The Sturgeon valley, after leaving the north boundary of the township of McNish, preserves pretty largely the same general features about as far up as the Parsons' Lake portage. It consists essentially of a sandy plain, attaining a width of several miles in places, notably where it is joined by the valley of the Obabika River..

In the valley around Wawiagami Lake, however, the soil is of a much more loamy nature, and considerable areas would be suitable for agriculture.

The valley of the Obabika River is of a similiar nature but much narrower than that of the Sturgeon River. The river is very winding and the current swift, varying from about two miles per hour near the mouth to perhaps three near Obabika Lake. The channel for canoeing is frequently interrupted by accumulations of driftwood, which a moderate amount of work would remove. A short distance above the mouth of the river a large stream comes in from the north which must drain a considerable lake area.

The Sturgeon River, throughout that portion of the valley described above, is quite swift; and meanders between sandy banks from fifteen to fifty feet high. The only notable breaks in this distance are the upper and lower Goose Falls, both of which could be economically developed into good powers.

The watershed of the river here extends far to the north and east, and all the large streams come in from that side; the Metagamasing basin being comparatively close on the west.

Kibble's Creek drains the Emerald Lake district. Plum Creek is quite small.

A large creek comes in just above Upper Goose Falls, whose valley nearly parallels that of the Obabika River for some distance, and can be seen stretching far away to the north.

Between Parsons Lake portage and Lyman Lake,above Kettle Falls,the valley of the river becomes very much contracted and rocky, the river itself consisting chiefly of a series of rapids and falls, alternating with small lake expansions; a few miles of comparatively quiet river intervening, however, in the neighborhood of the Stouffer Lake portage.

The most important of these falls is called "Kettle" or "Pothole" falls about seventy-five feet high, making an important water power. Just above the fall, on the east bold masses of rock rise to a height of about four hundred feet.

Other falls of twenty-five and twenty feet occur just above and below mining location W. R. 90. The rapids below range from eight to thirty feet, but are usually too long to be economically developed.

The portion of the river between Halleck Lake and Stouffer Lake portages is very little used for travel, the somewhat more circuitous lake route on the west being much less difficult.

Bobs' Creek is the main tributary of this part of the Sturgeon River. It is a swift, shallow stream of considerable magnitude, forming part of an old and disused trappers' route to Smoothwater Lake, but practically impassible at low water.

From Lyman Lake to the lake below Eagle Nest the river is very swift and shallow, making canoeing difficult. The banks change to gravel here, and the valley widens out to half a mile or more. Above this the main river seems to be much more nearly in the centre of its watershed, large tributaries coming in on either side. At the foot of Shakoba or Ghoul Lake, the river flows through two gorges, forming the "Twin Falls", having a drop of about forty feet.

Above Shakoba Lake the valley becomes flat and sandy again (changing to clay, however, above Bowland's Falls), and gradually widens out, becoming two to three miles in width and with arms extending up the tributaries. The river becomes much slower and very tortuous.

Stull's Branch, also called the East Branch, with several large lake expansions, forms the route to Smoothwater Lake, the headwaters of the East Branch of the Montreal River. This route was formerly well travelled, but has been very little used of late. The valley is from two to three miles in width and bordered with hills from five hundred to eight hundred feet high.

Bowland's Branch is a large tributary coming in from the west, some distance below Eagle Nest Lake. Trappers' routes lead from this branch to the Wahnapitae River, but were not explored. This stream is a dark reddish brown, and is believed to have its source west of the District Line, in the large spruce swamp areas existing there to the north of the Wahnapitae River.

Above Stull's Branch the country becomes much flatter and the river distributes itself among many smaller tributaries, reaching out east, north and west, into large swamp areas. In the upper reaches of the river travel was much impeded by driftwood.

Lady Evelyn River is peculiar from the fact that its valley lies east and west, or nearly at right angles to the general direction of most of the other prominent river courses in this region. It forms a route from Lady Evelyn Lake to Smoothwater Lake, but is very little used except for a short distance at the lower end.

Except at Lake Florence, the Sturgeon Watershed extends quite near to Lady Evelyn River all along the south side; and all the main tributaries come in from the north.

Leaving Apex Lake, which empties both into Smoothwater Lake and into Lady Evelyn River, the route follows a small branch some four or five miles east, to where the main branch comes in from the north.

Lake Florence is a fine, large body of water and a short portage at the south end takes one across the divide into a small lake, which is believed to be the headwaters of Bobs' Creek, a tributary of the Sturgeon River.

East of Jack's Lake the river forms a large double elbow, at the northern bend of which a branch comes in fully as large as the main stream up to this point.

Indians describe a very large lake (called Gray's Lake on the map), as being at the headwaters of the large stream called Gray's River, which empties into Macpherson Lake from the north.

At the small lake below Macpherson Lake, the river divides into two branches, one running east through several smaller lakes into Willow Island Lake, and the other nearly south into Suckergut Lake. On the former occur Helen's Falls and Rapids, with a total fall of about one hundred feet, the only power of any importance on this river.

This route from Lady Evelyn Lake to Smoothwater Lake is very difficult travelling at present, on account of the many rapids and rough portages where the valley is rocky. and the dense growth of alders blocking the channels in those reaches where the river is slow and tortuous; but a comparatively small amount of work would make it quite feasible for the more energetic class of tourists, for the river teems with brook or speckled trout, and is the only place where they were found in the whole field of the season's exploration.

Proceeding now to the route northward from Metagama Station lumbering operations were found in progress as far north as the vicinity of Mosquitoagema Lake.

A considerable stream finds its way from the main branch of the Spanish River across to the east or Mattagami branch. through what is called the Lost Channel. The east branch is very straight running almost directly north for many miles, and consists largely of lake expansions. A dam constructed by the lumber company gives quiet water to their storehouse at the foot of Mosquitoagema Lake.

The country thus far is chiefly rocky and broken, with small stretches of flat or swampy lands. North of Mosquitoagema, however, the hills disappear and at the height of land the surface consist of vast sandy plains and swamp areas.

A large stream coming in from the east in this level area forms a canoe route across to the Onaping waters.

Crossing over the height of land into Dividing Lake, we enter the west branch of the Mattagami River. Following the river northward, a few prominent ridges of rock cross the valley; but the country generally is very flat with long stretches of swamp and marsh. A long portage over a sandy plain leads into Minisinocwa Lake and avoids several miles of the river containing many rapids and not used at all for travelling.

From Minisinocwa Lake to Mattagami Lake the river is mostly quite wide and the current slow, with, however, one long heavy rapid having a fall of about one hundred feet.

The whole route from Mattagami Station to Fort Mattagami is kept in excellent condition for travelling by the Hudson's Bay Company, who use it for taking all their supplies in to Fort Mattagami. The portages, though somewhat long, are for the most part nearly level and are kept in splendid order.

Fort Mattagami, which occupies a beautiful site on a point in the lake of the same name, has a fair-sized farm in connection where potatoes and the ordinary garden vegetables of splendid quality were seen, as well as many varieties of flowers.

Mattagami Lake and Kenogamissee Lake, another expansion of the Mattagami River, are separated by Kenogamissee Falls and Rapids, which have a total fall of about forty feet, where a splendid power could be developed. At the front of this lake the channel narrows again and the river forms a very heavy rapid, but difficult to utilize as a water power as it is about a mile in length.

About two miles below this rapid the Kapiskong River enters from the east; and from its junction down to Niven's base line of 1898 the river is of a more uniform character, from three and a half to five chains wide, quite deep and having a current of nearly two miles per hour interrupted by only one small rapid.

Much of the country northward from Fort Mattagami is suitable for agriculture, especially beyond Kenogamissee Lake, where the river has a wide level valley, the soil being a rich sandy or clay loam with clay subsoil, very similar to the lower reaches of the Sturgeon River.

Patches of potatoes and other produce of the finest quality were found at numerous points along this river where cabins are shown on the map,and also along the Kapiskong River and Grassy River; and as late as the first of October they had not been touched with frost. The Indians stated that all ordinary vegetables would have quite sufficient length of season to become fully matured if they were planted at the proper time. They, however, leave all agricultural matters in abeyance until the finish of the spring hunt late in June.

The Kapiskong River for several miles above its mouth is shallow, with bottom of small boulders or gravel and an extremely swift current, against which in many places it is impossible to paddle ; and recourse must be had to poling or wading and towing.

Higher up the current becomes somewhat slower but broken frequently by short sharp rapids, until about eleven miles from the mouth, by the river, or about six miles in a straight line, a series of rapids and falls with a total drop of about one hundred and fifty feet is encountered.

Above this for several miles the river is slow with small lake expansions and the banks marshy, until a a series of small rapids is reached, terminating in a fall of about thirty feet, at the foot of Mesinocwanigwahiganing or Peter Long's Lake.

The valley of the Kapiskong River near the mouth is defined by high gravel or sand banks, becoming rapidly lower as the river ascends towards the marshy plateau.

At lake Mesinocwanigwahiganing (so-called from an outcrop on the shore of slate, the rock which can be written upon), we found an old Indian living named John Jarbeau (otherwise Bags) who stated that this is the real Peter Long's Lake, the long lake to the south bearing the same name as the river. He pointed out a line (shown on the map) which was run many years ago, and said he had never heard of any other in that locality. This line was traced a considerable distance both ways from the lake and runs very nearly east astronomically. From its position, as located by survey and checked by latitude observation, it seems very improbable that this can be Duncan Sinclair's Line of 1867. No other line was found, however, farther south.

Around Peter Long's and Kapiskong lakes the surface is somewhat rocky and broken.

Shetagami or Sinclair's lake is a very pretty sheet of water, connected with Mattagami lake by a long difficult portage now very little used. From here to Little Hawk lake the country is very flat, a few high isolated hills standing out like sentinels. The most prominent of these is called Sansawau Mountain.

The valley of the Kebswashising or Grassy River is wide and marshy, and the channel meanders among willows and reeds. The lake of the same name is also very shallow and reedy.

A very imposing elevation on the north side of Little Hawk Lake marks the divide between the Hudson's Bay and St. Lawrence waters ; but the actual divide is level, and at a very low elevation, probably not more than ten or twelve feet above the water on either side.

The four portages from Hawk Lake to Opishgoka or Pigeon Lake, known as the "Hawk Portages," have at one time been well cut out and much used, having formed a part of the Hudson's Bay Company's route between Fort Matachewan and Mattagami, but at present they are badly choked by successive windfalls.

From Opishgoka Lake through Obaunga and the other lake expansions of the Upper Montreal River, the country becomes more broken again, with high rocky ridges alternating with swamps or stretches of marsh.

Between Suwamnakong Lake (connected by portage with Obushkong on the east branch) and Mistinigon Lake the river becomes narrow and swift, with a series of rapids, several of which can be run by skilful canoemen, but none of which are of any economic importance.

Matachewan Falls and Rapids, however, at the Great North Bend where the river empties into Matachewan Lake, have a fall of about forty feet and constitute a very important waterpower, as the length is short and the site comparatively easy of development.

Agricultural Lands.

A summary of the districts where land suitable for cultivation is found is about as follows :—

The most important areas are along the valley of the Mattagami River principally below Fort Mattagami, and in the valley of the Montreal River from the junction of the East Branch down to Bay Lake. The latter seems to be a westerly extension of the large agricultural area to the northwest of Lake Temiscaming. Smaller and more scattered areas lie along the lower Kapiskong River and the Grassy River.

The Sturgeon valley below the Obabika River is another area of some importance, but the soil is rather light in many places. The Obabika River and Round Lake areas are smaller but quite worthy of notice.

In fact, wherever there are well defined valleys or plateaux, there are larger or smaller patches of land suitable for cultivation; except in those places where the soil is wholly sand or gravel; notably around the height of land between the Spanish and Mattagami Rivers and most of the Sturgeon River above the Obabika River.

GAME.

Large game is found over the whole of the area explored. Moose and caribou are most numerous in the Sturgeon and Obabika valleys and westward to Chinicoochichi Lake and the lakes to the north; their numbers apparently gradually diminishing further north along the Upper Montreal and Mattagami waters.

Red deer are very scarce on the northern slope, and are not found in any considerable numbers, until, coming southward, the neighborhood of Wahnapitæ Lake and the Lower Sturgeon River is reached.

Bear seem well distributed over the whole area, as also small game, such as partridge and rabbits.

Wolves were heard only in the Sturgeon valley. They always frequent the habitat of the red deer.

The smaller fur-bearing animals, such as mink, marten and fisher, are found only in small numbers, as they are kept pretty closely hunted by Indians and other trappers.

Beaver and otter are very scarce, and were found only in localities which have been left for a number of years undisturbed.

FISH.

Salmon and grey trout, bass, pickerel, pike and whitefish are found in abundance in most of the lakes, though usually not all these varieties in one lake. Individual lakes seem in many cases to have a predominant variety.

Pike and pickerel are the species most frequently found in the rivers, and these were always of excellent quality, even in the hottest summer months.

Splendid speckled trout were found in abundance in the waters of Lady Evelyn River. There are said also to be other streams equally as good in this respect to the north-east of Temagami Lake.

MINERAL DEVELOPMENT.

The most important mineral bearing area, as far as known at present, in the territory covered by this report, is the district around Net and Vermilion Lakes, north east of Temagami Lake, and extending westward to Kokoko Lake. Large bodies of auriferous pyrrhotite and mispickel have been partially opened up on a number of claims already surveyed in this neighborhood. Westward towards Kokoko the mineral formation changes to magnetite associated with jasper. Small bodies of hematite have also been found.

South of Lake Temagami also, at Austin Bay, a deposit of magnetite has been found close beside magnetic pyrites.

At Emerald Lake a number of locations have been surveyed. Auriferous sulphides of iron in quartz are here associated with jasper, of similar appearance to that found in the Kokoko district.

North of the Township of McCarthy, at Pedro and Nick's Lakes, some free gold quartz veins in a slate and diorite formation are being opened up.

Several quartz veins have been staked out on Chinicoochichi and Frederick Lakes, and on the Upper Sturgeon River near Kettle Falls.

A group of locations has been surveyed at Mattawapika, the outlet of Lady Evelyn Lake, and some development work done in quartz veins.

Other claims are located on islands in the North Arm and in Muddy Water bay of Lake Temagami and in Cross Lake. Several claims containing iron pyrites and galena were described by Indians as being in the neighborhood of Fort Matachewan.

As the country is mostly heavily timbered prospecting has been difficult, but now that several main leads have been located, systematic prospecting will doubtless be carried out on a larger scale, and with every prospect of success

Lines of communication and transportation are all that is necessary to insure rapid development of the deposits already known to exist.

A number of observations for latitude were taken during the season at various points; but as the surveys made were largely between points already established, the results obtained were used merely as approximate checks on the work.

Azimuth observations give the magnetic variation as ranging from about 6° 30′ west in the neighborhood of Chinicoochichi Lake, to 7° 25′ west along the Kapiskong River.

Heights of more than a few feet were measured with aneroid barometer and are therefore only approximate.

We have the honor to be, sir,

your obedient servants,

DEMOREST & SILVESTER,
Ontario Land Surveyors on Exploration Survey Party No. 3.

To the Honourable E. J. DAVIS,
Commissioner of Crown Lands, Toronto, Ontario.

LAND AND TIMBER ESTIMATOR'S REPORT OF EXPLORATION SURVEY PARTY NO. 3.

---

50 ISABELLA STREET, TORONTO, Ontario.
Jan. 7, 1901.

To the HON. E. J. DAVIS,
Commissioner of Crown Lands.

SIR,—I have the honor to transmit herewith my report upon the exploration work performed by Exploration Survey Party No. 3 under my control during last season.

In accordance with the instructions received I have endeavored to acquire as much general knowledge as possible regarding the districts allotted me for exploration. These districts comprise the territory lying east, west and north of Lake Temagami and on either side of the Montreal river and its tributaries east and west of the district line between Nipissing and Algoma, and I am pleased to state that we have been successful in covering this entire country in as thorough a manner as time would allow.

The districts explored by us proved to be, in all respects, much more valuable than we had expected to find. On this side, or south of the Hudson's Bay watershed, the timber and minerals are the most valuable assets of the province, although we find localities containing large areas of good agricultural lands.

The territory on the Hudson's Bay slope, or north of the height of land, which is well adapted for agriculture, is now principally covered with large quantities of pulpwoods of the most valuable kinds, spruce and balm of Gilead predominating.

A very noticeable feature of all the territory outlined for us to explore was the complete network of water-ways, rivers, lakes and creeks that drained the entire country. The even distribution of these water-ways has obvious advantages, inasmuch as the country is thus perfectly drained, and the timber will be cheap to operate. The rivers are also well-supplied with water powers that can be utilized for all kinds of manufacturing enterprises.

The healthy condition of the flora specimens that we gathered, as well as of those that were observed at different points during the season, proved conclusively the absence of early frosts. In fact, at no time or place did we notice the slightest indication that even the most delicate flowers or plants had been blighted before their maturity.

The scarcity of fur-bearing animals was very noticeable throughout the entire district traversed by us, and is due, no doubt, to the immense quantity of fur exported yearly by the Hudson's Bay Company.

Not alone will this vast territory with its many resources furnish employment for the lumber and paperwood man, work for the prospector and miner and homes for those who wish to make their livelihood by farming, but it will prove a paradise for the followers of the rod and gun and a resort for all lovers of the picturesque in natural scenery.

A large part of the territory south of the height of land between the Hudson's Bay and St. Lawrence River waters is of a rough rocky nature and will prove valuable only for its mineral and timber wealth. Practically all the country drained by the east and main branches of the Montreal River, that is, northwest and southwest of the juncture of these waters, as well as nearly all the territory lying between the Sturgeon and Wabuapitae Rivers, is unfit for cultivation. Another uncultivatable district is situated between Lady Evelyn and Smooth Water Lakes, comprising the watershed between the Montreal and Sturgeon waters. I may mention that small areas of good land were found in these localities, but nowhere were they extensive enough to be suitable for settlement.

The agricultural territory south of the watershed is generally confined to the river basins. The soil is for the most part of a uniform nature, sandy surface and clay subsoil. A large tract of good farming land was found on both sides of the Montreal River, extending from the outlet of Lady Evelyn Lake to Kekopacagawan or High Falls, a distance of about forty miles. All the territory lying between the surveyed townships of Sharpe, Robillard, Bryce, Henwood, and Hudson and the Montreal River is good cultivatable land suitable for settlement. The belt of good land on the west side of the river varies in width from three to ten miles. This whole district, comprising over five hundred square miles of territory, is all arable land of the finest quality, mostly clay subsoil and sand loam surface.

Oats, potatoes and other vegetables were seen growing at several points in this locality and appeared to be in a very healthy condition. "Rennabester," or Mowat's farm, situated on the Montreal River at the outlet of Lady Evelyn Lake, was growing a flourishing crop of oats and potatoes. Mr. Mowat informed me that during his residence in that section of the country, a period of fifteen or sixteen years, he had experienced no frosts that would be injurious to the successful growth of any farm products.

On this same Montreal River belt of land, about forty miles north of Mowat's farm, was found a small farm situated within half a mile of Kekopacogawan or High Falls. This farm has been cleared and settled by a halfbreed who is the owner of three or four head of cattle and one horse. It was the latter part of September when we visited this settler and as we found him digging and pitting his potatoes at that time we concluded that any root crops would mature as early in that locality as in any settled district in northern Ontario. It was noticed that his crop was a very large yield and that the potatoes were of good quality and size.

The valley of the Sturgeon River north of the surveyed townships contains a considerable area of good clay loam soil. There is a very limited quantity of arable land forming the valley on the west side of the river, as the rough country extends very close and parallel with it The east side of the river contains quite a large district of good farm lands. This tract of land, which commences at the line of the surveyed townships, extends north along the Sturgeon River to a point about six miles above the outlet of the Obabika River and east to Round Lake. This district, embracing the Obabika and Sturgeon Rivers' valleys, probably contains two hundred or two hundred and fifty square miles of good arable land.

At Fort Matachewan, on the Upper Montreal River, we found potatoes, corn, pumpkins and cucumbers all looking healthy and strong in the latter part of the month of July. The factor at this Hudson's Bay post informed us that some seasons frost injured the growth of his garden products. This may be accounted for by the fact that Fort Matachewan is located very close to the height of land. I found this to be a correct theory, after investigating the growth of garden products on the Hudson's Bay slope.

Two-thirds of the territory lying between the Hudson's Bay watershed and Niven's base line, which was the most northerly point reached by our party, is good arable land suitable for settlement. The southern portion of this district, or that situated between the height of land and Sinclair's line, is, for the most part, rough and rocky, although sections of country on the Kapiskong or Grassy River, not far distant from the watershed, are well adapted for farming purposes. We found quite a large patch of potatoes growing on the Grassy River. They were fully matured in September, of excellent quality and large size, and it took less than two hills to fill a large patent pail. The owner of this potato crop was a white man of French extraction, Resteuil by name, who has lived for twenty-five or thirty years among the Indians in this locality. He informed me that he had watched the changing of the seasons very closely and could not say that he had ever experienced any frosts that would damage the crops, or injure their growth.

The territory in the vicinity of Sinclair's line is somewhat broken, but a gradual improvement is noticed on travelling north, until a large level tract of clay land is found some distance to the south of Niven's base line and extending as far north as we explored, using the Mattigami River as a base.

Along all the canoe routes in the territory wherein our exploration operations lay, we found numerous patches of potatoes. These were planted doubtless by the Indians and, although they had received no attention whatever since they had been put in the ground, we found them, without exception, in a thriving condition.

The areas of pine, in all the territory explored by us, lie very noticeably in belts. The general nature of the country after leaving the waters running south, is more favourable at the present time, to the growth and sustenance of other woods than pine. Still we found several patches of pine north of the height of land, where it shows, in sections, clumps, each containing about four or five million feet. This condition of things may be accounted for by the circumstance that, probably one hundred years ago, this territory was exposed to the ravages of fire, which left these clumps of timber as survivors of the pine riches that at one time existed. Judging from the age of the surviving trees, it is probable that this whole territory was once a pine forest and that, when all the country has been thoroughly explored, some extensive pine belts will be found elsewhere on the Hudson's Bay slope. There are two or three of these patches of pine lying between the Kapiskong or Grassy River and the Mattagami River and north of Peter Long's Lake. In these sections there are twelve or fifteen million feet of pine, of which seventy five per cent. is white pine. This pine was of very good size and quality, some being large enough for board timber. We also traced a belt of pine commencing on the east side of Peter Long's Lake and ending at Shebandowan or Duncan's Lake on the head waters of the Montreal River, covering a distance of about twenty-five miles in length and varying in width from two to eight miles. This belt of pine will average with our well timbered territory on the Sturgeon or Wahnapitae Rivers.

More or less pine, both white and red, is found on nearly all of the country lying south of the height of land, between the Hudson's Bay and St. Lawrence river waters. Although this territory, which is drained by the Montreal, Sturgeon, Wahnapitae and Matabitchewan Rivers, is a pine growing district, we found numerous sections where other woods predominated. Generally speaking, however, the nature of this country is favourable to the production of pine.

The territory lying east of Lake Temagami and south as far as the surveyed townships, with the exception of small portions that have been burned, is all a pine forest. The sections that have suffered most from the ravages of fire are the districts north and west of Net Lake, the south end of the Rabbit Lake and along the Matabitchewan River, as far as the boundary of the timber limits already under license. Notwithstanding the fact that fire has done considerable damage to parts of this territory we found quite large quantities of timber still standing uninjured on the ground that fire had visited. The balance of this territory is almost one solid block of pine, and especial mention may be made of the country adjacent to and lying between Hanging Stone and Cross Lakes. This section may be classed as a board pine district, the timber being large and of excellent quality.

The territory west of Lake Temagami, or situated between the Sturgeon River and Lake Temagami, is fairly well timbered with pine, some portions of the district being heavily timbered, while in others it is not so abundant. A very large belt of pine continnes from Cross Lake west across the south-west arm of Lake Temagami, passing Gull Lake west to Lake Manitou-pee-pa-gee, then north crossing Emerald Lake and reaching Round Lake and the Obabika River. This timber also shows in large quantities on Obabika Lake and the north west arm of Lake Temagami. The pine in this belt is standing very thick and will be operated very cheaply. Although it is not of a very large size it may be estimated to cut about ten or twelve sixteen foot logs to the thousand feet. I would say that half, or possibly three-fifths, of the territory bounded by Obabika Lake and River, Lake Temagami and the Sturgeon River is well timbered with pine, yet in no place could we find where this timber extended to the Sturgeon River proper, although it will all come down the Sturgeon waters.

All the country lying between the Sturgeon and Wahnapitae Rivers north of the surveyed townships is largely a pine growing district. In some sections the timber is small, rough and scattered, but the greater portion of this locality is well timbered with pine of good quality and medium size. The territory east of Chinicoochichi, Button, Dougherty, Frederick and Stouffer Lakes is well timbered with pine of good size and quality. The pine on the west of this chain of lakes, or more properly that lying between the Wahnapitae River and these waters, is more scattered. We found the good pine in clusters or sections of a mile or two square, with the exception of a large extent of territory west of Frederick Lake, which apparently extended to Wahnapitae River. The pine described as lying east of Lakes Stouffer, Frederick, Dougherty, Button and Chinicoochichi was found to be the commencement of a large belt that continues northeast across the Sturgeon River and stretches as far north as Smooth Water Lake and south to Lady Evelyn Lake. The timber of this district, as far north and east as the watershed between the Sturgeon and Montreal Rivers, will be floated down the Sturgeon River and its tributaries. All around Lady Evelyn Lake there is quite a sprinkling of pine, but, upon exploring the country eastward, we found that it had been burned, and is now covered with the usual second growth of jackpine, birch, poplar, etc.

The large belt of pine, described as reaching Smooth Water Lake on the north and Lady Evelyn Lake on the south, does not extend as far north and east as the Montreal River, and very little pine of commercial value is found on the basin of the main Montreal. In fact we located no pine on this river worthy of mention, excepting one district situated south of Mountain or Round Lake, which contains an area of twenty-five or thirty square miles fairly well timbered with pine. The greater part of the country drained by the east and main branches of the Montreal River contains no very large quantity of pine. This district, with the exception of the spruce and tamarac lands, is mostly a brulé of about thirty-five years' growth.

After closely comparing the respective quantities of pine standing on the various districts located and defined by us, I would say that the average square mile of the territory described as pine lands would contain one million seven hundred and fifty thousand (1,750,000) feet of good merchantable timber. By careful calculation, the area of pine lands on the territory we explored was found to contain sixteen hundred and fifty (1,650) square miles, and therefore the total estimate of the pine located by us during the past season will amount to two billion, eight hundred and eighty-seven million, five hundred thousand (2,887,500,000) feet.

Spruce, whitewood, jackpine and other pulpwoods are found on nearly all the territory where the pine does not exist in quantities. This condition of things is very noticeable in the change of timber growth after crossing the height of land between the Hudson's Bay and St. Lawrence slopes, and especially may it be seen in the health and progressive nature of the various classes of timber. While we found the pine thriving on the St. Lawrence slope, the spruce and balm of Gilead flourished on the Hudson's Bay waters.

The spruce on the Lower Sturgeon and Temagami districts is found chiefly in swamps, ranging in extent from ten to five hundred acres. It was also found growing with the pine and, though scattered, was larger than that which grew on the low lands. We also found spruce on the river valleys, but no large areas in any one locality south of Smooth Water Lake and the upper waters of the Sturgeon River. These last mentioned districts are well adapted for the growth of spruce and balm of Gilead. The territory which comprises the upper waters of the Sturgeon River east and west of the Algoma-Nipissing boundary line, extending east as far as Smooth Water Lake and the east branch of the Montreal River, and comprising about one hundred and twenty-five square miles of country, is undoubtedly the best spruce territory we found south of the height of land. This district also carries quantities of whitewood and poplar, a very noticeable feature of these timbers being that they are invariably found growing in a brotherly way to the

Kettle Falls, Sturgeon River. Party No. 3.

Rapids on Upper S urgeon River. Party No. 3.

Smoothwater Lake looking south from outlet. Party No. 3.

Lake expansion on east branch of Montreal River. Party No. 3.

Indians at Fort Matachawan. Party No. 3.

Kekopakagawan Fall or High Fall, Montreal River. Party No. 3.

Dan O'Connor's House. White Bear Lake. Party No 3.

John Thompson's family. Mattagami River. Party No. 3

Mattagami Post. Party No. 3.

Kenogamisse Falls, Mattagami River. Party No. 3.

John Jarbeau and family. Peter Long's Lake. Party No. 3.

From top of Mountain south-east of 4th lake above Clearwater Lake. Party No. 3.

spruce, although the spruce is better nourished where the soil is moist and the lands low. As the ground rises and escapes the constant moisture so favourable to the spruce, the same soil is in the proper condition for the sustenance of poplar and balm of Gilead, and consequently we invariably found spruce and balm of Gilead growing in close proximity, the spruce on the lower and more level ground, the poplar and balm of Gilead on the higher and drier soil round about or intersecting the spruce districts. The territory on the upper part of the main Montreal River is dotted with small spruce swamps, but no great quantity of this timber was seen in any one place. This same feature was noticed in the country round about Lady Evelyn Lake, and all the creeks and rivers running into it. The lower Montreal River is chiefly wooded with jack pine, cedar, poplar and small birch.

The territory north of the height of land is a spruce and balm of Gilead district. Almost as soon as we crossed the watershed, by way of the Grassy and Mattagami waters, the spruce and balm of Gilead appeared larger, cleaner and more healthy. It is not unusual to see spruce trees that would make one hundred foot spars and whitewood as large, standing as thick as any beech and maple woods have stood in Southern Ontario. The country situated between the watershed and Niven's Base line, after leaving the broken section in the neighborhood of the height of land, is largely a pulpwood growing district.

Banksian or jack-pine is the most abundant wood of the pulp variety south of the Hudson Bay watershed. It is found almost any and everywhere, but usually in abundance on rough and hilly country. This variety of wood added to the list of pulpwoods will double the estimate of pulpwood on the territory south of the height of land. Jack pine is also plentiful on the Hudson's Bay slope, showing in quantities on the higher ground. We found a large amount of this wood on the broken territory near the watershed, but, as we travelled north and the country became more level, this timber became scattered and less abundant. Jack-pine is a sturdy, persistent wood and will grow where any other timber would starve. It seems to prosper under all conditions and is suitable to any climatic changes we have in Ontario.

Balsam, which usually grows on the low lands, is often found in quantities in the spruce and tamarac districts and will materially add to the aggregate estimates of pulpwood.

Although the territory south of the Hudson's Bay watershed will produce large quantities of spruce, it is a very noticeable feature of this entire country that, with few exceptions, there are no very large areas of this wood in any one locality. This pulpwood usually grows in the river and creek valleys, or on swampy districts which vary in size from two to five hundred acres. This condition of its growth makes it very difficult to arrive at an accurate estimate of the total quantity of spruce without close examination of these numberless swamps. The even distribution of the other pulpwoods, such as jack pine, poplar, etc., makes it much easier to arrive at the aggregate amount of these woods that the territorry will produce.

After careful consideration of the extent of territory within the limits of our exploration in which the different varieties of pulpwood are found, the entire district may be calculated to contain about four thousand five hundred (4500) squares miles, or two million seven hundred and eighteen thousand (2,718,000) acres. Considering that an average acre will produce two cords, the sum total of pulpwood in the entire district will amount to five million four hundred and thirty-six thousand (5,436,000) cords.

Although nearly all the early spring flowers had gone to seed or died before our exploration work commenced, in so far as we could note, they were almost indentical with, although probably a little latter in maturing than, those of southern Ontario.

During nearly all the month of July, numbers of blue flags of the ordinary type were noted along the shores of the lakes, where the waters were most quiet. In places where the bottoms of the lakes were suitable, white and yellow water-lilies frequently occurred. Beyond these specimens no other aquatic plants were seen in the open water, with the exception of one which was so mature that it immediately fell to pieces upon touching it, and we were therefore unable to make a satisfactory description of it.

Hazel bushes were noticed along a great many of the trails, and on the shores of Frederick Lake raspberries were found in small quantities. During the latter part of the month of July the low bush blueberries began to ripen, and from that time until late in August were found in very large quantities. These berries were especially noticed where forest fires had taken place. It seems to be the nature of the blueberry bush to follow the moss which almost immediately appears after the visitation of fire.

At Fort Matachewan the Hudson's Bay post, which is situated almost on the height of land between the St. Lawrence and Hudson's Bay waters, and is probably the most exposed to the attacks of frosts of any of the districts we visited, a number of vegetables, including potatoes, cabbages, cucumbers and carrots, flourished in the small garden which was kept by the Factor. In his flower garden we noticed pansies, sunflowers, morning glories and the ordinary marigolds.

The Factor of the Hudson's Bay post at Bear Island had all kinds of vegetables growing in his garden, in sufficient quantities to supply his needs during the winter. On the pasture lands surrounding the post nearly all the common field weeds of southern Ontario were noticed, including yarrow, shepherd's purse, ox eye-daisy and wild rose besides a number of the members of the compositæ family.

Around the shores of the lakes and on the portages were noted the ox-eye-daisy, several specimens of the golden rod, the wild honeysuckle and the dandelion in all it various stages. A few buttercups were seen on the Sturgeon River, and the winter-green grew in profusion throughout all these sections. On Lady Evelyn Lake we found an old Indian clearing upon which corn and raspberries were growing.

The season was so far advanced when we explored the territory on the Hudson's Bay slope and adjoining the Mattagami and Kapiskong branches of the Moose River, that we were unable to make much note of the flora. Some of the members of the compositæ family were the only surviving representatives of the plants and flowers.. Quantities of pin-cherries and high-bush cranberries were seen along these rivers, and the low-bush cranberries were quite plentiful on the marsh lands. We also noticed a number of pitcher plants growing in these swampy districts.

The majority of the botanical specimens were collected in two main centres, namely, Stouffer Lake and its vicintiy and Lake Temagami district, these points being selected as average localities of the whole country. While we did not search particularly for rare specimens, we gathered such as we thought would give a fair idea of the nature of the soil and climate of the regions explored by us.

The entire territory explored by us is an excellent field for lovers of sport and without a doubt, when this country is known to the sportsmen, it will be invaded by them "en masse," and districts hitherto untrodden by the foot of man will become the haunts of the pleasure-seeking Nimrod.

Fish of every description are plentiful in the rivers and lakes, the most abundant varieties being trout, pickerel and pike. The waters of nearly all the lakes and rivers are clear and cold, and some are so transparent that the bottom is plainly visible at a depth of twenty feet. The fish are very firm and of exceptionally fine flavor, due doubtless to the low temperature and excellent quality of the water.

Trout were very abundant in all the Temagami waters. Especial mention may be made of Gray's River, the outlet of Florence and Gray's Lakes, which empties into Lady Evelyn Lake, where the waters of the river actually teem with beautiful brook trout of very large size.

Pickerel were generally plentiful in all the lakes and rivers. This hardy fish was small in size but of a very fine flavor. It was easily caught with trolling hook and line and afforded great sport for us, owing to its presence everywhere throughout the entire season.

Pike were also caught in numbers. This particular fish of the northern waters is beyond comparison with those of its kind in southern Ontario, being firmer and of finer flavor.

Bass were found in quantities in nearly all the waters traversed by us during the season.

On the Grassy River we encountered a band of Indians who were well equipped with fish nets. The chief of the party told us that in season, or just before the formation of ice on the lakes and rivers, whitefish might be caught by this means in enormous quantities. By setting one net for two days and two nights and repeatedly running it, they have caught eleven hundred whitefish during the forty-eight hours. These fish were found to be of excellent quality, and would compare favorably in size with our Georgian Bay or Lake Huron whitefish. The Hudson's Bay Factor at Fort Matachewan informed me that whitefish were netted in quantities in the Montreal River near that point, during their spawning season.

The forests abound in game of all kinds, chief amongst these being the moose, bear, red deer and caribou.

Mention may also be made of the feathered tribe, of which the partridge and duck were found to be the main representatives. Partridge were seen almost everywhere and duck of various kinds were abundant on the lakes and rivers over which we canoed. Among the species of duck, we noted were the mallard, wood-duck, merganser, blue-bill, shelldrake or saw-bill teal, widgeon, spoon-bill or horse duck and whistler. The above mentioned birds were all breeding in these districts, as we found a representative female of each variety with a brood of young ducklings. In addition to these, almost all varieties of birds seen in other parts of Ontario were found to be inhabitants of these virgin wilds.

The fur-bearing animals of this territory are without a doubt but a remnant of the numbers which at one time inhabited these districts. There is scarcely a spot where the Indian or half-breed trapper has not found these fur-coated animals, studied their habits, and, by his native cunning killed the greater number of them and secured their valuable skins. The most numerous of the fur-bearing animals are, the bear, marten, mink, fox and muskrat. The beaver, otter, fisher and wolf are very scarce. Especial mention may be made of the beaver which are fast becoming extinct. They were never seen in large families, and as evinced by their work, have been driven from their homes and are now scattered broadcast.

The moose are the largest, and without a doubt the most abundant of any of the deer species. We found this animal in all the different districts over which we travelled. The upper waters of the Sturgeon River and the east branch of the Montreal River seemed to be their favored haunts. These sections were literally covered with hoof marks, and in some places their paths formed very good roads, which could be followed for miles through the woods. They would prove easy prey to the hunter, for they are not timid nor of a suspecting nature like the red deer. They are easily followed as their tracks are plainly visible even on hard, dry ground, owing to the enormous weight of their bodies. As many as eight individual moose have been sighted by myself in one day at separate places. Later in the season, when they had commenced to herd, it was not unusual to see ten or fifteen at the same time.

The red deer were found on nearly all of our territory but were much more plentiful in some districts than in others. Owing to the fact that the deer in these parts have been left unmolested, they are much tamer than those of their kind in other parts of Ontario, where they have been annually chased by dogs and hunters. The southern portions of the country, that is the Wahnapitae, Sturgeon and Obabika Districts, were the most thickly inhabited by these animals. As we travelled north they became more scattered and less abundant.

The caribou is also an inhabitant of this territory. We found signs of them every where. but did not see a great number. They are a much more wary and timid animal than the moose, and are consequently harder to see, and would prove much more difficult to capture. When seen, they are usually in bands or droves of various numbers. The country lying south and east of Smooth Water Lake, and north-west of the Wakenika Lake and River, seemed to be the best suited to these cautious animals, as it is a rough and hilly country. In these districts the caribou dwelt in greater numbers than in other localities which we explored.

The general character of the territory explored by us south of the Hudson Bay watershed may be described as that of an undulating rocky plateau. The surface of this elevated plain is by no means uniform, consisting as it does of a succession of more or less parallel rocky ridges. The valleys which intervene between these ranges of hills are occupied by swamps of varying extent and lakes of different sizes. The valleys of the Montreal and Sturgeon Rivers form notable exceptions to this general rule. The basins or valleys of these rivers are largely composed of clay formations of very great extent.

The rocks of the country which is drained by the Montreal and Sturgeon Rivers, are of Huronian formation, which in Ontario, is productive of our most valuable economic minerals. At several places ore was encountered throughout this formation. The most northerly point where we located mineral was on the Montreal River, near Fort Matachewan The country around Peter Long's Lake and generally that territory covered by our party on the Hudson Bay slope, is composed of Laurentian rocks, which, as is well known, does not contain important economic minerals.

There are very few prominent bills, the highest very seldom attaining a greater altitude than four hundred feet, while throughout most of the district, hills of from seventy-five to one-hundred feet are conspicuous topographical features. The two highest ranges of hills on the whole area are situated about sixty miles apart. The more northerly range was found on the height-of-land where it is crossed by Duncan Sinclair's exploration line of the year eighteen hundred and sixty-seven. The highest peak of this range is Mount Sinclair, which is fifteen hundred feet above the sea level. The highest land in the whole area is situated immediately to the west of Lady Evelyn-Lake, where a range of hills, of which Maple Mountain is the highest peak rises to the height of about two thousand feet above the sea.

The entire district is traversed by numerous rivers, and lakes which serve as inlets and outlets to these rivers. The Sturgeon and Montreal Rivers occupy deep and important depressions in the somewhat uneven plateau. The Sturgeon is the larger of these two rivers, draining probably three thousand five hundred square miles of territory, while the Montreal drains an area of about twenty-five hundred square miles. The Wahnapitae, a large and important river, and the Matabitchewan, a much smaller stream, are also worthy of mention. Lake Temagami, during the freshet season or in the early portion of the summer, drains both northward by Lady Evelyn Lake into the Montreal River, and southward by way of Temagami River into the Sturgeon. The north outlet is usually dry during the latter part of the summer, and can only be utilized in the spring season at the time of high water.

The southern, or Temagami outlet is the larger and no doubt the main stream draining these waters. Lakes Ohinicoochichi, Adelaide and Button are drained by the Maskinonge River, one of the largest tributaries of the Sturgeon, which enters the main stream from the west in the Township of Janes.

WATER POWERS —The Sturgeon and Montreal Rivers are furnished with numerous water-falls, some of which are large enough to furnish power for extensive manufacturing enterprizes. There are four waterfalls on the Sturgeon River north of the surveyed territory, the smallest having a fall of over twenty-five feet. Ascending this river, the first fall is found about three miles below the mouth of the Obabika River. The banks are rock on the west and clay on the east side of the river. The banks at the head of this fall are about seventy-five feet apart, and there is a drop of forty feet in a distance of seventy-five yards.

The next fall is situated about three miles above the outlet of the Obabika River and is called Goose Falls. The banks are rock on both sides and the river is sixty-five feet in width and has a drop of about twenty-five feet. There is a good chance here to increase and improve the natural power of the falls. The third fall is situated about two miles below the Stouffer Lake Portage, and about nine miles above Goose Falls. The river at this point is about sixty feet wide and has a drop of twenty-eight feet. Kettle Falls, situated about four miles above the Stouffer Lake portage, is a nice clear fall of over forty feet. The river is sixty-five feet wide and the banks are high rocky bluffs. This water-power could easily be improved so as to have an artificial head of forty or fifty feet more than its natural fall.

The Main Montreal River, north of the outlet of Lady Evelyn Lake, has two large falls. In ascending the river, the first of these is Kekopacagewan or High Falls. Here the river is about seventy-five feet wide and has a drop of about thirty-five feet in seventy-five yards. The banks are high and rocky and form a good opportunity for increasing the natural power of the falls. The second valuable water-power on this river is Matachewan Fall which has a drop of over forty feet. This fall is situated above Fort Matachewan at the Great North Bend. The banks are high and rocky, and the river which is about fifty-five feet in width, could be easily improved.

A great many minor falls and rapids are found on the east branch of the Montreal River. The principal one of these, which has a fall of thirty feet, is situated south of Abushkong Lake. Both banks are of a high, rocky formation and the river at this point is about forty feet in width. The Mattawapika Falls, situated at the outlet of Lady Evelyn Lake, have a drop of about thirty feet. The banks are high and of a rough rocky nature.

The most important falls on the Mattagami River, south of Niven's base line, are the Kenogamissie Falls, situated at the foot of Lake Mattagami. This fall has a drop of about forty feet, and the banks are rocky and high enough for any purposes of improve ment.

The Kapiskong or Grassy River from its outlet to Peter Long's Lake has a probable drop of two hundred feet. This entire distance of about twenty-five miles is almost one continuous stretch of swift water and rapids, with the exception of two falls of twenty-five or thirty feet each, which can be utilized for water-powers.

It was found on examination of the rivers and creeks that drain the entire territory that comparatively little improvement would be necessary to enable all the timber to be transported or floated down the respective water-courses.

On reviewing the information acquired by myself and party regarding the soil, timber and minerals, of this district, I beg to state that the immense natural resources of the newly-explored tract of country are bound to contribute largely to the growth and prosperity of the Province. When the territory has been opened up by a railroad, a nucleus will be formed around which the lumbering, mining and agricultural industries will develope with rapidity, thereby inducing the settlement of these regions which now support only a few scattered families of Indians.

I have the honor to be,

Sir,

Your obedient servant,

GEORGE R. GRAY.

I, George R. Gray, of the City of Toronto, in charge of Party No. 3, make oath and say that the contents of this report are true to the best of my knowledge and belief.

Sworn before me this 10th day of January, 1901.

GEORGE R. GRAY.

G. B. KIRKPATRICK,
A Commissioner.

TORONTO, Dec. 31st, 1900.

To GEORGE R. GRAY, Esq.,

In charge of Exploration Survey Party No. 3,
Toronto.

DEAR SIR,—Acting under instructions, received early in June, from Mr. Archibald Blue, late Director of the Bureau of Mines, I accompanied Exploration Survey Party No. 3 as Geologist. The territory assigned to this party is described as follows: "From Lake Temagami northward to the Montreal River, and upon the east and west sides of the Algoma-Nipissing line."

In company with Mr. George R. Gray, who is in charge of the party, I left Toronto on June 24th, travelling by Canadian Pacific Railway to Sudbury. There we met the remainder of our party and obtained our supplies and canoes. From Sudbury we drove out to Lake Wahnapitae, a distance of twenty-one miles, then crossed this lake, (ten miles) on the tug "Maid of the Mills," which landed us at Crystal Mine, on the north east bay of the lake.

From this place our trip was made in canoes. We followed a chain of twelve lakes with portages, northward to the Sturgeon River, which we then ascended to the forks. Taking the north branch from here we passed through a chain of ten lakes with portages in a direction north-northeast to Smoothwater Lake, which is at the head of the eas branch of the Montreal River. We then descended this river to its junction with th main Montreal River, about seven miles below Hudson's Bay Company's post at Matachewan.

From this point we descended the Main River to Mattawapika, where the northern outlet of Lake Temagami empties into the Montreal River. Here we turned south, passing through Lady Evelyn Lake, Lake Temagami, Lake Obabika, and Obabika River, to the Sturgeon River which we descended to the north boundary of McNish Township, where we turned west through a chain of lakes to Lake Wahnapitae and descending the Wahnapitae River reached Wahnapitae Station on the Canadian Pacific Railway

From this point we travelled by train to Metagama Station, where we again set out in canoes, ascending the Mattagami branch of the Spanish River to the head waters. We then crossed the height of land between St. Lawrence River and Hudson Bay waters and entering the head waters of the Moose River (Mattagami branch) descended it to the point where it is crossed by Niven's base line from the one hundred and twentieth mile of the Algoma-Nipissing line.

Returning a few miles we ascended the Grassy River to Little Hawk Lake, the head waters of one of its branches, and recrossing the height of land entered the head waters of the Montreal River, which we descended to Mattawapika River, turning south from here we again reached the Sturgeon River, which we explored and mapped northward from Obabika River to the Stouffer Lake portage, also several chains of lakes east and west of this part of the river.

At the close of the season's work the party returned to Sudbury by Wahnapitae Lake and thence by Canadian Pacific Railway to Toronto.

Before commencing my report proper, I would state that the route we followed up the Sturgeon River, down the east branch of the Montreal River, and up the Main Montreal River, was covered by Dr. Bell of the Ottawa Geological Survey in 1876. As frequent side trips were made by our party I have a good deal to add to what Dr. Bell has said, however, where I have no new information I shall simply refer the reader to the above report.

DEWDNEY LAKE.—The route followed from Crystal Mine was up the northwest arm of Lake Matagamishing, a distance of six miles, then through three small lakes (with portages) into Dewdney Lake, which is crossed toward its upper end by the north boundary of the Township of Mackelcan. Dewdney Lake receives a stream from Lake Chinicoochichi which enters at the head of the lake falling about ten feet over quartzite near its mouth. The quartzite is coarse-grained, light green in color, and contains grains of reddish felspar scattered through it. The Lake is bounded on each side by high ridges of this quartzite, which have a general direction, a little west of north.

LAKE CHINICOOCHICHI.—A short distance east of the inlet, a thirty chain portage, running northwest, leads into Lake Chinicoochichi (meaning the lake with big bays). This Lake is six miles long and contains a large island in the centre of the main part. The rock around the shores and on the island is light colored quartzite similar to that just described. A bluff on the west shore opposite the island is 560 feet high (aneroid), composed of quartzite, which becomes coarser-grained toward the top, where in places it passes into a conglomerate containing pebbles of quartz several inches in diameter.

From the northwest bay of this lake a chain of small lakes leads westward. The first portage (55 chains) crosses the height of land between Sturgeon River and Wahnapitae River waters; from the west end of this portage two small lakes and two ponds connected by trails lead west, a distance of one and one-half miles to a larger lake, on the east shore of which quartz is exposed, similar to that already mentioned.

Another chain of small lakes runs northwest from the north narrows of Lake Chinicoochichi, and on this route the quartz is again found to be the prevailing rock.

N. E. BAY OF LAKE CHINICOOCHICHI.—The entrance to the northeast bay of the lake is by means of a narrows, leading southeast from the north arm. The rock is a dark reddish quartzite along the north side of this channel, and greywacke of the typical variety on the south side. Toward the southeast extremity of this bay several small islands occur composed of massive diabase, on the north side of the larger of which the rock is a good deal weathered, and stained by decomposing pyrites.

CANOE ROUTE TO STURGEON LAKE —From hte north end of this bay a 40 chain portage north 85 degrees east, through a spruce swamp, leads to Laura Lake. This lake which is about four miles long and varying from one-quarter to one mile in width has a general east and west direction, and is one of a chain of five lakes, which with portages, is an old Indian canoe route to Sturgeon River and thence to Hudson Bay Post on Lake Temagami. A small stream enters Laura Lake near the end of the portage about three quarters of a mile east of the portage, on the south shore. Here an exposure of slate occurs: it contains specks of muscovite and small crystals of pyrites and approaches a schist with its wavy cleavage. This rock forms a ridge along the south shore of the lake, which 3 miles east of the portage rises to a height of 50 or 75 feet, where a talus of boulders has broken off and fallen down. The South shore is wooded with second growth spruce and poplar while the north shore is thickly covered with a grove of pine. Slate also outcrops at the narrows near the east extremity of the lake.

A 26 chain portage (south 60 degrees east) leads from Laura Lake to the second lake of the chain, near the centre of which was observed an exposure of greywacke striking north and dipping at a high angle.

The second lake is roughly crescent-shaped measuring 2 miles from portage to portage, the direction of the lake being first south-east then north-east. It receives a stream at its south-eastern extremity which falls 10 feet over a ridge of slate, at its mouth. This stream drains a couple of small lakes toward the south east on which fresh beaver cutting was observed. Between these two lakes a small vein of quartz was noticed in the slate, having a width of three inches, in which, however, no mineral was apparent.

From the foot of the second lake a 32 chain portage (north 15 degrees east) leads to the third lake. The portage is between the two outlets of lake No. 2 and is for the most part over a ridge of slate, in which the cleavage runs north and south. A fall of about 12 feet occurs on each of the streams, which empty into the third lake within 5 chains of each other.

No rock exposures were observed on lake No. 3 which is narrow and about one mile in length. The remains of an old beaver dam were noticed at the foot of the lake. The 25 chain portage (south 60 degrees east) out of this lake, on the left of the outlet, crosses a ridge of greywacke near the centre. This rock which contains small pebbles of felspar approaches a greywacke conglomerate.

The fourth lake which is quite shallow in places is 1½ miles long and is contracted near the centre, to a very narrow neck. Opposite the narrows a stream enters from the north which drains two lakes in that direction. The outlet is at the north-east end of the lake and runs east to a small lake No. 5. A 12 chain portage passes a rapid on the outlet.

The fifth lake is marshy and about half a mile wide. The outlet leaves the northeast of this lake and is about half a chain wide. One mile below the lake it becomes rapid, and a 68 chain portage is necessary, which leads to the Sturgeon River. The first 20 chains of this portage is through rolling country wooded with spruce and jack pine which to the east gives place to a sandy plain, the original forest of which has been burned and it is now covered with very small poplar.

The shores of this chain of lakes are everywhere green, the timber being pine (both red and white) spruce, poplar and birch.

A chain of lakes extends southward from lake No. 2 described above. A trail 52 chains in length (south 15 degrees west) commences at the south side of this lake and leads to a small lake half a mile long; from the west shore of this about the centre is a lift over into another lake half a mile in length, from the south end of which there is a well worn trail 20 chains in length leading to another small marshy lake. This trail crosses an exposure of slate near its north end, and an outcrop of greywacke on the shore of the last lake of the chain.

Another chain of waters extends from the south-east of the north-east bay of Lake Chinicoochichi. There are five lakes having a general north and south direction. The prevailing rock is a very massive quartzite and as this runs south to the north boundary of the Township of Mackelcan and is thus directly east of and parallel with Lakes Chinicoochichi and Dewdney the quartzite is probably continuous between these two waters.

The description now returns to Lake Chinicoochichi. From here we paddled north through a narrows into Sawhorse Lake which is small and marshy. A 30 chain portage from this to Adelaide Lake is over a level sandy plain. This portage crosses a watershed; the lakes to the south draining into Lake Matagamishing and thence to the Sturgeon River through Lake Maskinonge, while Adelaide and Button Lakes are part of a chain which drain into the Sturgeon River in a north-easterly direction.

Adelaide Lake is half a mile in length, with shores thickly wooded with pine and spruce. It drains north into Button Lake. The 10 chain portage is through a marsh, at the north end of which is an exposure of greywacke which has been glaciated, leaving a smooth surface.

BUTTON LAKE.—Button Lake is three-quarters of a mile long from portage to portage, with a small island near the centre, and heavily wooded shores. A rocky ridge runs along the west side. A sample of this rock examined under the microscope proved to be quartz diabase which is very massive in its arrangement. On the north and north-east shores the rock which is also a basic eruptive, was found on the examination of a thin section to be hornblende porphyrite.

This lake drains north-east into a long narrow lake. The outlet which leaves the north east end of the lake is a small stream, only navigable with canoes for a short distance. Hornblende porphyrite forms its north bank, while on the south is a spruce swamp. The portage leaves the south-east bay of Button Lake and runs due east for 62 chains, the last ten chains being through a spruce swamp, which forms the east shore of the lake. This lake is 3½ miles long and quite narrow, with a general north-north-east direction. Several exposures of massive quartzite occur along the shores and fresh beaver cutting was observed on the west shore near the inlet. The outlet flows out of the foot of the lake into the Sturgeon River and is about one mile in length. A sixty chain portage on the left bank of this stream runs north 40 degrees east. Several exposures of massive diabase occur along this trail forming high bluffs toward the north-east. At a distance of 37 chains from the Sturgeon River the trail crosses a vein five feet wide, nearly vertical and striking north 10 degrees west. The vein matter is white quartz and spar with a good deal of iron rust, a sample gave on assay a trace of gold and 1.22 oz. of silver (60c).

The description now returns to Button Lake. The canoe route which we followed was by a 22 chain portage out of the west end of this lake, which crosses a ridge of quartz diabase 150 feet high. This portage is the watershed between the chain of lakes of which Button Lake is one and a chain of waters to the north.

DOUGHERTY LAKE.—The next lake is Dougherty Lake, which is 3 miles from portage to portage. The rock around the south end is quartzite which is almost white on the weathered surface. Several small islands on the main part of the lake are also composed of this rock. The west shore of the lake is composed of a ridge of diabase from 50 to 75 feet high, thickly wooded; the rock outcropping at many places along the water's edge. The north shore is composed of massive quartzite which also frequently outcrops along the water's edge.

The outlet of this lake is a short stream flowing north, out of the north-west arm, with an eight chain portage on the left. No rock is exposed on the portage.

FREDERICK LAKE.—Frederick Lake into which Dougherty Lake drains, is 3 miles in length. It is quite shallow with numerous small islands. The west shore is a ridge of diabase, which is a continuation of that described on the west shore of Dougherty Lake and takes the form of a succession of hills from 200 to 300 feet high (aneroid). The east shore is composed of quartzite of the ordinary variety. The shore line is very rocky, the quartzite being in a vertical position and striking north 35 degrees east. All the shores are thickly wooded with pine, spruce and poplar.

STOUFFER LAKE.—An 8 chain portage leads from Frederick to Stouffer Lake on the west of the outlet. This lake is 1½ miles long, with a large island about the centre. Directly west of this island a narrow neck leads into a large bay at the upper (north) end of which is a fresh beaver dam, where I saw a beaver at work. To the south west of this bay is a hill of diabase measuring 400 feet high (aneroid), from the top of which the valley of the Wahnapitae River is visible in the distance to the west, the whole stretch appearing as a mass of green bush. A trip inland from the east shore of this lake, to the Sturgeon River showed the rock to be a succession of ridges of quartzite striking in a general north and south direction, separated by valleys and all thickly wooded with pine, spruce and poplar. The aneroid barometer registered 150 feet as the height of one of these ridges near the shore of the lake. The outlet is a small stream running out of the north east arm of the lake, into the Sturgeon River. The portage is a few chains to the east and is 40 chains in length. Quartzite is exposed to the right of the portage and is a continuation of the ridge above noted.

Sturgeon River.

Below Stouffer Lake portage.—At the point where Stouffer Lake portage touches the river it is running with a good current and is about one chain in width. One quarter of a mile below is a marshy expansion half a mile in length with a high bluff of quartzite on the west bank. For two miles below this, the current is steady and the banks level and wooded with red and white pine and spruce. Here the river flows around a large island and then falls over chute caused by an exposure of slate, crossing the river in a direction south 85 degrees east. The river falls 20 feet between high walls of slate, passed by a 10 chain portage on the right. Five chains further down is a second chute having a drop of 8 feet and passed by an 8 chain portage on the right. The east bank is solid rock 200 feet high and almost vertical, the rock having broken off in rectangular blocks and fallen in the form of a talus to the water's edge. Below this second chute is a short rapid, then a small lake expansion one quarter of a mile in length. Following this is half a mile of swift water, then a rapids where the river passes through a narrow gorge between vertical banks of slate. The next half mile is swift water with a rapids at the end opening into a small lake expansion, on the east bank of which is an exposure of massive, light colored quartzite containing specks of pyrities, which have rusted on the weathered surface. A quarter of a mile below is a fall of 10 feet over very tough greywacke, which forms a small island at the head of the falls, and a 2 chain portage passes over it. The banks of the river for some distance below are composed of this rock, which has been very much disturbed, making the river rough and the scenery wild and rugged. Ten chains below is another rapids passed by a five chain portage on the left; opposite the south end of the portage, across a small expansion, has recently occured a rock slide, about 3 chains in width, which in falling from the top of a hill 200 feet high carried all the trees before it, leaving the bare rock exposed, while on either side pine, spruce and poplar trees of good size are thrown into relief. This rock is greywacke, which, crossing the river causes a rapids, passed by a 12 chain portage on the left. Ten chains below, an exposure of quartzite occurs on the east bank, and just below it, an exposure of greywacke containing considerable iron pyrites and quartz, which causes a rapid here. A short distance inland this same rock rises in a sheer bluff to a height of 75 feet, striking north and south and dipping eastward at a high angle. A trip of a mile inland to the east, from this point showed the rock to change to light-colored quartzite, which continues eastward and forms a very rocky difficult country. Fifteen chains below, *i.e.* about 7 miles below the Stouffer Lake portage, what is known as the North Branch of the Sturgeon River, or Bob's River, enters the main stream. A trip was made up this branch. It enters the main river at a chute which falls over massive quartzite, this is passed by a 10 chain portage on the left. The country for 4 miles above this is very rocky on each side of the river, but the immediate shores are marshy and several small lake expansions occur. About six miles from the mouth, an exposure of sericite schist occurs. This rock causes a chute which is passed by a difficult portage of large rhombohedral blocks of schist having broken off and fallen down, and these covered with moss make very dangerous walking. Half a mile above is a small lake on the shores of which three or four new beaver houses were observed, indicating the presence of several families of beaver.

For a half mile below Bob's River, the main Sturgeon River is swift with large rounded boulders in the bottom. At about this point however the country suddenly changes from the very rocky, rough character described above, and becomes more or less level with but few rock exposures, the soil being represented by sand. About one mile further is an outcrop of chlorite schist striking north 20 degrees west, dip west 50 degrees. Two miles below we passed the portage from the chain of lakes described, from Lake Chinicoochichi. Here the soil is sand covered with a few inches of decayed vegetable loam. The portage crosses a young brulé which extends eastward on the lett bank, with however occasional patches of green bush. Below this the current becomes slower and the river wider and deeper, and at a distance of three miles below the Ohinicoochichi portage we reached Upper Goose Falls, with a drop of 25 feet over a ridge of slate. one mile below this the Obabika River joins the Sturgeon from the east. Three miles further down is the Lower Goose Falls passed by a 10 chain portage on the left over a sand hill.

For 20 miles below, there is not a portage and no rock exposures were observed. The river winds its way, with a steady current through a sandy country, which forms the immediate basin of the river, the country inland soon becoming rocky and broken. In some places the sand and clay banks are 30 and 40 feet high; these are wooded with spruce, poplar, pine and occasionally some small ash, this latter indicating good agricultural land. The remaining part of the lower Sturgeon River has been mapped and described in several reports* of the Geological Survey to which the reader is referred.

As will be seen from the above description the south part of the Upper Sturgeon is rough and difficult and for this reason is but little travelled. The easier route which is always chosen, is the chain of lakes described in an earlier part of this report, from Lake Matagamishing to Stouffer Lake. And this route can also be recommended for the tourist on account of the beauty of the scenery, as well as the abundance of game. Moose, red deer, caribou, partridge and duck are all plentiful.

UPPER STURGEON RIVER.—The description now returns to the Sturgeon River at the Stouffer Lake portage from which point the ascent of the river will be described. A short distance above this is a small rapids passed by a 3 chain portage on the left, on which quartzite of the ordinary variety is exposed. A few chains above is another rapid and short portage caused by the river passing over diabase. This rock outcrops at several points along a small lake expansion, where it is striking north 60 degrees east. Ten chains above the portage I examined what is known as the Leroy location. It is a quartz vein in country rock of greenstone (diabase). The vein is 5 feet wide, strikes east and west and dips south 60 degrees. It contains iron and copper pyrites and galena. A sample taken across the mineralized portion of the vein gave the following results on assay:— 02 oz. gold (40c), 23.28 oz. silver ($11.64), 5.64 per cent. copper ($11.28) 25.74 per cent. lead ($10.29). A littlework has been done on this location consisting of several test pits, and stripping of the vein for a short distance. A quarter of a mile paddle, through Perkins Lake brought us to another rapids and portage on the left.

The rock at this portage is a dark, green-colored greywacke or quartzite. Large blocks have broken off and the portage which is over these is rendered very difficult and rough. A paddle of half a mile through another small expansion Renfrew Lake, leads to another rapids and short portage on the left, on which the rock is similar to that just described. At the head of this rapids is another small expansion into which the river flows from the north, over what is called Kettle Falls.

KETTLE FALLS.—Here the river drops forty feet (aneroid) in a succession of chutes over quartzite, situated in a vertical position and striking south, 70 degrees east. There is a very steep portage on the left of the foot of the falls which leads almost vertically up for 100 feet, then descends on the other side to the river. This precipitous bluff is composed of dark greenish-red colored quartzite. Several small pot holes have been worn in the rock near the head of the falls from the presence of which the falls have probably been named. A fine grained basic eruptive (diorite, probably) outcrops on the west shore opposite the foot of the falls, which I think may be the continuation of the eruptive dike noted as forming the west shores of Dougherty, Frederick and Stouffer Lakes. Directly east of the head of the falls is a high hill, wooded with jack pine, birch and poplar, while the bluff on the west side is almost bare, the whole producing a very romantic bit of scenery.

* Dr. Bell's Report of 1891-- (Sudbury Mining District.) Mr. A. E. Barlow's Report on Lake Temagami—Vol X. New Series, Part I.

For half a mile above Kettle Falls the river forms a narrow lake expansion called Lyman Lake, the upper end of which is marshy and the west shore flat. Above this light green, fine-grained quartzite is exposed and about one mile above the falls, on the east bank, occurs an outcrop of reddish syenite, containing a little quartz, a large amount of flesh-colored orthoclase and varying quantities of hornblende. Light colored quartzite again outcrops above this ; and one and a quarter miles above the lake is an exposure of fine-grained greenish rock, a specimen of which examined under the microscope proved to be quartzite (arkose). Half a mile above this quartzite of the ordinary variety again outcrops and crossing the river at this point, it causes a rapid which is passed by a lift over on the right. Half a mile below Eagle Nest Lake there is a contact exposed on the east shore between greenstone and dark reddish gneiss; the line of contact runs due north and south and is very distinct. A trip inland to the west from this point showed the gneiss to continue for that distance. A quarter of a mile above, the gneiss crosses the river causing a rapid. Above this are two small expansions, followed by a large one which we named Eagle's Nest Lake on account of an eagle's nest having been conspicuously placed on the top of a dead pine a short distance east of the lake. The east shore of the lake is composed of a ridge of dark reddish, fine-grained gneiss seventy-five feet high ; no exposures of this rock, however, occur along the shore, as it is thickly wooded to the water's edge. At the head of the lake (which is one mile long) there is a short stretch of swift water and one quarter of a mile above this a rapid, passed by a three chain portage on the left, then a small lake expansion thirty chains in length which is at the foot of the Twin Falls. Here the river flows around an island of gneiss. This rock forms a barrier and causes a fall in each channel with a drop of about fifty feet. The rock, which is reddish grey gneiss also outcrops on the east shore of the lake expansion above the falls, where it strikes south, 65 degrees east. The island is crossed by a ten chain portage. A quarter of a mile above the island is a narrows where the river flows with a strong, deep current between high walls of rock. On the northeast bank is an almost vertical wall of micaceous schist 150 feet high, from which immense rectangular boulders have broken off and fallen down. A few chains above is a beautiful lake expansion called Paul's Lake or Ghoul Lake.

This lake is one and a half miles long and about half a mile wide, having a general direction north-northwest. A low ridge of reddish grey gneiss forms the east shore, while the country to the west is flat, and no rock exposures were observed on it. The shores of this lake, as is the case all the way up the Sturgeon River, are heavily wooded with jack pine, spruce and poplar, with scattered groves of red and white pine.

At the upper end of the lake to the west of the inlet is an old clearing some twenty-five acres in extent with the remains of an old Indian village and graveyard.

For three and a half miles above Paul's Lake the river is quite narrow and winds its way through a black alder swamp in a general direction south-southeast. This swamp is flanked on either side (a short distance inland) by high wooded ridges. At a point three miles above the lake, on the west bank, is a bold exposure of dark green diabase, about seventy-five feet high. It is striking north, 15 degrees east, and diagonally across the river is continued in the same direction to the east, the dip is to the southeast.

Upper Forks of the Sturgeon River.

About four miles above the lake occur the upper forks of the Sturgeon River, where a trapper has built his cabin, One branch comes from the west with dark brown water, the other from the north with clear water called Clear Water Creek.

No opportunity occurred of following the west branch, however, I have been told by an engineer* who made the trip to the head waters of this branch, that the river is extremely tortuous with occasional exposures of light colored quartzite. Twenty-five or thirty miles above the forks it breaks up into three small branches which take their origin in a spruce swamp, very level and muskegy, and many square miles in extent.

Route to Head Waters of North Branch.—The route we followed was up the North branch to the head waters by a chain of ten lakes, Light colored quartzite and coarse,

*Lake No. 1 is about one and one-half miles from portage to portage.—Mr. W. W. Stull, Sudbury, Ont.

dark green diabase are exposed on the east bank at the forks. About one mile up the branch it becomes too shallow for canoes and it is necessary to make a 78 chain portage, due north to a small lake. The portage is through a sandy plain, wooded with jack pine, spruce. birch and whitewood. A windfall through this bush has made the portage very rough. No rocks were exposed but several float boulders of quartz were observed.

The outlet (Clearwater Creek) leaves from the south west bay at which point is an exposure of light colored quartzite containing some small pebbles of quartz and a few of red jasper. The west shore of the lake is composed of a ridge of similar quartzite, while the east shore is low and flat. This lake receives two short streams from a large lake to the north—Stull Lake. There is a drop of 15 feet between these lakes, and they are also connected by a two chain portage just to the right of the west stream.

Stull Lake.—On the south shore of Stull Lake, east of the portage, massive diabase is exposed, striking north-east dip south 60 degrees. This lake is five miles long consisting of a stretch of 3 miles, east and west, and an arm 2 miles long running north from the eastern extremity. At the north-west end of the lake, diabase is again exposed. The country to the north of the main part of the lake is level with high hills in the distance. On the north shore toward the east, are the remains of an old trading post and Indian clearing ; the soil at this point is a sandy loam. The rock to the east and south is quartzite similar to that described above.

Our route was east from Lake No. 1 through two small lakes with portages to Lake No. 4, which is 1½ miles long and a direction north-north east, the east shore of which is a ridge of massive light-coloured quartzite. This lake is connected by a short stream with Lake No. 5, 1½ miles in length, to the north. It is surrounded by high hills of quartzite. The highest hill, that to the south-east, measured 650 feet high (aneroid). A portage of three-quarters of a mile leads north east to Lake No. 6. which is very small and with beautifully clear water. A two-chain portage leads to Lake No. 7, which is 1¼ miles long. On the north-east shore of this lake is an exposure of a banded, purple and slate-coloured rock which strikes north-east and dips north west 30 degrees. A specimen of this rock, examined under the microscope, was found to be a slate.

Lake No. 7 is connected by a short portage with Lake No. 8, which is narrow and 1¼ miles long, and this by a short portage with Lake No. 9, which is one-half mile in length.

A 52-chain portage leads from Lake No. 9 to Lake No. 10, running north-north-east. The portage is very rough, and we had to chop it out before taking over our supplies. Light coloured quartzite outcrops about the centre. The portage crosses the height of land between the waters of the Sturgeon River and those of the Montreal River (east branch), Lakes 1 to 9 all draining south. Lake No. 10 is three-quarters of a mile long, direction north-west. The portage from it to Smoothwater Lake is 30 chains in length and passes over a high, narrow ridge of sand and gravel, which is apparently of morainic origin and is probably an Esker.

Smoothwater Lake.—Smoothwater Lake is four miles long, with an average width of about one mile. The immediate shore on the east side is level and sandy, but half a mile inland there are a series of high hills composed of light-coloured quartzite, from the top one of which a good-sized lake is visible to the eastward. The aneroid registered a difference of level of 75 feet between these two lakes. This same rock composes the south-west shore The rock on the north point of the west shore is fine-grained and green in color, with fine-grained patches of a light green color. A specimen of this rock proved to be an impure talc schist, when examined in thin section under the microscope. The remaining rocks on the east shore of this lake are completely dealt with by Dr. Bell, as are those of the east branch of the Montreal River. On account of this river having been surveyed and mapped some years ago our party travelled more rapidly than on the above noted waters and consequently but little time was available for the study of the rocks. Wherever I am able to fill in details of the geology I shall do it below.

East Branch of Montreal River.

The east branch of the Montreal River has its head in the small lake to the south of Smoothwater Lake. However, Smoothwater Lake receives several good sized streams and the river leaving the north arm of it is about two chains wide and flowing with a strong current. About two miles below the lake the high ridges which were noted at the height of land drop off gently to the north, the country becomes level and several lake expansions occur.

About 9 miles below the lake there is a 60 chain portage on the right past a rapid No rock outcrops on this portage; about the centre it passes through a swamp and at the north end crosses a ridge of sand. About 4 miles below there is another rapid and short portage on the right. Here the rock is a massive reddish grey arkose, containing small pebbles of quartz and occasionally pebbles of red and black jasper. For one quarter of a mile below the 2nd portage the current is very swift, after which the river becomes very crooked, as it passes through a tamarac and alder swamp. To the east and west of this swamp are low ridges which gradually approach one another, and about 3½ miles below the portage the river becomes straight and runs for a short distance between well defined walls which lead to a rapid. This is passed by a 40 chain portage on the left (3rd portage). This portage is over bare rock for the most part, the rock being similar to that at the 2nd portage.

A few hundred yards below this portage the river opens into a large lake, called by Dr. Bell Lady Evelyn Lake, about three miles in length. The river flows out of the north of it, and a rapid occurs just below the lake, which is passed by a 20 chain portage on the right, leading to a small lake. The rock here is greywacke. The aneroid recorded a drop of 30 feet between the two lakes. This small lake is about 1 mile in length. The river flows out of the north of it, and another small rapid occurs with a short portage on the right, although we were able to run it with our canoes, light.

Lake Obushkong.—Another lake, 1 mile long, leads to the 6th portage, which is 4 chains in length. Here there is a rapids and falls as the river enters Lake Obushkong. This lake is 5 miles long and contracted to a narrows at about the centre. There is a low ridge forming the east shore while the country to the west is rolling. A high bare peak forms the north-east shore, and a high mountain (probably Sinclair's mountain) is visible in the distance to the north-west. The rock on the south end of the portage into this lake is a light grey quartzite, at the north end it is massive diorite. This eruptive rock seems to be a narrow dike, which crosses the river in a direction north 30 degrees east, causing the falls at the foot of the rapids. Quartzite is found in contact with this rock both to the north and south, and this quartzite, which in places passes into arkose containing rounded pebbles of quartz and angular pebbles of jasper, is found on both the east and west shores of the lake. About half way up the west shore of the north part of the lake, the rock is greywacke, and the rock composing the peak to the north-east is greywacke conglomerate.

On leaving Lake Obushkong, the river turns east for about 3 miles. Hornblende (actinolite) schist outcrops along the river during the first two miles, then the rock changes to reddish granite. Four miles east of the lake, the river turns north again flowing between high rocky hills of reddish granite which in places passes into syenite. A mile and a half below the turn, there is a rapid and 12 chain portage on the left (seventh portage) the rock being reddish syenite. Half a mile below is another rapid with a 15 chain portage on the right, over similar rock to above. About three-quarters of a mile further the river turns sharply to the right (east) expanding into a lake. The rock on the south shore just east of the bend is hornblende porphyrite. The lake is short and the river at the foot of the lake flows around an island. This forms a rapid passed by the 9th portage 6 chains in length on the right. Below this the river again turns north, and half a mile from the bend there is a rapid and fall passed by a 6 chain portage on the left. This portage which is the 10th and last, passes over a bare outcrop of greywacke of the typical variety. Just below this a quartz vein outcrops down the face of the high rock on the west shore.

Gold and Copper.—The quartz shows iron and copper stains and a sample gave a trace of gold with 0.11 per cent. of copper.

The river then enters a lake 2 miles in length, which being surrounded with thickly wooded hills, makes a beautiful bit of scenery. Below this are two small lake expansions, after which the river enters a stretch of swift water, 2 miles in length, which empties into the expansion at the junction with the main Montreal River.

The description now turns to the Mattagami and Grassy Rivers and Peter Long's Lake. From these we crossed to the head waters of the main Montreal River, and this river will afterwards be described without interruption.

Leaving Metagama Station on the Canadian Pacific Railway we put out our canoes in a small stream a little east of the station and paddled 1½ miles west parallel with the line of the railway, the creek winding its way through a marsh, between low ridges of pink granite. At about 1½ miles from the station the creek expands into a small lake, from the west of which the creek runs and turning northeast crosses the railway near the depot camp of the Metagama Lumber Company.

We then paddled northeast about one mile across a lake to a short portage on the right, where we were able to draw our canoes up with a line from the shore. Here the rock is pink granite varying in texture from fine-grained to a pegmatitic variety. From this point we paddled northeast up stream, for about 4 miles, passing through three small expansions. Here a branch of the the Spanish River enters from the north, while what is called the Lost Channel turns southeast for 2½ miles and joins the Mattagami Branch of the Spanish River, thus forming a large island. At a distance of 2 miles down this channel there is a 5 chain portage on the right over granite, the glacial striæ here observed having a direction north 10 degrees east. A quarter mile below is a second portage 20 chains in length on the right, around a rapid; the river running between walls of granite 50 to 75 feet high, which leads to the Mattagami branch.

MATTAGAMI BRANCH.—Turning north-north-east, we paddled through a lake for about three quarters of a mile to the dam of the lumber company, where a short portage on the left was necessary. From the dam our course was through two small expansions. There are ridges of sand and gravel on either side, but these give place to pink granite about two miles above the dam. The timber has been cut and a fire has been through this part recently so that the rock is exposed in many places. A paddle of nine miles north through a series of small expansions of the river, brought us to one of the Metagama Lumber Company's camps. The rock at this point is reddish grey gneiss.

Above this is Lake Mosquitowagemaw, which is quite narrow and about 4 miles long, on which the pine is now being cut. To the north of this are two small lakes which are bounded on each side by terraced ridges of sand. For the next three miles the river is very crooked, flowing between sandy banks; above this is a small lake. From this we paddled one quarter of a mile up a small creek, with very brown water, pushed through some reeds into a small clear water pond, at the north side of which is a 14 chain portage leading northward over a level, sandy plain to Perch Lake, which is about half a mile in length and contains beautifully clear water. From the north end of this lake a 30 chain portage leads to Blue Lake, about 1½ miles in length, to the north of which is the height of land between Hudson Bay and St. Lawrence River waters.

HEIGHT OF LAND —The Height of Land portage is 100 chains long and is over a level sandy plain of very considerable extent, This has been burned over recently and now is covered with a small growth of jack pine. Near the south end of the portage is an outcrop of massive pink rock on each side of the trail. A thin section of this rock examined under the microscope showed it to be a crushed granite. About the middle of the portage is a clump of large jack pine. At about 75 chains from the south, the trail crosses what is apparently a kettle hole about 200 acres in extent and 75 feet deep. The portage ends at the south end of Dividing Lake, the water in which is quite brown and at the time of our visit in September was much colder than that in Blue Lake.

MATTAGAMI RIVER.

We paddled north about one and one-quarter miles the length of Dividing Lake, then turned south east down the Mattagami River, or the Mattagami branch of the Moose River. At about two miles from the lake an exposure of dark greenish-colored granite was observed. Just below this point the river turns northward into a lake expansion three miles in length. The river flows out of the north of this lake and bends off to the north-east half a mile from the lake, becoming narrow and rapid, with two portages. The Indians have cleaned out a channel in this part, by removing the stones, so that we were able to let our canoes down with a line. For two miles below the river flows north-east, through a flat sandy country, and turns east for one mile. At the end of this distance there is a 110 chain portage due north, to avoid a bad part of the river. The first part of the portage is over a level sandy plain, while for the last 40 chains the country is rolling and more thickly wooded; no rock exposures occur. The portage is well taken care of by the Hudson Bay Company's Indian voyageurs. This portage ends at the south-east bay of Lake Minnisinaqua, which stretches off to the north for a distance of seven miles, then turns eastward for about three miles, and turning to the north again passes into the river. The shores of the lake are rocky, but no time was available for a thorough examination. One mile from the portage on the west shore the rock is a dark greenish-colored diorite. Three miles further north on the east shore is an exposure of dark green gneiss.

The north shore of the east arm is level and sandy, wooded with small spruce, poplar and jack pine; the south shore is rocky, with pink gneiss exposed at the head.

The rock at a point one mile below the lake on the west bank of the river is a coarse-grained, reddish-grey gneiss. In the next four miles frequent outcrops of granite and gneiss occur. Five miles below the lake is a rapid and sixty chain portage on the right, where the rock is fine-grained gneiss. The river for the next four miles is deep and from two to three chains in width, and flowing with a steady current. At the end of this distance it opens into Lake Mattagami, on which the Hudson Bay Company's post, of the same name, is situated. At the time of our visit, about the middle of September, there were excellent potatoes growing at the post, which we had the pleasure of sampling, owing to the kindness of Mr. Miller, the post-keeper. Mr. Miller's son rides a bicycle along the level sandy shores of the lake, and these bicycle tracks were the subject of much speculation by the members of our party, before we learned that this mark of modern civilization had found its way so far north into New Ontario.

LAKE MATTAGAMI.

Lake Mattagami stretches off for 12 miles south of the post, while the part to the north of the post is 18 miles in length and varies from one quarter to one mile in width. Pink granite and gneiss are exposed frequently along the shores with intervening stretches of undulating country, composed of sand and gravel with more or less vegetable loam on the surface. Thirteen miles north of the post on the east shore is an outcrop of very much weathered diabase, on which the glacial striæ run north and south.

North of this the west bank is level, with a sandy beach; the east bank is rocky but thickly wooded with spruce, poplar and jackpine, with some red pine.

The river flows out of the north end of the lake, over a fall and rapid, Kenogamissee falls, having a drop of 30 feet (aneroid); Kenogamissee portage is 45 chains long and leads to Kenogamissee Lake.

Kenogamissee Lake is about 22 miles long but is quite narrow in places, and has a direction almost due north. The country is flat and sandy to the west and rocky to the east, the rock being granite and gneiss. A trapper named Joe Moore has a house and clearing near the north end of the lake. We met a party of Indians in camp near the north end of the lake. There were about 30 in all, who had come from Moose Factory, and were on their way to hunt for the winter near Nighthawk Lake.

There is a fall and rapid at the north end of this lake which is passed by an 80 chain portage on the right (east). Toward the south end of this portage there is an outcrop of soft chlorite schist, which strikes northwest and southeast and dips at high angles. This same rock also outcrops near the north end of the portage. From the end of this portage the river runs north-east with a good current for a mile and a half to a point where the Grassy River joins it. It then turns north.

GRASSY RIVER.

A trip inland from the mouth of Grassy River, eastward, revealed a rolling sandy plateau, 50 to 75 feet above the level of the river, covered with moss and wooded with large jack pine and spruce with some birch and poplar. This character continues for 6 or seven miles up the river. The course of the Grassy River, near the mouth is from the north, but half a mile above it gradually swings through 180 degrees, after which the general course is from the south. The river winds its way in a very circuitous manner down the valley, which is flanked on each side by high banks about a quarter of a mile apart. These banks are composed of sand and gravel, which have been worn down to the depth of 50 to 75 feet, thus forming a valley of denudation. In this process an enormous quantity of sand must have been carried down the Moose River toward James Bay. The river has an average width of about 1 chain and runs with a very swift current, over a gravel bottom. It is nowhere very deep and in the wide parts quite shallow (2 or 3 feet). The current which in places runs at the rate of 5 or 6 miles an hour often rendered it necessary for us to wade and tow our loaded canoes.

About nine miles from the mouth the river becomes straighter, and one mile above, the first outcrop of rock occurs, a reddish-grey gneiss. Half a mile above this the river is very swift, which is at the foot of a fall. This fall is over reddish gneiss and is composed of a double chute and rapids above. The first fall is 20 feet, the second 25 feet. This is passed by a 34 chain portage on the right over fine sand and clay. The aneroid carried over the portage recorded a drop of 150 feet (probably too great).

Above this the river broadens out and is quite sluggish, the country is flat and the immediate shores quite marshy. About one and a quarter miles above the portage is an outcrop of crystalline diorite. There are low ridges of this rock on each bank of this part of the river, and also quite a grove of white pine. This same rock again outcrops half a mile to the south. One mile further dark greenish-grey granite is exposed. Half a mile above reddish-grey gneiss outcrops cause a short rapid, and similar rock was also observed at several points during the next mile and a half. Beyond this for half a mile the current is very slow and the river marshy and three or four chains in width, at which distance there is a fall and rapid called Nonwakawan, one quarter of a mile above which was observed a bluff of reddish gneiss on the west bank.

The fall is passed by a 30 chain portage on the east side. Ten chains east of the middle of this portage is a small lake about 30 chains in width, which drains into the Grassy River by a stream which crosses the portage at 13 chains from the south end. On the north shore of this lake fine-grained, dark greenish gneiss is exposed. Fresh beaver cutting was observed around this lake and stream.

For one mile above this portage the river is wide and forms a lake-like expansion. At this distance it flows through a narrows caused by high banks of reddish-grey gneiss. The next three and a half miles is a long narrow lake called Mazanokwanokwahagaming Lake, which is the Indian for "the lake where the green rock is, which you can write on." Gneiss similar to that above noted outcrops at several points on each side of this lake during the first mile and a half, after which it gives place on the east shore to a basic eruptive Huronian rock. This rock varies from fine-grained near the water's edge to a very coarse-grained variety a short distance inland (eastward), this latter resembling a syenite in appearance; a thin section, however, examined under the microscope, showed it to be quartz diorite. The fine-grained variety, which weathers a bright green, is exposed all along the east shore, on account of the bush having been burned two years ago, and it is from this rock that the Indian name of the lake has been derived. A quarter of a mile south of the contact between the granite and diorite which occurs on the east shore, the line referred to on the map as an "old line" crosses the lake. This line which was run a number of years ago is now almost obliterated.

We met two families of Indians camped on the lake—Sandy Green and Joe Jarbeau—who treated us in a most friendly manner, giving us much information about the country. Jarbeau who is an old man, said he had not seen white men there since Sinclair and his men ran the line in 1866. He stated that the lower part of the Grassy River is never used by the Indians, and this is easy to believe, as we did not see an axe mark on it for ten or twelve miles after leaving the mouth, although for the trip from Matachewan to James Bay, it would be found much easier, than the usual route, over a six-mile portage to Lake Mattagami, as the current runs very swiftly, and could be easily travelled down stream.

The rock on the west shore of the lake is reddish gneiss.

The jack trout or pike shown in the foreground of the photo, was caught in the lake and was 3½ feet long, and weighed 20 lbs. White fish are also numerous as well as pike, pickerel and black suckers.

The Grassy River enters Lake Mazanokwanokwahagaming from the west three quarters of a mile north of the south arm of the lake. One mile above the lake there is a rapid passed by a 12 chain portage on the left. Above this the river is very sluggish and marshy and gradually merges into Peter Long's Lake. Sandy Green's cabin is at the north end of this lake.

Rennebester or Mowat's Farm, Lower Montreal River.
Party No. 3. (See page 90.)

Lake Kabinakagami showing islands. Party No. 4.

Camp on 5th portage on Kabinakagami River, showing fish caught by Indians. Party No. 4.

Foot of rapid on Kabinakagami River. Party No. 4.

Rapids on Kabinakagami River. Party No. 4.

Lake Wanzatika looking east, 3 miles east of Missanabie River. Party No. 4.

Lake Wanzatika. Party No. 4.

Lake Wanzatika. Party No. 4.

Foot of rapid on Kabinakami River. Party No. 4.

At Long Lake. Party No. 5.

Long Lake House. Party No. 5.

On Long Lake in H. B. Co.'s sail boat. Party No. 5

PETER LONG'S LAKE.

This lake is six miles in length and at no place over one mile in width. It is really composed of several lake expansions, all having a general north and south direction. There are, however, frequent bends and narrows so that about three miles is the longest stretch of straight water.

The east shore of the lake at the north end is formed by a low ridge, wooded with pine and small spruce and poplar. One mile to the south of the north end or foot of the lake is an outcrop of greenish slate. Half a mile further south reddish gneiss is exposed on the east shore, in this rock flesh-colored orthoclase predominates. The west shore is quite flat and level for the first three miles. At this distance, however, at a narrow part of the lake, similar gneiss to that just described outcrops on both shores.

At this narrows the lake turns slightly to the east, then stretches off to the south for three and a half miles with a width of half to three-quarters of a mile. Gneiss similar to the above was noticed at several points along the west shore. Near the south end of this expansion a stream enters from the west, on which there is a fall one-quarter of a mile from its mouth. The rock around the mouth of the stream is greyish gneiss. On the east shore opposite this the gneiss is pinkish in color on account of the predominence of pink felspar.

One mile south of this point this expansion narrows to about three chains, with a short turn to the east, then a turn to the southeast for half a mile with quite a noticeable current. Above this it stretches off to the south again in another lake expansion, one mile from the foot of which is another narrows which we named Moose Narrows. From this point we made a trip inland to the east for one and a half miles and found the country to be rolling and wooded with spruce, birch, poplar and whitewood. The soil is black vegetable loam for two or three inches and underneath a fine, light-colored sand. A quarter of a mile from the shore we crossed an exposure of granite. A short distance south of Moose Narrows granite is exposed on the east shore, and a half a mile further is an outcrop of slate, containing specks of pyrites and lying in an almost horizontal posision. One mile above this another exposure of the same rock was observed, and one mile further south the Grassy River enters from the east, while Sinclair's Lake or Lake Shatagami stretches off to the south and west. This lake is about five miles from north to south and two and a half miles across. It is in the form of a "cul de sac" at this point, where Grassy River turns north. From its west shore a six mile portage leads westward to Lake Mattagami and this forms part of the route used by the Indians from Fort Mattagami to Fort Matachewan.

GRASSY RIVER.—On the south shore of the Grassy River, two chains east of Sinclair's Lake, there is an exposure of graphitic slate. A quarter of a mile up stream is a fifteen chain portage on the left around a rapid. This portage, which passes through a pine, spruce and poplar bush, terminates at the west end of Net Lake, a small expansion of the river. From this lake, which is about one mile in length a short narrows leads into Loon Wing Lake. At the west end of this Daddy Restoul has his cabin. He is an old French halfbreed, eighty-one years of age, and has lived here among the Indians for twenty-five years. He has quite a clearing around his cabin and a good patch of potatoes which he was digging on September 23rd, and up to that time there had been no frosts to injure them. Two families of Indians had their summer camp near this point.

Loon Wing Lake is about one mile long and and receives the inlet from the southeast A quarter of a mile above the lake there is a rapids passed by a five chain portage on the left, no rock is exposed and the country is sandy and well wooded.

For half a mile above there is a strong current, then a marshy lake expansion. For two and a half miles the river winds its way slowly through the marsh, then expands to to form another marshy lake called Lake Kopskajeeshing. From the outlet of this lake a high hill is visible to the east, called Mount Sansawaju, meaning square mountain. It is in the form of a rectangular prism with a plateau on the top. This lake is about one mile long. From its southern extremity another Lake Waaquakopskajeeshing (meaning the head of Lake Kopskajeeshing) stretches off for four miles to the south.

Grassy River enters this lake at the southeast and near the entrance on the east bank a rock exposure occurs in the midst of the marsh, consisting of greywacke conglomerate containing pebbles of varying size, up to 2 inches in diameter, the larger ones of granite

Ten chains above this there is an Indian cabin where an old squaw lives. The Indians of the neighborhood supply her with any food which she cannot procure for herself. Ten chains further brought us within half a mile of Mt. Sansawaju to the south. The river here is sluggish and very crooked, flowing through an extensive marsh. For the next six miles above there is no change in the river. At this distance, a basic eruptive rock is exposed on the north bank. This rock carries a considerable per cent. of magnetite so that the compass was strongly attracted when brought near it. This rock shades off to a coarse-grained diorite of the ordinary variety to the east, and it drops into a swamp in a distance of 3 or 4 chains. At a point 4 miles further east, i.e., about 14 miles above the last portage, the river receives a branch from the south-east.

The course of the main river above this branch is from the north-east. A trip 7 miles up it showed the river to be 1 to 2 chains in width and flowing with a steady current through an extensive marsh, similar to that to the west. This marsh is dotted with small outcrops of rock—chiefly greenish ash rocks—which are wooded with spruce, poplar and jack pine, which give them the appearance of islands in the midst of the surrounding marsh.

The canoe route which we followed was to the south-east up the branch referred to above. We ascended this for about 1 mile then turned east into a short stream which drains Little Hawk Lake.

LITTLE HAWK LAKE—This lake is about 3 miles long, with a general direction north-west and south-east. The east shore is formed by a high rocky bluff composed of crushed granite.

A 25-chain portage leads from the east bay of Little Hawk Lake to a small lake, one of the chain of three leading to Pigeon Lake, and called the three Little Hawks. This portage, which is quite level, is the height of land between St. Lawrence River and Hudson Bay waters.

The small lake is three-quarters of a mile long, it drains by a small stream southward into the second Little Hawk Lake. This stream is navigable for canoes for a distance of only 10 chains. It is then necessary to make a 110-chain portage, crossing a recent "windfall" very rough and difficult. At the north end greenish slate, dipping slightly eastward, is exposed. The second Little Hawk Lake is one-quarter of a mile across. From it a 120-chain portage leads to the third Little Hawk Lake, crossing a small muskeg near its north end. About the middle is an exposure of slate, striking north and south and dipping eastward. The third Little Hawk Lake is one-quarter of a mile across,* a 40-chain portage from the south shore of which, terminates at Pigeon Lake. This chain of lakes forms part of the head waters of the Main Montreal River.

On account of the Main Montreal River having been previously surveyed, our party travelled rapidly down stream and but little opportunity was given of observing the rock exposures.

Pigeon Lake is a long narrow sheet with several small islands near the north end. One mile south of the portage a bay runs off to the west. The lake is 6½ miles long. Near the foot it becomes quite narrow, and the current of the river becomes again apparent near the head of a fall of twenty feet, composed of a double chute. This is passed by a twenty chain portage on the left (east) called by the Indians Omeemeeomigamee (Pigeon Portage). At the north end of this portage there is an exposure of a distinctly bedded reddish rock, in beds eight to ten inches in thickness, dipping eastward at a slight angle; a specimen of this rock examined in thin section under the microscope showed it to be arkose. At the south end of the portage the rock is slate, dipping east at twenty degrees.

On this portage we met an Indian named Joe Chimigon, who lives on the Grassy River, and who was just returning from examining his four bear traps, which he had set twelve days before. He had caught two black bears and one fisher, and the fourth trap had been set off, but the bear had escaped, leaving its foot. He considered this a very good catch, and it seems probable that fur-bearing animals are still numerous in this section. Below this portage the river again expands into a narrow lake. Half a mile below the falls Duncan's Lake stretches off north-north-east for a distance of about fifteen miles. These two lakes come together in the form of a Y, the stem of the Y being two miles in

* Referred to by Mr. Burwash, in Bureau of Mines Report, 1896, p. 176.

length and a continuation of the lake expansion with a direction south-east. At a distance of two miles, the river flows out of the lake and turns sharply north-north-east, forming what might be fittingly called the Great Southern Bend. At this point similar rock formation to that noted on the last portage outcrops in sheer bluffs on each side of the river, causing a narrows with a short piece of swift water. The course of the river for the next thirty-five miles is north north-east, parallel with Duncan's Lake, the river being composed of a succession of lake expansions connected by the river proper.

At a distance of six miles below the bend an outcrop of slate occurs on the west bank. Seven miles further is a rapid with a short portage on the right, which, however, we were able to run with our canoes. One mile below is an exposure of pinkish gneiss. Three miles further on is a rapid where the river flows over an exposure of slate; the portage on the left is twelve chains in length. Three-quarters of a mile below is another portage fifteen chains long, on the right. A few chains further is a lift-out and three-quarters of a mile below this a five chain portage on the right. A paddle of ten miles brought us to the south point of the large island on Lake Mistinigon (meaning the lake with the large island). Here the rock is greenish slate with a strike north-east, and in a vertical position. This island is about two miles long. Three and a half miles below the foot of the island we ran a rapid, around which, however, there is a portage on the left. Two miles below this is Matachewan Falls and the Great Northern Bend, where the Montreal River plunges over a ridge of diorite, with a drop of forty feet, into a long narrow lake, called Lake Matachewan. This lake has a direction north-north-west, which is nearly at right angles to the direction of the river above.

The Hudson Bay Company's post—Fort Matachewan—in charge of Mr. Stephen Lafrican, is situated four miles below the falls, at the foot of the lake, on the east shore, at which point there is a short rapid. The post consists of a store, the factor's house, and a couple of other buildings.

At the time of our first visit, in July, there was an Indian village near the fort, in October, however, this village of canvas had vanished; the Indians having dispersed to their hunting grounds for the winter. The Hudson Bay Company's factors arrange and allot these hunting grounds for each family of Indians, so that everything seems peaceful and prosperous among those we met. In fact most of the Indian braves were better dressed than any of our party.

Mr. Lafrican showed me a sample of magnetic sand which had been found by an Indian near the post. This contained a considerable per cent. of finely divided magnetite. He also showed me a piece of red hematite associated with quartz which an Indian had discovered about six miles from the post. These I was unable to visit, the Indians who had discovered them being away. Mr. Lafrican also stated that the sand and clay near the post pans gold, with, however, only a few colors to the pan.

I examined a quartz vein on the south shore about one and one-half miles below the post. It outcrops in a steep rocky bank about fifty feet high, and is exposed down the side. The vein is five feet wide, almost vertical, and strikes east and west. The quartz, which is rusty, carries pyrites, with a little copper pyrites and galena. A sample on assay gave .02 oz. gold, .6 oz. silver, .14 per cent. copper, and a trace of lead. A sample of the country rock, examined in thin section under the microscope, proved to be porphyrite.

Half a mile below is a rapid which we were able to run; in low water it is passed by a 10 chain portage on the right. The river turns west at the rapids, and just below these it turns south again. A short distance south-east from this bend a small vein or stringer of quartz and spar with pyrite was observed, a sample of which on assay gave a trace of gold. The country rock for this vein was shown by microscopic examination to be red, quartzless porphry, which forms narrow dikes in the greenstone.

Four miles below this the east branch joins the main Montreal River. A description of the east branch is found in an earlier part of the report.

The river for eighteen miles below the junction flows with a strong current, with several pieces of swift water. At this distance occurs Kekopakagawan Falls, the river here falling over reddish granite with a drop of 30 feet (aneroid).

This is passed by a 10 chain portage on the right. In the above distance several outcrops of reddish granite were observed, but the country is comparatively free from rock, and is composed of coarse sand and gravel for the first 10 miles, then it changes to sand with clay sub-soil, the sand banks being from 30 to 40 feet high in places.

An Indian half-breed has quite a clearing half a mile above the falls and with his family he works his farm in the summer, and hunts and traps in the winter. We bought some excellent potatoes from him on October the first, which he was just digging at that time, no frost having occurred to injure them. We met this enterprising fellow in July, on the Montreal River below, bringing up his farm stock, which consisted of a horse and three head of cattle. He had a bark canoe lashed on either side of a small scow, on which was a platform enclosed by a railing for the cattle, while he, his squaw and family of small children paddled in the bow and stern of the canoes. He was only able to make a few miles a day, up stream, and we were glad to note that his perseverance had been rewarded with success.

Below the falls for 10 miles the river is crooked, flowing through a level wooded country, composed of light sandy loam and clay subsoil. At this distance the river broadens out into a narrow lake—Elk Lake—9 miles in length, having a south-east direction similar to that of the river. There are several small cabins and clearings toward the south end of this lake. and we met a settler, bringing up some lumber in a large canoe to finish a new cabin. This he had brought from the saw mill at Haileybury on Lake Temiscaming.

From the foot of Elk Lake, a short stretch leads into Mountain Lake, which is about three miles across. There is a succession of high wooded hills forming the south shore of this Lake. On the north bank of the river as it leaves this lake occurs an outcrop of very coarse-grained diabase, at which point also is an old Indian burying ground.

The river then bends to the north-east, and one mile below occurs a rapid and chute with a drop of 12 feet. This is passed by a ten chain portage on the left. Seven miles below a large stream joins the river from the south, and one mile further the river expands into Indian Lake which is about 1½ miles in length and half a mile in width. The surrounding country is level and the soil is sandy loam with clay subsoil. Ten miles below this is what is called Rennebester Settlement where a Mr. Mowat lives. He was for many years in charge of Fort Matachewan, and after leaving the employ of the Hudson Bay Company settled at this spot, where he has a comfortable cabin, a very substantial barn, several head of cattle and quite a large clearing.

One mile below this the Montreal River receives a large branch from the south which is the northern outlet of Lake Temagami, and drains also Lady Evelyn Lake and Lake Mattawapika. The route followed by the party was southward, through these lakes. The geology of this section of country has however been worked out by Mr. A. E. Barlow of the Geological Survey Department of Ottawa, and published in Vol. X, New Series, Part I., so that no mention of it need be made here. I shall however mention the occurrence of economic minerals, which I noticed in the region.

A short distance south of Mattawapika Falls Mr. Klock of Klock's Mills, has done some work on a quartz vein. Some good samples of copper ore were seen, but the vein, where cut by a short tunnel at the foot of the hill, seems to break up and disappear.

A quartz vein was seen outcropping at the north narrows of Lady Evelyn Lake, 2½ feet wide, in a country rock of quartzite. An assay of a sample of this quartz showed a trace of gold.

On Lake Temagami, 2 miles north-north-east of Bear Island is a large island known as Terry's Location. The country rock is greenstone and it is crossed by a quartz vein striking north 65 degrees east dipping at a high angle. A test pit has been sunk on the vein at about the centre of the island and the vein is 10 feet wide at this point and is mineralized with copper pyrites. A sample taken across the width of the vein gave on assay:— .02 oz. gold per ton,
.52 oz. silver " "
2.36 per cent copper " "

Emerald Lake is situated south west of Lake Obabika. The geology of this section is also included in Mr. Barlow's report and nothing will be said of it further than to give the result of an assay of mineralized quartz taken from Location W. D. 154 on the north east shore of Emerald Lake, which was as follows:—
.14 oz. gold ($2.80) per ton.
.32 oz. silver (.15c.) " "

In addition to the above I visited a mining location on the north-east shore of Lake Wahnapitae. which is being opened by Mr. McVethe of Sudbury. There are three parallel veins striking east and west and dipping southward. The south vein of the three is 12 to 18 inches wide and carries free gold easily visible to the naked eye and frequently small nuggets. A test pit has been sunk to a depth of 10 feet on this vein, which shows the vein to be widening. The second vein a few yards to the north is several feet wide, no work however had been done on this at the time of my visit. The third or north vein is a wide deposit of pyrites. A surface sample of this gave an assay of 32.42 per cent sulphur with traces of gold, silver, nickel, cobalt and arsenic.

CONCLUSION.

The rocks encountered during the summer belong to the Archean Age, and include those of both the Laurentian and Huronian formations. The trip from Metagama to Niven's Base Line, was through Laurentian rocks with only a few exceptions and these continued eastward to Peter Long's Lake. Around this lake, however, this formation gives place to the Huronian which extends eastward and southward throughout the Sturgeon and Montreal River basins.

Economic mineral was found at nine different places during the summer, indicating the presence of gold, silver, copper, lead and iron, distributed through the region examined, as far north as Fort Matachewan. The quartz veins with one unimportant exception, were found in a country rock of greenstone, which rock seems therefore to offer the best reward to the efforts of the prospector.

Many valuable water powers were noticed, among which might be specially mentioned, Upper and Lower Goose Falls, Kettle Falls, and Twin Falls, on the Sturgeon River; Matachewan Falls, Kekopwakagewan Falls and Mattawapika Falls, on the Montreal River; Kenogamissee Falls on the Mattagami River; and Nonwakawan Falls on the Grassy River. The number of feet of fall in each of the above cases may be found in the body of the report.

Finally I want to express my sincere thanks to Professor A. P. Coleman of the University of Toronto, for assistance and advice in the preparation of this report, and also to yourself and the following other gentlemen for kindness and information received during the summer :— Mr J.F. Black of Sudbury; Mr. Stephen Lafrican, officer in charge of Fort Matachewan; and Mr. G.E Silvester, O.L.S. of Sudbury.

J. L. ROWLETT PARSONS, B. A.
Geologist with Exploration
Survey Party Number 3

I, J. L. Rowlett Parsons of the City of Toronto and County of York, solemnly declare that the above report dated Dec. 31st, 1900, and the facts and statements therein contained, made by me, are true and correct.

Declared before me this 11th day of January, 1901, in the City of Toronto and County of York. } J. L. ROWLETT PARSONS.

GEO. B. KIRKPATRICK,
A Commissioner, &c.

SURVEYOR'S REPORT OF EXPLORATION SURVEY PARTY NO. 4.

LEAMINGTON, ONT., January 5th, 1901.

SIR,—I have the honor to submit the following report on the track survey along the Kabinakagami, Matta-wish-guaia and Missinaibi Rivers in New Ontario, and the exploration of the section of the country on each side of the said streams, which was performed during the past season by Exploration Survey Party No. 4, assigned to my charge, under instructions from your Department dated 12th June, 1900.

I left Leamington on the 18th of June and proceeded to Toronto, where I met Mr. Stewart, the geologist, and Messrs. Laidlaw and Miles, three other members of the party.

On the 20th of June we left for Grasett Station on the Canadian Pacific Railway, from near which point, or Loch Alsh, near Missinaibi station, I was instructed to go into the work. The party was joined at Huntsville by Mr. McConachie, the land and timber estimator.

At Missinaibi I engaged an Indian guide and also purchased a birch-bark canoe for the use of the party, and was informed that the best and most direct route to get into the work, was by way of the Magpie, Shell or Esnogami River

I decided to take the route by way of the Magpie River, and after some delay caused by the non-arrival of our supplies we proceeded to a point on Magpie River where it is crossed by the railway, about five miles east of Grasett Station, pitching our tents there for our first camp.

Our supplies arriving the following day, and having obtained the services of an Indian to guide us to the portage over the height of land, from Lake Esnogami to Lake Kabinakagami, we started to move the supplies on the 25th of June up the river, and made a track survey along our course, which I was instructed to do in order to connect the starting point proper of our survey "the head of the River Kabinakagami," with some known point on the railway.

Starting the survey at the Canadian Pacific Railway on the northwest side of the Magpie River, at a distance of sixteen chains and fifty links northwesterly from mile board seven hundred and five of the railway, I proceeded up this river and through its expansion, Magpie, Shell or Esnogami Lakes, to and across the portage, over the height of land into Lake Kabinakagami, and through this last named lake to the head of the river of the same name, which really was the commencement proper of the survey and explorations. Having made a careful micrometric track survey to this point, in order, as before stated, to connect it with the railway, I proceeded to make an accurate survey of the River Kabinakagami, and have the tract of country on each side of the same, for a distance of twenty miles on each side of it, explored. Surveying past the rapids at the head of the river by way of the portage leading past the same, I continued the survey along the river and across the portage leading past the rapids along it, to the fifteenth rapids, which are about forty-four miles from the head of the river.

The river here becomes very rapid and shallow, and my Indian guide, who claims to know it well, informed me that from these rapids north to its junction with the Kenogami river at a place called Mammattawa, this river could not be travelled with a canoe except in very high water in the early part of the season and then only in short stretches, it being one continuous shallow, and rapids with no portages cut out, from there to near Mammattawa. I examined the river below these rapids for about a mile and found it in this distance and as far as could be seen further down much as represented by our guide with no sign of any portage past the rapids. The guide and other Indians of the party refusing to attempt to go further down river, in such low rapid water, I was unable to proceed further with the survey of this river. I concluded to cross over to the Matta-wish-guai-a River, which I ascertained flowed into the Missinaibi River, but did not flow out from or join the Kabinakagami River as was supposed. A portage trail about a mile in length leads from Kabinakagami River about a quarter of a mile above the fifteenth rapids on it, east to the Mattawa-wish-guai-a River.

Planting a cedar post with the date on which it was planted, "the 18th of August, 1900," and my name marked thereon at the end of the trail where it leaves the Kabinakagami River, so that should it be necessary to continue the survey of this river at any future time further north, this point could be readily found, and having connected this post with my survey of the river, I ran a line due east from it, to the Matta-wish-guaia River, which followed closely along the portage trail, and along this line I measured the distance between the two rivers. At the end of this line where it strikes the last named river, I planted another cedar post, with the date on which it was planted, "the 20th of August, 1900," and also my name marked thereon, and from this post I made a micrometric survey of the Matta-wish-guaia River to its junction with the Missinaibi River, and on each side of it the country was explored similarly to that along the Kabinakagami River.

The water in the Matta-wish-guaia River was very low at the time we reached it; and it was with great difficulty we succeeded in getting our canoes and supplies through it.

From the junction of the Matta-wish-guaia River with the Missinaibi River, I continued the survey on down the last named river six and one-quarter miles, exploring the country on each side of it to this point, with the intention of continuing the survey through along this river to Moose Factory as instructed. The water still continued very low as in the Matta wish-guaia River, and the canoe bottoms having become worn very thin they could not be prevented from letting the water through. In consequence our provisions and baggage were wringing wet every night, and as the water was getting cold our Indians became discontented with the work of wading and portaging, in the rocky beds of the rivers, and refused to proceed further with the work. The other members of the party were not willing to attempt the prosecution of the work further without a guide and Indians to assist us through the rapids. None being procurable short of Brunswick Post, about one hundred and fifty miles distant, or at Missinaibi Station, a distance of about two hundred miles, and finding I could not induce or persuade our Indians to continue the work further it was decided to return to Missinaibi Station.

Planting a cedar post on the east side of the river with the date on which it was planted, "the 17th of September, 1900," and my name marked thereon, at the point where the work was left in order that it might be readily continued at any time, we on this date commenced the homeward trip.

The return journey was made up the Missinaibi River to Missinaibi Lake and through this lake, Crooked or Mosogonosh Lake and Dog Lake to Missinaibi Station, which was reached by the entire party without accident on the evening of the 8th of October, 1900.

On the way up the Missinaibi River, the line run by Mr. Niven from the east was passed at two and a half days' journey up from the point where our work ceased.

General Description.

Magpie River.—The Magpie River flows southward to Michipicoton River and thence into Lake Superior and is sometimes called Shell or Esnogami River. The most direct and shortest route to the Kabinakagami River head, from the Canadian Pacific Railway, is up this river and through the lake of the same name from which it flows, and through Lake Kabinakagami. At the Canadian Pacific Railway this river is three chains wide, but varies greatly in width, as it is ascended, until it widens out into the lake about three miles from the railway. South of the railway it is very rapid and a strong current is perceivable for a short distance up from there, and but little further up, until where it emerges from the lake, where it becomes very narrow with a stronger current. It varies greatly in depth, being from three or four to ten or twelve feet deep, without any rapids to the north of the railway, and for this portion, at least, easily navigable with canoes. Its shores consist of rocky bluffs of grey Laurentian gneiss, in places reaching fifty or more feet in height, the shores in the deep indentations being low and marshy. At the Canadian Pacific Railway a large bank of coarse gravel eighty feet high, adjoins this river.

Lake Esnogami.—Lake Esnogami of which the Magpie or Esnogami River is the outlet, also varies greatly in width and consists of a number of large indentations or bays, towards the head of which the shores are low and swampy, but otherwise the shores of this lake are a series of rocky bluffs. It has no apparent current and attains a considerable depth throughout, from sixteen to twenty feet or more. This lake is easily reached from Grasett Station on the Canadian Pacific Railway, by a portage trail leading to it from Mud Lake, which approaches the railway about a mile east of this station.

About half way up its length the rocky shores attain the greatest height, and approach close together narrowing up the lake for some distance into a narrow rock-bound passage, but the shores again diverging, the lake widens toward its head, and the shores there become very low and swampy. The rocky shores are principally the same as along the river, "grey Laurentian gneiss."

TIMBER.—The region of country along the river and around this lake has all been burnt over, and the timber which is principally spruce, tamarac, cedar and birch, is blown down over extensive areas, but in many of the swampy sections, which have escaped the fires, some good spruce and tamarac were found, suitable for pulpwood and railway ties, and where the burning has not reached, recent fine growths of poplars are to be found.

On the east shore of the lake about three and one-half miles from the railway, an Indian burying ground is located, with many of the head boards decorated with ornaments and trophies, and in close proximity to this burying ground, was found an Indian hut, with a small patch of potatoes and raspberries adjacent in a fine healthy condition.

HEIGHT OF LAND PORTAGE.—From the head of Lake Esnogami at its north west angle the height-of-land is crossed by a portage trail leading to a small marshy river named Papakapesage by the Indians, which flows into Lake Kabinakagami. This portage is three hundred and ninety-six chains in length, and for the first two and a quarter miles from Lake Esnogami the trail passes through a dense tamarac and spruce swamp, very wet, with a small swift stream running through it.

At the end of this distance is a ridge of high land eighty feet high and ten chains wide, again descending into the neck of this swamp, nine chains wide, over which is a tract of high, dry land, nearly one mile wide, north of which the trail again passes into a low, marshy tract of country, through which pass two small streams. Towards the north end of the trail there is another small tract of dry land, but at the north end the trail leads through a dense, marshy thicket of alders. On the higher land some fine, large valuable timber, such as poplar, birch, cedar, tamarac, spruce, white-wood, and scattering pine are to be seen along this trail. In the swamps the timber is dense, but small, a large quantity of it being suitable for railway ties and pulpwood. The high land along this trail would apparently make excellent farming land when cleared up, as also the swampy land, which can be readily drained by the creeks and streams running through it, the soil being a sandy loam, covered in the swamps by a considerable depth of black muck and mossy peat.

This portage was found very difficult to get over, and in order to expedite the work, I found it necessary to engage the services of some Indians to assist us to portage the supplies over the same.

LAKE KABINAKAGAMI.—Lake Kabinakagami is entered from the River Papakapesage, which is a small sluggish stream, about two rods wide where the trail over the height of land joins it, but expanding as it reaches the lake, which it joins a mile and a half further down. The southern end of Lake Kabinakagami is shallow with a sandy bottom. It is diversified with numerous rocky islands, varies greatly in width, and contains numerous indentations or bays being very similar in this respect to Lake Esnogami. Its shores are for the greater part rock-bound, with the same grey Laurentian gneiss as Lake Esnogami, with occasional outcrops of high bluffs of yellow sand on a layer of greyish-blue clay, but the rocky shores do not reach to the same height as on this other lake, and at the head of most of the bays, the land is low and swampy.

TIMBER AROUND THE LAKE.—The timber around this lake is principally standing, and not nearly so much burnt as around Lake Esnogami. In the swamps at the head of the bays and upon the numerous islands which have escaped the fires there are large quantities of small sized spruce, tamarac and cedar suitable for ties and pulp, mixed with some tall poplar and birch. On the northerly side of this lake where it turns to the westward a little more than half way up from its head, at the end of a deep bay on a high sandy hillside, is a beautiful grove of red pine, clear of underbrush, which Mr. McConachie, the timber estimator, estimated to contain about a million feet of timber. From this bay along the hillside a portage trail about a quarter of a mile long leads to a deep indentation or bay of Oba Lake, an expansion of the Oba River, which is used by the Indians when travelling up this river to and from the head of the Kabinakagami River.

From this bay of Oba Lake an exploration trip was made along the Oba River by Messrs. McConachie and Stewart, who report finding some ridges of good red pine, though the country was generally burnt over. Through this bay a canoe trip was also made into Oba Lake, and from this lake down the river to its outlet into Kabinakagami Lake, on which three rapids were passed with a fall of ten feet in so many chains each, two of which required to be portaged around. This river, as far as gone over, is deep and sluggish, except on approaching and leaving the rapids. Its banks are generally low and swampy with small spruce, tamarac, cedar, balsam, poplar and birch, mostly burnt over adjacent to the river.

This river enters Lake Kabinakagami at the end of a deep bay where the lake turns to the westward a little more than half way down the lake from its head. The soil along this river is a rich sandy loam, well adapted for farming purposes. On the bank of this river near the Oba Lake, a small log house has recently been erected, and a small clearing made adjacent to it. About half way down Lake Kabinakagami from its head, a long peninsula projects from the east, and across a neck of this a portage is made when going from one end of the lake to the other, greatly shortening the distance, and avoiding rounding the point.

The lake to the north of this peninsula being much deeper, is consequently not so much affected by wind as the more southern portion, and easier of navigation in stormy weather.

Kabinakagami River and District.—The Kabinakagami River is the outlet for and flows from the northerly end of Lake Kabinakagami. At the head of the river the waters rush out of the lake through a rocky pass, forming rapids which extend down the river, a distance of about half a mile, with a fall in this distance of thirty-five feet. It is passed by a portage 42.50 chains long, going in from the lake. Along this portage the land is high and dry, with a sandy clay loam, the timber being all burnt and mostly blown down. Some fine raspberries are found along the trail.

An exploration trip was made to the westward from the northerly end of this portage. The country is reported as having been burnt over, with some high land alternating with lower land which approached muskeg in character. The burnt timber is principally red pine, tamarac, spruce, and cedar. In the low lying land the timber is not burnt, and consists of a dense growth of black and white spruce and tamarac of small diameter, and of healthy and vigorous growth. The soil, as almost everywhere in the region passed through, is a light-colored sandy clay loam. Very little rock was found away from the outcroppings along the river, and where found was low-lying and principally of the Laurentian and Huronian series, with which the geologist of the party, Mr. Stewart, deals fully in his report.

The Kabinakagami River, which flows north-easterly, is a fine large stream, for the distance traversed by the party at least, averaging from three to five chains wide, and is easily navigable with canoes to the point where the survey of it was ended, "in the neighborhood of forty-four miles from its head." It will average in depth from ten to twelve feet in the distance gone over, and has in this distance fourteen rapids, which have a fall of from six to forty-five feet, all of which, with one exception, have to be portaged past. The portages, of from one-half mile to about a mile in length, are generally good. The rapids are all located on the plan, and the length and estimated height of each, as also the length of each portage, are shown in the field notes which accompany this report. Another heavy rapid occurs where the survey on this river ended, and at each of these rapids are excellent sites for milling purposes, the rapids forming valuable natural water-powers. Only for a short distance above and below the rapids is this river a rapid running stream, the intervening stretches being rather sluggish.

Along the banks of this river, which are high throughout, is a belt of high and dry land varying in width, from a quarter to two or more miles wide, and occasionally, where two streams come together, making an angle between them. This land is also of the same nature as that along the river, "high and dry," and this may be said of all the land along any of the streams met with. From this high belt of land along the river, as the interior of the country is reached on either side, the land becomes slightly lower and is generally swampy, rising gradually again where a stream is approached.

The soil along the river is of a sandy clay and gravelly nature, and everywhere in the country a light-colored sandy clay loam. In the low-lying lands in the interior of the country this is covered over with a depth of from two to four feet of muck and moss. The soil seems to be well-adapted for farming purposes, as all kinds of wild berries, currants, cherries and grasses grow in profusion throughout the country, where there are patches clear of timber. The low lying parts have a heavy fall to the north-east, as noticeable by the numerous small streams running therefrom, and when the streams become cleared of the numerous jams of fallen timber, etc., in them, those sections of the country can be readily drained and converted into excellent farming lands.

TIMBER ON THE RIVER KABINAKAGAMI.—The timber of the high land along this river is practically the same throughout, and consists of fair-sized tamarac, spruce, cedar, balsam, poplar and birch. Along the river edge where the shores are low, are dense clusters of alders, willows and dogwood,and on some high sand banks,a number of which are seen along the river varying from fifty to ninety feet high, are dense and healthy growths of jack or pitch pine, with scattering small red or white pine. In the interior of the country back from the high land along the river where the land becomes lower, the balsam, cedar and birch are not seen, and the poplar only in small clusters where a slight rise in the land takes place. The predominant and only timber of the lower land of the interior is spruce and tamarac, of less diameter than that on the higher land, but of dense growth and unlimited quantity, getting smaller as we recede from the river and smaller streams. Immense quantities of spruce, tamarac and cedar of the high and low lands of this section of the country are suitable for pulp, railway ties and building material. From a trip made into the interior, southeast from the river, at the foot of the third rapids, five and a half miles from the head of the river, it was found that the country had been burnt over for a considerable area, and on this trip was found a belt of red pine, surrounding a small lake about eight miles in, estimated to comprise half a million feet.

The portage trail passing the fifth rapids on the river is reached by ascending a small stream a short distance, which empties into the river above the rapids from the west. This portage is nearly a mile in length (seventy-three chains) and it passes over a fine high level tract of land, through a beautiful grove of red pine, with about a mile square of jack pine, which no doubt the timber estimator mentions and estimates in his report. A number of small streams enter the river from both sides, but none of these could be utilized for exploring the country, as near their mouths or a short distance up, they were found to be either blocked up with timber jams, or so shallow they could not be navigated, and all explorations made into the interior from the river had to be made on foot. Three in all of these trips were made from this river, beside the two already mentioned; one was made to the west of the eighth rapids, one to the east at the twelfth rapids and one at the head of the fifteenth rapids, to the west where our work on this river ended. No burnt country was found to the north of the trip made from the third rapids, and generally the timber north of this was in a fine, healthy, growing condition. On the trip made eastward at the twelfth rapids, the upper end of the Matta-wish-guaia River was discovered and crossed near its source "a small lake," and established the fact that it did not flow out of, and from the Kabinakagami River as had been supposed. It was ascertained that the Kabinakagami River was not a branch of the Missinaibi River as was surmised, but that it flowed north and was a branch of the Kenogami River.

MATTA-WISH-GUAIA RIVER.—The Matta-wish-guaia River flows to the north-east, taking its rise in a small lake some miles to the south-east of where it is joined by the portage trail that leads from the Kabinakagami River, near the fifteenth rapids. It flows in a north-easterly direction and joins the Missinaibi River just to the west of Lower Skunk Island, about four miles further up this river than is shown on the maps up to the present.

This river is a narrow, shallow and rapid stream, and at the time our party reached it, the water was very low, making it very difficult to navigate. Its bed is covered with boulders and ocks, and its banks are generally muddy. In high water the Indians say this river is good for canoes, and can be run from the portage to its mouth in a day or two, while it took our party from the 19th of August to the evening of the 4th of September to make the trip. It was, at the time we went down it, full of small rapids and shallows over which for days we had to wade, portage and drag, the canoes and our supplies. The banks show that the water in the early part of the season is good in this river, being about five feet higher than we found it. A foot in depth more water would have saved our party many hard days' work, and time in getting through. The banks of this river are much lower than those of the Kabinakagami River and of a light-colored clay, much similar to the other river but less sandy. Many little streams flow into this river from both sides, draining the low, flat country inland.

Along the banks of this river there are some good spruce, cedar, tamarac and large quantities of small poplar, whitewood, balm of Gilead and some balsam. Where the rock exposures occur they are all Laurentian gneiss. For some distance down this river the land is high and dry, and from a trip made northward from the river eighteen and a half miles down from the portage, it was reported that eight miles of this high land was passed through and an inland lake was found which was surrounded by large balsam, poplar, cedar, spruce and tamarac, this timber being some of the best seen along the rivers. From where this trip was made as we decended the river, the country inland becomes more swampy until the high land narrows down to a quarter of a mile wide. The swampy land continues along the river in close proximity to it, to near the lake of the same name as the river, about half way down from the portage, where the high land widens out again.

Good high dry land with large timber was also found on the east side of this river, from a trip about five miles down from the portage, extending fourteen miles inland from the river, where the land became low and swampy. About half way down this river to the Missinaibi River on the southerly side of it, are three lakes, one of which approaches the river at one point to within five chains, and two of these lakes are connected by small creeks with the river, and discharge their waters into it. Along these creeks and in the vicinity along the river were seen some swamp elm and small swamp ash, the only trees of the kind met with in the country. The upper one of these lakes could not be entered or navigated, the water being too low, as also the creek leading from it. The middle one of these lakes was crossed with difficulty in a canoe, and the land on the far side of it from the river, as also on the upper lake was found to be high and dry, extending well inland and well timbered. Around these lakes is a great hunting ground for the Indians, and a portage leads from one lake to the other, and from each to the river, and our guide informed me on our way out that a portage trail also led from the lower one of these lakes to the Missinaibi River. The shores of the lake next to the river are low and swampy. A good stretch of navigable river extends for some distance above these lakes to a short distance below them, which at the time we traversed it, may be said to have been the only navigable part of this river.

Above the junction of this river with the Missinaibi River twelve and three-quarters miles, is a big rapid or falls with a descent of thirty-five feet that requires to be passed by portaging. This is the only portage required upon this river when the water is high, and is valuable as a water-power for milling or other purposes.

From the lakes above referred to, going down the river, the belts of high land along its banks widen out inland from the river, and as the Missinaibi River is approached, fine large areas of high, dry, rolling land are found, timbered with a fine growth of poplar, spruce, birch, tamarac, balsam and whitewood. The soil is of the same nature throughout the entire section of the country—light-colored, sandy clay. Wild cherries, cranberries and red and black currants grow profusely along the banks of this river, and the wild grasses are very rank in the open spaces, the wild pea also being frequently found. The land through the section of the country which this river drains inclines to the north east and the low-lying lands can be readily drained by the clearing out of the numerous small streams which lead out to the river, and converted into excellent farming land, as the higher lands along the river now are.

THE MISSINAIBI RIVER.—The Missinaibi River is a magnificent, rapid stream with many rapids and portages, and is the natural highway to and from Moose Factory on James Bay. It is divided into two channels by the Skunk Islands, and our party entered the west channel from the Matta-wish-guaia River at the lower island. The east channel is the one usually travelled in going up and down the river. From the trips made inland from this river a few miles below Lower Skunk Island, it was found that the high land which forms its banks extended inland only about a quarter of a mile, when the land became low muskeg, the land rising again where a stream or lake was approached. It was ascertained from our guide that the country north from there to James Bay was much the same throughout.

Two miles below Lower Skunk Island, where a trip was made into the interior to the west of the river, a rapid running stream fifty feet wide was crossed nineteen and a half miles inland. This stream is called by the Indians Pewabiska River, and is said to rise in a small lake and run north-easterly, discharging into the Missinaibi River some miles below where our survey ended. The land rises as this river is approached and the elevation extends for about two miles after crossing the stream, when it again descends into a low flat section with scrubby spruce, tamarac and poplar. On the east side of the river three miles below Lower Skunk Island, is a portage trail, which our guide informed us had been cut out by him some years ago. This trail is about three miles long, and ends at a lake called by the Indians Wawzatika, the shores of which are rock bound. On the west shore the same Laurentian gneiss appears as along the rivers. The guide informed us a slate out-crop was to be seen along a portion of the east shore of this lake and along other small lakes and streams leading into the Opazatika River. Not having a canoe at hand we were unable to explore the east shore of the lake. The portage trail along which this trip was made to the lake, passes through a low muskeg, except for about a quarter of a mile near the river, and also the same near the lake. Two small streams cross the trail and the land rises slightly in close proximity to these.

This muskeg is covered with small, scrubby, scattering spruce and tamarac, and where the land rises slightly it is covered with a dense growth of poplar with scattering spruce and tamarac of fair size. Another portage trail leads to the south-west corner of Lake Wawzatika from the Missinaibi River, three-quarters of a mile below Lower Skunk Island.

Tho soil of the section of the country traversed along the Missinaibi River is much similar to that already described along the other rivers, with the exception that it contains more clay and less sand.

From where the survey proper started "the head of the River Kabinakagami" to the point where it was ended "on the River Missinaibi" the land taken as a whole throughout the territory traversed, and as far as could be seen, may be termed level, with a heavy inclination or fall to the north-east, rising slightly along the rivers, streams and lakes, and there being undulating and rolling. The soil is of the light-colored sandy clay mixture frequently mentioned herein, covered in the lower land with a depth of from two to four feet of boggy peat and moss. Rock exposures occur but seldom, except in the vicinity of the rapids on the rivers, and the shores of some few lakes, where they are generally low-lying and covered in most places with a sufficient amount of soil to admit of being farmed over. The timber of the section of country explored by my party may be said to be one dense forest of spruce and tamarac. The other timbers consisting of poplar, cedar, birch, balsam, white-wood and balm of Gilead are found in narrow belts along the rivers, streams and lakes, and in isolated patches in the interior, and in these belts and patches are to be found some of the largest and finest spruce and tamarac.

The majority of the spruce and tamarac, which cover immense areas, is tall, but small in diameter, but sufficiently large to be utilized for pulp-wood and railway ties, and of great value, if saved from destruction by fires, so soon as a means of transportation is given to carry it to the points were it can be utilized and manufactured.

This section of the country may be said to be north of the pine limit, no pine of large size being found. The only areas worthy of note are mentioned in a former portion of this report. The land can be easily cleared, and when cleared will make excellent agricultural land, much of it in fact equal to some of the best farming sections of the older and now settled portions of the Province. Much of the country that now looks swampy, when cleared of the timber, will, with the natural fall of the land, and many small streams, secure perfect natural drainage. With the mixing of the sub-soil of sandy clay, which it has, with the deep muck now covering it, it will prove to be rich and very productive. The country is well watered with numerous springs.

On our way home to the shore of Lake Missinaibi at Brunswick Post, one of the stores of the Hudson Bay Company, we saw some excellent crops of potatoes and other roots. Here good crops of hay, oats and barley had been raised from soil similar to that in the district explored by our party. At this post we saw also a number of fine fat cattle pasturing on the nutritious natural grasses which grow where a clearing is made, showing that for stock-raising this country is also adapted.

A number of views were taken with the kodak of sections of the country passed through, notably views of the height-of-land and timber, rapids and falls on the rivers, and catch of fish taken near one of our camping grounds; the films with the views taken, and a list of these views you will find accompanying this report.

A collection of some of the plants of this region was also made and pressed between folders, which have been sent in to your Department, a list of these stating where they were collected also accompanying this report. Plants of many varieties grow in profusion, and are found in many places throughout this region.

The fur bearing animals of this section of the country are the black-bear, moose, caribou, red deer, red, white, and coon fox, fisher, otter, beaver, skunk, mink and muskrat, but with the exception of large quantities of muskrats, but few of the others were seen, one black bear, three otter, and some mink being all that the party saw. Fresh tracks and paths of bears, moose, caribou, and red deer were often seen inland, and around the streams and lakes, and moose were frequently heard at night. Fresh work of the beaver was but seldom seen, but bear seem to be very numerous. This section of the country is a great hunting ground of the Indians, from the vicinity of Moose Factory, Missanabie and Brunswick Post.

The rivers and lakes of this section of the country are great breeding grounds of wild duck, which are very numerous, and many were seen. Wild geese were often seen towards the end of the survey, flying southward, being very numerous at Moose Factory, as our Indians informed me. Very few partridges were seen and very few birds. The Canada bird and chic-a-dee were the only birds noticed and heard. In the Magpie River and Lakes Esnogami and Kabinakagami, pike and pickerel abound, and a few perch were also caught in these waters. At the head of the River Kabinakagami where the waters start over the rapids, our Indians made a haul of twenty-five fish, pike and pickerel, in about half an hour, another fine haul of fish consisting of white fish, speckled trout, herring, and suckers, was obtained at the mouth of a small stream which leads up to the fifth portage on the Kabinakagami River, but at no other point in that river or the Mattawish-guaia was fish found to be very plentiful. Some attention was given by Mr. Stewart, the geologist of the party, to a collection of the insects of the country, which are enumerated in his report.

Astronomical observations were taken frequently during the progress of the work, the details of which accompany the field notes in connection with this report.

The Kay taffrail log did not work satisfactorily. The first one I received was lost in Lake Esnogami, while a test of it was being made. The indicator of this recorded backwards, and the indicator of the second one I received failed to record and could not be made to work, and therefore its use had to be abandoned. Readings of the aneroid were made daily, a record of which is submitted herewith, also my diary kept during the progress of the work.

The weather throughout, during the time occupied in the field, was remarkably fine.

In conclusion, I am sorry to say that I was unable to perform the entire work assigned to me and my party, but for the reasons hereinbefore stated it was impossible for us to do more than was done, and it was too late in the season to risk returning to finish the work after a new outfit could be procured. Should it be necessary to have the work completed next season, or at any other time, I would recommend that the party performing it be upon the ground not later than the first of June, in order to be able to take advantage of the high waters in the rivers and thereby save a great amount of time and labor.

Submitted herewith are the field notes and plan of the survey, and upon the latter are shown a number of small lakes, streams and portages not met with in the survey, but information of which was obtained from our guide, who had a great knowledge of this section of the country, he having hunted over a great portion of it.

I have the honor to be, Sir,
Your obedient servant,

A. BAIRD,
Ontario Land Surveyor.

To the Honorable E. J. DAVIS,
Commissioner of Crown Lands,
Toronto, Ontario.

I, Alexander Baird, of the Town of Leamington, Ontario Land Surveyor, make oath and say that the above report is correct and true in all its particulars.

Sworn before me at the Town of Leamington this 7th day of January, A.D. 1901. } ALEX. BAIRD.

JOHN MCR. SELKIRK,
J. P. in and for the County of Essex.

LAND AND TIMBER ESTIMATOR'S REPORT OF EXPLORATION PARTY NO. 4.

HUNTSVILLE, Ontario, December 29th, 1900.

Mr. ALEXANDER BAIRD, O. L. S.,
Surveyor in Charge of Exploration Survey Party No. 4,
Leamington, Ont.

Sir :—

I have been instructed by the Honorable the Commissioner of Crown Lands to make out my report as timber and land estimator for your party No. 4, which you will find as follows :—

June 25th. Went up Mud Pine River accompanied by Mr. Laidlaw and an Indian guide, thence into Esnogami Lake, which extends up to the height of land, a distance of twenty-five or thirty miles from the Canadian Pacific Railway up the Mud Pine River. The land is very low and the timber is all burnt, except in the very low swamps, where it was too low for the fire to run. The timber in those places is principally spruce, mixed with tamarac and cedar of a medium size. We went up Esnogami Lake to the height of land. The shores of this lake are very rocky with some swamps. The swamps are very well timbered with spruce and tamarac, but the high grounds are principally all burnt. The timber in these swamps, which extend for miles back from the shores of the lake, are spruce, tamarac and cedar. The head of the lake is at the height of land. The distance over the height of land is five and a half miles. The first three miles are very low and swampy and thickly covered with spruce and tamarac. Then we come to a high ridge of land about two hundred feet high and two miles wide. On this high land the timber is of a larger size and more of a variety, viz : whitewood, cedar, spruce and tamarac, poplar, white pine, and the remainder of land is good clay loam. The spruce timber on the height of land is larger than it is in the swamps, and would average about ten inches and cut about thirty cords per acre, and mixed with whitewood and poplar which would cut, net, about five thousand feet, board measure, per acre and also with a few tamarac which would make good piles, averaging from fourteen inches up to two feet and sixty feet long on the flat lands. On the height of land there is some fine cedar which would made good shingles. The white pine is the scarcest timber, but what I did see was of a good quality and good fair length and size.

July 24th. We left camp on Lake Kabinakagami to prospect the county lying between the lake and Oba River, a distance of about ten miles. The first half mile we came to a high ridge of sandy land covered thickly with red pine, about one million feet, the average being about sixteen inches in diameter and fifty feet long. There was no other timber growing on this land of any account. We then went in a south-easterly direction about nine miles till we struck the Oba River passing through a large tract of forest. All the high lands are mostly burnt and the low lands are heavily timbered, principally with spruce, which is not burnt. The spruce timber is not very large, but very thick and will cut about fifty cords per acre. The average size would be about seven inches on the stump. We came to a high ridge of sandy loam where there were found about half a million of nice red pine of an average size of about twelve inches and about sixty feet in length. We came to a mountain of rock about eighty feet high. I climbed a tree and saw to the north east of us some large groves of red pine about three miles distant. No man can estimate the timber in this country as regards spruce, it is unlimited. The only thing the timber estimator can do is to estimate about how many cords it will cut to the acre. I came across some nice groves of balm of Gilead, poplar and birch, not in very large quantities, but of a nice size, averaging about fourteen inches. I noticed that where the fire had been a few years ago, a nice second growth of timber is springing up, and if the fire is kept out of it for a few years it will be very valuable timber. On the way up the shores of the Kabinakagami Lake are very rocky, with some flat land extending far back into the country, heavily timbered with spruce and tamarac of medium size. The high lands and rocks are all burnt.

July 27th. We went inland at the first rapid at the head of Kabinakagami River about ten miles on the west side of river in a north-west direction, and came to a river about five miles from the head of the river. The land is those five miles is very low and heavily timbered with spruce, tamarac and cedar. The cedar is of a good size and has a good healthy appearance. The spruce also is of an average size of about seven inches, and of a good average length, and would cut about thirty cords per acre. The tamarac is scattered but what there is there is very large and about sixty or seventy feet long. The river we came to is about forty rods wide at the point where we struck it, and has low, flat shores. We had to change our route and went west about five miles, and about three miles of the five as regards the timber, are about the same as the first five miles, and the last two miles we travelled were principally burnt over and the timber that is left standing is mainly birch of a very good size. The greater part of the ten miles is good farming land of clay loam. Going farther we came to where the land was more level and not burnt. I climbed a tree and as far as the eye could see the land was low and flat and very heavily timbered, being about the same as the first eight miles we travelled.

July 31st.—Went down river to about ten miles from the mouth, then went inland on the east side of river in a south-easterly direction. The first mile is high land and all the timber is burnt. The timber that is burnt is red pine, balm of Gilead and birch. The land is sandy loam with a mixture of clay. We travelled through three miles of low, flat land, thickly timbered with spruce and tamarac, about half and half. The average size of the spruce is about seven inches in diameter and it would cut about twenty cords per acre. The tamarac is a very good size and would average about two hundred feet B.M. per tree. We came to a high range about eighty rods wide and half a mile long covered with red pine all burnt, the clay being sandy loam. We then went through a swamp three quarters of a mile wide, thickly timbered with spruce and tamarac about the same size and quality as stated above. We came to a half mile of burnt land. The principal wood burnt was red pine. I notice that where the red pine grows the land is a sandy loam, and where spruce and other timber is found, the land is clay loam. We came to a creek half a chain wide running north-east and then crossed a beaver meadow half a mile wide and two miles long. We then came to a creek, half a chain wide, running south-west, succeeded by a swamp one mile wide thickly timbered with spruce, tamarac and cedar. The spruce is about the same average as the last. The cedar also is a very good quality. The land is a clay loam. We found a small lake on our left, running south-west, about one mile wide with swampy shores. A high range covered with red pine contains about one-half a million feet, not burnt. We went through one mile of swamp thickly timbered with spruce, tamarac and cedar, the average size the same as the last. The soil is clay loam well adapted for agriculture.

August 1st.—We continued our journey south east, going through two miles of burnt red pine, growing on sandy loam, and half a mile of swamp, well timbered with spruce and tamarac about the same average as last. We came to a ridge of sandy loam, covered with red pine comprising about half a million feet not burnt, the average size of which is fourteen inches in diameter and sixty feet long. This was succeeded by some high land covered with burnt timber, such as balm of Gilead, poplar and birch, which gave place to one mile of swamp well timbered with spruce, tamarac and cedar, all of a good size and about the same average as we have passed. We came to a lake or large river half a mile wide running north-east and south-west, to avoid crossing which we made a detour to the south, arriving at a swamp, one mile wide, thickly covered with spruce, tamarac and cedar. The spruce will average about seven inches and cut about forty cords of pulpwood per acre. There will be about ten trees of tamarac to the acre which will cut two hundred feet to the tree B.M., and the cedar will yield about the same. We went through one mile of swamp thickly timbered with spruce, tamarac and cedar, with about the same average as the above. A large tract of burnt country then presented itself. I climbed a tree and could see nothing for four or five miles but burnt land. The land was well adapted for cultivation. I noticed that on the east side of this river there is more burnt land than on the west side, as this land lies between Kabinakagami River and Oba River, and the Indians hunting and camping on these rivers set fire to a large portion of this part of the country.

August 4th.— From camp we went down the river about four miles to four rapids. Then we went inland on the west side of the river. I noticed while going down the river some red pine and jack pine of a fairly good size, and also some good balm of Gilead, poplar, spruce and tamarac, with good high land. The banks also were good soil and farming land. All the timber down the river is a good size, and I would judge about twelve or fourteen inches average. We went inland about one mile, where we saw a nice grove of red pine to our right. On examination we found about two and a half million feet of nice red pine, the average size being fourteen inches in diameter. We came back to the river, and went down four rapids, a distance of five miles. Then we went up a small creek flowing west a distance of about a mile to reach a portage in order to avoid a bad rapid. We saw some good timber all the way down. For the last five miles travelled down the river and creek the shores were covered with red and jack pine, averaging from one-quarter mile to one mile back from the shores. The soil is loam and fit for farming. Both sides of the creek are covered heavily with red pine. We crossed over the portage about one and a half miles and found a large grove of red pine on a portage, containing about three million feet of a very good quality, the average size being twelve inches. The trees will run about sixty feet in length up to the limbs. The soil, a sandy loam, is well fitted for farming.

August 8th.—We went inland from the eighth rapids on the west side of river with Mr Baird and chained five miles in a westerly direction. The first two miles we went through was mostly low, flat land, timbered with spruce and tamarac of a medium size and very thick on the ground. The spruce average about six inches in diameter. The tamarac are much larger, will average about ten inches in diameter and are of good length. The spruce will cut about forty cords per acre. I saw some good poplar and balm of Gilead, and also some good red pine on both sides of this creek. The creek was running north and was about one-quarter of a chain wide and two feet deep. I estimated the pine at about half a million feet B.M. When about three miles on our way we came to a beaver meadow about one mile long and a quarter of a mile wide. We went on about one-quarter of a mile further, and came upon another beaver meadow of about same size as the first one. The rest of the five miles were flat lands, covered thickly with spruce and tamarac, averaging about the same as already mentioned. The land is all good clay loam and fit for farming.

August 9th.—We continued our exploration, going ten miles farther west. Mr. Blair had to return back to camp to carry on the work on the river. I found on those ten miles one continuous tract of flat land covered with spruce and tamarac of a good size. The spruce would cut thirty cords of pulpwood per acre, and there would be about twenty trees of tamarac that would yield two hundred feet b. m. per tree. The land is good clay loam and fit for farming. We came to a bunch of spruce timber mixed with tamarac, both of good size the tract being about half a mile wide and one mile long. All the spruce was dead and the tamarac was quite green. I could not find the cause of this.

August 11th.—We moved camp down the river about ten miles. On the way I noticed some very nice red pine and jackpine growing on high sandy banks. I went up and examined it and found it to extend inland from a quarter of a mile to a mile. The land is sandy loam.

August 13th.—We left camp at the eleventh rapid, and went down the river to the twelfth rapid. About three miles of the land is low and flat and covered with small timber, viz., poplar, spruce, cedar and tamarac not averaging more than five inches diameter. We went inland on the east side of the river in an easterly direction. The first

Grove of spruce and balm of Gilead near Devilfish Lake north of Long Lake. Party No. 5.

Spruce and jack pine, Devilfish River, about 40 miles from Long Lake House. Good timber in this district. Party No. 5.

Rapids on Kenogami River, north of Pine Lake. Party No. 5.

Falls on Kenogami River. below Kepeesowatin Lake Party No. 5.

On Little Pine Lake, a tributary of Kenogami River, about 50 miles from Long Lake House. Party No. 5.

Sample of spruce 12 in. to 24 in. in dia., a few miles interior from mouth of Moose River. Party No. 5.

On Kenogami River near Mouse River. Rapids, and heavy spruce and poplar woods. Party No. 5

Near Mouse River. Party No. 5.

Rapids and portage, No. XI, Kenogami River. Party No. 5.

Little Pine Lake. Party No 5.

Brulé, south of Pine Lake. Good Land. Party No 5

Red Rock Falls on Little Pine River, above Little Pine Lake.
Party No. 5.

mile I found to be good high land of sandy loam mixed with clay, well timbered with spruce, poplar, birch and tamarac. The spruce is very thick on the ground but not large, its average size being about six inches and it would cut about fifty cords per acre. The tamarac is too small for present use, as is likewise the poplar and birch. The land is very good clay loam. The timber and land of the next mile were about the same. We proceeded three miles further and found the land more level and well timbered with spruce, tamarac, poplar, birch and balm of Gilead of a good size. The spruce will cut twenty cords per acre, and the remainder of the timber is large and fit for lumber and and would cut about one hundred and fifty feet per acre. This land is clay loam and good farming land. We came to a small creek about ten feet wide and two feet deep flowing north, supposed to be the head of the Matta-wish-guaia River. At the end of the next mile we struck high land which was very heavily timbered with spruce, balm of Gilead, tamarac and birch of a good size. It will cut thirty cords of spruce per acre, and all the rest of timber will cut two hundred and fifty feet per tree. The land is good clay loam. We came to a small creek about fifteen feet wide and two feet deep flowing south-west, then to low, flat land, well timbered, about the same as above. Going on we struck a ridge of rock about fifty feet high, then crossed the same creek as before following north-east. The timber is principally spruce of a good size which will cut sixty cords per acre, and the soil of a good clay loam. The first half of the sixth mile was stony, the second half low, flat land, thickly timbered with spruce and tamarac which would cut about twenty-five cords of pulpwood and about three thousand feet of tamarac lumber b. m. 7th mile. Low and swampy land, timber, spruce and tamarac which will cut about same as last mentioned. Good clay soil and very mossy bottom. 8th mile. Low and swampy land, timber spruce, not very large. Would cut about ten cords to acre. 9th mile. Same as last.

August 14th.—10th mile. We went in same direction (easterly) through one mile swamp. Same kind of timber as last mile. 11th mile. First half of this mile swampy, timber, spruce and tamarac. Last half stony and land covered with spruce and tamarac only smaller. Would cut ten cords per acre. We went through two hundred yards alder brush swamp. 12th mile. Low and swampy land, timber a little larger than last mile. Would cut about fifteen cords per acre. At the end of 12th mile we came to a stony ridge two hundred yards wide, running north and south. 13th mile. Low and flat land, small timber. Very thick, mossy bottom, kind of muskeg. 14th mile. Same as last mile only good farming land, sandy loam. 15th mile. I climbed a tree looking east about a mile. Just the same kind as last mile, looking like some kind of swamp for a distance of two or three miles, looking north about three miles the land was also swamp but with larger timber. I judge it would cut an average of twenty cords per acre.

August 17th.—We left the camp at the sixteenth rapid at the portage over to Matta-wish-guaia River, going inland on the west side of the river, where we found no timber except alder bush and small scrubby spruce, and very wet swampy ground for the first mile. 2nd mile. Same as last. 3rd mile. Same. 4th mile, Same. 5th mile. We came to a river running northeast, thirty feet wide, four feet deep, muddy bottom. We built a raft and crossed the river. 6th mile. We came to a little higher land covered with short spruce and poplar, which will cut out about ten cords per acre. Soil clay loam and about one and a half feet peat. I took samples of soil at this point. I think it will make a good farming country. 7th mile. Not quite as high land as last mile. Small timber spruce and poplar. 8th mile. First half mile same as last mile. On the second half we came to a creek the same as we crossed before running south. We crossed over the creek and found timber the same as on the first half mile. 9th mile. Low, flat, swampy land. Timber principally spruce and tamarac. The timber is a little longer, it will cut about ten cords per acre pulpwood. I climbed a tree and could see about three miles. The same kind of swamp and timber prevailed.

August 18th.—We came back to the starting point. We crossed the Kabinakagami River. We went across portage about one mile to camp on the Matte-wish-guaia River. We found good farming land, with good large timber consisting of spruce, tamarac, balm of Gilead and poplar. The soil is clay loam.

August 20th.—We went down the Matte-wish-guai a river five miles, and thence went inland to the east. On the first mile found good dry land fit for farming, sandy loam mixed with clay. Timber, spruce, tamarac, poplar, and balsam, spruce will cut thirty cords per acre. Tamarac, poplar and balsam will cut about seven thousand feet per acre. 2nd mile, same as first. 3rd mile, same. 4th mile, same as last, land high and dry. 5th mile, same as last mile. 6th mile, same as last only land a little lower. 7th mile, same as last mile. 8th mile, same as last only more thickly covered with brush. 9th mile, higher land. Timber a little larger. 10th mile, same as last, land a little higher. Land good for farming, soil sandy loam, clay bottom. 11th mile, same as last mile except that we came to a high stony ridge about fifty feet high running north about one hundred yards wide, the remaining portion of the mile is dry land covered with large timber spruce, tamarac and poplar. The spruce will cut twenty-five cords per acre, the tamarac and poplar about ten thousand feet per acre. 13th mile, same kind of soil and timber as last two miles. 14th mile, same as last. 15th mile, low and swampy and larger timber. Very nice spruce, would cut about twenty cords per acre. 16th mile, low and swampy with good spruce and tamarac. 17th mile, same as last mile. I climbed a tree and as far as the eye could see it appeared to be the same kind of country. The five miles we paddled down the river, the land looked like good farming land. The water in the river is very shallow and the bed stony. I noticed that timber was larger on the east side of the river than on the west. How to account for this I cannot say.

August 24th.—We went down the river about five miles from camp and inland northwest. 1st mile, high dry land. Good soil, sandy loam, thickly timbered with spruce, tamarac, poplar and balm of Gilead. The spruce would cut about twenty cords per acre. Poplar, balm of Gilead and tamarac would cut from ten to fifteen hundred feet per acre. 2nd mile, same as last. 3rd mile, dry, same as last. 4th mile, very dry land, timber very large. 5th mile same as last, very large poplar, balm of Gilead and spruce. 6th and 7th miles, same as last. 8th mile, we came to a large lake with high clay banks about three miles long and one and a half miles wide, surrounded with good large timber, such as spruce, tamarac, poplar and balm of Gilead. Good farming land. Examined soil and found it to be good clay loam.

August 27th.—We went down the Matta-wish guaia River about ten miles and came to a river flowing into it which we went up in a southerly direction. We found a large swamp covered with water. It would be a lake in the spring of the year. We could not get into it, as the mouth of the lake or swamp was full of weeds. The land down this river is very low and swampy and not much timber. I noticed some large elm and black ash along the shores of the river. The lake is about four miles long and one mile wide. By looking across the lake the land seemed to take a rise and is much higher and thickly covered with timber such as spruce, poplar and tamarac. When we were coming down the Matta-wish-guaia River these ten miles, I noticed that the shores were covered very thickly with timber, viz: black ash, spruce, poplar and tamarac. The last two or three miles the land back from the shores was very flat and low and not much timber.

August 28th.—We went down the Matta-wish-guaia River about three miles and came to a river running into Matta-wish-guaia River which we went up in a southerly direction. We came to a lake at the head of the river about one mile wide and about five long, shallow with low shores. Crossing it in a canoe we found the south shore of the lake very stony. The timber all around the shores is very thick, consisting of spruce, tamarac, balsam and cedar. We went inland about four or five miles and found good dry land well timbered. Good soil, fit for farming. The spruce will cut about thirty-five cords per acre. Tamarac and balsam timber will cut about seven thousand feet B.M. per acre. The water of the lake was not over two feet deep in any place.

August 29th.—We went down the Matta-wish-guaia River about five miles, seeing good timber along the shores, consisting of spruce, poplar, balm of Gilead, tamarac, balsam of a large size. The water down the river is very low. I noticed some very fine birch, the land being good clay loam.

August 30th.—We went in on the west side of the river in a northerly direction and found nice level dry land, thickly timbered with spruce, tamarac, poplar and birch of a good size. We came to a marsh running east and west about one-quarter of a mile wide. No timber on it. 3rd mile. Found good level land fit for farming, well timbered with spruce, tamarac and poplar of good size. Spruce cut about eighteen cords per acre, tamarac and poplar about fifteen thousand feet per acre. 4th mile. Same as last mile, very dry land. 5th and 6th mile. Same as last. Soil good clay loam.

August 31st.—Went inland in the same direction north. 7th mile. More high land. Good spruce, tamarac and poplar. 8th mile. We came to a river running east, twenty feet wide, stony bottom, and reached higher land, timber poplar and balm of Gilead of good size, and will cut out about twenty thousand feet b.m. per acre. 9th mile. Low and swampy land. Timber good size, principally spruce and tamarac. 10th mile. We went through low flats, swampy land with small scruby spruce and tamarac. At the end of the 10th mile I climbed a tree, from which I could see about four miles. Saw same kind of swamp and timber, the poplar being very large. Soil, sandy loam mixed with clay.

September 3rd.—We continued journey down river about one mile, and there came on a very heavy rain storm which continued all day and night. We camped for the night.

September 4th.—We started to explore the country between Matta-wish-guaia River and the Missianaibi River on the east of Matta-wish-guaia River, going inland in a south-easterly direction. 1st mile. Nice high rolling land, well timbered with spruce, tamarac, poplar and birch of a large size. We came to a small creek running north-east which we crossed. 2nd mile. Wooded same as last. 3rd mile. Little higher land, well timbered with spruce, balm of Gilead and balsam. 4th mile. A little lower land, more swamp, more scrubby timber, spruce and tamarac. 5th mile. Same as last. 6th mile. High land, good timber, spruce, tamarac and poplar of large size. Good soil, clay loam, 7th and 8th mile. Same as last, very dry land and good timber. 9th mile. Low and swampy land. Good soil, clay loam and well timbered with spruce, tamarac and balm of Gilead. Spruce will cut about twenty cords per acre, and the rest of the timber about five thousand feet b.m. 10th mile. Same as the last except some fair or dead timber. At the end of 10th mile we struck the Missinaibi River.

September 6th.—We came back to Matta-wish-guaia River and camped all night.

September 7th.—We started down Matta-wish-guaia River about five miles until we struck Missinaibi River. We had to wade the river and carry the canoe two or three miles of this distance, when we met the rest of the party and camped. I noticed coming down that the banks of the river were covered with poplar and spruce.

September 8th.—We went down the river about two miles. We went inland on east side of Missinaibi River in an easterly direction. The river banks are about fifteen feet high. 1st mile, nice high land well timbered with spruce and poplar. 2nd mile, same as last. 3rd mile, low swampy muskeg with small scrubby spruce and a few large tamarac mixed in. Fourth mile, south of our range a large bluff of poplar; the rest of the mile same as last mile. 5th mile, same as fourth mile; swampy and deep marsh and very thickly timbered with scrubby spruce and a few scattered tamarac of small size.

September 10th.—We started out from the Missinaibi River on the west side in a north westerly direction. For the first ten chains we found high land with some scrubby brush. We came to good poplar and balm of Gilead of good size, which would cut out about fifteen thousand feet per acre, and good soil, clay loam. Next ten chains same as last. Next ten chains a little more spruce timber. Next ten chains we came to ridges of rock fifty feet wide, running north and south. The rest of the ten chains was the same as last only more spruce. The next ten chains was low flat land, a kind of muskeg a little wet, thickly timbered with spruce, not very large, would cut off about ten cords of pulp per acre. Next twenty chains same as last ten chains. Coming to a ridge of rock running southwest and northeast we found a little larger timber running around the ridge. Next ten chains same as last. Next ten chains higher land and larger poplar, spruce, tamarac, balsam; will yield thirteen cords spruce per acre, tamarac and balsam about fifteen thousand feet per acre. Next ten chains same as last. Next ten chains little lower land, very deep moss, and scrubby timber of spruce and tamarac, Next ten chains same as last.

September 11th.—We stayed forenoon for rain, in the afternoon returned to head-quarters on the Missinaibi River. One of our men, W. Laidlaw, was taken sick.

September 12th.—I continued the trip myself with two Indians. 1st mile, low and swampy, covered thickly with what I would call spruce, the Indians calling it "sisecantik" timber, so I have taken a sample of it. It is very small, averaging about five inches, and very thick on the ground. The moss is very deep and dry. The soil a good clay loam under the moss. 2nd mile, partly much the same. 3rd mile, I noticed a slight difference between the regular spruce timber and the tree which the Indians call "sisecantik ". The wood looks the same but the trees or the branches are a little different under the bark and a little rougher. The Indians call the common spruce "Menahick." I find that the other kind of spruce grows densely where the muskeg is, and very small, most of it averaging only 4 inches in diameter. 4th mile, we came to high land, on which grew ordinary spruce, only much larger. The sisecantik spruce I find, does not grow very large and grows principally in low flat muskeg land. 5th mile, same as last mile with a little deeper moss, and timber a little smaller. 6th mile, same as last mile, very dry with small scrubby timber, no water to drink, muskeg low and flat, poorly timbered. 7th mile, low and flat muskeg, scrubby spruce averaging four inches. 8th mile, same as last, dug a hole in the moss and got something that looks like water, and camped for the night.

September 13th.—We went in the same direction exploring. First mile same as last mile of the 12th of September, muskeg covered with scrubby spruce and a few tamarac, not healthy looking. 2nd and 3rd mile, same as last. At the end of third mile came to a ridge of rock, thirty feet wide, running north and south. 4th mile, same as last, some scrubby timber and deep moss. 5th mile, same as last, with some scattered tamarac. 6th mile, we came to another ridge of rock, running north and south, about twenty feet wide, a little rougher timber, another ridge of rock one hundred feet wide running north and south; the timber on the ridges was mostly poplar and spruce of a good average size and healthy looking. 7th mile, a little higher land and larger timber, mostly spruce. 8th mile low and muskeg. We came to a small creek fifteen feet wide running north and south with larger timber, principally spruce, which would cut about twenty-five cords per acre. 9th mile, higher land and larger timber. A ridge of rock runs north and south about sixty feet wide and about six feet high above the level of the land. At the latter end of this mile we came to a small creek running east, and further on another ridge of rock fifty feet wide running east and west. 10th mile, high, good land. Timber principally poplar, mixed with spruce and balm of Gilead of a good large size. The spruce would cut about thirty cords per acre and other timber about fifteen thousand feet, b.m., per acre. We came to a river running north-east about fifty feet wide and two feet deep with high banks and stony bottom. The Indians call this river "Pewabare." This river flows out of a lake called "Pewabare." This lake lies south west from where we struck the river. I noticed running into Missinaibi River a river shallow and running very fast. I noticed by the banks that the water rises about eight feet higher than at the present time. We crossed the Pewabare river on the north side. The land rises about twenty feet higher than the river and is thickly covered with nice large timber, principally poplar and spruce.

September 14th.—We continued in the same direction, north-east. 1st mile, found good timber, poplar and spruce. The spruce would cut about twenty cords per acre, poplar about ten thousand feet, b.m. The ground began to ascend. 2nd mile, good timber for first half-mile, spruce and tamarac. Last half same as first half. Coming to muskeg swamp, climbed a tree and as far as the eye could see, for three or four miles, there was nothing but muskeg and small spruce with a tamarac here and there of a good size. Looking up and down the river the land is higher for about half a mile on each side and better timbered. We returned home, got half way, then camped.

September 15th.—We started in the morning for camp. It was bad walking over deep moss and muskeg and raining hard, but we walked all day, and got home to camp on shore of Missinaibi river tired and hungry. Found rest of party well and sick man improving.

September 17th.—We broke up camp and all hands started up the Missinaibi river for home. I noticed coming down the river some very nice cedars of a large size, also some good spruce, tamarac, balm of Gilead, poplar and yellow birch, with some good red pine. The red pine was not in large quantities, but the rest of the timber seemed to be plentiful and has a good healthy look.

I have the honor to be, sir,
Your obedient servant,

JOHN McCONACHIE,
Land and Timber Estimator on
Exploration Survey Party No. 4

Toronto, January 8th, 1900.

I, John McConachie, of the Village of Huntsville, District of Muskoka, Land and Timber Estimator, make oath and say that the above report is correct and true to the best of my knowledge and belief, so help me God.

Sworn before me at the City of Toronto, in the County of York, this 8th day of January, A.D. 1901.
GEO. B. KIRKPATRICK,
A Commissioner, etc.

JOHN McCONACHIE.

GEOLOGIST'S REPORT OF EXPLORATION SURVEY PARTY NO 4.

TORONTO, Jan., 1901.

To the Hon. E. J. Davis,
Commisioner of Crown Lands,
Toronto :

Sir :— I have the honor of reporting as follows on the country covered by No 4. Party:—

No. 4 Party consisted of a land surveyor, Mr. Baird ; a timber estimator, Mr. McConachie ; and three others besides Indians. The party assembled on the 21st of June 1900 at Grasett Station on the C. P. R., a point 574 miles from Toronto, or 347 from North Bay, and almost directly North of Michipicoton. Esnogami or Magpie River passes near this station, and the party left the railway up this river. The Kabinakagami River was the point where the survey really commenced, but the intervening lakes had to be mapped so as to connect the survey with the C.P.R. and accordingly short notes on their general features are included in this report.

ESNOGAMI AND KABINAKAGAMI LAKES.—The Magpie River flows southward from a lake that is known to the whites as Shell Lake, and to the Indians as Esnogami. This lake was mapped by micrometer, and then it was necessary, in order to cross the height of land into the Kabinakagami Lake, to cross the Esnogami portage, which portage is about five miles in length,

COMMENCEMENT OF SURVEY PROPER.—At the lower end of Kabinakagami Lake the survey proper began, and according from this on trips were made by Mr. McConachie and myself to each side of the river, extending for about twenty miles on either side. These trips had to be made overland as the small rivers are everywhere impassible owing to timber jams.

KABINAKAGAMI RIVER.—The Kabinakagami River flowing out of the lake proved to be a fine large stream for its upper half at least, and easily navigable by canoe. It has a fair number of rapids, most of which must be portaged, but between them the river is deep, and not excessively rapid. Dr. Bell has surveyed Lake Kabinakagami and about a third of the river in 1882 and his map was the one used by us in our work.

KENOGAMI RIVER.—The Kabinakagami River was descended for something more than half its length, when it became very shallow and rapid. We ascertained that it flows off north-westward into the Kenogami River, a tributary of the Albany, meeting the Kenogami at a place called Mammatawa as suggested by Dr. Bell, and does not connect with the Mattawishguaia River, which flows across into the Missinaibi or Moose River.

MATTAWISHGUAIA RIVER,—The Mattawishguaia River is narrow, and in the summer has broad mud banks on each side of it. It was reached by a portage from the Kabinakagami River and proved to be very shallow and stony for almost its whole length In June it is a good river with only one rapid that requires portaging, but in August when our part went down it was almost dried up and, except for a short deep stretch in the middle, it was obstructed by numerous rocky and rapid shallows, over which it was necessary to either make a portage or to wade beside the canoes.

MISSINAIBI RIVER.—The Missinaibi River was entered at Lower Skunk Island. This magnificent river has formed a canoe route to James Bay from time immemorial, and its broad sweeping waters gave excellent canoeing, offering a striking contrast to the little stony rapids of the Mattawishguaia. The survey was continued for a short distance down the Missinaibi River, but owing to the lateness of the season, our Indian guide refused to go on to Moose Factory, and we had to return up the Missinaibi, passing through Missinaibi Lake, Crooked Lake (Wowogonosh) and Dog Lake to Missinaibi Station on the Canadian Pacific Railway. The line run by Mr. Niven's party (No.2) from the 198th mile of the Algoma Nipissing Line eastward was passed on the way out.

GENERAL FEATURES OF THE REGION.

MAGPIE RIVER.—The readiest way of reaching Kabinakagami River from the C.P.R. is by the Esnogami River, Esnogami Lake, and Kabinakagami Lake. Esnogami or Magpie River crosses the C.P.R 5 miles east of Grasett Station. The river flows southward to Michipicoton Inlet on Lake Superior. This river, and the lake of the same name from which it flows, were surveyed with micrometer northward from the railway so as to connect the survey with the C. P. R.

LAKE ESNOGAMI —A short distance north of the C.P.R. the river widens into Lake Esnogami or "Shell Lake." A current is perceived, however, for some distance up the so-called lake, and no exact point can be taken where the lake may be said to end, and the river to begin. Just south of the railway track the river again widens into a small lake about half a mile wide, with a very narrow outlet where the water is so rapid as to be paddled up only with difficulty. Mr. Willmott's paper in the Bureau of Mines Report for 1899, shows that south of this the river becomes very rapid.

North of the C. P. R. the lake in a few miles loses all trace of current and gains a considerable depth. It is very variable in width, its shape being governed by the ridges and bluffs of rock that are numerous all through the region. In the bays the shores are almost invariably low-lying; often swampy in the deep bays, but at the points there are bluffs of grey Laurentian gneiss often 50 feet or more in height. At the C. P. R. on the east side of the river is a great hill of coarse gravel 80 feet high. The lake narrows up about half way to its head into a close rock-bound passage, and north of this again widens into another large expansion.

TIMBER.—The timber around Esnogami is much burnt, and often blown down over large areas. It consists mainly of spruce and tamarac with some cedar. Where the burning is not too recent a thick growth of poplar has sprung up. A characteristic feature of the scenery both here and throughout the region is the close fringe of red-osier dog-wood that lines the shores, and whose hanging clusters of white berries serve as the main food for the bears after the blue-berries and raspberries are over.

INDIAN GRAVES.—This lake appears to be much used as a burial place by the Cree Indians of the district, as numerous graves are found on its shores and islands. The largest burying place is at the outlet of the first lake expansion north of the C. P. R. where a little cemetery is placed with graves in crude imitation of those of the white man, and having the headstones decorated in many cases with ornaments worn by the dead. Next to this ground a Missinaibi Indian has, with characteristic callousness, chosen a site for a small log house, in front of which he has planted a good patch of potatoes and raspberries. This was the only instance of the kind seen by us, but I believe that near Missinaibi Station some of the more advanced of the Indians have built themselves houses even more pretentious than this.

ROCKS OF THE LAKE.—The only rock observed on Lake Esnogami was Laurentian gneiss of a gray color. Two dikes were seen cutting each other on a burned-over bluff near the mouth of a long bay that goes in off the west side of the lake and turns south. These dikes are of diabase containing a little quartz. The larger of the two is eighty feet wide and strikes north 20 degrees west right up the bluff, and the other, thirty-five feet wide, cuts it striking north 70 degrees east. This smaller dike is much more finely grained owing to rapid cooling, but the two dikes fuse together at their junction showing them to be of the same age. They are closely jointed along and across the strike, and also horizontally. The dip in both cases is nearly vertical. At the narrows before the upper lake expansion are three similar dikes that, owing to their power of resisting erosion, are left as ridges standing forty feet above the level of the adjacent country. Two of these dikes are about fifty feet wide and 250 yards apart, they strike north 35 degrees east. The other is somewhat smaller, and strikes north 35 degrees west. In this upper part of the lake there are a number of smaller dikes of the same kind and the same was observed near the upper end of Lake Kabinakagami, the country appearing to be more broken as we approach the Height of Land

Height of Land Portage.

The Height of Land north of Lake Esnogami is crossed by a portage leading inland from the northwest corner of the lake. This portage is nearly five miles in length 396 chains. For two miles and a quarter from Lake Esnogami the trail passes through muskeg covered with tamarac and spruce. The trail is very wet and is crossed at 37 chains by a rapid stream, but the timber is dense though small. At the end of this distance is a ridge 80 feet high and some ten chains wide. At some places on this hill the country rock can be seen by scraping away the moss that covers it. It is a fine hornblende schist, probably Huronian, and is very similar to those hornblende schists aftewards found on Lake Kabinakagami. From this ridge for a mile and a half the land remains fairly dry and has some very good timber: poplar, cedar, birch, balsam, etc., with scrubby pine. From here on, the land becomes again swampy, and is crossed by two more streams. At the northern end there is a little more dry land leading into a dense alder thicket at the northern end of the portage.

Lake Kabinakagami.

Lake Kabinakagami, situated on the northern side of the Height of Land, has a different type of scenery from that of Lake Esnogami on the southern. The shores are flatter, the rock exposures lower, and the sand banks larger and much more numerous. Also the timber is not nearly so much burnt as that round Lake Esnogami.

River Papakapesage.—The lake is reached from the Height of Land portage by means of a little marshy river called by the Indians "Papakapesage." This river where the portage strikes it is about two rods in width, but nearer to Lake Kabinakagami it expands to some 30 chains.

Upper Part of Lake.—The upper end of Lake Kabinakagami is shallow and has a sandy bottom. The rock exposures are of gray Laurentian gneiss with eruptive diorite in places. They are numerous, but often small, the greater part of the lake being surrounded by sand. The ice-shove in winter, however, banks up these sandy shores and builds a natural breakwater of stones against them, giving them a more rocky appearance than they would otherwise have.

Rock Formations.—The geology of this lake and of part of the Kabinakagami River has been described by Dr. Bell in the Canadian Geological Survey for 1882. The lake is for the most part Laurentian, but is crossed near its middle by two narrow bands of Huronian rocks lying roughly north-east and south-west, and parallel to one another. The most southern of these bands consists mainly of hornblende schists, the more northern band has in addition a rock containing tailed fragments that was described by Dr. Bell from near the Oba River. A rock probably identical with Dr. Bell's, but differing somewhat in this place from his description, was found on the western shore of the lake about opposite the mouth of the Oba River. It consists of a diorite gneiss of fine grain, containing tailed pebbles that have much epidote and quartz. Dr. Coleman suggests that the rock is an agglomerate.

About half way down the lake is a long point, or rather peninsula from the east, over which the Indians usually portage when going from one end of the lake to the other, and so save rounding the point in the open and often stormy lake. Below this portage (i.e. to the north of it) the lake is deeper, und consequently is not so much affected by wind as is the upper part, making it much easier of navigation.

Timber.—The timber on this lake is burnt in most of the high places, but it is not nearly so bad in this respect as Lake Esnogami, perhaps because its shores are for the most part lower than those of Esnogami. Most of the way round the lake the timber consists of spruce and tamarac, with small quantities of cedar and birch, balsam poplar in the high places, and thick small poplar where the country has been burned over.

RED PINE AT OBA RIVER.—There is some red pine near the north end of the lake especially near the mouth of the Oba River. This mouth is shallow, with low marshy banks, and with three rapids that require portaging a short distance from its mouth, and the Indians have a portage leading out from Lake Kabinakagami, and cutting the Oba some distance from its mouth. This portage is about a quarter of a mile long, and passes along a hillside with no underbrush, where there is a splendid grove of red pine. Mr. McConachie estimated that this grove contained about a million feet, board measure.

An inland trip was made from the Oba Lake, running a little east of south, until the Oba River was met with. All the high land on our line was burnt, but some ridges of good red pine were seen three to four miles away. The hornblende schist of the most northern of the bands crossing Lake Kabinakagami was again found here four and one half miles inland. This schist has the appearance of a metamorphosed diorite.

KABINAKAGAMI RIVER.

The Kabinakagami River flows from the northern part of the lake of the same name. At the head of the river is a rapid with a fall of thirty-five feet. This is passed by a portage going in from the lake. This portage is 42.50 chains in length.

HURONIAN BELT AT HEAD RAPID.—At this rapid is a little patch of hornblende schist mapped by Dr. Bell. The same rock was again found four miles away in a north-west line inland from the foot of the rapid. Here it formed a ridge fifty yards wide, running north-east. The planes of cleavage strike north, 45 degrees west and dip 58 degrees easterly, and lie parallel to the line of contact with a diorite to the westward. The ridge contains some stringers and bands of coarse white and pale brown quartz running with the cleavage, and the hornblende-schist itself is very quartzose in character.

COUNTRY TO THE N. W. OF HEAD RAPID.—An inland trip from this rapid showed besides this rock already mentioned high land with burnt timber, of red pine, tamarac, spruce and cedar, alternating with lower land that approaches muskeg in character, having sphagnum moss for footing with shallow peat below, and supporting a dense growth of black and white spruce and tamarac. These trees are small in diameter, but grow very close together, and are healthy and vigorous. The soil here is, as almost everywhere in the country, a light-coloured, sandy clay loam. Only the one ridge of rock was seen on this side trip, and indeed very little rock of any kind was found in the bush, all the country inland being low-lying and generally swampy. A man who never leaves the river gets a very wrong impression of the country owing to the narrow border of high land that forms the river banks everywhere.

Below the head rapid there are occasional exposures of grey Laurentian gneiss, but, except for this, the shore consists of stratified sandy clays and gravels. A few scratched stones were found in these beds, but no fossils. The river here is about three chains wide, and does not vary greatly in width to the end of the survey of it.

TIMBER.—The timber of the river is practically the same throughout; along the banks is a belt of high land with tamarac, spruce, balsam, cedar, poplar, and birch, with alder, willow and dog-wood at the river edge where the shores are low. This belt of high land along the rivers varies from a quarter of a mile to two miles or more in width, and occassionally where two rivers flow together so as to make an angle between them, this land will all be of the same character as that of the river banks. Inland from this the country becomes slightly lower and sphagnum moss and lichens come in, Then sarracenias are seen among the moss; and orchids—cypripedium, habenaria, and spiranthes—make their appearance. The cedar, poplar, and birch are no longer seen, the poplars appear in isolated clumps only, and the spruce and tamarac become dense though not so large, getting smaller as we go inland. The river is fairly good for canoeing purposes. The rapids are frequent, but are for the most part short with good deep water in between, and they can be passed by good portages, though not so good as those on the Missinaibi River. This refers to that part only of the river covered by our survey; lower down the river is reported to be very much worse.

Below the third rapid a trip was made inland going south east. At the rapid the river had a bank of sandy, pebbly clay that goes up sheer to 80 feet. For a mile from the river the land was burnt, with fallen timber lying in all directions. This burnt land has the same sandy clay soil noted before, and now that the trees are burnt a wonderful growth of berries has sprung up. Beyond this the land as usual becomes lower, with close spruce and tamarac, and the usual orchids and other swamp plants among the moss. Another small area of large timber (burnt except half a million feet of red pine) surrounds a lake at the eighth mile. This lake covers about a quarter of a square mile, and is bordered by a broad moose path, showing that the big game is still far from being driven out, although the Hudson's Bay factors at both Brunswick House and Missinaibi Station say that they get fewer skins of bear and moose every year. With the exceptions noted the land was all low with dense spruce and tamarac, and peat up to the sixteenth mile, where the land begins to rise ; and at the eighteenth mile a great burnt country was entered that extended as far as could be seen from a tall tree.

Two ridges of rock were seen, one at the 4th and one at the 15th mile. Both were pink Laurentian gneiss. The deepest peat on the whole trip was at about the 14th mile. Here it would average four or five feet, the whole being much more shallow than would be expected from its extent, so that it seems probable that much of this swamp could be drained and converted into farm land.

Red Pine and Jack Pine.—The fifth rapid on the Kabinakagami River is the longest on the river. The descent is only about forty feet, but it extends over a mile and a half. The portage is on the left side and is 73 chains in length. It passes over high land, with a beautiful grove of red pine, together with about a mile square of jack pine.

From here the shores become higher, with many hills still of stratified clayey sand, rather powdery when dry, firm and clayey looking when wet. Some are 70 feet high, and for two miles above the 10th rapid they would average more than 50 feet. For the rest the banks are of the same soil but quite low, the low and the high land alternating all along the banks.

Huronian Rocks.—The first Huronian rock found after that at the head rapid was in the form of a single exposure below water level, and not visible when the river is high. It is about nine and a half miles from the head of the river. It is a fine grey mica gneiss with orthoclase and much quartz, and lies nearly horizontal. It is apparently sedimentary in origin and resembles Lawson's Couchiching rocks. Again at the 7th rapid eighteen miles down is a grey syenitic gneiss, with biotite, hornblende and some titanite and apatite. It has little felspar in the rock itself though it contains a few stringers of orthoclase so that it approaches a siliceous mica schist in character. The same rock comes in again at the 9th rapid, nineteen and a half miles down the river, the intervening part being obscured by the sandy clay deposit.

At the 10th rapid, twenty three and a half miles down the river, is another micaceous rock, but this again has the Couchiching appearance of that first mentioned. It is, however, much more siliceous and the mica is more evenly disseminated through it; not in layers as in the other. This tenth rapid is caused by two eruptive dikes of coarse pegmatite. These run with the cleavage of the mica gneiss (*i.e.*, strike north 80 degrees east, dip 80 degrees southerly). One is just under the water where the rapid falls over it, the other is ten feet further down the river, and larger, being twelve feet wide, and above water in summer though probably covered earlier in the year. It juts out 100 feet from the shore, two-thirds of the way across the river. This rapid descends 30 feet and is passed by a portage of twenty three and a half chains.

Eruptive Contact of Huronian and Laurentian.

A quarter of a mile below this rapid the same mica gneiss comes in again, and occasional exposures are seen as far as the eleventh rapid, at which there is a contact with a Laurentian gneiss. The gneiss is coarse and much contorted. It is chiefly grey, but has streaks of red. It has much felspar but very little mica. It has every appearance of having erupted through the mica gneiss, as pieces of the latter, often several feet in area, are included in it, and appear fused at their edges. The whole contact is very irregular and broken. Exposures of the same Laurentian gneiss are found from here to the Mattawishguaia portage, where our survey of the river ended. There is a large exposure at the 12th rapid and at all other rapids and shallows are smaller exposures of the same kind. For the rest it is sandy clay land, chiefly low, dry near the river but soon passing into muskeg, usually in about a quarter of a mile.

The Lower River.—For about a mile below the Mattawishguaia portage there are very large Laurentian rocks, with piles of great boulders, making the river very rapid and stony. We could find no portage here, and the Indians state that, though the river flows into the Kenogami River at a place called Mammatawa, it is of no value as a canoe route, and an Indian travelling from Mammatawa to Brunswick House on the Missinaibi River will portage across a chain of small lakes to the northward rather than drag his canoe up the stony shallows of the Lower Kabinakagami River.

Three more inland trips were made, as well as that already mentioned, from the Kabinakagami River. An account of these will give an idea of the country inland from the river. The first was in a westerly direction from the first falls or eighth rapid 18⅛ miles down the river. A fringe of poplar, birch, cedar, spruce, and scattering small red pine and jack pine extended inland for twenty chains. Then we entered a great muskeg country, and went for fifteen miles finding hardly any change. The soil was all sandy clay loam with one to three feet of moss and lichen. Some of the timber is scrub, but generally it is of good fair size, large enough for pulp, and consists of spruce and tamarac, the tamarac being slightly larger than the spruce but not so numerous. All of the timber was healthy except a patch about a mile square at the end of the fifteenth mile in which the spruce was all killed while the tamarac remained healthy. We could discover no cause for this blighting of the spruce. The wood appeared sound, and no insects or galls could be found on the bark or leaves. Three small streams were crossed on this strip.

The second inland trip was made to the east going in below the twelfth rapid. Here the land was better drained. The soil was the same, but the timber for twelve miles was of good size, and having poplar with the sprnce and tamarac. A small lake was passed at one and a half miles, and the Mattawishguaia River crossed our line three times between the third and fifth miles. At this place the "river" was only ten feet wide and two feet deep with a bottom of soft clay like that at the portage from the Kabinakagami River to the Mattawishguaia River. The current was rapid. All the land near the Mattawishguaia River was dry. It must be rapidly drained by the river, this being shown by its banks which in spring would be more than twice as wide apart as they are in August. After an exposure of Laurentian gneiss at four and a half miles the land becomes gradually lower and more swampy, until at the end of twelve miles it is a low scrub muskeg. This low muskeg must be of great size as on climbing a tamarac of larger size than usual at the fifteenth mile it was seen to extend to the horizon in front and on each side. The same thing was seen on other occasions, and it is possible that this low, scrubby muskeg forms the interior of all the country wherever the land is ten miles or so from any river.

The last trip from the Kabinakagami River was made opposite the Mattawishguaia portage in a westerly direction. Here the land becomes low and wet ten chains from the river, and soon the timber thinned out into small spruce with stretches of alder swamp. This country extended for about five miles when a creek 30 feet wide crossed our line. This creek had a very soft muddy bottom, making it necessary to cross on a raft. Two and a half miles on the far side of this creek were fine dry land, on which, however, the timber was small and scattered. At the end of this distance the creek crossed our line again and the land in consequence became more swampy as the creek was left behind.

Appended is a list of the rapids on the Kabinakagami River, with their estimated fall and their distance in miles from the head of the river:

1st rapid, falls	35 feet in	½ mile,	0 miles down the river.
2nd "	25 "	5 chains,	⅞ " "
3rd "	15 "	5 "	5½ " "
4th "	20 "	5 "	8½ " "
5th "	40 "	1½ miles,	11 " "
6th "	12 "	6 chains,	15½ " "
7th "	25 "	6½ "	18 " "
8th "	8 ft. in a perpendicular fall,		18⅛ " "
9th "	10 "	"	19½ " "
10th "	30 "	23½ chains,	23½ " "
11th "	25 "	13¼ "	31½ " "
12th "	6 "	2 "	33 " "
13th "	45 " (rapid and falls)	2 "	33¼ " "
14th "	10 feet in	4½ "	38½ " "
15th "	17 "	"	43¾ " "

MATTAWISHGUAIA RIVER.

The Mattawishguaia River is reached from the Kabinakagami River by a short portage that leaves the Kabinakagami just above the fifteenth rapid. It rises in a small lake some miles to the south of the point where it is entered by the portage, and flows in a north easterly direction reaching the Missinaibi River opposite Lower Skunk Island. Previous maps put it four miles lower on the Missinaibi River, probably owing to a confusion between it and another river that does flow into the Missinaibi at that point.

LOW WATER IN SUMMER.—At the time our party reached the Mattawishguaia it was a narrow, shallow stream, with a broad bank of soft mud on each side, and strewn throughout its bed with boulders, chiefly of gneiss. Early in the summer, however, the Indians say that it is a good river for canoes, and that it can be descended from the Kabinakagami River to the Missinaibi River in a day, whereas in August it was full of little rapids and shallows so that our party was sixteen days on the river. The banks show that the water must rise at least five feet at the time of the spring floods, and another foot of water would have made a great difference in the case of navigation.

CHARACTER OF RIVER BANKS.—The shores of this river are lower than those of the Kabinakagami. The entire bed is of light clay. Over this many little streams flow into the river bringing good cold water that has filtered through over the clay from the Muskeg island. There is good spruce in patches along the river and a great quantity of small poplar. The shores are all lined with the red osier dogwood. Where rock exposures do occur they are all of Laurentian gneiss; the river flowing through the middle of the Laurentian country north of the great Huronian country to the eastward.

For some distance down the Mattawishguaia River the land is high. On a trip made northward from the river, 18½ miles from the Kabinakagami portage, eight miles of high land was passed, and we came on an inland lake surrounded by large timber, balsam, poplar, cedar, spruce, tamarac, etc., and having a broad moose-path beside it. This timber was the best seen after leaving the lakes. From here on, however, the country inland becomes more swampy as we descend the river until the scrub timber gets as close as a quarter of a mile to the river. This country continues as far as the upper Mattawishguaia Lake, where the improved drainage again gives rise to very good dry land with large timber. On a trip on the south side of the river five miles down the same good high land was found for fourteen miles, when it became low and wet. The timber was large, and the soil a light clay apparently good for farming. At the eleventh mile of this trip a moss-covered ridge of Laurentian rock was found. This consisted of a fine gneiss with irregular masses, probably dikes of a felspathic granite. This had very little quartz or mica A microscopic examination of a sample showed that the felspar was all microcline and what little mica there was, was muscovite. A very similar rock was found on the portage between the Missinaibi River and Wanzatika Lake.

THE MATTAWISHGUAIA LAKES.—Off the right bank of the Mattawishguaia, and about half way down to the Missinaibi are three lakes, two of which are each connected with the river by a small creek. On these creeks are some trees of swamp elm; the only elm seen in the country, and looked upon as a great curiosity by the Indians.

The upper of these lakes was approached by means of a little winding creek two miles long. The second mile of this passes through a willow swamp, dry in August, but evidently flooded in spring. The entire lake, although four miles long and a mile wide, was too shallow to float a canoe, and we were unable to cross it. On the far side the land was not swamp, and had good-sized timber—tamarac, spruce, balsam, poplar and and birch.

The middle lake was explored with more success. It can be entered either by a short portage from the river, or by a little creek a quarter of a mile long. It is somewhat larger than the upper lake, and is shallow in August but could be crossed in a canoe though with some difficulty, as before the shore of the lake next the Mattawishguaia River is a willow swamp, but the far shore is high and well timbered for a long way inland. There was a hundred yards of bare shore all around this lake where the water had dried up, and this was strewn with boulders of various kinds of rock including much Devonian limestone from the north. Similar limestone boulders are found all down the Mattawishguaia and the Missinaibi Rivers from this point to the end of our survey. These three lakes are connected with each other by portage trails, and form a favorite hunting ground for the Indians of the district. For a short distance on each side of the lake the river is deep and easily navigated. The deep water on the eastern side extends to the one big rapid of the river. Here a land portage has to be made on the left side. The rapid is caused by an eruptive mass, probably a dike, of diabase, that rises up out of the stratified sand and dams the river. The rock contains bands and stringers of coarse grey gneiss that run for the most part south-easterly. The rapid drops by two falls, one of fifteen and the other of twenty feet, and is always impassable by canoes. From this rapid to the Missinaibi River the river is again stony and shallow, with boulders of rock in the bed. Laurentian exposures are seen at long intervals and all are low. For the rest the shores are of the same stratified sandy clay.

Inland from the lower part of the river the land becomes again fairly dry with occasional ridges of Laurentian rock. The timber is the same as before. In the angle between the Mattawishguaia and Missinaibi Rivers where the drainage is good, some dry rolling land was found.

Missinaibi River.

The Missinaibi River was entered from the Mattawishguaia River at Lower Skunk Island, and was surveyed for a short distance. The river itself is fairly well known and needs little notice here. It is excellent for canoeing purposes and, as it forms a natural highway to James Bay, has well-cut portages over every rapid. A trip was made to the west by Messrs. McConachie and Laidlaw. They entered muskeg in less than a quarter of a mile and went through it for twenty miles when a small stream (the Pewabisko River of the Indians) crossing the line, drained some two miles of land giving better timber. Beyond this another great muskeg country was entered. A sample of rock brought back from the Pewabisko River is a coarse pegmatite rock with biotite and large pieces of garnet. It is probably from a dike.

On the eastern bank of the river about three miles below Skunk Island is a portage trail cut out by our guide some years ago, and leading to a lake called by him "Wanzatika." This portage proved to be about three miles long. Two small streams cross it, the second about half way to the lake. These however cause very little change in the drainage, the portage being almost entirely through scrub muskeg except for a quarter of a mile at each end. The soil was heavier than usual, a stiff clay of a yellowish color being found near the first stream. The lake is about a mile and a half across from east to west, and some $3\frac{1}{4}$ miles north and south.

Rocks of Wanzatika Lake.—The rock on the west shore of the lake (that next the Missinaibi) is Laurentian gneiss, and the same was found at one place on the portage. Another bluff on the portage five-eighths of a mile from the Missinaibi was a very felspathic granite, probably an irregular dike in the Laurentian gneiss.

Slate.—The most remarkable rock seen was a fine grey slate that cleaved readily. This was found in boulders only on the west shore of the lake, and it was not found possible to explore the other shore in search of it. Our guide, who knew this region well, stated that the slate formed a part of of the far shore of the Wanzatika Lake and that it was also found on various rivers and lakes across to the Opazatika River. In view of the fact that the slate boulders are found nowhere on the Missinaibi River, and that the region mentioned by the guide is all Huronian it seems probable that he was right about the rock.

Timber and Soil.

The main part of the survey was north of the pine limit. No large pine was seen north of Lake Kabinakagami, and on the Missinaibi River the most northern pine was at the portage to Brunswick Lake. However, now that the pulp-wood industry has become so greatly developed in Ontario, the dense spruce and tamarac forests of the Moose Basin must be regarded as of great value. This spruce, though not large, is very dense, and covers an immense area. It may be said that our whole survey was through spruce and tamarac forest, the other timber being merely as islands among the spruce and as strips along the rivers.

Peat.—From the spruce and tamarac we come to the consideration of the peat on which they grow. Wherever this peat was poled it was found to be less deep than would be expected from its area, about one to five feet with an average depth of three or four feet. It is practically all formed of sphagnum moss. The moss itself is about a foot deep, then we have a foot or so of half decomposed moss of a light brown colour, and below this is the dark brown peat. Below the peat we invariably found a yellowish sandy clay. The upper part of the peat when sun-dried would probably form excellent moss litter. As a fuel it suffers from the fact of its not being deep, and from the thick timber making it hard to remove. In the sun-dried form it contains some 22 per cent. of water all of which cannot be driven off at boiling point, so that some artificial means of drying would have to be resorted to.

The value of the soil below this peat is increased by the shallowness of the latter, as it is probable that much of this peat-covered land could be drained and converted into an agricultural area. The soil consists universally of a yellowish or gray sandy clay loam. Beside the rivers, notably the Kabinakagami, there are cliffs of stratified material, and here the sand predominates over the clay, but these never go far inland from the river. The Hudson Bay factor at Brunswick House on Missinaibi Lake, has shown that a similar soil is capable of raising very good crops, especially roots and barley.

Animals of the Region.

The country covered by the survey still continues to bring in good returns of furs to the Hudson Bay Co. at Moose Factory, at the northern end of Missinaibi River. The quantity has probably not greatly decreased, and at the southern stations, although the returns are greatly less than they used to be, still a good hunter often gets several bears in a winter, and such ordinary game as red fox and crossed fox is trapped in numbers. Moose, caribou, red deer, bear, fisher, otter, fox (red, white and crossed), with a few beaver form the chief game of the Indians, together with small animals such as mink, skunk and great quantities of muskrats.

Tracks of bear were very often seen on our trip, especially on the Mattawishguaia River, and one bear was seen on the Missinaibi River. Some of these tracks were very large (a large one measured nine inches from heel to claw) and, though one bear makes many tracks, must show that the bear is common in these regions. Moose paths were often seen, and in the bush the moose were often heard at night, while caribou and red deer tracks were also common. Marks of beaver were also seen on many of the small creeks, notably on the Mattawishguaia River, near its source, above the portage from the Kabinakagami River. Wolves are seen only very occasionally and in winter, and the porcupine, so much valued by the Indians further south, is almost unknown.

Wild duck are fairly numerous, and in the fall a flight of geese would often be seen going south-west. These, the Indians told us, are very numerous at Moose Factory. Apart from this birds of all kinds are few, though far back in the bush the Canada bird and the Hudsonian chickadee can often be heard.

Good fish were found in places. Lake Esnogami and Kabinakagami had many pickerel and pike. Just above the head rapids of the Kabinakagami River our Indians caught 25 in less than an hour, but the Kabinakagami and Mattawishguaia Rivers both have few fish. Some speckled trout and some whitefish are caught. The Oba River is a noted place for the latter.

Special attention was paid to the insects of the region, as they probably affect the settlement of a country more than all the larger animals together. Not many enemies of timber were seen. The dreaded larchsaw fly has not yet spread as far north as this, but cimbex Americana was found together with several species of uroceidæ, but they do not seem to have done much damage. Near the Height of Land the dreaded boring beetle monohammus scutellatus was taken, but it was not very numerous. Very few borers were found in the burnt timber about the lakes. The old enemies of the settler, mosquitoes and black flies, were in their usual numbers, together with the little "no-see-um," the "bruleur" of the French Canadians. All of these continued till killed off by the frost in late September. A list of some of the insects is appended.

LIST OF INSECTS.

Coleoptera (kindly identified by Mr. H. F. Wickham, of Iowa City.
Carabus Servatus, Say.
Bembidium littorali, Oliv.
Psephenus lecontii, Lec.
Sericosomus incongruus, Lec.
Buprestis maculiventris, Say
Podabrus modertus, Say.
Podabrus (probably) peniphilus, Esch.
Thauasimus dubius, Fabr.
Xylotrechus undulatus, Say.
Acmæops pratensis, Larch.
Amæops protens, Kby.
Leptura Subargenlata, Kby.
Leptura 6-maculata, Lim.
Leptura chrysocoma, Kby.
Leptura vibex, Neum.
Monahammas scutillator, Say.
Donacia emarginata, Kirby.
Orsodachua atra, Ahr. var.
Bassareus maminifer, Newn.
Pachybrachys (near) Atomarius Melsh.
Diachees catarius, Snffr.
Adoxus obscurus, L. var. vitis. Fabr.
Chrysomela bigsbyara, Kirby.
Galerucella decora, Say.
Upis Ceramboides, Lium.
Macrobasis unicolor, Kirby.
Dendroctônus rufipenuis, Kirby.
Orthoptera (kindly identified by Mr. E. M. Walker, of Toronto).
Circotetlia veruculatus, Kby.
Nulauoplus extremus, Walk.
Nulauoplus islandieus, Blatchley.
Podesina glacialis, Scudder.
Lepidoptera.
Argyniis Atlantis, Edw.
Brentlies charichlea, schneider. Common especially northward. A circumpola species.
Vauersa j-albun, Boisd. and Lec.
Vauersa milberti, Godart.
Vauersa autiopa, L.
Grapta Famnes, Edw. Common especially northward.
Grafta progne, Cramer. Not common.
Basilarchia arthemis, Drury. Common throughout.
Oeneis macouni, Edw. One spec. at C.P.R. A species described from N. shore of L. Superior.
Chrysophanus epixanth, Boisd. and Lec. Common on burnt highlands northerly in August.
Lycalua Fulla, Edw. (?)
Colias eurytheme, Boisd. Common.
Colias interior, Scudder. Common.
Papilio turnus, L. A few especially southerly.

I have the honour to be, Sir,
Your obedient servant,

GRAEME M. STEWART, B A.,
Geologist of Party.

I, Graeme M. Stewart, of the City of Toronto, Geologist, make oath and say that the foregoing report is true to the best of my knowiedge and belief, so help me God.

Sworn before me at the City of Toronto, in the County of York, this 5th day of January, 1901.

GEO. B. KIRKPATRICK,
A Commissioner.

GRAEME M. STEWART.

SURVEYORS' REPORT OF EXPLORATION SURVEY PARTY No. 5.

SARNIA, ONTARIO, December, 1900.

SIR,—I have the honor to submit the following report on the exploration of the Kenogami River and adjoining country by Exploration Survey Party No. 5, under my charge, performed under instructions received from your Department, dated the 12th day of June, 1900.

The party left the railroad at Jack Fish station and crossed Jack Fish Bay to Empress Landing, where we took a wagon road to McQuaig's Lake, near the Ursa Major mine. A portage thirty chains long leads from McQuaig's Lake to Duncan Lake, sixty chains long. From the west end of Duncan's Lake, we crossed a two-mile portage to Owl Creek, a shallow, sluggish stream forty links wide, which we descended forty chains to Black River. Six miles up Black River we entered Lower Trout Lake, sixty chains long and passing through a narrows five chains long with swift current, reached Middle

Trout Lake, two and a half miles long and one-half a mile wide. Passing a narrows fifty links long, Upper Trout Lake, one and three quarters miles long is entered. Above the Trout Lakes we passed eight small lakes with short portages between them and reached the height-of-land portage, one hundred and eight chains long, at the end of which we came to Long Lake. The country passed through between Lake Superior and Long Lake is hilly and rocky, the soil generally being sandy and the timber small and scrubby. Long Lake is about fifty-four miles long and from one mile to three miles wide. The water is deep and fairly clear. Along the south part of the lake the banks are high and rocky, some of the hills being four hundred feet high. Towards the north, the banks become lower until at the north end of the lake, the country is quite flat.

There is a Hudson's Bay Company's post one and three quarter miles from the north end of Long Lake. About one hundred families of Indians bring their furs to this post.

The Kenogami River flows out of the north end of Long Lake, following a northerly course to the first portage, which is five and three-quarter miles from Long Lake. Three and one-half miles further down the Little Long Lake River enters from the west. Two miles further down, the Devil Fish River comes in also from the west. We ascended the Devil Fish River and Devil Fish Lake and other small lakes and a branch called Mink Creek to Head Lake, twenty-seven miles from the mouth. Here we crossed a portage one and one quarter miles long to Fleming's Lake, on the Kawakaska River, a branch of the Albany. We descended this river a distance of about forty miles. Two miles below Fleming's Lake we reached Kawakaska Lake, two miles long and one mile wide. Two miles lower there is a portage sixty chains long leading to Lake Wawong, a beautiful sheet of clear water three miles long by two and one-half miles wide, surrounded by a rolling sandy country, timbered with scattered small jack pine and poplar. Ten miles below Kawakaska Lake there is a portage of five chains. There is a fall of twenty feet at this portage ; eighteen miles further down there is a portage of five chains past a jam of logs. Five miles below there is another fall of twenty feet. These are the only portages on the Kawakaska River as far as explored. The Kawakaska River below Fleming's Lake is from two to five chains wide and generally deep with a sluggish current. The banks are usually low and covered with spruce, but occasionally sandy hills with poplar and jackpine are seen.

To the northwest of Lake Wawong there is a large lake called Eskinagagaini, said to be twenty miles long, the outlet of which is another branch of the Albany River. Between this lake and the Kawakaska River there is a belt of large poplar and spruce, some of the spruce being three feet in diameter four feet above the ground.

Messrs. Neelands and Proctor went up the Kawakaska River from Fleming's Lake as far as Mountain Lake, where they found a country very similar to that below Fleming Lake. Below the mouth of the Devil Fish River the Kenogami River flows north for four miles, and then turns north-east. Seven miles below the bend Twenty-mile creek is reached. This creek enters the Kenogami River from the north a short distance above the fourth portage. Twenty-mile creek is about seventy-five links wide; there is a slight rapid near the mouth with a fall of four feet, above which there is very little current for several miles. Two and a half miles up is Lake Waukouika, four miles long and one mile wide. Five miles above Lake Waukouika the creek became shallow and rapid and was not ascended any further. Half a mile above Lake Waukouika a branch creek twenty chains long leads to Lake Truax, two miles long and half a mile wide. From the north end of Truax Lake a creek leads through several ponds to Jemar Lake, two miles from Truax Lake. Jemar Lake is one mile long and one mile wide. The creek continued upward from the north end of Jemar Lake and Indians report two more lakes above Jemar Lake.

The country along Twenty-mile Creek and these lakes is generally flat, but around the lakes there are some rocky hills from fifty to one hundred feet high. The shores of the lakes are generally rocky. Below Twenty-mile Creek the Kenogami River runs nearly east to Pine Lake, which is twenty-six miles from Long Lake. This stretch of the river is generally about four chains wide, but narrows to about one chain in many places where there is a swift current. Pine Lake is ten miles long and one and one-half miles wide at the widest place. South of Pine Lake there is a chain of lakes and portages leading to Granite Lake and McKay's Lake at the head of the Pic River. Wawong or Clearwater Lake, three miles long, is the largest of the chain. There is also a chain of small lakes to the east of the south end of Pine Lake. This was followed to the fifth lake. The country around the south end of Pine Lake, and as far inland as explored is hilly. Towards Clearwater Lake the hills are of coarse sand, from two hundred to three hundred feet high. Towards the north end of the lake the country is more level and the soil better. The river leaves the north end of the lake and flows east three miles to Arm Lake, a shallow lake two miles long, surrounded by a slightly rolling country. Five miles further down is Kapessawatin Lake, a small lake with several small, flat islands and low shores covered with spruce and tamarac of good size. Little Pine River flows into Kapessawatin Lake from the south. This was ascended eight miles to Little Pine Lake, which is four miles long. Three and a half miles south of Little Pine Lake there is another lake one and one half miles long, the shores of which are rocky, and it contains several small islands. The country south of Little Pine Lake is hilly and rocky, the soil being mostly coarse sand. Around the north end of Little Pine Lake and along the Little Pine River below the lake there is a rolling clay country which extends back as far as explored. Three and a half miles below Kapessawatin Lake the Mouse River enters the Kenogami River from the south. This is a very shallow, rapid stream, and could not be ascended by canoes.

Below this for twenty miles, the river has a swift current with many rapids, at eight of which are portages. Sixty-six miles from Long Lake the Atick River is reached. It is shallow and rapid and not navigable with canoes.

From the Atick River to Pembina Island, ninety six miles from Long Lake, the Kenogami River is nearly all shallow rapids. Below Pembina Island the current was swift but water deeper as far as explored.

From Pine Lake down as far as explored, the banks of the rivers are generally clay from twenty to fifty feet high. Back from the river the country is level or slightly rolling with good clay soil as far as explored, but more swampy below the Atick River.

Table of Distances.

The following is a table of distances measured on the Exploration Survey.

Distances from Long Lake, following River.

Miles.		
0	Portage 1	5¾ miles
2¾	L. L. L. River	8½ "
1½	Devil Fish River	10 "
2¾	Portage 2	12¾ "
¼	Head of Long Rapids	13 "
4	Portage No. 3	17
3	Camp Aug. 15	20 "
1½	20 Mile Creek	21½ "
½	Portage 4	22 "
1	Portage 5	23 "
1	Portage 6	24 "
1½	Portage 7	24½ "
1½	Pine Lake	26 "
7¾	Outlet of Pine Lake	33¾ "
3	Arm Lake Camp	36¾ "
5¼	Kapeesawatin Lake	42 "
1	Kapeesawatin Camp	43 "
3½	Mouse River	46½ "
4½	Portage 10	51 "
1½	Portage 11 (Long Portage)	52½ "
5¼	Portage 12	57¾ "
¼	Portage 13	58 "
½	Portage 14	58½ "
1½	Camp, Sept. 7th, just above Portage 15	60 "
2	Portage 17	62 "
4	Atick River	66

Falls on Kenogami River below 16th portage. Party No. 5.

A grove of spruce and tamarac, average pulpwood, near the 16th mile portage, Kenogami River. Party No. 5.

Water power on Kenogami River. Party No. 5.

Falls and rapids on Kenogami River. Party No. 5.

On Kenogami River. Party No. 5.

Rapids on Kenogami River. Party No. 5.

Rapids and portage on Kenogami River. Party No. 5.

Long rapids. Party No. 5.

Kenogami River near Flint River. Party No. 5.

Rapid a second portage, Ombab ka Ri er. Party No. 6.

Kenogami River, 11 miles north of Pembina Island. Party No. 5.

Rap d a first portage Ombab ka River. Party No. 6.

DISTANCES FROM LONG ISLAND LAKE, FOLLOWING RIVER.—*Continued.*

Miles.			
3½	Portage 18	69½ miles	
½	Portage 19	70 "	
2	Camp, September 14	72 "	Rainy Camp.
8	Limestone Rapids	80 "	
2	Flint River Camp	82 "	
8	Camp, September 20	90 "	
6	Pembina Island	95 "	
4	Farthest North	100 "	Neelands & Howard.

FROM MOUTH OF DEVIL FISH RIVER.

0	Devil Fish Lake	4½ miles
7½	Outlet Devil Fish Lake	12 "
1½	Round Lake	13½ "
2	Arm Lake	17¼ "
4½	Mink Creek	21¾ "
2¼	Muddy Lake	23 "
3	Outlet Muddy Lake	26 "
1	Head Lake Portage	27
1	Fleming's Lake	28
1	Outlet Fleming's Lake	29 "
3	Kawakashgama Lake	32
2	Outlet Kawakashgama Lake	34 "
2	Wawong Portage	36
3	Camp August 2nd	41
1	Wawong Creek	42 "
2	Falls	44
10	Camp August 3rd	54
8	Log Jam Portage	62 "
5	Howard's Falls	67

I have the honor to be, Sir,
your obedient servant,

WALTER S. DAVIDSON,
Surveyor in charge of Exploration
Survey Party No. 5.

To the Hon. E. J. DAVIS,
Commissioner of Crown Lands,
Toronto.

LAND AND TIMBER ESTIMATORS' REPORT OF SURVEY EXPLORATION PARTY NO. 5.

WALKERTON, Ontario, December 8th, 1900.

SIR:—In connection with Exploration Survey Party No. 5, under your charge, I started my explorations from the north end of Long Lake House, where I found the country rocky, with some good farming land, parts of which were good clay loam. In the neighborhood of the Hudson's Bay Post they grow all kinds of vegetables—potatoes, turnips, carrots and other garden productions. I saw a most beautiful field of hay. It has been seeded for fourteen years and was a good crop. The manager of the Long Lake House rears cattle with which he tills his land. His potatoes, which were of different varieties, were very good and equal in size to any grown in central Ontario.

Leaving Long Lake House I went down the Kenogami River to the first portage which took us past the first falls, which are called the Kenogami Falls, about five and a half miles from the mouth of the river. In this stretch the land gradually became flatter and there was no timber to speak of. It was small scrubby spruce and tamarac. The land in this locality is fairly good and there is plenty of fall for drainage purposes.

From here I made a side trip to the east for a distance of about six miles, and from a tree-top could see the country for five or six miles further. The land was light, covered with a few scattered swamps. The timber was small spruce and tamarac, from six to twelve inches in diameter.

Returning I followed the river down for five miles where I came to the mouth of the Devil Fish River, a river coming in on the left. The Kenogami River widens here to from six to eight chains. In this stretch I found the land fairly level with a few swampy places. Parts, however, would be suitable for farming purposes, but considerable drainage would have to be done.

Starting from camp at first falls on this river (about half a mile from the mouth) I made a side trip of three weeks. On this side trip the land for the first six or seven miles was low and marshy, and would take considerable drainage to make it profitable for farming. The timber is small and scrubby.

The first lake I came to on this side trip was the Devil Fish Lake, a beautiful sheet of water in the form of an inverted "L," the length from end to end being about fifteen miles by one to two miles wide. This is one of the finest lakes in the district.

The fire which swept through this country some fifty-three years ago left only small timber of about four to eight inches in diameter. The shores of this lake are principally rock with a great number of exposed quartz veins.

The next lake I crossed was Round Lake, about two miles further where I found the banks fringed with small jack-pine, spruce and poplar. To the west, however, the land was very rocky. The land from this timber fringe is inclined to be swampy. I went inland for about five miles and climbed a tree whence I could see the country for miles and it was apparently the same. On the south side the land is not so swampy but light and of a fair quality. There is considerable poplar in this section.

Leaving this lake and following the river for a distance of two miles, I came to Arm Lake No. 1, the timber around which was about the same as I have just described, inland for about four miles being just the same. The land was generally low and swampy with some rocks on the north bank.

Leaving this lake I found the banks of the river fringed with small spruce and tamarac, also some poplar and jack-pine. I went up the river here for over eight miles where I found the above timber in the first four miles. Then I came upon a nice grove of spruce and tamarac. The land was generally marshy with a few sand ridges. I spent two days on this stretch of country and made a very minute survey. In the neighborhood of the camp I found a grove of jack-pine embracing over fifty thousand feet, also good timber for pulp-wood.

Time would not allow me to go any further up this fine river, which is a nice clean stream of two to four chains wide and quite navigable. From a tree top I could see the country for miles, and the river, while being winding, appeared to be much the same as the part just gone over, the land appearing to be the same also, which, with drainage, would be quite suitable for farming.

Returning I went up a small winding creek called Mink Brook which took me to Mud Lake, which is three miles long and very shallow with a mud bottom. This lake was fairly alive with wild fowl. On either side of Mink Brook the timber is small, the land low and marshy, while on the north east side of Mud Lake, the timber (composed of spruce, tamarac and poplar) is better in size and quality and is more healthy looking. The outlet of Mud Lake, a small creek running through a marsh of three-quarters of a mile brought me to another lake called Head Lake. The country around here looked better. The timber, though small, is very healthy looking There are some good spruce, tamarac, balm of Gilead and jack-pine, while the land, being somewhat light, is good, and farming could be successfully carried on. At the next lake, Fleming Lake, our party divided, the geologist and I taking our canoe across Fleming Lake, where we found conderable rock together with some good timber and land.

From Fleming Lake up the Kawakeshkagama River, I found for the first two miles timber of jack pine, poplar and balm of Gilead from four to eight inches in diameter. Further on I came to some sandy ridges, finding moss on top of rock on one side, while on the other side there was very fair land.

The next lake, Island Camp Lake, gave me only small poplar and white birch from three to six inches, the land being marshy with sandy ridges. Some two miles further I came to a small lake, not named on the map, around which was small timber, the land being light and sandy with a few sections of gravel together with a few rocks.

Up this river about two and a half miles I came to Egg Lake, where I found small timber of spruce, tamarac, jack-pine and poplar, which continued the same all around the lake. Sandy ridges were on three sides of the lake, north, east and west, while on the south it was low and marshy.

This lake is a fine sheet of water, and would prove a grand thing for a settlement; in fact nearly all the lakes and streams would be very beneficial for a settlement.

Following up the main river in a south westerly direction for about nine miles, I entered Mountain Lake, three and a half miles in length. The banks of the river leading into this lake are fringed with black alders, while further inland the timber is small spruce, jack-pine and tamarac.

Traces of a recent fire were quite apparent, and from a tree top I could see in the distance green timber which looked healthy. The land was generally low with a few sandy ridges, but fairly good soil.

In the neighborhood of Mountain Lake there are some hills rising from 100 to 200 feet above the general surface, the most remarkable being Granite Mountain on the south side of the lake, which is composed of granite and massive gneiss, and has an elevation of about two hundred feet above the lake. This portion of country is overspread with a yellowish sand, beneath which a considerable thickness of gravel is found, while back from these lakes the land is nice and rolling and of a sandy character. The timber around Granite Mountain Lake consists of white birch, tamarac, spruce, balsam, jack-pine and white cedar.

I learn from good authority that the same conditions prevail for many miles inland to the westward to Lake Nepigon and eastward to New Brunswick House on the Moose River, a country which will, without doubt, in the near future, become a very valuable section to the settler. With a railroad in this direction it will make as fine a portion of country as can be found in the whole Province of Ontario. Mountain Lake being the terminus of this extra side trip, I returned over the same water course, and after re-crossing Fleming Lake, I made note of the following large blocks of good timber, comprising approximately:—

Spruce, about	1,037,800 feet.	Board measure.
Tamarac	34,400 "	
Poplar	22,500 "	
Balm of Gilead, about	24,200 "	
Pulpwood	5,000 cords.	

This block of timber, which escaped the fire, lies between Fleming Lake and Head Lake, and is very valuable.

Leaving this lake I went up the Kawakashgami River to a lake of the same name, the timber along this stretch being small spruce, poplar and tamarac about four to six inches in diameter, and the land light and sandy, but very good.

From this lake up the same river about two miles, I came to a portage three-quarters of a mile long to the north east, which led me to a beautiful lake called Wawong Lake, six miles long, dotted with many islands and having many deep bays. Crossing this portage I came to a jackpine plain, then spruce swamp and again jackpine, the soil being quite sandy. A portage in the northern part leads to another lake, a small one called Duck Lake. Some of the bays on Wawong Lake are so closely connected that a portage of just twelve feet saved me a paddle of two miles. The land generally around this lake is sandy with some slopes of 50 to 100 feet high, while in other parts it is generally flat and fairly dry. The timber is spruce, tamarac and poplar of a medium size.

Coming back again and continuing down the Ka-wa River, I found a fall which is a fine water-power, with about fifteen feet head, natural fall, all rock. Then for about sixteen miles, the stream is nice and clean. Along this stretch of river I found some good spruce, while further inland the timber is not so good, the soil a black muck, but further down the land became gradually better, as did also the timber, especially the jack-pine and spruce; and where I turned back (being forced to do this for want of provisions) I found the prospects for good land and timber really good. The soil is inclined to be sandy near the water's edge, but further in is much better.

I here estimated a large block of timber to be as follows: Of spruce, about one million feet, board measure; of jack pine one million feet, board measure; of red pine one million feet, board measure, with similar blocks in the distance. Generally speaking, this seventy miles of exploration showed me a very fair portion of the country, a great deal better than some portions in the older parts of Ontario.

Arriving back at the mouth of the Devil Fish River (where I had my supplies cached) I made a short side trip up Little Long Lake River, which is about one mile up river from Devil Fish Lake, where I found an excellent waterpower, one fall of twenty-five feet of head, and another of thirty-five. The timber was fair, of spruce, tamarac and balsam. The land was stony near the river but very good in the interior, the soil being fine clay loam. A few sandy ridges could be seen in the distance, Returning again to the Devil Fish River, I continued down the Kenogami River, where for a distance of three miles I found the land good, a clay loam. Almost anything could be grown here, as the climate is similar to that of Old Ontario. The timber along this portion of the river is spruce, cedar, tamarac and balm of Gilead, some spruce being from eight inches to two feet in diameter. The largest trees are scattered.

About ten miles below the Devil Fish River, there is another excellent water-power with about twenty-five feet of head. The land is rocky for a few miles along the banks, while in the interior it is good clay loam. Another stream enters the Kenogami River, about twenty miles from Long Lake, which I went up for over ten miles, where I found, two and a half miles up, a nice lake which I named Pettypiece Lake. The fire has made great havoc through here, but the land is good and fairly high, the timber being second growth and small. Later on I discovered another lake, which I called Truax Lake. There are very high rocks around this lake, some being over eighty feet high. On the north side there is some good spruce. I also discovered two or three other small lakes. While on this side trip I noticed a number of very large trees, but they were scattered. The next stretch of river took me to Pine Lake, which is almost twenty-five miles from Long Lake. I met with much swift current and many rapids. A number of portages were crossed over on this stretch. There were many huge boulders in the stream and the banks were rocky. The timber along banks is good, measuring from six to fourteen inches in diameter. The land is rather rocky close to the river but very good further in. Pine Lake, a beautiful sheet of water, twelve miles long, is surrounded by high hills and rocks with numerous sand beaches. The timber is composed of groves of poplar, spruce and pine, all in a good healthy condition and of a fair size. The land is light and sandy. I made an inland trip to the east of Pine Lake which took me up a chain of lakes which I named Haggart, Montague, Pardee and Hyslop Lakes. The timber is small but with a good promise in the future. On the east shore of Pine Lake I discovered two very good copper showings. The mineral resources of this lake appear to me to be good. Arriving at the extreme end of Pine Lake, I found more high hills and rocks. The timber is small and the soil light and sandy.

From the north-eastern extremity of this lake, the country becomes gradually flat, and it puts on a different appearance, the land being clay and red muck. On Arm Lake No. 2, three miles further down Kenogami River, I found the timber small. The land is fairly good continuing so to Kapeesawatin Lake. The shores of the river are of gravel. From this lake I made a side trip up Little Pine River to a lake of the same name. The land being a beautiful clay loam up to and beyond Little Pine Lake, a distance of over fifteen miles, this section would be excellent for agricultural pursuits. One side of Little Pine Lake has a number of mineral exposures of copper and iron. The prospector and the farmer would find this a good place to locate. This is a beautiful lake, five miles long, clean and very clear. At the south end of the lake there is a beautiful natural grove, a park in fact, of spruce and balm of Gilead. The land is level and good and the soil of a clay loam. Some portions of the country are low but quite suitable for hay, etc. For ten miles the land was excellent clay loam. The neighborhood would be a good place for a settlement as I believe almost anything could be grown here. The large timber is scattered.

I made a side-trip from the south-west end of Little Pine Lake for over four miles over a trail, which led me to a beautiful lake which was afterwards named Proctor's Lake. The timber is small spruce, tamarac and pine. I came across a few rocks which had some quartz exposures. Upon my return I made a more minute examination of the country. I went into the interior on both sides for five miles and found the timber much the same as on the shores, small but good for pulpwood. The soil was "number one," a nice clay loam, quite suitable for farming purposes.

Arriving back to Kapeesawatin Lake, I continued down the Kenogami River, where I found the country very good to the Atic or Deer River,and these conditions continued the same to the Mouse River, three miles further down from the Atic River. The timber is very good for pulpwood and the land is a nice clay loam. A good settlement could be formed in this stretch of country, as everything necessary for the formation of one can be found here.

I made a side-trip from the mouth of Mouse River and found the land good. Then three miles further down I made another side trip of seven miles where I found the land first-class. It could not be better. There were a few scattered portions of swamp, but good soil generally, high and rolling. Timber, spruce, poplar, tamarac and balsam, promising to be a good size in the future. From a high tree-top could be discerned in the distance many spruce and poplar groves, comprising I should judge, from twenty-five to one hundred acres, while various small streams intersect these valuable lands.

Continuing down the river, I found the land becoming more flat with a few swamps. I found many good water-powers ranging from fifty to thirty-five feet head. The timber is good for pulpwood.

I made an inland trip from the Atic River of five miles, where I found the first half mile fairly good timber of spruce and tamarac, poplar and balm of Gilead. The land is good sandy loam, further inland the country became more flat. I met with some swampy places which could be drained. The timber is small. This portion, while not being quite as good as further up, is fair, and with a little energy and enterprise could soon be converted into good farming lands.

Leaving the mouth of the Atic River, I found the fringe of timber on the banks very good. The land for half a mile inland is good, while after that distance it gradually becomes flat and inclined to be swampy.

From the Atic River to the Flint River, some seven miles, the timber along the banks was good and the land fairly good. There are portions in this stretch of good clay loam, very suitable for farming. I encountered a fair block of timber in this stretch, which I estimate as follows:—

500,000 ft. board measure, spruce for saw logs. 100,000 ft. board measure, white wood saw logs. 100,000 ft. board measure, tamarac, saw logs. 10,000 cords of pulpwood. Some of these trees are three feet in diameter.

I made a side trip up the Flint River, a river flowing in from the east,and discovered a considerable quantity of good spruce, average in size of six to twelve inches in diameter, also some good tamarac. The soil is a good clay loam, while in other parts there is a good black muck soil. There is jack-pine and cedar on either side of the river. The land is generally flat. The same conditions appear to prevail for miles. From the mouth of this river to Pembina Island, where I turned around to come home, I found the country flat with good opportunities for farming. The timber is generally small but good for pulpwood; in fact there are portions of country where thousands of acres of pulpwood could be taken out. From Pembina Island camp I made a side trip on both sides of the river where I found the land very good clay loam The country was generally flat and the timber was of a fair size and would make good pulpwood.

Estimated Distances from Long Lake, following river:

	Portage I	5¾	miles.
2¾	L. L. River	8½	"
1½	Devil Fish River	10	"
2¾	Portage II	12¾	

Estimated Distances from Long Lake, following River.—*Continued.*

¼	Head of Long Rapids	13	miles.
4	Portage III	17	"
3	Camp, Aug. 15	20	"
1½	20 Mile Creek	21½	"
½	Portage IV	22	"
1	Portage V	23	"
1	Portage VI	24	
½	Portage VII	24½	"
1½	Pine Lake	26	"
7¾	Outlet of Pine Lake	33¾	"
3	Arm Lake Camp	36¾	
3¼	Kapusawatin Lake	42	"
1	" Camp	43	"
3½	Mouse River	46½	
4½	Portage X	51	
1½	" XI Long Portage	52½	
5¼	" XII	57¾	
¼	" XIII	58	
½	" XIV	58½	
1½	Camp Sept. 7, just about Portage XV	60	
2	Portage XVII	62	
4	Atic River	66	
3½	Portage XVIII	69½	"
½	" XIX	70	"
2	Camp Sept. 14	72	
8	Limestone Rapids	80	"
2	Flint River Camp	82	"
8	Camp Sept. 20	90	"
6	Pembina Island	96	
4	Farthest North	100	"

From North of Devil Fish River.

	Devil Fish Lake	4½	
7½	Outlet D. F. River	12	
1½	Round Lake	13½	
1¾	Outlet Round Lake	15¼	"
2	Arm Lake	17¼	
4½	Mink Brook	21¾	
2¼	Muddy Lake	23	
3	Outlet Muddy Lake	26	
1	Head Lake Portage	27	
1	Fleming Lake	28	
1	Outlet Fleming Lake	29	
3	Kawakashgami Lake	32	"
2	Outlet Kawakashgami River	34	"
2	Wawong Portage	36	
5	Camp August 2nd	41	
1	Wawong Creek	42	"
2	Falls	44	
10	Camp August 3rd	54	"

From Mouth of Devil Fish River.

8	Log Jam Portage	62	"
5	Howard's Falls	67	"

The length of the territory explored from Howard's Falls to Pembina Island is about one hundred and twenty miles, and the average width explored (including what was observed from hills and tree-tops) about twenty-five miles, making the total area explored three thousand square miles, of which half is good soil and one-third timbered. That is fifteen hundred square miles, or nine hundred and sixty thousand acres of good land, and one thousand square miles, or six hundred and forty thousand acres of timbered land,and half of this, or three hundred and twenty thousand acres good pulpwood or timber.

This is a fair estimate of timber and land observed on the district.

I am, sir, your obedient servant,

DANIEL PROCTOR,
Land and Timber Estimator.

W. S. DAVIDSON, Esq,

Ontario Land Surveyor,
In charge of Exploration Party No. 5,
Sarnia, Ontario.

I, Daniel Proctor, of the town of Walkerton, in the County of Bruce, Land and Timber Estimator, make oath and say that the foregoing report is correct and true to the best of my knowledge and belief, so help me God.

DANIEL PROCTOR.

Sworn before me at the town of Walkerton, in the County of Bruce, this eleventh day of March, 1901.

OTTO E. KLEIN,
A Commissioner.

GEOLOGIST'S REPORT OF EXPLORATION SURVEY PARTY NO.5.

LINDSAY, Ont, Jan. 3rd, 1900.

Sir,—In accordance with instructions received from the Director of the Bureau of Mines, I left Toronto on June 21st, 1900, and proceeded via Owen Sound and Fort William to Jack Fish Station on the C.P.R., where I joined Mr. W. S. Davidson, O.L.S., who was in charge of party No. 5, for the exploration of the Kenogami River Basin. By steam-tug and wagon we reached a point 5 miles north-west of the extreme end of Jack Fish Bay, on the Government road to the Ursa Major Mine, and thence proceeded by a chain of small lakes to Owl Creek, a tributary of the Black River, which is affluent to Lake Superior.

Ascending the Black River and its lake expansions, we arrived at the head of Long Lake on July 13th Owing to continuous head winds and a double trip with the boats, we were unable to reach our initial point, the outlet of Long Lake, till July 24th.

We descended the Kenogami River 12 miles to the mouth of the Manitounamaig or Devil Fish River, which we used as a base from which to explore the country north-west of Long Lake House. Ascending this stream and its lake expansions, Manitounamaig, Round, Arm, Muddy and Head Lakes, and crossing a portage a mile and a quarter in length, we reached Fleming's Lake on the main water system of the Kawa-keshkagama, a stream said by Indians to be tributary to the Albany River. From this lake we explored the river southward to its source in Bell's and Mountain Lakes, and north-westward to a point about thirty-five miles below Fleming's Lake. Beyond this point the river was not followed, and the party returned to the Kenogami River, the furthest point reached in this expedition being about fifty miles in the straight line from that river.

From the Devil Fish River we descended the Kenogami River to a point about thirty chains above the Fourth Portage, when we paused to explore Twenty Mile Creek, which entered here from the north and including lakes is about 15 miles in length. Continuing the descent of the Kenogami we arrived at Pine Lake on July 25th. From the South-east Bay exploratory trips were made south-west and south-east for about six miles.

Below Pine Lake the examination of the river was continued to Arm Lake, the country north and south of which was explored on foot. About four miles below this the river enters Kapeesawatin Lake, from which two main trips were made; a line was run in 7 miles on the northwest side of the river, and on the south-east side of Little Pine River was ascended for eleven miles to a lake bearing the same name. Beyond this a three-mile portage was crossed and a lake tributary to the same water system was examined.

Below Kapeesawatin, the last of the lake expansions of the Kenogami, the river was followed to a point eleven miles below Pembina Island, side trips being made at numerous places along the route. The country was carefully examined on both banks, the Mouse, Dead Sucker, Mindidino, Atic and Flint Rivers being used for that purpose.

On Sept. 24th the ascent of the river was begun, the party arriving at Long Lake House on Oct. 9th. Here a division was made, Mr. Davidson with the main party journeying to the railway via Long Lake and the Black River, and myself and one man descending the Pic to Heron Bay. The object of this was to examine the country north of McKay's Lake and upper Pic, which was reported by Indians to be copper bearing. After spending several days prospecting in this vicinity, we began the descent of the river and reached the railway at Heron Bay on Oct. 22nd after an absence of about four months.

The Kenogami River.

The instructions of the Crown Lands Department demanded an examination of the geology of the Kenogami River basin and some of the country adjacent to it, but not actually drained by this river, the whole district to be explored being upwards of eighty miles in width at the outlet of Long Lake. At this point the Kenogami River is from two to three chains in width, and is deep and sluggish for 7½ miles to the first portage. With the exception of a sandy bar at the outlet of the lake, this whole stretch of the river is navigable for steamers of light draught. Clay along the banks, the soil inland soon becomes sandy, and in many places swampy. Fire has at some time swept the country, and only small scattered spruce and tamarac fringe the bank.

A few exposures of gneiss occur at the north end of the lake, but no rocks are seen on the river down to the first portage. Here the river plunges over a ledge of gneiss, approaching syenite, striking northeast and dipping south 80 degrees, forming Kenogami Falls.

Below the falls the river is swift and deep, with clay banks well timbered with spruce, poplar, birch and tamarac, up to 15 inches in diameter. At distances varying from five to fifteen chains from the river the country changes to sandy, sparsely timbered with small jack pine. About three miles from the portage the Kenogami-shiesh, or Little Long Lake River, enters from the west, and a mile lower down, the Manitounamaig, or Devil Fish River. Both streams are nearly two chains wide at the mouth, and below them the Kenogami River has an average width of about four chains.

The next stretch of the river, to the second, or Island portage, two and a half miles in length, is swift, and in one place occurs what might be fairly called a rapid. Numerous exposures of coarse-grained Laurentian gneiss occur along the banks, striking north 70 degrees east, and having a roughly perpendicular dip. Below this portage the river is wide for about twenty chains, and then narrowing suddenly to about a chain and a half, runs swiftly through a kind of gorge with high rocky banks of gneiss or syenite. Near the foot of this rapid the river bifurcates around an island, the eastern channel being the best for canoes. This rapid may be avoided by a portage of sixty chains on the the west bank over dry, level ground.

For the next mile and a half the river is fairly sluggish, but becomes swifter at the head of the next rapid, which is about six chains long over Laurentian holders and may be run with a small load. One mile further down the river is a fall twenty-five feet high, over a ridge of interbanded gneiss and mica schist. This necessitates the third portage.

Below the falls the river is about eight chains wide and very shallow, with reedy banks and several marshy islands, for about one mile, and then narrows to about three chains and runs swiftly between high banks of a fine-grained gneiss, striking generally northeast. Spruce, jack pine, poplar, birch and balm of Gilead, up to twenty-four inches in diameter, occur as a belt of timber on both banks, which vary in width from a few chains to half a mile. Inland the country is sandy except in the immediate neighborhood of running water and is usually burnt.

Several small dikes of a dark and very fine-grained trap, cutting across the strike of the gneiss at an angle of about 90 degrees and dipping north 85 degrees were observed in this stretch, which is terminated by the fourth portage. Twenty chains above this, Twenty Mile Creek enters from the north. The portage avoids a rapid about twelve chains long over boulders. At the foot occurs an exposure of gneiss striking and dipping 18 degrees north.

From this point to the seventh, or Thirty-four Chain portage, the river is swift and there are portages at two points, but these rapids can be run with care. A few exposures of gneiss were observed but they in no way differ from those previously described.

One mile north of the portage the river enters Pine Lake, a large open body of water about nine miles in length, which is divided into two parts by a channel about ten chains wide three miles from its outlet. The shores of this lake are usually high with sand and gravel beaches, and sometimes, especially on the south shore, high rocky bluffs.

On the shore of the South West Bay occur some fine-grained binary granites, and in one place underlaid by a fine-grained gneiss is a small exposure of pale green dolomite. In the southeast bay the eastern shore rises abruptly to a height of from twenty to sixty feet in hills of red gneiss in which occur several dikes of dark green trap, similar to those described on the Upper Kenogami River, striking north, eighteen degrees east, and dipping west eighty-five degrees. On the west shore is an exposure of a very fine-grained diorite schist, distinguishable as such only under the microscope, the contact with the eastern rocks being lost in sand beaches. South of this lake, except in the hollows, the whole country was burnt for at least five miles and probably far beyond that distance.

About two miles from the south end of the lake and immediately opposite the mouth of the Kenogami River, in a dike of fine-grained trap, occur three distinto parallel veins of iron pyrites, with some chalcopyrite. The first vein is twenty-seven feet wide, the second, eighty feet north, nine, and the third, one hundred and fifty feet north, fifteen feet. The strike and dip of both the dike and the veins is similar, north, thirty degrees west, and north, forty degrees. The surfaces were considerably weathered, and the vein, could not be traced inland owing to the soil covering the rock. Several samples from these deposits gave on assay a trace of gold and nickel.

From a short distance north of this point to the narrows between the lakes a beautiful graphic granite prevails, though on some islands mica schist is seen. The country inland on both sides of the lake is burned and is growing up with jackpine on the high ground and spruce and tamarac in the swampy parts.

No exposures occu r north of the narrows except some gneiss at the extreme north end near the outlet of the lake.

For about a mile below the lake, the Kenogami River has a strong, swift current to the eighth portage, some gneissoid rocks occurring on this stretch. Immediately north of the portage is Arm Lake, the main body of which lies at right angles to the general course of the river and is about three miles long.

Gneiss outcrops occur in a few places on the northeast shore. The country to the southeast of the lake is low and swampy, with spruce and tamarac trees. Further inland some ridges of poplar and white birch appear. The shores are high on the east and west sides of the lake and are well timbered with spruce, tamarac, balsam, jackpine, poplar, white birch snd balm of Gilead, up to fourteen inches in diameter.

Below Arm Lake the river is wide and sluggish with marshy banks for about two miles, where it narrows to about three chains, and a shallow rapid about a mile and a half long occurs, terminating in Kapeesawatin Lake. This rapid may be run without danger, with the exception of a short chute which necessitates a portage of about five chains over rocks. About seventy chains below its head is a portage sixty chains long leading to a bay on Kapeesawátin Lake. Laurentian gneiss occurs on both banks of the river at the rapid, striking generally, northeast.

Kapeesawatin Lake is described as a shallow reedy lagoon, with several islands, formed by the junction of the Little Pine and Kenogami Rivers. It is in no place over half a mile wide, and is about two miles in length. No rock is seen anywhere on the lake, unless a few boulders of gneiss or granite at the bottom of the rapid be so regarded. The shores are well covered with a fringe of timber, but the country inland, of which a description will be given latter in this report, is universally burnt, except in the swamps.

North of Kapeesawatin Lake the river flows sluggishly for about a mile, and breaks into a series of small rapids and chutes over boulders. On the bank at one of these chutes is a small exposure of a fine grained gabbro carrying some traces of iron pyrites. The river here is about five chains wide, and the banks are fairly well wooded. Near the foot of the rapid the Wahigano, or Mouse River enters from the east

Coarse-grained mica schists outcrop frequently to the 10th portage, where the river plunges suddenly through a short gorge, the walls of which are composed of very coarse-grained granite or gneiss.

About a mile below this is the 11th portage. The river between these portages is broad and shallow, with numerous islands. The rock along the bank is a coarse-grained, much weathered syenite, approaching diorite. Below this portage, which is about a mile and a half long, the river is swift for several miles to the 12th portage, the rock being a coarse hornblende syenite. This rock prevails all the way to a point about three miles below the 15th portage. The river throughout this stretch differs in no way from that immediately above, except that its width increases to an average of about five chains.

Between the 16th and 17th portages, and half a mile below, the Atic River, which enters the Kenogami River here from the west, occurs an exposure of sandstone on the left bank of the river. The rock is calcareous, dark green in color, and is composed of coarse-grained quartz material in a fine ground mass. In this bed occur numerous small orthoceratites, closely crowded together. About three miles below the Atic is the 17th portage over coarse-grained gneiss or syenite, and about half a mile further down the 18th, which is the last portage on the Kenogami River, and below which there is no interruption to navigation all the way to the sea.

The last exposure of gneiss occurs about three quarters of a mile below the 18th portage, and the character of the river alters considerably and bears a marked resemblance to the Lower Moose and Abitibi River. The width is here increased to about ten chains and the river is shallow enough to be fordable almost anywhere. The banks are from ten to twenty feet high, and are crowned with a thick fringe of poplar, balm of Gilead, white and red birch, spruce, tamarac and balsam, which usually extends inland about twenty chains or half a mile. Between the timber and the low water level the bank is usually grown up with thick alders, coarse grasses, etc. The bottom of the river is composed either of flat lying, unaltered, paleozoic strata, which are seldom seen above water, or of pebbles and fragments of the same material. Beyond the belt of timber on the banks is an almost continuous muskeg which prevails over the whole country, except in the immediate vicinity of running water. The current is swift and strong, and in many places extensive shallows and bars make fairly hard rapids.

From the 18th portage to the most northerly point reached is over thirty miles, and in this stretch the Mindidino and Watistiquam rivers enter from the west,and the Pewona or Flint from the east. The best idea of the river at this point will be given by quoting an average example of the original notes, below Pembina Island.

20.00 Low banks, spruce and tamarac, 12 inches in diameter. Rapid 3.00 long, fall 18 inches.

30 00 Strong current; deep water; river 7.00 wide.

30.00 Very strong current; limestone exposure; well marked horizontal strata, dull yellow in color, not fossiliferous.

80.00 River 15 00 wide,high banks,large poplar, birch, spruce and tamarac, soil sandy.

60.00 High limestone cliff on north bank, pale greenish-grey in color, not fossiliferous; all flat, shallow rapids; fall 8 feet; island 15.00 long.

No fossils were found in any of the limestone exposed, but in the shingle and gravel on the bars and banks many fossils, mostly silicified, were seen. They were usually too worn to be recognizable, but among them occurred specimens of paleozoic corals, such as Favosites and Halysites, similar to those which are described later in this report as occurring on Flint River in limestone deposits, which are undoubtedly contemporaneous with those on this part of the Kenogami River.

The last exposure seen was in a cliff fifty feet high about eleven miles north of Pembina Island. The rock here was a soft greyish-green argillaceous limestone having well defined horizontal strata. The country inland was burnt, and growing up with poplar, birch, spruce, balsam and tamarac.

Little Long Lake River.

About three miles below Kenogami Falls the Kenogami-Shiesh, or Little Long Lake River, enters from the west. At the mouth it is about two chains wide and has a strong current. About forty chains up a rapid over a ledge of very coarse-grained mica schist necessitates a portage. The trail is on the north bank, and is not well cut out, owing to the fact that the river is little used by the Indians, as they prefer a more direct route from Little Long Lake to Long Lake House *via* a chain of small lakes.

The river was ascended four miles to Proctor's Rapids, and is rough and shallow all the way. With the exception of the first, all the rapids in this stretch can be poled. At Proctor's Rapids the river runs through a narrow gorge about fifteen chains long, and a portage must be made. Above this rapid the upper reach of the river bears off to the south-west.

The rock above the first portage is an extremely cleavable hornblende schist striking north 35 degrees east, and dipping south 45 degrees, iron pyrites occurring in minute particles in most exposures examined. The banks are low and sandy, and are well timbered with spruce, tamarac, jack pine, poplar and balm of Gilead.

Country North-West of Long Lake.

The Manitounamaig or Devil Fish River, enters the Kenogami River about eleven miles below Long Lake. Close to the mouth navigation is interrupted by a rapid which is evaded by a twenty-six chain portage on the north bank. Above this the river is rough and shallow, with numerous small rapids, all of which can be run to Devil Fish Lake. Gneiss outcrops here and there along the banks, the average strike being north sixty degrees east, and the dip south 70 degrees.

On Devil Fish Lake gneiss occurs everywhere, usually banded with coarse-grained mica schist. The strike and dip, with a few local exceptions, are the same as in the rocks on the river. From the outlet of the river the course is about north-west for five miles. Here the main body of the lake reaches off to the south for about twelve miles, and the river enters at the angle formed by the two arms. The shore line is irregular, and in places the country is burned, but generally the shores are well timbered with spruce, jack pine, tamarac, etc. In no part of the country explored were so many evidences of Indian occupation seen as here.

Above the lake the river is a tortuous stream leaving Round Lake about a mile and a half above Devil Fish Lake. An exposure of coarse red gneiss, generally similar to those on the lower river, was observed about midway between the two lakes.

Round Lake is about one mile wide and about two miles long. The north and south ends are marshy, but some exposures of gneiss and mica schist occur on the east shore. On the south-west side of the lake are extensive gravel beaches, composed largely of paleozoic pebbles, with many fragments of silicified orthoceratites. The country to the east of the lake is a thick swamp in which no bottom could be found at eight feet. The timber was small scrubby spruce and tamarac.

Above the lake the river winds its way through two miles of swamp to Arm Lake, a shallow lagoon about a mile long. North of the lake a ridge of fine-grained gneiss about fifty feet high occurs, striking north east. The country inland on both sides of the lake was burned for many miles, the soil being sandy and growing up with young birch, jack pine, etc.

One mile above Arm Lake is a small chute over a ledge of gneiss, striking north seventy-five degrees east and dipping north seventy degrees, and about three miles further up Mink Brook enters from the east. The main river was followed for about four miles farther through a flat sandy country, which may be described as old brulé growing up with young jack pine.

Ascending Mink Brook the only interruption to canoe navigation is a small chute near the mouth. Here the gneiss strikes north seventy-five degrees west and lies a vertical dip. Four miles above this the river leaves Muddy Lake, a shallow body of water about a mile and a half across. A creek half a mile long joins this lake with Head Lake, and from a point half way up the west side of the latter a portage one hundred chains long leads to Fleming's Lake, an expansion of the Kawa-kesh-kagama River, which is a tributary of the Albany River.

Fleming's Lake is about six miles long and two miles wide, but very irregular in shape. The banks are usually high, with sand beaches and sometimes bluffs of gneiss striking about west. The country in the vicinity of the lake is well timbered with large spruce and jack pine.

For about two miles the upper continuation of the river passes through level country, clay along the banks but sandy farther back, to Duck Lake, which is an open body of water lying at right angles to the general course of the stream. A few bluffs of gneiss are seen along the shores. Above the lake is two miles of river to Egg Lake. No rock occurs in the stretch, except the gneiss so characteristic of the district. Egg Lake is about two and a half miles in length, and two rivers enter at the south-west end. On a point between these occurs a coarse-grained mica schist and some decomposed gneiss. A short distance from the mouth of the western stream is an outcrop of fine-grained diabase and about a mile and a half above this is Bell's Lake. This lake is about a mile and a half long, with sandy shores, but no rocks are seen. Above this the river was followed for about five and a half miles till canoe navigation was finally interrupted by thick alder bushes in which the river loses itself. From a tree top the country was seen to be burned for a long way in every direction.

Ascending the southern river, about ten chains brings us to a small marshy lake, above which, with the exception of one small rapid, the river is deep and sluggish for five miles to Mountain Lake. Near the outlet of the lake the shores are rocky, and inland the country is very hilly. The only rock is a fine-grained granite. The total length is about five miles. From a bay south-west of the outlet is a portage leading to a large lake, which, according to Mr. Godchere, the officer in charge of Long Lake House, flows westward to Lake Nepigon.

Below Fleming's Lake the river is rapid for about fifteen chains, and then flows quietly for about three miles to Kawa kesh-kagama Lake. The river is deep with marshy shores generally, but in one place occurs an exposure of mica schist and gneiss striking north 65 degrees east, and dipping north 80 degrees. On Kawa-kesh kagama Lake the course is down the left shore for about two miles. Gneiss is found on the east shore and on several islands, striking about north 55 degrees west.

Two miles below the lake a portage leads to Wawong Lake, a beautiful and very irregular sheet of water, with sand beaches but no rock exposures. From its extreme north end a portage runs to a small lake from which a short canoe route leads to Oskanaga or Bare Bones Lake, which is said to be twenty-five miles long and flows north to the Albany River.

Below the Wawong Portage the river was followed for about twenty-five miles. The country is burned, with large sandy hills covered with small jack pine, and some extensive exposures of red gneiss for about five miles. Several small rapids occur in this stretch, which terminate at Rupert's Falls. A short distance above this occurs an outcrop of Huronian chlorite schist, and similar rocks are seen as far as the river was explored. The strike of this series of rocks was about north-west, and the dip north 60 degrees. Below the falls the river is swift, with high clay banks, well timbered with large spruce, poplar, tamarac, birch, balm of Gilead, up to thirty inches in diameter. Numerous exposures of chlorite schist occur everywhere in this stretch. Some samples from small quartz veins in one extensive exposure striking north 15 degrees west, gave a trace of gold on assay.

There are two more rapids in this part of the river where portages are necessary. The first, Driftwood Portage, is eight chains long, the trail being on the right bank, and the second, Howard Rapids, with the trail on the left bank. Here the river narrows to about ten feet and plunges through a miniature canyon for fifteen chains, forming a beautiful and picturesque series of chutes and waterfalls.

Owing to the limited time at the disposal of the party no attempt was made to descend the river farther; but it is said by the Indians to afford a good canoe route from the Albany River, owing to its freedom from rapids.

At several points along the north-east bank of the river portages lead to Lake Oskanaga, and on the south-west is a canoe route via a tributary stream to Lake Nepigon..

Twenty Mile Creek.

Twenty chains above the Fourth Portage on the Kenogami River, Twenty Mile Creek enters from the north-west. It is about one chain wide at the mouth, and is the outlet of several small lakes. A short distance from the mouth navigation is interrupted by a small chute, and above this is a fairly continuous succession of small rapids for about three miles to Pettypiece Lake. No rocks "in situ" occur in this stretch, the rapid being formed by Laurentian boulders and gravel bars. The country is universally burnt, and inland is sandy or swampy.

Pettypiece Lake is about 3½ miles long, and varies in width from twenty to sixty chains. The banks are usually low, but in places bluffs of red gneiss rise to a height of eighty feet. In the north end of the lake are many islands, most of which are entirely gneiss banded with coarse mica schist striking north 70 degrees east. Inland the soil is sandy, and for a short distance from the lake is well timbered with spruce, jack pine, tamarac, poplar and birch up to about 15 inches in diameter. Numerous sand beaches occur, and in some places coarse gravel, composed largely of pebbles of paleozoic limestone and some undistinguishable silicified fossils.

Above the lake the upper continuation of the river winds through a flat marshy country for about half a mile, when it divides, the north branch constituting the outlet of several small but beautiful lakes which fill the valleys of a rough Laurentian district, and the western branch passing through flat level country, frequently cedar swamp, finally becomes too small for navigation about six miles above the junction. No exposures of rock were seen on the western river, and the gneiss and mica schist of the northern in no way differed from similar rocks, already described, on Pettypiece Lake. The timber of the whole country drained by the upper rivers is that usual in the country, jack pine on the high land, and in the wetter portions spruce, tamarac, etc.

South of Pine Lake.

From the south-east bay of Pine Lake a portage 70 chains long leads to Haggart's Lake, a pretty sheet of water one mile square. A few chains of creek connect it with Montague's Lake, which is about twenty-five chains long. From this a portage about three chains long leads to Pardee's Lake, which is about one mile long. Four chains south of this is Hyslop's Lake forty chains long, and from it a portage fifty chains in length runs to another lake, which was not explored.

On the west side of Haggart's Lake are gneiss hills ranging from forty to one hundred feet in height, burnt and quite bare. The country on the opposite shore is similar in character but is free from burn. Where timbered, spruce, tamarac, poplar, and birch occur in large quantities and of fair size. The strike of the rocks is generally about north east and the dip roughly vertical.

Southwest of Pine Lake the country is hilly and is burned everywhere. It is growing up with small jack pine and birch. Everywhere the soil is sandy, except where hills of gneiss appear. No traces of mineral were observed in the district, except some fragments of hæmatite among the gravel on the beaches. A chain of small lakes is said to be the source of a creek flowing into this bay.

Little Pine Lake.

The Little Pine River enters Kapeesawatin Lake at its south-eastern extremity. It was ascended to its source in Little Pine Lake. No exposures of rock are seen throughout its entire length of eight miles. The banks are clay, which extend for twenty chains to half a mile inland and then change to sand. The timber is spruce and tamarac near the banks, and inland largely jack pine of small size.

About twenty chains below Little Pine Lake the river passes through a shallow lagoon half a mile wide, and from this point to Kenogami River there is no obstruction to navigation.

Little Pine Lake is about four miles in length, and averages perhaps one mile in width. The banks are high and the soil clay loam when not in the immediate vicinity of rocks. About 2½ miles from the outlet on the west shore a fine-grained hornblende schist occurs, striking north-west and dipping south-west 15 degrees. Twenty chains south is a considerable exposure of pegmatite, which reappears at the next point and covers a large area around the south end of the lake.

On the east side of the lake a cliff of hornblende schist occurs, the strata being lifted at the point of contact by the pegmatite. A little farther north the hornblende again occurs with well marked strata, in the folds of which are small deposits of iron pyrites with actinolite. No signs of this mineral existing in any quantity were observed.

A portage from the southwest end of the lake about three miles long, over sandy soil and rocky hills, leads to a lake one and a half miles long. The rock in this neighborhood was all a pale flesh-colored pegmatite. A river which constitutes the outlet of the upper lake flows parallel to the portage. The timber about the lake is of fair size and similar to that on the river.

Kapeesawatin Line.

North of Kapeesawatin Lake a line was run in for about 4½ miles, and the country was explored for several miles farther. Except in the wetter parts the whole country was burned. In the swamps small scattered spruce and tamarac were the only timber.

Two small exposures occur near the line, both of fine grained mica schist. About two miles farther north the country is swampy, and outcrops of pale grey gneiss or sheared granite occurs. The soil near the lake is a clay or sandy loam, and is possibly capable of producing light crops.

Atic River District.

From the mouth of the Atic River a line was run northwest for 4¼ miles. The soil is a sandy loam near the river, but after about half a mile the country was swampy as far as explored. Spruce, jack pine, poplar, tamarac, etc., occur near the river, and inland small spruce and tamarac. From a tree top at the head of the line the country as far as could be seen was a great muskeg, open in places, with small scattered trees. No rock was seen in this region.

Flint River.

The Pewona or Flint River enters the Kenogami River about ten miles above Pembina island. At its mouth it is over two chains in width and its water is of that characteristic red hue which is usual in streams drawing their water supply from the muskegs. In the lower part of the river the banks are low and at a very short distance from the river the ground becomes swampy and soon merges into muskeg. Numerous small, shallow rapids occur throughout the whole distance explored, but all may be ascended by poling.

The timber along the bank is sometimes of a fair size, spruce, tamarac, poplar, birch and balm of Gilead occurring up to twenty four inches in diameter. Trees of this size, however, do not occur at any distance from the river, except in rare instances and over small areas, the timber a short distance from the water being small, thick and scrubby. About five miles from the mouth an extensive growth of jack pine was observed on a sandy bluff, none of the trees being over twelve inches in diameter.

Gravelly banks prevail in low water, the pebbles being composed mainly of limestone, with some irregular fragments of schist and some representatives of the gneiss rocks further up stream. About five and a half miles from the mouth occurs an exposure of light brown limestone approaching dolomite. The strata were well defined and roughly horizontal. In this deposit were found the following fossils :

Zaphrentis.
Cornicula prolifica.
Favosites species.
Orthis species.
Cyathophylloid coral, probably Heliophy Elum.
Syringopora perelyaris.

These fossils are sufficient to determine this deposit as of Upper Silurian age and it is probably contemporaneous with the Niagara limestones or Guelph dolomites. This was the only rock exposure seen in the thirteen miles of the river which was explored.

A fair idea of the nature of the country adjacent to the river may be given by quoting the original notes on an average mile :—

8.00—Considerable cedar 20 inches in diameter, but knotty and twisted. Poplar and balm of Gilead up to 18 inches in diameter.

18 00—Rapid 2.00 fall 2 feet ; banks 6 feet high ; spruce and tamarac 15 inches on the bank, inland a few chains average 4 inches.

20.00—Creek .40 wide enters from west banks ; gravelly limestone and schist pebbles ; sandy bank 10 feet.

15.00—Rapid fall 2 feet 3 inches long.

20.00—Banks low ; spruce and tamarac 12 inches in diameter ; coarse grass, "Blue Joint," alders ; strong swift current.

The water volume of this river, like that of all others in the district, is subject to enormous variations, the high water mark being in places ten feet above the water level on September 18th. The stream has two banks, a flood bank often ten or twelve feet high, and a low water bank about two feet high. The ground between these two varies from a few feet to a chain in width and is frequently covered with alder bushes.

Pic River District.

About one and a quarter miles south of the outlet of Long Lake the Making Ground River enters from the south. It is about one chain wide at the mouth, and is used for about eleven miles in following the canoe route to Lake Superior. Two small rapids occur in this stretch, but portages are not necessary. From this river a portage over the height of land one and one and three-quarter miles long leads to Mud or Shallow Lake. A few chains north of this and parallel to the trail is the rollway used by the Hudson's Bay Co. in bringing their batteaux from Lake Superior. Mud Lake is a shallow lake with high rocky banks, extending for about four miles in a north-easterly direction. North of the lake, and connected with it by a small creek, is Hollow Rock Lake, which is about two and a half miles long. From the north end a trip was made inland for some distance, the country being rolling and rocky and sparsely timbered with jack pine, spruce and tamarac.

From Mud Lake a creek one and a quarter miles long leads to Granite Lake. There is one short portage in this stretch. Crossing the southern end of Granite Lake another creek fifty-five chains in length leads to Yankee Bay on McKay's Lake. This lake is about twelve miles long and is very irregular. From the north-west bay a canoe route via a chain of small lakes leads to Powgutchewan Lake, eight miles to the north. This lake is tributary to the Kenogami River, and the river draining it enters the main stream about twenty-five miles below Pembina Island. From the north-east end of McKay's Lake flows the Pic River, a shallow, winding stream seldom exceeding a chain in width. This stream, which flows into Lake Superior,, affords a good canoe route to the railway at Heron Bay.

A massive red granite occurs towards the mouth of the Making Ground River and also on the south-east side of Mud and Granite Lakes. About the north end of McKay's Lake is a greyish syenitic rock which passes over to the north end of Granite and Hollow Rock Lakes. On a point one mile south of the river mouth on McKay's Lake is a dark fine-grained mica schist containing small garnets. At summit portage, over the height of land, and about the south-west end of Mud Lake, occur dark green fine grained hornblende schists, striking about north 70° east and having roughly a vertical dip. On some of the islands in Mud Lake a dark green diorite schist was observed, which may be regarded as porphyritic owing to the presence of crystals of hornblende. Several trap dikes were observed in this region, especially in the vicinity of McKay's Lake, striking north 20° east. Below McKay's Lake the country around the upper Pic River is generally sandy a short distance from the banks, and may be regarded as unfit for agricultural purposes. Numerous hills of light-colored gneiss are seen, the general strike being north 80° east. Everywhere the country is burned, and growing up with small jack pine except in the neighborhood of the river and in the wetter valleys, where spruce and tamarac grow to a diameter of 24 inches.

Geological Features.

Laurentian rocks, consisting of gneiss, granites, mica schists, and syenite, occupy the country to the north of Long Lake and north-west to the lower Kawa-kesh kagama River. On the Kenogami River these rocks, with the exception of some gabbro and fine grained hornblende schists immediately below Kapeesawatin Lake, extend as far north-west as the eighteenth portage. Huronian rocks, mainly chlorite and other soft green schists, occur on the Kawa-kesh-kagama River from about four miles below the Wawong portage to the northern limit of exploration, and also on the Little Long Lake River. Here, however, the rock is hornblende schist. In the Pic River country, rocks of Huronian age, fine hornblende and diorite schists, occur on the Making Ground River and about the west end of Mud Lake. About half way down the lake a belt of massive red granite occurs, which extends across to the south end of Granite Lake, and north of this is a belt of greyish syenite, which extends from McKay's Lake across to Hollow Rock Lake. About two miles north of the eighteenth portage on the Kenogami River unaltered Silurian strata appear, and sedimentary rocks occupy the whole district from this point to James' Bay. From the fossils observed there is no doubt that the rocks in this district belong to the Niagara period, and possibly are contemporaneous with the Guelph dolomites.

Mineral Resources.

No mineral deposits existing in such conditions as would make them of economic importance were found in the district. Iron pyrites occurs in considerable quantities on Pine Lake, but it carries only very small traces of gold, nickel and copper. Traces of this mineral were also observed on Little Pine Lake, but assays revealed nothing of value. The most prominent district is the country on the Kawa-kesh-kagama River below the Wawong portage. Here Huronian exposures are numerous, mostly chlorite and other soft green schists. Several samples from small quartz veins in this district showed traces of gold, and it might be that careful prospecting in this district would be rewarded.

Timber Resources.

Much of the whole region is burned, and as, according to Mr. Godchere, the whole country was fire swept fifty-three years ago, no timber of great size is to be found in the district, except in parts where nature in some way has protected the trees. The best timber district seen was between the Kawa-kesh-kagama River and Lake Oskanaga, where extensive groves of large spruce and tamarac up to thirty-six inches in diameter occur. Generally, it may be stated that the timber everywhere along the river banks is of fair size, though liable to be very knotty, but inland there is, practically speaking, nothing of commercial value. The poplar, which occurs everywhere along the river, is usually of fair size, and is singularly free from "black heart," which should make it very valuable for pulp-wood.

Agricultural Possibilities.

According to Mr. Godchero of Long Lake House, the climate of the district is similar to parts of the Northwest lying in the same latitude. Frost is very uncommon during the summer months, and all the ordinary garden vegetables are grown without difficulty. Barley and oats mature successfully, but wheat has not been attempted. A tomato patch was observed on the shores of McKay's Lake near a deserted Indian camp, but whether a success or not I am unable to say, as when examined it was partly covered with snow and the crop had been picked. Good land does not occur in large areas, but in the neighborhood of the streams there is usually a strip of arable land of excellent quality. In the northern part of the district I am of the opinion that the muskeg country is valueless, as it is too flat to be drained. Generally there seems no reason why the clay areas should not under cultivation support industrious settlers.

Powitik River below Summit Lake. Party No. 6.

About 4 miles below Alma Lake on the Kapikotongwa River. Party No. 6.

Burton's Falls, north branch of Ogoki River. Party No. 6.

Indians running rapids, Ogoki River. Party No. 6.

From a tree top, up the Ogoki River. Party No. 6.

Second portage, Ogoki River. Party No. 6.

Running rapids on Ogoki River. Party No. 6.

Right hand side of upper or main falls, Amy's Fall, Ogoki River. Party No. 6.

Amy's Falls, right hand side of upper or main fall. Party No 6.

Amy's **Falls**, main fall. **Party** No 6.

Amy's Falls. Party No. 6.

Rapids at foot of **Amy's Falls, Ogoki River. Party No. 6.**

FUR AND GAME.

Most of the small fur-bearing animals found in Northern Ontario occur in this district, including mink, otter, fisher, marten, fox and muskrat. Beaver are becoming very rare, and few traces of their work were seen. Only two or three moose are killed in a year by the Indians who trade at the local post, but they are said to be coming into the district from the east. Caribou are fairly common, especially in the neighborhood of Pine Lake. Bear is the commonest big game, over a hundred being killed annually in the Long Lake House district. Ducks and partridge are very numerous, though the former are shy and hard to obtain. Geese do not breed in the district, but are plentiful in the spring and autumn months. Pike, pickerel, salmon trout and suckers are found in most of the lakes and streams.

WATER POWERS.

The water volumes of the rivers in the district, while subject to great variations are in many cases sufficiently great to develop considerable power, and will probably some day constitute an important supply of energy for the commercial enterprises of the lake shore towns. The following may be quoted as falls important from this standpoint :—

	Falls.
Kengami River—	
Kenogami Falls	20 feet
Falls at 3rd Portage	20 "
Falls at 7th Portage	32 "
Falls at 10th Portage	12 "
Falls at 11th Portage	70 "
Little Long Lake River—	
Lower Rapids	20 "
Proctor's Rapids	20 "
Devil Fish River—	
Lower Rapid	30 "
Kawa kesh kagama River—	
Fleming's Lake Rapids	15 "
Rupert's Falls	18 "
Howard's Falls	20 "
Pic River—	
Sandy Hill Falls	140 "
White Otter Rapids	27 "
Lake Superior Falls	40 "

Before concluding this report I wish to express to Mr. Davidson my sincere thanks for continual assistance afforded me during the summer, and also to Mr. Peter Godchere of Long Lake House for most generous hospitality and valuable information. My gratitude is also due to Professor Coleman, Geologist of the Bureau of Mines, and to W. A. Parks, Ph.D., of Toronto University, for valuable assistance in the preparation of this report.

Your obedient servant,

E. V. NEELANDS,
Geologist, Party No. 5.

W. S. Davidson, O.L S.,
In charge of Exploration Survey Party No. 5.
Sarnia. Ont.

TORONTO, January 7th, 1901.

I, the undersigned, do declare and say that to the best of my knowledge and belief the above report is correct in all particulars.

Sworn before me at Toronto, in the County of York, this seventh day of January, A.D. 1901. } E. V. NEELANDS.

GEO. B. KIRKPATRICK,
A Comm'r, etc.

LAND AND TIMBER ESTIMATOR'S REPORT OF SURVEY EXPLORATION PARTY No. 6.

UTTERSON, MUSKOKA DISTRICT, ONT.,
December 18th, 1900.

SIR,—I beg to submit to you my report on the timber and land examined by me in northern Ontario in connection with Exploration Party No. 6, in charge of Ontario Land Surveyor J. M. Tiernan, who died of typhoid fever since his return home. The examination of the district allotted to our party commenced at the north-east shore of Lake Nepigon. We went up the Ombabika River and we used the canoe route to the Albany River as our base, exploring and examining the timber and land on both sides, and then taking the Ogoki River from what is called the French channel, exploring both sides until we again reached the Albany River some distance below Martin's Falls, and we took for our base the Albany River, exploring the south side only until we reached the Kenogami or English River. Owing to the late date of our party in getting started from Lake Nepigon, and the extensive country to explore and get out of again before navigation closed on the inland lakes and rivers, I took the advice of my chief that a distance of about ten miles on either side of our base would be as far as we could examine, but of course I saw a great deal more of the country outside of this ten-mile limit at different points and times. It would be presumption on my part to attempt to give an accurate estimate as to the quantities of timber or areas of farming land, as nearly the whole of our long journey across this wonderful country was through vast forests of great commercial value and large areas of good farming land; and in examining the timber, I decided that the only way would be to estimate about how much spruce and jack pine only it would yield in pulpwood per acre. Carefully examining this timber, I am safe in saying that as regards 40 per cent. of the country examined by me the timber is suitable for saw logs. I regret to report that I found some large areas of burnt country. One of these, the largest, extends from a chain of small lakes south of the Ogoki River as far north as Lake Kaginogame. I find in these burnt districts in every case a new crop of timber growing in a healthy condition. I also found two small fires started by careless Indians in camping, both of which I succeeded in extinguishing. I have drawn a map showing location of burnt districts or sections, lakes and rivers and courses travelled in my explorations. I also send my field book with notes entered each day, the distance in every case being estimated.

In commencing our work we went up the Ombabika River and, as we ascended this river, as far as we could see from our canoe, both banks are well timbered. The land along the banks is mostly sandy; and about ten miles up this river from Lake Nepigon I was instructed to make my first exploration at right angles from the river, and in this trip for the first mile was rolling sandy soil timbered with small white birch, spruce and poplar, and then a rocky country evidently an old brulé, as it is now grown up with small jack pine and scrub spruce, and the timber in this exploration would only cut out about ten cords of wood per acre. On the north-west side of the river, and some three or four miles further up stream in this exploration, the timber is much better, and the land rolling, with some nice sandy loam flats, broken by rocky ridges, and I put the pulpwood, jack pine and spruce only, at about twenty cords per acre. The balance of timber is white birch, some poplar and balsam. Further up the Ombabika River and south we found a splendid spruce and jack pine growth around Robinson Lake. This lake is about eight or nine miles south-east of the river and flows into the Ombabika River by a stream called Robinson River. This stream flows through a valley of low marshy land, with a rolling rocky country back from the river and well timbered. Two streams flowing from north-east are tributary to Robinson Lake, with splendid spruce along both streams as far as I saw them. The land in this exploration is not farming land. I put the cut of pulp wood at about thirty-five cords per acre. The balance of the timber is small tamarac and poplar. Ascending the river to Summit Lake and exploring both sides of the river, no farming land is found. There are some small flats along the banks, but they are low and swampy, and produce some fine large thrifty tamarac. Back from the river the land is rolling and rocky in the low places. We found splendid spruce, and some poplar, and on the slopes and tops of hills white birch and jack pine. From the forks, that is from the mouth of Robinson River, to Summit Lake, a cut of thirty-five to forty cords per acre is about what we would get there. Then we have a fine lot of good tamarac, the remainder of timber being balsam, poplar and white birch.

Leaving Summit Lake the water flows north and we found that we were getting into a better country. We saw some fine clay land along the Ponitick River and with the exception of some burnt places, the timber is good. Some distance down this river flows through the end of a lake (see map). In going out from there I got into a fine jackpine country as far as explored. It is fine long straight timber, some of the largest sixteen to eighteen inches in diameter. Here we did not find much rock, but about the lake, and a small one further inland, are sandy rolling hills. Just here we left this fine timber, turning north, but before doing so, I went down this river about nine miles north-east through a splendid forest of spruce and jackpine. I should have liked to explore this stream down to a greater distance, as in my opinion it flows through one of the most valuable timber districts in the north. From the most reliable information which I afterwards received, I am confirmed in my opinion that this river flows through a vast forest of spruce and jackpine until it joins the Kenogami or English River at a distance of nearly two hundred miles, and in this whole distance it is said by my informant that there is not two miles of burnt country.

Turning north we followed the canoe route and at some distance we came to a chain of small lakes. At the first one we met the country is burnt and rocky east and west of this lake for miles. I should say that this burnt district has been a finely timbered country, as large parts of burnt trees are standing or lying, measuring from ten to sixteen inches in diameter, and I notice that in all of the burnt districts a new crop of young timber is growing, so that at the end of forty years, if protected, we shall again have a fine forest instead of burnt and charred stumps.

We continued on through burnt country to the Otter River and further on the Ogoki, and we saw some fine clay flats at the forks, but burnt and rolling rocky land in the back ground. We again turned and left the Ogoki River and went north to Lake Mahamoigama. The country about this lake is rough and rocky and burnt for miles. This burnt district continues on to Lake Kaginogame and at the south end we found the end of the fire district. But it follows up the north-east side about four miles further. This is a fine lake, teeming with lake and speckled trout and contains several large islands with valuable timber. In this lake section we again found splendid spruce, jackpine, balsam, poplar and some white birch. West of this lake is Trout Lake, a fine sheet of water and surrounded by grand timber, but the land in this section is rocky with small sandy flats. Further north above we entered the Cat tail River, a small stream, and in going up this river a short distance and then walking on through a rocky piece of country, we got into some extremely large jackpine, we continued on down stream until we arrived at Sucker Lake. The timber in this section is the best I have met with on this route; there is a slight brulé on the north-east side, but it does not go inland far. The spruce in this section, and from lake Kaginogame down to Eagle Rock Lake, is rich. East and west from these lakes as far as examined by me, is one grand spruce and pine forest. The land is not farming land, except some small flats and some of these are clay, but they are not extensive, and I put the estimate of pulp-wood, spruce and pine at about forty five to fifty cords of wood per acre. I may say a few words about the fishing in this section or chain of lakes. I was on the Nepigon River for two seasons some twenty years ago and saw some wonderfully fine trout, but I never saw them so large and plentiful as they are in this stream. At the time our party was going down this stream to the Albany River, and, returning, we agreed not to say how large the trout were, as if we did it would be taken as a fish story. But from Lake Nepigon to the Albany River the streams and lakes are full of fish of large size. From Eagle Rock Lake or just a short distance the river flows into a long lake with the Albany River flowing through the north end, and the timber on both sides of this lake through to the Albany River is second growth, small spruce, jackpine and poplar, with dead spruce of good size standing in the low ground. Both shores are rocky and rolling and, except the flat along the Albany River the country is not farming land.

Our explorations of the Ogoki valley commenced at the French channel. We followed this great water-course northeast and as we went down we noticed that the valley of the Ogoki River became more uniform with some fine clay flats. On the southeast side we found green timber, but north is a continuation of the brulé. A fine water power is some distance down from the channel. This power is capable of great development, with splendid sites for mills on either sides, with a fall of about twenty-five feet. The country undergoes a change. We had rocks behind us and found a wide valley of splendid sandy loam for some distance. This is really excellent land. We saw no more brulé but the timber is second growth for some distance below this fall increasing in size as we went on. The river flows through splendid forests of spruce and through a fertile country of good land, until we came to a large lake and, just at the inlet, and for some little distance along the shores of the lake, it is a low and sandy shore, though further on in places there are some rocky points. Now, from the fall near the French channel to this splendid lake we have a valley from twenty to thirty miles in width of rich, fertile land well suited for settlers. The timber in this stretch will cut from sixty-five to seventy cords of wood per acre, besides a considerable percentage of large saw-logs, as I examined the country on all sides of this lake. We named this Lake Tiernan. I was surprised at its rich soil and valuable timber. The south and west shores of this lake, as far as my exploration went, comprise one rich valley, except a few points jutting out in the lake. On the north and west sides it is more rolling but not rocky, with a small brulé in a deep bay, but outside of this brulé the timber is good until we came to the outlet of this lake. We found a good deal of poplar. There are two channels flowing out of the lake, but they join into one about a mile down stream, and just at the outlet in a deep bay there is a canoe route from this lake to Long Lake. The Indians say they go through to Long Lake House in five or six days' light. As we passed on down through splendid forests and rich land we found in several places (but they are not extensive), some old brulés of years ago, and in these localities the timber is smaller, and some distance below Lake Tiernan the river again divides. We took the south channel where we passed some good water powers. This division forms an island about thirty-five or forty miles long and ten miles wide, with excellent land and timber throughout the whole length of the island, except the lower end of a recent brulé which extends from one side to the other, and another, I should say an old one,—at the extreme end. I found in exploring the south side of this channel that we were still in this wide valley with clay and fine timber, but, as we got further down, it would seem as if the country back from the river became lower and wet, as the moss that we found, more or less in all of this north country, becomes deeper and we found in some places bogs or open swamps of good size. Again we found, perhaps in our next inland exploration on the same side, a much higher country and such we found in going south from the end of Long island. Some distance in from the river we ascended a hill of about thirty feet higher than the banks, and saw south of this a rolling country for some distance. On the north side of the river about one mile from the forks there is a splendid water power with a fall of twenty feet, and in going in from these falls it appears that there was a fire of some years ago. It left some fine large spruce standing along the banks of the stream and some belts inland. In this fire belt I made a careful examination of the soil. It is an easy matter, for as you go along you may kick the rich clay loam up with your boot and if there had been a deep moss on this side it must have burnt. We found the same kind of land and timber on down to the mouth of the Jackfish River, a stream flowing into Ogoki River from the northwest. In examining up this stream I found that it flows through a clay country until where another stream comes into it and at the forks are some sandy points, but inland we got clay. We found some fine timber along this stream as far as explored, and southeast from this point inland from the river, is low and wet land and the timber small. I found in putting down a pole some deep moss or peat down to the depth of eight or nine feet. I also found old timber away down and covered up with moss. Now, I believe that at the first this whole valley has been one flat country, that the ice which forms in winter on this river to a good depth in breaking up in the spring jams the narrow places, causing the water to rise higher than the original banks, and that when this water subsided it left a sediment on the banks. This keeping on year after year the result to-day can be plainly seen, as both banks are certainly higher than the country inland, for in walking in from the shore you go down an incline. This sediment and rubbish has completely stopped up the entrance to the river of any small streams, thus the water in the background cannot get away except by soaking through the banks, and of course this is slow and the result is that miles back in this flat country it is wet. From time to time timber falls, as it does in any evergreen timbered country, with the deep snows, and as the fallen trees sink deeper the moss grows over them, so that we have in these wet places a woody matter of moss, peat and water to a good depth. Now, sir, I venture to say that when this great country is opened up that if there were deep drains cut through these flats we would see a splendid result, and I believe that the land back along the Albany River is in the same state. There is a good fall in all the streams, the banks are high just below Jackfish River, the banks of the Ogoki River are rocky in some places, but the rocks disappear as you go inland. There is clay soil on both sides down all the way to the Albany River, but the timber below the last fall is not nearly so good. A good belt of spruce and poplar grows on either side, but inland it is small and as we joined the Albany River it is of no value; and as we examined the country south of the Ogoki River we did not find any timber in the limit of ten miles of any commercial value, nor did I see any land, except in some two or three spots that is not covered with moss or peat to a depth of in some places twelve feet.

In concluding my short report, I am pleased to say that from the mouth of the Ombabika River to the Albany we have a rich country of timber: that in the Ogoki valley we have got a rich country both in timber and land; and that the area of this rich land is so extensive as to be nearly beyond belief. Here we have homes for the millions with great timber resources, a country teeming with fish and game, also water power to turn the wheels of a nation, and a climate that if we take this year as a criterion is fine as I have ever seen either in Muskoka or Parry Sound districts or even the county of Simcoe, and in proof of this our party saw no frosts until the 25th of September. When we were on James' Bay we saw in a garden at Fort Albany potatoes, their tops as green as they would be in August. In this part of Ontario they grow all garden produce that we grow here where I live, and Fort Albany is nearly one hundred and forty miles north of Moose Factory. At this latter place on October 3rd they had not removed their garden produce for the winter, and as I walked through their gardens and saw such fine vegetables I thought I was in some other part of the country. Here they grow tomatoes, cucumbers and everything we grow further south. The Hudson's Bay Company have a fine farm and grow enough fodder to feed fifty cattle and seven horses, and they do not know how or pretend to farm. All through October it was fine and warm, as in September in this country; we only having two nights of frost up to the 28th of October.

Estimate of Pulpwood.

The following is an estimate of pulpwood timber on the Ombabika River, etc.:—Ten miles each side from Ombabika Bay to the Albany River, a distance of about one hundred and forty miles as per route taken and explored by Exploration Party No. 6:

The first sixty miles of this route and twenty miles wide, I estimate the pulpwood at thirty cords per acre all through. 768,000 acres at 30 cords per acre making 23,040,000 cords.

The next twenty-four miles is burnt country with no timber on it.

The remaining fifty six miles of this route I estimate at forty seven cords per acre. 716,000 acres at 47 cords making 33,652,000 cords.

The whole, 116 by 20 miles wide of this route will average about 38 cords per acre, or the 1,484,800 acres explored will cut out about 56,346,400 cords of pulpwood.

Estimate of pulpwood on the Ogoki River.—Ten miles each side commencing about ten miles northeast of French Channel and following the river through to the Albany River, a distance of some 140 miles, as per route taken by Exploration Party No. 6: I estimate that the pulpwood along this river will cut on an average 44 cords per acre, or 1,792,000 acres will cut 78,848,000 cords. I also estimate that out of the above 1,792,000 acres explored, that 1,500,000 acres of it is good farming land.

SUMMARY OF ESTIMATES OF PULPWOOD.

	Acres.	Cords pulpwood.
Ombabika River ·		
	1,484,000	 56,346,400
Ogoki River :		
	1,792,000	 78,848,000
Total acres	3,276,800	Total cords....135,194,400

Trusting this short report will be satisfactory,

I am, yours respectfully,

J. L. HANES,
Land and Timber Estimator,
Exploration Survey Party No. 6.

HON. E. J. DAVIS,
Commissioner of Crown Lands,
Toronto, Ont.

Sworn before me this eighteenth day of December, 1900, at Utterson.
W. G. STIMSON, J.P., } J. L. HANES.

GEOLOGIST'S REPORT OF EXPLORATION SURVEY PARTY, NO. 6.

PETERBOROUGH, Ont., Dec. 27th, 1900.

Sir,—Under instructions from Mr. Archibald Blue, director of the Bureau of Mines, I accompanied as geologist Ontario Government Exploration Party No. 6, Jos. M. Tiernan, Esq,, O. L. S., surveyor-in-charge. The country examined was that along the Ombabika canoe route between Lake Nepigon and the Albany River, the Ogoki River from where the above route leaves it at the French Channel to its junction with the Albany River, and the south side of the latter from the Ogoki River to the Kenogami River. My instructions were to report on the geological and typographical features of the district to be explored, and also, on its timber, soil, water powers and in short on everything of economic or scientific interest.

Leaving Nepigon Station on the Canadian Pacific Railway, on June 28th, our canoes heavily loaded with supplies, camp outfit, etc , we proceeded by way of Nepigon River and Lake Nepigon to the mouth of the Ombabika River, where our work was to commence. The Ombabika River flows into the lower, or southerly end of the bay of the same name at the north-east corner of Lake Nepigon, and here we arrived on July 14th.

We enter Ombabika River by a wide, shallow, marshy mouth, which however, soon narrows down to two chains while the low bushy shores give place to banks of clay 30 feet high and thickly wooded with small spruce, tamarac, poplar and balsam This continues up to the first portage which is twenty chains in length on the left bank of the river. The portage is past a rapid where the river narrows to some thirty feet. It has a fifteen foot fall, would be easily dammed, and there was water enough going over to make it a good water power. Here there is an outcrop of medium-grained grey gneiss cut by several narrow dikes of diabase. At the contact of the dikes with gneiss most of the dark minerals in the latter are altered to light yellowish-green, probably epidote.

Scarcely a quarter of a mile further up the river is portage two, four hundred paces long, also on the left bank, past a rapid eighteen feet wide with twenty-five feet fall and a splendid water power. Grey gneiss is seen here again, but is more schistose than below.

For the next mile and a half the river is from one and one-half to two chains wide, with low clay banks showing gneiss at the water's edge, and sloping up gradually back from the river. Then comes portage 3, one-fifth of a mile long on the right bank past a fall of 25 feet and a good water-power. A dark grey mossy micaceous gneiss much folded and crumpled and cut by numerous veins of pink granite occurs at this portage. The strike is north, 85 degrees west.

For four and a half miles above the third portage, the river continues with slack current, through two small lake expansions, between banks usually low and rocky. The rock along here is a light grey gneiss with darker, more micaceous bands, and striking a little south of west. At the upper end of the second lake expansion is a massive grey granite or granite gneiss with two sets of very perfect vertical joints crossing each other nearly at right angles. The country on both sides is low, broken and rocky with a thick surface covering of moss, and timbered with rather small spruce, tamarac, poplar and balsam, a lot of which would, however, cut pulpwood.

At the end of this stretch a rapid with a fall of 4 feet necessitates a portage of fifty yards on the left bank over a band of dark fine-grained hornblende schist, broken up and apparently enclosed by a coarse-grained reddish granite.

Another mile of river, also with two small lake expansions, brings us to a portage of sixty-four chains on the right bank. This, the fifth portage from Lake Nepigon, cuts off about eight miles of river in the form of a long, narrow bend. The trail runs easterly from the river over ridges of reddish coarse-grained gneiss striking south 80 degrees west, with troughs of muskeg between. About half way across the portage a dark green trap dike, at least 300 yards wide, occurs. One edge is covered by the swamp, but the irregular contact of the other can be traced for some distance. The contact the dike is very fine-grained and compact, but towards the centre is lighter-colored and much coarser-grained and forms the body of a hill 40 feet to the right of the trail. The timber along the trail is scrub, and the country rocky. At the upper end of the portage is an Indian grave covered with a tomb of logs 2½ feet high, with, at the head, a cross, a flag-pole with small flags of colored cloth, a little piece of firewood, birch-bark vessels and torch, a clay pipe and a black bottle. A trip inland here shewed the country to be substantially the same as along the river, i.e., granite gneiss and muskeg, much small poplar, birch and tamarac mixed with spruce and pitch-pine.

Above, the river banks are mostly of banded gneiss, bands of light and dark grey, for the two or three miles to the next portage. This is 50 yards long on the right bank past a small, shallow rapid. Grey and pink gneisses much folded and contorted and carrying bands and fragments of a dark, finer-grained hornblende schist occur here. There is also a small burned tract at this portage. The included pieces of schist are often badly broken and sometimes faulted with the interstices filled with gneiss, and both gneiss and schist are cut by small veins of a coarse-grained granite and of glassy barren quartz. The strike of the gneiss is west-south-west. A trip into the bush on the left bank showed the same sort of country and timber as at portage five.

Shortly after leaving the sixth portage there are two small rapids up which canoes must be poled or waded, but the remainder of the two and a half miles to portage seven is slack current with small lake expansions. Exposures of gneiss are quite frequently seen along this part of the river, and the general aspect of the country is rocky and broken with very fair spruce, tamarac, poplar and pitch pine. The seventh portage is one-tenth of a mile long on the left bank and leads over clay land through a good spruce bush. The obstruction to canoeing here is a rapid, made up of a four foot fall at the top over a ridge of gneiss with a boulder rapid below.

Just above the seventh portage a branch almost as wide as the main stream enters the Ombabika River from the right. This is the outlet of a small lake about seven miles up. Two rapids have to be passed between the Ombabika River and the lake. The first of these, one mile from the mouth, is passed by a sandy portage on the left bank. The second, two miles further up, is a long, shallow, broken rapid, part of which must be portaged and the rest waded. Grey gneiss outcrops at the first rapid, and also at several places on the river's banks. The latter are, however, usually low and inclined to be marshy. At the second rapid is grey gneiss, folded in places, and carrying included fragments of dark, finer-grained schist, and cut by irregular veins of granite and felsite. From the second rapid to the lake there is no fixed rock and the banks are low, wet and covered with small spruce and tamarac. A short distance back from the river, however, the quality of the timber improves greatly and plenty of spruce and jack pine up to 8 inches or ten inches in diameter are found. The lake, of which this river is the outlet, is about two and a half miles long, and a half mile wide, with the long axis running north 15 degrees east. The westerly shore is an extremely hard, coarse-grained, reddish granite, showing a very perfect jointage. The easterly shore is low and sandy or marshy. Back from the lake the country is broken by low hills of rock and sand, but is well covered with good spruce, poplar, pitch-pine, tamarac, etc. The best timber is that around the north end of the lake where two small, drift-wood choked, creeks enter it. There is spruce here up to 16 inches diameter. Numerous caribou tracks were met with a short distance west of the lake but none of the animals themselves were seen.

Returning from this lake to the Ombabika we continued on up the latter one and three-quarter miles to the next rapid which has a fall of seven feet. The portage is a quarter of a mile long on the right bank and over rock and muskeg. From here to Pigeon Lake the river flows through an extremely level piece of country, apparently alluvial flats. It is very crooked, sluggish and often obstructed by sand bars and fallen trees. This is an excellent piece of farm land but of limited extent. Spruce and tamarac on these flats frequently attain a diameter of 10 to 20 inches.

The west shore of Pigeon Lake consists of a hard grey and reddish granite gneiss, well jointed. About three quarters way up the lake a band of hornblende schist 20 feet wide and striking southwesterly occurs. Near the north end of the lake is a small dike of diabase porphyrite striking south-east. The eastern shores of the lake are low and sandy or bushy. There is no good timber around the shores.

We leave Pigeon Lake by a narrow, weedy channel, winding through marsh for half a mile to a portage one-tenth of a mile long on the left bank, leading over level clay land. Above the portage the river continues the same to Goode's Lake.

Just before reaching Goode's Lake, however, a crooked, shallow, weedy stream enters the Ombabika River from the left. This we named Sound River. It flows from Mud Lake, a small lake half a mile in diameter, a few miles up. The lake is surrounded by hills of gneiss and granite, well wooded with spruce and pitch pine.

The west shore of Goode's Lake consists almost entirely of grey and red gneiss. The easterly shore, like that of Pigeon Lake, is low and sandy. Green spruce, pitch-pine and poplar surround the lake.

We leave Goode's Lake by a portage of 37 chains on the right bank of the river, past a shallow boulder rapid. An outcrop of coarse-grained, reddish, porphyritic granite gneiss with porphyritic crystals of felspar often 1½ inches long occurs at this portage. The boulders in the river bed along here are also worthy of mention. They consist of green stones and boulders of a dark, fine-grained compact rock with numerous hemispherical holes. The green stones no doubt come from the Huronian rocks at Cross Lake, but the dark rock, which proved to be arkose, was not found in place at all, though large sub-angular boulders of it were very numerous.

From Goode's Lake to Cross Lake the river flows through a low, flat country timbered with poplar, spruce and tamarac. There is one more portage. This is the eleventh from Lake Nepigon and the last before crossing the height of land at Summit Lake. Above this portage, porphyritic gneiss outcrops occasionally up to Cross Lake.

Bluffs of red and grey granite form the southerly shore of Cross Lake but around the rest of it are found only green Huronian schists and diorites. These green rocks strike north 65 degrees west to north 70 degrees west and dip almost vertically. They consist of soft fine-grained green mica schists, hornblende schist, hard quartzite schist, diorites and dark fine-grained slatey rocks. On the west shore of the lake these schists are cut by a number of small, irregular veins of quartz. A sample was taken of the best looking of these, but gave no results when assayed. About a mile inland to the west, is a high, bare ridge of pink binary granite, running in a general north-easterly direction. Most of the country around the lake has been burned over and is grown up with small scrub poplar and birch and also young spruce and pitch pine. Pike and pickerel are plentiful as, indeed, they have been all along the route. Near the inlet to the lake are the remains of an old Indian house and what was probably at one time a small clearing.

The river from Cross Lake to Summit Level Lake is quite wide, but so grown up with wild rice as to be almost impassible for canoes. There is no perceptible current.

Summit Level Lake or Shoal Lake is an extremely shallow lake almost filled up with a thin greenish slime or mud over which paddling is very difficult.

On the west shore is found a light grey rock of white felspar and biotite and sometimes quartz. Several fissure veins of glassy looking quartz 15 feet to 20 feet wide and continuous for considerable distances were found. One sample from here yielded 80 cents gold per ton. The country around Summit Level Lake is level and rocky and timbered with spruce, poplar, tamarac, etc. Four and a half miles west is another small lake, three-quarters of a mile long and half a mile wide, with its longest axis north-west and south-east. Its outlet is at the south-east end and is probably the Little Pawitik River. At the south-east angle of Summit Lake a small stream, named the Lily River, enters. This stream is said by the Indians to rise in a lake of considerable size five or six miles up. We ascended it for about three miles to the head of the first rapids beyond which it was too shallow and broken for canoes. About two miles up from Summit Lake a vein of quartz 8 inches wide cutting diorite was noticed and a sample taken for assay. It yielded $2.80 of gold per ton. The vein could not be traced any distance as it ran nuder the river on one side and under the thick moss and roots on the other. The Lily River as far as examined was about 1.5 chains wide, the water very dark and swampy, and the current slack. The mouth is completely hidden by wild rice.

Leaving Summit Lake we started down the Albany slope by way of the Pawitik River which flows out of the northerly end of Summit Lake by a muddy and often mud-choked channel.

There is, however, more water flowing northward out of Summit Lake than there is flowing south. Two small rapids occur just after leaving the lake, both of which can be run; then comes a third down which canoes may be run light, while a mile and a half farther on is another portage of six chains on the right bank. Very little rock is to be seen and this is green Huronian schist, until we reach portage fourteen, a short distance below the last. A felspathic granite occurs here, consisting principally of white felspar with a little quartz and muscovite, and is very coarse-grained. There are also rusty brown coarse-grained mica schists with small lenses of quartz striking east north-east. From Summit Lake down the country is level and covered with green spruce, tamarac, balsam and poplar.

Six miles from Summit Lake the Pawitik joins the Kapikotongwa River. At the junction the latter is 4 chains wide, 4 or 5 feet deep but weedy and sluggish. We ascended the Kapikotongwa River for some distance in a north-north-west direction. Alternately narrowing and widening it follows along the foot of a range of sand and gravel hills 40 feet to 50 feet high on the westerly bank. The other bank is lower and more inclined to be swampy. Both are well wooded with good pitch-pine, spruce and tamarac, the former especially, on the sand ridges. There is no fixed rock to be seen in the bed of the river, but about five miles up from the junction with the Pawitik River and about one and a half miles inland, there is a bare hill of very coarse-grained red granite carrying included fragments of dark schist. The surrounding country is made up of conical sand hills and ridges of granite and gneiss, mostly well timbered with pitch-pine and spruce. A considerable burned tract extends in to the north-east however.

Returning to the mouth of the Pawitik River we now continued down stream on the Kapikotongwa. Three and a half miles below the mouth of the Pawitik River a coarse grey granite outcrops, and five miles down gneiss with veins of granite. The next portage, portage fifteen, is past a rapid with a fall of ten feet, before reaching it however two rapids must be passed, down which light canoes may be run but loads must be portaged. The country is fairly level and there is considerable good clay land. The banks at portage fifteen are of clay 10 feet high. A brulé starts a short distance above the same portage and continues on both sides for twelve miles down. Before reaching portage sixteen is a shallow boulder rapid which is canoeable. The rapid at portage sixteen has a fall of eight feet and is passed by a portage of six chains. The burned district ends about four miles below this, at a stream entering the Kapikotongwa River from the west.

Going up this branch we follow along the base of a range of sand hills for half a mile, but after that the country becomes very flat on both sides. The stream is narrow, shallow and crooked with low mud banks covered with good timber. Some distance up it is blocked with drift wood and becomes impassable. A portage of three-quarters of a mile, however, leads to a small lake to the north-west. This shows evidence of being a well travelled canoe route. The only rock seen on the creek was dark, micaceous gneiss, three quarters of a mile from the Kapikotongwa River.

Below this branch or rather from portage sixteen to Mokoke Lake, the Kapikotongwa River follows a swampy valley about a mile and a half wide, but expanding in places to three miles. On each side of this valley are low sand hills covered with pitch pine and spruce, intermingled with groves of poplar. The river itself is 3 chains to 4 chains wide, deep, with a slack current and low swampy shores. Half way down a small winding creek enters the river through a marsh on the left. Its banks are 4 feet high of stratified peaty matter mixed with clay—evidently river deposit. The valley of the creek is a mile wide and runs south-east from the river. The southerly side is bounded by sand hills 50 feet to 60 feet high. There are also hills to the north but smaller and more scattered. Both sides are covered with good green spruce, etc. The following outcrops of rock were noted between this creek and Mokoke Lake :—

A small island of dark micaceous gneiss striking east-north-east.

Light and dark grey banded gneiss, contorted, and with included fragments of darker finer grained schist, strike east north-east.

Light grey gneiss.

A mile above Mokake Lake a deep bay runs in on the right hand side and parallel with the river, from which it is separated by a long narrow point. At the west end of this bay a channel 3 chains wide, between two sand knolls, leads to a small lake surrounded by sand hills named Alma Lake. Another narrows leads to a second, smaller, sandy lake. The pitch-pine on the sand hills around these lakes will easily average ten inches diameter.

The route to the Albany leaves the Kapikotongwa River at Mokoke Lake by way of a small stream called Mokoke River. We continued ten miles further down, however, for exploratory purposes. Half a mile below Alma Lake the river expands, and flows by several channels through a wide marsh. These reunite to form a lake one and a half miles long and three-quarters wide with high gravel and sand shores covered with pitch-pine. Below the lake the river narrows to three or four chains and continues for eight miles with low sandy and weedy shores. The country on both sides is almost a dead level and thickly wooded. The land is sandy clay and no rock is to be seen. The Indians say the Kapikotongwa River continues its easterly course until it reaches the Kenagami River which it enters under the name of Little Current River.

We now return to the Mokoke River, a shallow, muddy, marshy creek entering Mokoke Lake from the north. Its mouth is completely hidden by weeds, and canoes are taken up with difficulty. Only one exposure of gneiss was noted to Spider Lake. This lake is the source of the Mokoke River and receives its name from its very irregular shape. It is a fair sized lake surrounded by hills of coarse sand, gravel, gneiss and granite. The easterly shore is banded grey gneiss much contorted. The westerly shore is a more massive grey gneiss, less disturbed than the former, striking north 75 degrees east, and cut by large veins of granite. East of the lake there is considerable green timber, but the rest of the surrounding country is burned over and partly grown up again with poplar, birch and young spruce and pitch pine. The burned district, which starts from here continues through to Lake Kaginogame.

A portage of 40 chains from the bay in the north-west corner of Spider Lake, leads to a smaller lake some 30 chains diameter. A second portage of 5 chains from the other end of this lake leads to Slime Lake, a lake 1½ miles long and widening from 30 chains at its easterly to 40 chains at its westerly end. From the west end of Slime Lake the route is down a small, crooked, choked-up creek to the Lanb schquay River. Slime Lake is a shallow lake filled up with the same greenish slime as Summit Lake, but more so. We found it necessary to dig a passage for our canoes through this mud with our paddles, in order to reach the outlet. This in its upper part is not much better than the end of the lake, but further down becomes deep and easily navigable. The creek is sometimes avoided altogether by making a portage from Slime Lake to the Lanb schquay River.

The rocks noted between Spider Lake and the Lanb schquay River are, light grey gneiss and syenite, red gneiss sometimes garnetiferous and sometimes hornblendic, reddish granite strike nearly east and west.

The Lanb schquay River, from the mouth of the creek to where it joins the Ogoki, follows a wide marshy valley, flanked by bare rocky hills. The river itself is is 4 to 5 chains wide, deep, and with a very slack current. A large number of otter were seen in this 4½ miles of river

The Ogoki from the mouth of the Laub schquay River to the French Channel is from 6 to 8 chains wide and very deep, and has no perceptible current, the left hand side is mostly wide marshes and lagoons with bare hills in the back ground. The other shore is drier and inclined to be rocky, and along it there is a belt of good timber of some extent, though the rest of the country is burned. The rocks met with on this part of the Ogoki River are light grey granite-gneiss, grey banded gneisses, often crumpled and folded, striking north-west and south-east.

From the Ogoki River the route turns north-west up the French channel, a narrow weedy stream, winding through a marshy valley to French Lake. This is another shallow lake nearly filled up with slime and about three-quarters of a mile long and half a mile broad. It is surrounded almost entirely by bare hills of gneiss and granite with a few clumps of green timber. On the north easterly shore is a sand beach. The rocks on the lake shore are grey gneiss banded with reddish hornblende gneiss. There is much folding visible and included pieces of dark fine-grained schist are to be found. The strike is north, 80 degrees east. From French Lake a portage of 40 chains over gneiss and biotite schist cut by granite veins—these latter sometimes with veins of epidote one-half an inch wide—leads to Mahamosagami River or Lake Mahamo.

Lake Mahamo is about seven miles long, one mile wide in the widest place and surrounded by bare hills of rock and sand. At the lower end there are quite a number of small patches of good green timber scattered over the country. The rock around the lake is pink and grey gneiss striking east-north-east to north east, and massive granite. There are a great many sand hills and ridges in the vicinity of the lake, but no good land. The route leaves the end of Mahamo Lake by a creek about 15 feet wide at the start. It is at first very shallow and crooked and obstructed by logs and sandbars and flows through a level, swampy country broken by an occasional sand hill. About half a mile down it expands to a small lake 20 chains in diameter with a sand hill covered with jack pine on the right hand shore. Below this expansion for some distance it is again shallow and obstructed. It gradually deepens and widens, however, to flow through a wide open marsh with scattered scrub tamaracs out of which it enters Lake Kaginogame.

Lake Kaginogame is a fine and large lake with a number of islands and about 8 miles across from inlet to outlet. Bays on each side are almost as large as the lake itself. The water of Lake Kaginogame is beautifully clear and transparent and is in marked contrast to the dark-colored water of the previously mentioned lakes and rivers. It is quite possible on a bright day to see the pebbles and boulders on the bottom in 25 feet of water. There is a large amount of green timber around the lake especially to the north and west, and the brulé that started at Spider Lake ends here. The easterly shore consists of bare granite ridges. On the west shore a light grey massive granite is found with two very perfect jointages. Sand beaches are also common at the bottom of the bays. The Indians report another very large lake west or south-west of this, which they say can be reached by a portage of a mile and a half from the bottom of a large bay that runs in to the south.

The outlet of Lake Kaginogame is at the bottom of a deep narrow bay at the northerly end of the lake. The bay gradually narrows down to a river with a good current, and two to three chains wide. After passing through a small lake expansion surrounded by gneiss striking north, eighty degrees west, we came to a boulder rapid which is passed by a portage of thirty chains on the right bank. Below this are two more portages, one of thirty five chains and the other fifteen chains, both on the right bank. Below the last portage the river is rather shallow, weedy and with a sandy bottom, and shores alternately swamp and sand. Good spruce is found all along between the two lakes, but especially fine timber occurs below the portages. Outcrops of gneiss occur occasionally. Below the last portage a stream, the Cat-tail River, enters from the left This is three chains wide at the mouth in a cat-tail swamp, but it soon narrows to ten or fifteen feet, where it flows as a broken rapid at the bottom of a ravine sixty or seventy feet deep, with almost perpendicular walls of gneiss, cut by large veins of coarse-grained granite. Though the stream is very small it is apparently an outlet of a lake, as its water is quite clear and shows none of the dark color of swamp water.

The shores of Sucker Lake are chiefly grey gneiss striking from north, eighty degrees west, to east and west, with small areas of sand, gravel and clay. It is burned all along the east shore of the lake but the brulé does not extend far inland. Back of it and on the west shore are splendid green spruce, pitchpine and poplar. Many individual spruces along this shore measure two feet and upwards in diameter. A small river enters this lake at its northwest corner. The mouth is a shallow stony rapid past gneiss cut by granite veins. One mile of rapids with a portage of ten chains on the left bank half-way down leads from Sucker Lake to Eagle Rock Lake.

At the southerly end of Eagle Rock Lake, walls of light grey granite-gneiss, twenty to forty feet high and cut by irregular veins of very coarse reddish granite, rise from the water's edge. At the northerly end of the lake the shores are low and often sandy. A good portage over clay land, covered with large spruce, poplar and balsam, thirty chains long, leads from Eagle Rock Lake to Lake Abagotikitchewan, an expansion of the Albany River. For three miles along the shore of Lake Abagotikitchewan there outcrops a grey gneiss, striking north, seventy degrees west. Then follows a coarse-jointed massive diorite, which continues to the Albany River. The Albany River, where it enters the lake, is three chains wide with swift, dark, muddy water and boulder shores. The timber line at the time of our visit was three chains back from the water line.

When we had reached the Albany River by the direct route we turned back to the Ogoki River, and followed it from the mouth of the French Channel to its junction with the Albany River.

OGOKI RIVER.

For half a mile below the mouth of French Channel the shores are low and swampy, then there is a straight course of north, thirty degrees east, for four miles, with steep shores of grey and pink gneiss. Below this to Amy's Falls, five miles further down, the river is five to six chains wide with low clay banks and the rock just showing at the water's edge. A short distance back from the river the country breaks into low rocky ridges. The right bank is well wooded with spruce, poplar, tamarac and balsam, the left is an old brulé grown up with scrub. Nine miles below the French Channel the river turns sharply to the right over Amy's Falls, then northerly again on the rapids below.

Amy's Falls consists of two falls of twelve feet each, with a rapid below. The upper face extends right across the river, but the lower one is broken by an island in the centre, to the left of which the main body of water passes. This fall would make an excellent water power. It is passed by a portage of sixteen chains on the left bank. The rock at the upper fall is grey gneiss, striking west northwest. At the lower fall and rapids there is very coarse-grained granite, probably a dike, also grey gneiss and a band of fine-grained syenitic rock jointed into small diamond-shaped blocks, the joints often showing slickensided faces.

There are six rapids between Amy's Falls and Tiernan or Ogoki Lake, all of which can be run, though very stony and often shallow. For half a mile below the falls outcrops of gneiss are frequently seen on the banks. They then disappear for six miles until we reach a rapid across a bank of gneiss. The rapid is three quarters of a mile long, very stony, and spreads at the bottom into wide shallows. Two-thirds of a mile further is a second rapid, a short chute over a ridge of gneiss. For eight miles then the river continues with a good current through a level country of sandy clay and sand. It spreads out and becomes very wide in places. At the end of this eight mile stretch is a long, crooked, stony rapid of one and a half miles. Less than three-quarters of a mile further is a fourth shallow, stony, rapid, three miles long, around a long bend in the river. Rapids five and six are short, deep rapids and the last before reaching Tiernan Lake. Below the last rapid the river divides into several channels for a short distance. Below the falls it will be seen the river loses its lake-like character. Its width becomes very variable, and its current swift and mostly shallow. Large islands, also, occur frequently. From Amy's Falls to the lake the country is extremely level, and thickly wooded with spruce, jackpine, balsam and poplar. This is second growth and not large, but much would do for pulpwood. The soil is clay, sandy clay loam and sand, the last becoming more prevalent as we approach the lake. Most of it would make good farming land. No rock is met with below the third rapid.

Tiernan Lake is a sheet of water eight or ten miles long from inlet to outlet, and four to five miles wide, with large bays running in to the north-east and south-west. The northerly shore is low and sandy, with small spruce, tamarac, poplar, etc. One smal elm was noted here, also, the only one seen during the summer. The country to the south and east is better timbered and more broken by low hills. On the northerly shore the following rocks were observed :—At a point three miles easterly from the mouth of the river, light grey gneiss cut by a dark green trap dike seventy-five feet wide and striking west, north-west. There were glacial markings on the dike almost due north and south. Half a mile farther east, a large dike of red granite cutting grey granite-gneiss. Other exposures of light grey gneiss occur. There are sand beaches in most of the bays and at one place a bed of marl and clay which had become hardened and almost rock-like occurs at the water's edge. The southerly shore is light grey gneiss with darker micaceous bands, striking west, south-west, to south eighty degrees west, and large dikes and knobs of a very coarse-grained graphic granite.

At the outlet of the lake there were noted :—Grey gneiss striking south, eighty degrees west, and dipping at a steep angle cut by a trap dike, one foot wide, striking north, fifty degrees west. A quarter of a mile further down a dark grey massive coarse grained rock, extremely friable, and weathering to a dark green sand, forms one wall of a fine-grained diabase dike, sixty feet wide, striking north, north east, with three sets of joints, two cutting the rock into diamond shaped sections, the third running from wall to wall and cutting the diamond sections diagonally. The easterly wall of the dike is an extremely coarse, dark-green, massive pyroxenite. A short distance down the river the rock exposed is a grey hornblende granite, hard and medium grained.

Shortly after leaving Tiernan Lake there is a slight rapid, then two miles of swift current and a second rapid, three-quarters of a mile further a rapid five chains long, while a fourth crooked shallow rapid two miles long follows at the end of one and three quarter miles. Down this last rapid only light canoes can be run, loads must be portaged part way. Two more rapids bring us some thirteen miles from Tiernan Lake. The country along this thirteen miles of river is very level clay land with only an occasional exposure of gneiss or granite on the river's bank. The rapids are full of large angular boulders of the same rocks, as indeed is a large part of the river bed. In many places low wet swamps and lagoons run back from the river. At the end of thirteen miles, an examination of the country showed in one of these swamps gneiss striking south, sixty degrees west, folded and crumpled and with included fragments of dark schist, coming to the surface like low islands.

Fifteen miles below the lake is another rapid, down which light canoes may be run, but a portage of fifteen chains on the right bank is necessary for the loads. Just above this portage the river branches and does not again reunite for many miles below. We followed the southern or right hand channel, but were informed by Indians we met later, that the north branch was the best and the one used by Indians travelling the river. Half a mile below the last portage is a rapid of five or six chains, then another of half a mile, three chains wide at the top, but spreading to eight chains and very shallow at the bottom. The channel is to the left of an island half way down. Half a mile further is a rapid ten chains long and two chains wide, the channel to the right of an island. After another slight rapid there comes a lake expansion two miles long and three quarters of a mile wide. Grey gneiss and granite were the only rocks on this stretch. The country continues level and well wooded with spruce, poplar, tamarac and jack pine.

On the small lake expansion just mentioned there was a camp of three or four Indian families fishing for white fish and sturgeon, of which there seemed to be a good supply. These Indians trade at Long Lake House which they say can be reached in five days by a route from the southerly shore of Tiernan Lake.

A mile below the lake is a rapid forty chains long. The upper part is run and a portage of ten chains made over the left end of an island of grey and rusty micaceous gneiss and red granite. The gneiss strikes west, south-west. There is a total fall of eighteen feet including a straight fall of six feet and good water power at this rapid. Another half mile brings a slight rapid of ten chains, a bad rapid fifteen chains, and then twenty-five chains of swift water and exposures of gneiss. After this comes four or five miles of almost continuous rapids with five short portages. Next follows a rapid of ten chains with a portage of one chain on the right bank, then two more rapids, at the second of which there is a fall of ten feet, a good water power, and a portage of three chains on the left. After a stretch of swift water we run the upper part of another rapid to a portage across an island past a fall of eight feet and also a good water power.

A grey granite-gneiss striking north eighty degrees east outcrops at this portage. Seven shallow stony rapids all of which can be run, are followed by a five mile stretch of deep water with slack current. The river banks are now of clay three feet to ten feet high. After the deep water stretch comes a broken fall twelve feet high and two chains wide, an excellent water power. A portage of two chains on the left bank leads over a bank of hard, siliceous, grey gneiss. The banks here are of clay twelve feet to fifteen feet high. A boulder rapid follows, then a long rapid with two portages over bands of gneiss. The first portage is two chains in length over an island, at its foot on the right bank is a dark green trap dike fifteen feet wide exhibiting columnar structure, the walls perpendicular. The second portage is three chains in length with another trap dike on the right bank striking east and west. Three-quarters of a mile below the two branches of the river reunite.

On the north branch, a mile up from the junction of the two branches, is Burton's Falls, a fall 1 chain wide and 30 feet high over dark fine-grained mica schist striking south 85 degrees west, affording a good water power. Lower down is grey gneiss and granite. A high clay bank at the junction shows in section, at the top a fine yellowish loam mixed with decayed vegetable matter, then stratified clay and sand, and at the bottom unstratified boulder clay. The stratified clay is in bands 1 inch to half an inch thick and does not lie flat but in waves or folds 3 inches from trough to crest. They are alternately white and chocolate colored, and the latter bands are quite hard and solid. A hill of white sand and gravel occurs here also, the only one in the neighborhood. From Tiernan Lake to the junction the country maintains a uniformly level aspect. No hills are to be seen, and the level clay soil is thickly covered with a heavy growth of spruce, poplar, balsam and tamarac, and this was also found to be the case on the expeditions inland.

Three-quarters of a mile below the junction is a rapid with a face of 8 feet over a ridge of crumbled grey gneiss, striking south 75 degrees west, and a portage of 1 chain. Then a short, deep rapid followed by a long rapid with several islands and a fall of 15 feet, a good water power. There is a portage of 1 chain over a small island of crumpled grey gneiss striking north 35 degrees east on the north side of the river. One hundred yards below is a portage of 2 chains over similar crumbled gneiss and grey granite gneiss, both carrying included fragments of dark fine grained schist. Next follows a rapid of 20 chains over reddish and grey gneiss with darker bands. There quarters of a mile below is a rapid with a fall of 11 feet and a portage of 4 chains on the left bank. The upper part is massive granite gneiss, below it is fine-grained and banded, striking north 75 degrees east. At the end of 400 yards is a portage of 6 chains on the right bank over banded gneiss. Next is a rapid with 8 feet fall and a portage on the right, and then rapids to a fall of 4 feet. The portage is 2 chains long, over the foot of a small island just to the right of a large island in the middle of the channel. Gneiss here strikes north 10 degrees east. Then a rapid to a fall of five feet and a portage of 2 chains on the left bank over banded gneiss. Two hundred yards of swift current follow and then a fall of 6 feet passed by a portage of 4 chains over an island of gneiss in the middle of the river. From here there is clear water to the Whitefish River.

The Whitefish is a stream 2 chains to 3 chains wide entering the Ogoki River from the north. The water is very dark, shallow, the current strong and the bottom pebbly. The banks are high and of clay. A belt of splendid spruce, tamarac and poplar from half a mile to one mile wide extends along both banks, but beyond this the country becomes wet and the timber poor and scattered. In the wet country a layer of peat and moss from 18 inches to 5 feet thick covers a hard blue clay Two miles up the Whitefish River from the Ogoki River is an exposure of gneiss striking north 55 degrees east. Three miles up there is a short rapid over a ridge of crumpled grey gneiss striking north 60 degrees east. Then, some distance farther, a second rapid and 100 yards above it a fall of 4 feet necessitating two portages, both over gneiss. We stopped at the foot of a fourth rapid with a fifth in sight.

Below the mouth of the Whitefish River, three portages and six rapids, in a distance of about five miles, flow out of the area of Archaean rocks and past the bad part of the river. Below this we are in Silurian country, and the river is clear sailing to the Albany River. The Ogoki River below the last portage is 4 chains or 5 chains wide, but this increases to 7 chains or 8 chains farther down while at its mouth it is a quarter of a mile.

The width in this lower stretch is much more uniform than in the upper rocky portion, where it frequently varies from 2 chains to 30 chains or more. The current is very swift, often approaching a rapid and always shallow. The bottom is pebbly, with the exception of one or two places where it apparently consists of flat limestone. The banks are 12 feet to 30 feet high, the top of yellow loam mixed with decayed vegetable matter, then stratified clay and sand, the base of tough blue boulder clay. On each bank of the river there is a strip of large poplar, spruce, balsam and tamarac, which however extends only one-quarter to half a mile back. Beyond, the country becomes wet and swampy, though not, apparently, lower, and the timber very small and scattered. At the mouth of the Ogoki a river 3 chains wide called the Blackwater enters it from the north west. A short distance up it narrows to 1½ chains and is about 4 feet deep with very dark coffee colored water and a good current. It maintains this width and depth as far as explored. The country along its banks is similar to that along the lower Ogoki River. One and a half miles from its mouth the first exposure of limestone was found. It is a small exposure of a shaly, yellow limestone containing a great deal of quartz sand and, as far as could be seen, non-fossiliferous. No other rock was found on the Blackwater River. Trees on the banks of this stream and the lower Ogoki River, 15 feet above the present water level, have been broken by the ice, showing that they are subject to high spring floods.

ALBANY RIVER

The Albany, between the Ogoki and the Kenogami River is a fine, large, deep river one-quarter to half a mile wide, with clay and gravel banks 60 feet to 80 feet high. The current in this portion is very strong, in many places approaching a rapid. It is full of gravel bars and low gravel islands of rounded pebbles, prominent among which are many of hematite and jasper. Only three streams of any size enter it from the south in this distance, the Ruby, the Wabimaig and the Upper Sturgeon Rivers, all of which are very shallow. An attempt was made to ascend the first of these, but it was too shallow and swift for any progress to be made. It was not thought worth while attempting the others from their appearance near their mouths. Three small outcrops of a blue concretionary limestone were noted, which, from the few fossils obtained is apparently Upper Silurian. At the mouth of the Upper Sturgeon beds of a buff-coloured indurated marl rise nearly to the top of the bank. The country along the Albany River is a dead level clay country. Just along the banks is a narrow strip of green timber and dry land but, from a few hundred yards inland, there is no timber but a few scattered black spruce about 1½ inches in diameter and 8 to 10 feet high growing in an immense peat bog. The peat and moss is at least 10 feet deep in most parts, how much deeper could not be ascertained. Were this peat and moss removed and the country drained the clay should make good farm land. This kind of country continues from the Ogoki to the Kenogami River, and in fact to James Bay.

The return trip from the forks of the Kenogami River was made by way of the Albani River, James Bay and the Moose and Missinaibi Rivers to Missinaibi station on the Canadian Pacific railway, which was reached on November 1st.

SUMMARY.

SOIL, ETC.—While a large portion of the district explored is rock or sand and valueless from an agricultural point of view, there are, nevertheless, large amounts of good land also. First there is a comparatively small area of black alluvial soil along the Ombabika River from the seventh portage to Pigeon Lake. On the other side of the watershed, from Summit Lake as far down the Kapikotongwa river as it was explored, there are many tracts of good clay of considerable extent, though a large portion of this district is made up of sand hills. The whole valley of the Ogoki River is a wide level tract of clay and to a much smaller extent sand. The upper portion of the Ogoki valley was the most promising agricultural land discovered. The lower portion, *i.e.*, below the portages to the Albany River and all down the latter is very wet and covered with peat bogs. It lies, however, at a considerable elevation above the river bed and should be drained without difficulty into the latter. This done and the peat cleared off the land should be fit for farming. Little is definitely known of the climate, but as far as could be ascertained it is much the same as that of the Lake Temiscaming district and not appreciably colder.

TIMBER —The timber consists of spruce, poplar, jack pine, tamarac, balsam, birch, with, in places, a few small cedars. Excepting in the burned districts, the principal of which is that from Spider Lake to Kaginogame Lake and along the Albany River spruce and jack pine of sufficient size and quality for pulpwood is plentifully distributed throughout the district. Mention should be made of the fact that where burned districts are again growing up, although apparently only with poplar and birch, there is, nevertheless, a vigorous young forest of spruce and pitch pine concealed among the former. Excepting the spruce and pitch pine, the timber is of little or no value. In the muskegs along the Albany the only trees are worthless scrub black spruce.

MINERALS.—No minerals of economic value were discovered, though a careful watch was kept for lignite along the lower stretch of the Ogoke River and down the Albany River. Traces of gold were found in the quartz veins in the Huronian rocks about Cross and Summit Lakes, two samples yielding respectively 80c and $2.80 per ton. This Huronian belt should be worth careful and thorough prospecting

WATERPOWERS —Good water powers are numerous 'Mention may be made of the first three rapids met with on the Ombabika after leaving lake Nepigon, the rapids between Kaginogame and Sucker Lakes and between Sucker and Eagle Rock Lakes, Amy's Falls on the upper part of the Ogoki River, and the almost continuous rapids and falls from Tiernan Lake down.

FUR AND GAME —Large game seems to be scarce throughout the district, with the exception, perhaps, of black bears. A few caribou tracks were seen, but not many. Ducks are said to be plentiful at Summit Lake in the fall, attracted, no doubt, by the wild rice. Very few partridge or rabbits were met with. The most numerous furbearing animals noted were muskrats, mink and otters, and a solitary beaver. Pike, pickerel and whitefish are generally distributed throughout the district. Speckled trout are numerous from Lake Kaginogame to the Albany River and also on the Ogoki River. Sturgeon are caught in the Ogoki and Albany Rivers.

GEOLOGICAL FEATURES.—From Ombabika Bay through to the Albany at Lake Abagotikitchewan the rocks are Laurentian gneisses and granites of various shades of color and texture, with the exception of a belt of Huronian schists some 10 or 12 miles wide. The southerly edge of this belt crosses the lower part of Cross Lake in a general east, north east direction; the northerly contact was not seen on account of overlying drift, but is on the Pawitik River shortly before its junction with the Kapi-ko-tongwa. The Ogoki River, down to the last portage before reaching the Albany River, flows entirely through Laurentian country overlaid with clay and sand. Below this to the Albany River, and along the latter from the Ogoki River to the Kenogami River, is, as far as could be told from the few fossils found, silurian limestone overlaid with deep beds of drift.

A. H. A. ROBINSON.
Geologist with Exploration Party No. 6.

Peterboro', Dec. 27th, 1900.

Hon. E. J. Davis,
Commissioner of Crown Lands,
Toronto.

I, A. H. A. Robinson, of the City of Toronto, in the County of the York, Geologist, make oath and say that the foregoing report is correct and true to the best of my knowledge and belief, so help me God.

Sworn before me at the city of Toronto, in the County of York, this 4th day of February, 1901. } A. H. A. ROBINSON.

G. B. KIRKPATRICK,
A Commissioner

Indians running rapids, Ogoki River. Party No. 6.

Tiernan's Lake. Party No. 6.

Party on portage 24, Eaglerock Lake. Party No. 6.

Indian fish trap at mouth of small tributary of the Albany River. Party No. 6.

Indian camp at mouth of Wabimaig River. Party No. 6.

At Moose Factory. H. B. Co.'s schooner in winter quarters, undergoing repairs. Party No. 6.

In Moose Factory. H. B. Co.'s storehouse. Party No. 6.

Camp at Moose Factory. Party No. 6.

Cottage hospital for sick and aged Indians, Moose Factory. Party No. 6.

Parsonage and stores, Moose Factory. Party No. 6.

SURVEYOR'S REPORT OF EXPLORATION SURVEY PARTY NO.7.

FORT WILLIAM, District of Thunder Bay, Ontario,

Sir,—I beg to present my report as Surveyor in charge of Exploration Survey Party No 7, the work assigned to which comprised the exploration of the land and timber on each side of the Portage Route going through the Albany River, in a north-westerly and northerly direction from Wabinosh Bay on Lake Nepigon, in the district of Thunder Bay, across the height of land, thence down the chain of lakes to the Albany River, then up the Albany River to lake of St Joseph, etc., etc.

Leaving the Canadian Pacific Railway at Nepigon station on July 5th. 1900, I proceeded up the Nepigon River in canoes to Lake Nepigon, taking the Lake Hannah route and thus missing the last four portages on the river, but being greatly detained, on the way, by wet weather—almost incessant rain,—I arrived only at Lake Nepigon on Saturday, July 7th, and camped at Flat Rock portage over Sunday.

The sail boat belonging to the Hudson's Bay Company arriving at Flat Rock portage about noon on Sunday, July 8th, to take Mr. Beatty's party (No. 8) across the lake to Gull River, I deemed it advisable to wait for Mr. Beatty and cross the lake with him.

We got away from Flat Rock portage on Monday, July 9th, at noon. We had then three days of alternate calms and headwinds, making progress very slow. On Thursday evening the 12th, I took to the canoes and arrived at the Hudson's Bay Company's post at 9.30 p.m.

Lake Nepigon, on account of its size and large stretches of open water, is an unreliable and precarious lake for canoe travelling. A breeze that on a small lake would not render canoe travelling difficult, even in a small canoe, on Lake Nepigon with a large canoe, necessitates lying up for hours and sometimes for days at a time. In using the Hudson's Bay Company's sail boats, high wind, unless it is a fair wind, also necessitates lying up, as the sail boats are provided with neither keel nor centre board.

The term "height of land" when used in this report refers to the height of land between the waters of the River St. Lawrence and those of Hudson's Bay. Subsidiary heights of land are always designated by giving them a particular description.

From the daily journal kept by the officer in charge of the Hudson's Bay Company's post at Nepigon House I found that it was impossible to obtain any reliable information as to the freezing or closing of Lake Nepigon in the fall of the year. In 1897 on December 1st the bays were frozen but the main lake clear. In 1898 on December 1st the channel in front of the Hudson's Bay Company's post was frozen over. In 1898 on December 14th channel frozen over. This information is so indefinite that I did not consider it advisable to go further into the matter.

As a rule however, Lake Nepigon freezes over from December 1st to 15th. Trips are made along the west coast on the ice, long before it is safe to cross the Nepigon River More accurate information has been obtained as to the opening of Lake Nepigon for navigation in the spring as follows:—

1887, May 14th.	1894, May 20th.
1888, June 8th.	1895, May 11th.
1889, May 11.	1896, No record.
1890, June 4th.	1897, May 16 th.
1891, May 10th.	1898, May 14th.
1892, May 28th.	1899, May 23rd.
1893, June 8th.	1900, May 8th.

From the same source I also found that barley, potatoes, radishes, rhubarb, parsnips, carrots, cabbages, onions and all small fruits ripen readily at Nepigon House.

From conversations I had with the officer in charge at Nepigon House and with Indians from the north country, I found that I would have to alter considerably the route laid down in my instructions. The canoe route, as so laid down, does not exist on the ground. After passing the height of land on the Wabinosh route the canoe route descends through various lakes and rivers to the river to which I have given the through name of Okokesibi—although known to the Indians and traders by a different name for each of the stretches between lakes—and which is actually the outlet of Wahbahkimmug or Savant Lake of the plans. The Okokesibi is the river called Ogoki by Dr. Bell in 1870, and flows easterly almost parallel to the height of land, emptying into the Albany River, about forty miles below Martin's Fall.

At Kenojiwan Lake, the first lake down this river, on the north-easterly outlet of Wahbahkimmug Lake, the route to Lake St. Joseph is encountered. I did not ascertain the name by which this river is known to the Indians, and have called it the Palisade River.

I started to work at the Indian houses at the head of Wabinosh Bay, Nepigon Lake, and traversed the various lakes and rivers as shown on the plan to the height of land. In 1870 Dr. Bell, of the Geological Survey Department, of the Dominion also made a survey of this route across nineteen portages. The only plan I received of it was the lithograph map of the country on the scale of four miles to an inch ; and as the various lakes, etc., shown on the plan, did not in any way resemble the lakes as they are on the ground, I made a complete traverse of them all and located the height of land on the 14th portage, or on the portage between Tunnel Lake and Rock Island Lake.

From Wabinosh Bay up to the 5th portage the country is very broken and hilly, but from the 5th portage to the north the hills are very low and the country flat. Hills two hundred feet in height seldom occur, and look like mountains in contrast to the flat surrounding country. No land suitable for agriculture is met with on the south side of the height of land.

From the height of land to Smooth Rock Island Lake all the country has the same characteristic features, low hills and small timber, although some good spruce and tamarac is met with at the south end of Tamarac Lake. Adjoining Boiling Sand River, there is a small tract of fair agricultural land, but I did not take time to explore the extent of it, expecting to come out that way in the fall.

After passing the 21st portage a very flat country is entered as evidenced by the low falls or rapids in the rivers, the number of lakes with more than one outlet, and the size and shallowness of the lakes themselves.

Smooth Rock Island Lake (Zhooshquabeanahmenis).

We arrived at this lake July 30th, and did not get through with the survey until August 16th, most of that time having two sets of discs working with the micrometer, measurements being also made with the log. It would be almost impossible to describe this lake without referring to the map, as it is so crooked and full of islands, large and small. At the south-west corner of the lake, and also in Lone Breast Bay some big hills are observed, but the general character of the shores and islands is low, rocky hills, burned recently in the south-easterly portion, but as a rule covered with a thick growth of small spruce and pitch pine.

There are several canoe routes coming into this lake, some of which were explored by the geologist and timber estimator. The canoe route coming into the northerly end of Lone Breast Bay, the one usually travelled by the Indians going to White Water Lake, passes through a "height of land lake" immediately before entering Smooth Rock Island Lake. There is another route southerly from the east end of east bay, which I am informed joins the route we travelled from Lake Nepigon at some place on Sucker Lake, also branching into Rocky Island Lake between the 14th and 15th portages.

This lake also receives part of the water of Wahbahkimmug and empties into the Ogoki River.

I have endeavored to show on the plan most of the islands that occur on this lake, but some are too small to appear on a map of a scale of two miles to an inch.

Although the distance between Smooth Rock Island Lake and Wahbahkimmug Lake is short, still there are four portages to be made. With the exception of the first portage, or the 22nd according to my numbering, the falls or rapids are mostly low and flat, low hills and small timber still being the distinguishing characteristics of the country.

Wahbahkimmug Lake.

This is evidently the Savant Lake shown on old maps. It actually consists of two lakes, but is known to the Indians, traders, etc., as one lake and I have considered it as such; calling the two parts Little and Big Wahbahkimmug. This lake is about seventeen miles long and about four miles wide at the widest place. It has two outlets, the southerly, flowing into Smooth Rock Island Lake, and the northern being the Ogoki River. About ten and one-half miles from the westerly end of this lake the canoe route from Sturgeon Lake enters. It was my intention on my return from Lake St. Joseph to have traversed and explored this route, if not to Sturgeon Lake at least as far as the height of land. On account of the lateness of the season, however, I had to abandon that project. About three-quarters of the distance up the lake on the north side there is a considerable archipelago of large and small islands, which would have taken too long a time to survey for the small amount of information that would be obtained thereby. At the extreme westerly end of this lake a canoe route comes in, which leaves the route I travelled to Lake St. Joseph at Pushkokogon Lake. I did not have this route explored, more on account of want of time than anything else, and I thought that I would be passing through a similar country in going up the Palisade River route to Pushkokogon Lake.

Ogoki River.

Between Wahbahkimmug and Kenojiwan Lakes the Ogoki River will average three chains in width, and is a succession of flat rapids with the river bed full of large boulders. This is also noticeable in the Ogoki River down as far as White Water Lake. The country adjoining the river is low, but fairly well timbered with spruce, tamarac, jack-pine, poplar and birch with some cedar.

There is a considerable stretch of fairly good agricultural land on both sides of the Ogoki river and lakes from near Pike Lake down the river as far as I surveyed about 40 miles on White Clay Lake. The soil is a sandy loam but the timber as a rule is small, especially on the north side of the route on account of old brulés. Some fair-sized tamarac, pitch pine and poplar is met with on the south sides of the route.

To Lake St. Joseph.

It was with great difficulty that I could obtain any reliable information from the Indians about this route. No one in the party knew the name by which the Indians called the lake, and "St. Joseph" is an "X" quantity to the Indians. I found however that there was a route entering Kenojiwan Lake from the north, and the guide I had was well pleased when he found that I had determined to take it, as he knew the route, but could not tell where it came from by any names I could understand. I did not know though where it was going to bring me out until I had my work roughly plotted and found by the map that I was striking in the direction I wanted.

There are several lakes of considerable size on this route, but the rivers and creeks are mostly small until Pushkokogon Lake is reached. The subsidiary height of land between the Albany River and the Ogoki River occurs either on the thirty-ninth or fortieth portage, the country is so flat it would be hard to determine where the height of land is. The pond between these portages has no outlet, the water seeming to flow in all directions through the swamps.

The route I followed from the lake south of Muskeg portage to Pushkokogon (Pick-oo-Ko-gon of the maps) is not the one usually travelled by the Indians and others, but it was the shortest route.

We encountered quite a number of mud lakes, or lakes almost filled with a vegetable deposit that will eventually develop into peat, but which at the present time makes canoeing difficult and slow.

Some years ago the Indians in this district made an attempt at farming on Pushkokogon Lake, clearings were made, houses erected, and a short distance down the river from the lake, root houses were built. They have abandoned their farms of late years and now grow their potatoes near Osnaburg House on Lake St. Joseph.

From Pushkokogon Lake I followed the river to the Albany River; but there is also another route from the west bay of this lake to Lake St. Joseph, though the portage is long and rough. The whole country, with the exception of the first few miles of the Palisade River, which is somewhat hilly, is very flat, low hills predominate. The rivers are small and the falls and rapids are low. The country is rocky and there are large tracts of muskeg or spruce swamp. In some of the swamps the spruce and the tamarac is of fair size, but on the high lands the timber is all small. On the Albany River some large sized timber is met with, principally spruce, poplar and birch.

On the 13th September I reached the Albany River, where it widens into a lake just below the outlets of Lake St. Joseph, and worked up it about two miles, but had to stop on account of wind and rain. Leaving camp the next morning I worked up the south side of the river into Lake St. Joseph and surveyed the route to Osnaburg House, the Hudson's Bay Company's post, finding thereby that I had entered Lake St. Joseph by what was called the "canoe route to James' Bay" on some of the maps. On the way I kept a sharp look out for the post planted by Thos. Fawcett, D. L. S., at the outlet of the lake but failed to find it. I enquired of Mr. Wilson, the Hudson's Bay Company's officer in charge at Osnaburg House, if he knew anything about the post. He informed me that he was at Osnaburg when Mr. Fawcett did the surveying in 1885, but he knew nothing of the post and mound referred to. He also informed me that the route by which I had come into Lake St. Joseph was the outlet of Lake St. Joseph and was always called such; it was not until I returned to camp that evening that the timber estimator and geologist told me of the other outlet further to the north. The next day being very stormy it was useless to try reaching the northern outlet, and as I had failed to get any quantity of provisions at Osnaburg, as the company do not keep much in that line at the inland trading posts, I could not waste any more time endeavoring to check the latitude given for Fawcett's astronomical station; but no doubt that has been attended to by Mr. Robertson, who, I understand, found the post.

I ascertained at Osnaburg that the officials of the Hudson's Bay Company make very little attempt at agriculture; potatoes are grown in some abundance by them and the Indians, but no other vegetables had been planted this year, although they are said to do very well. Mr. Patterson, now in charge of the post on Lake Nepigon, assured me that he had a good garden at Port Hope, Eabimet Lake, about one hundred and twenty miles down the Albany river from Lake St. Joseph, and could grow anything that would ripen quickly, the only thing that he had no success with being tomatoes.

Lake St. Joseph usually freezes over from the 15th to the 20th of November, and opens again in the spring from the first to the 15th of May, although some years it is well into the month of June before the lake is free of ice.

The Hudson's Bay Company keep a large stock or trading goods for the Indians, but very few provisions, I could have got any quantity of flour that I wanted, but it was by a favor that I got twenty pounds of pork. Of other provisions there was nothing. It was while camped on the Albany River that we had the first snow of the season on September 16th. On that date I started on my return trip to Kenojiwan Lake to take up the survey of the Ogoki River, (having decided that the season was too far advanced for me to attempt the survey of the Sturgeon Lake route, or the Albany River), and also try to carry tie-line through to connect with the work of Party No. 6.

I arrived at the caché near the thirty-first portage on the 18th September. Two days were spent here in repairing canoes and outfit and getting provisions cooked and on account of rain.

Where it is river proper, not lake expansion, the Ogoki is wide and deep with a fairly strong current and flowing through a country which I have no doubt will prove a good agricultural district, although the soil is sandy and timber at present, very light. This country also is not so flat as it is on the route to Lake St. Joseph, and around Smooth Rock Island Lake. A range of hills from two hundred and fifty to four hundred feet high runs in a northeasterly direction, starting from Lone Breast Bay, Smooth Rock Island Lake, and is cut by the Ogoki River near the east end of Long Lake; but to the north and east of that lake the range is very much broken, and only isolated high hills appear from the heights adjoining.

The first lake met with descending the Ogoki from Kenojiwan is called Long Lake; for the most part very narrow and about six miles long. There are two outlets to this lake, one leaving the lake at the south side near the east end, and the other at the extreme east end of the lake, joining together a short distance above White Water Lake. The south outlet also receives the waters of Wahbahkimmug Lake that flow through Smooth Rock Island Lake and the waters of the last named lake a short distance below the portage from Long Lake.

By referring to the map it will be noticed that the Ogoki river is very rapid with numerous small falls, from its leaving Wahbahkimmug Lake to where it empties into White Water Lake. There are numerous fine waterpowers which may be developed some time in the future, but the best powers seen on the whole trip lie between white Water Lake and White Clay Lake.

At the time I was in this part of the route the water was pretty high and still rising. On the sixteenth trip from Kenojiwan Lake to St. Joseph and return the water rose about fourteen inches in Kenojiwan Lake, and during the week we were camped on White Clay Lake the water rose nearly a foot. In going up the river to the height of land, south of White Clay Lake, I found the rivers and lakes full to high water mark. I cannot very well account for such a rise at that time of the year; although considerable rain fell, I do not think there was sufficient to increase the flood of water to such an extent.

It will be noticed on the plan of this route that the survey of the last two large lakes was not quite completed. The season was getting well advanced and supplies low; and I was very anxious to complete a tie-line from Wahbahkimmug Lake to the work of Party No 6 on the Ogoki River. Unfortunately on White Clay Lake we lost a full week through wet weather, rain, fog, mist and snow, and during this week I met with an accident which precluded me from doing any walking—or unnecessarily getting in or out of a canoe. I had, however trained some members of the party in the use of the instruments, and I was able therefore, to carry on the tie-line to Round Lake on the Pitikigouching River, to the point at which Mr. Lount stopped work in 1870. From Indians I met on White Clay Lake I obtained the information that Party No. 6 had not ascended the Ogoki River at all, and in my crippled condition and want of supplies I was reluctantly forced to turn south before completing the tie to the work above mentioned. Moreover, experience has taught me that it is unwise to be among small lakes so far north after the beginning of October, unless well rigged for winter work,

On the route to the Pitikigouching River I passed for some distance through comparatively flat country, having fair sandy soil with large stretches of muskeg in places. A rough country is entered about eight or ten miles south of White Water Lake, but which is mostly burned and very rocky. This rough country does not continue far and is passed before the height of land is crossed. This height of land I found to be shown very much out of place on the present map of the country.

The Pitikigouching River.

Only the upper stretches of this river were surveyed, that is, from the height of land to the north end of Round Lake. The river comes into the lake south of the height of land portage from a northwesterly direction, and is a stream of some size but little current. I could not ascertain where it took its rise. From the height of land portage the river flows through a tract of country that has been mostly burned, and is very hilly and rocky, to where it empties into Round Lake.

Round Lake is described by Mr. Lount as being about two miles wide and lying twelve miles in a general direction of north east by north from Lake Nepigon, but it is thirty-six miles from that lake by following the river. It is the first lake on the river from Lake Nepigon.

From Round Lake to Lake Nepigon the river is very crooked, with a strong current and flows through a high rolling sandy country, thickly timbered in the valley with heavy spruce and tamarac. Near Lake Nepigon high rocky hills are observed, but they do not extend any distance back from the lake. I arrived at Lake Nepigon on the return trip on October 16th. As I had not obtained an observation for latitude at Lake Nepigon when I started work in the summer I moved down to Wabinosh bay and camped on the north shore near the main shore line of Lake Nepigon, and there observed the sun for latitude, and also traversed the shore line from the Indian houses to this astronomical station. Some time was lost both at this camp and at Nepigon House by head and strong winds, but the railroad was reached on the evening of October 29th.

I have already delivered to the Department the box of botanical specimens which I and my assistants collected during the summer.

In Field Book No. 4 will be found the record of weather, temperature, barometer, wind, etc., during the season.

In regard to the fauna of the country passed through I can say very little, except from hearsay evidence. Otter, martin, mink, fox—silver-grey, red and black—fisher, bear, a few beaver, caribou, moose, skunk, muskrat, lynx and weasel are the principal furs obtained from the Indians by the Hudson's Bay Company.

The rivers and lakes are said to be full of fish. Unfortunately, after passing the height of land we saw very little of them. The principal fish seemed to be whitefish, lake and speckled trout, pickerel, pike, sturgeon, and a fish the Indians call Ohdoannebee, being something between a whitefish and a herring.

I have the honor to be, sir, your obedient servant,

H. B. Proudfoot, O.L.S.,
Surveyor in charge of
Exploration survey party No. 7.

The Hon. E. J. Davis,
Commissioner of Crown Lands,
Toronto.

LAND AND TIMBER ESTIMATOR'S REPORT OF EXPLORATION SURVEY PARTY No. 7.

Sudbury, District of Nipissing, Ontario,
December 28th, 1900.

Sir,—In accordance with instructions from the Crown Lands Department, I left Sudbury, June 23rd, 1900, by Canadian Pacific Railway for Nepigon to join Exploration Party No. 7, under your superintendence to explore the territory from Wabinosh Bay at the north-west of Lake Nepigon, up Wabinosh River across the height of land to Albany River, and sections of the Albany and Savant Rivers. My duties were to examine and estimate the timber on our route, and also report on the quality and nature of the soil with a view to agricultural development. At Nepigon I met the other members of our party, and after arrangements as to supplies were completed, we proceeded up the Nepigon River with four canoes and two Indian guides. The Nepigon River is well known for its beauty and as a fishing ground for speckled trout. It has several lake expansions, and a number of rapids (notably the White Chute) the difficulties presented by which are overcome by portages, the first and longest, Camp Alexander, being two miles in length. On the evening of the second day we arrived at Lake Nepigon by way of Lake Hannah and on the following day the Hudson's Bay Company's sailboat came down the lake from Nepigon House to the landing, of which we took advantage to make the trip to Gull River and Nepigon House, a distance of eighty miles over the lake. When we reached the entrance of Gull Bay, on the west shore, we went by canoe to Nepigon House, which is situate near the north-west corner of the lake on a beautiful rising slope, from which a good view of the lake can be obtained in clear weather. Close to the shore are two large islands, one of which is occupied by the Roman Catholic Mission, in connection with which there is an Indian school. The other island, which lies immediately in front of the House, is used by the Indians as a camping ground, and as it was near "Treaty Day" at the time of our visit, quite a number of wigwams were scattered over it as well as on the main shore. There are about five acres cleared, which is a sandy loam near the shore, but as it recedes the rock crops out. A promising patch of potatoes was growing, besides timothy hay of average length. There is a cemetery which seems to have been used for a considerable time by the appearance of the graves, which according to Indian custom are covered with birch bark or boards to form a roof. In the channel between the islands and the shore is an excellent fishing ground, and at the time of our visits we saw many fine specimens of speckled and lake trout, that are daily caught to feed the dogs used by the Hudson's Bay Company in winter. We proceeded to Wabinosh Bay, which is about ten miles north of Nepigon House, where our work of exploration began. This is a winter trading post of the Hudson's Bay Company, and at this point there were two or three small patches of potatoes growing in a little clay mixed with boulders on bed rock. A short portage brought us past the rapids into Wabinosh Lake. On the east side of this lake is a high ridge, reaching an elevation of about two hundred feet running north and south, dividing Lake Wabinosh from Lake Nepigon. The west side of it facing Wabinosh is very abrupt, but it gradually slopes towards Lake Nepigon. The timber on this ridge is scattered and consists of small birch, poplar, jackpine and balsam, with fair-sized spruce ten to twelve inches at the stump, and would average about two cords of pulpwood to an acre. The country to the north is level, drained by a small river about forty feet wide, but is not passable after the first two miles, owing to fallen timber. The timber along its banks is chiefly spruce growing thickly, averaging ten inches at the stump, but back from the banks is small birch, poplar and jackpine. There is also another river running into this lake on the south-west corner, from eighty to one hundred feet wide, which with Mr. Percy Dawson I followed about two miles, and found it winding its way between mountains. There was a nice current and no rapids as far as we went. It seems to be a regular canoe route as we met Indians in canoes returning from Nepigon House after "Treaty Day." Our route was up the Wabinosh River on the west side of the lake and over three portages to Round Lake (Wahweyah), the mountains becoming less lofty as we proceeded, and the timber small and scrubby with balsam not so plentiful, but birch, poplar and jackpine growing thickly along the banks. From Round Lake we continued up the river through two lake expansions, viz. : Sucker and Valley Lakes, and over several short portages, the last being very high and difficult. By this portage (12th) we leave the river and enter Clear Lake, which is the first of the chain of lakes which now lay on our route. Next to Clear Lake is Tunnel Lake. Proceeding out of this lake we cross the height of land or Fourteenth portage which is short, being only about fourteen chains in length. From this portage we go into a star-shaped lake called Rock Island Lake, from a large rocky island in the centre, and across Prairie Portage (15th) which was a trading post years ago, where the remains of the old buildings are to be seen. East of this portage and also east of Rock Island is Fawn Lake, which we explored on a side trip and is part of the portage route to Caribou Lake. Leaving the 15th portage we proceed by a chain of lakes to Tamarac Lake. So far the timber has been small and scrubby, consisting of small birch, poplar, jackpine, spruce and balsam, the last not so plentiful as we approach the height of land. On the south and east sides of Tamarac Lake there is a block of good spruce suitable for pulpwood, which would cut from fifteen to twenty cords to an acre. Its area would be about three square miles. There is some good tamarac growing on the swampy ground close by. The spruce grows on the rock, which is covered with a thick bed of moss, this seems to account for its healthy condition. This was the best timber that I saw on our route, considering area, above the height of land. From this lake we went north by a short, shallow river into Smooth Rock Island Lake, which is a lake of long, narrow bays and dotted with a great number of low, rocky islands covered with small timber, the country around it being also very rocky and flat. At the entrance to this lake there are several small brulés, as there are a number of Indian camping grounds. The northern and western sides of this lake are burned bare in some places, while in others the dead timber is standing. This timber appears to have been burned within the last ten years, and the same condition prevails for about fifty miles north. On the south side the timber is small and scrubby, and appears to be a second growth. From this lake the Geologist and I took a side trip easterly to Caribou Lake, a distance of about six miles. This is a large lake about four miles square and contains some large islands at the south side. With the assistance of our Indian guide we found a portage route which is not travelled much, as the route lies through several muddy ponds and muskegs which made our progress very laborious. This brought us to Fawn Lake as before mentioned. The timber is the usual jackpine, poplar, birch and spruce, all small and scrubby from the absence of soil to sustain vigorous growth.

From Smooth Rock Island Lake we proceed west up Wahbahkimmug Lake which we suppose is Lake Savant. This lake is divided by a long point of land into Upper and Lower Wahbahkimmug connected by two narrow channels. On the lower or eastern side there is an outlet, and a small creek emptying into the lake from a smaller lake. The south side of the lake is inclined to be swampy and is chiefly low and flat. On the north side the country is more elevated than that around Smooth Rock Island Lake which is flat. There is a mountain of about one hundred feet one and one half miles from the shore which gives a good view of the surrounding country. From this point the timber appeared to be the same as on the shores of the lake, jackpine being the principal tree, growing in a thin sandy loam covering granite rock. At the north east corner of the lake we saw two or three of the largest tamarac trees seen on our journey, one of them measuring twenty-eight inches at the stump, the others twenty-six and twenty-two inches. They were growing in a wet swamp and on this account had not been destroyed by fire as had been the case with timber on higher ground. The north side of Upper Wahbahkimmug Lake is all a brulé, while the rest of it is timbered with jack pine, birch, poplar and some balsam, which was the first seen since leaving the height of land. Another side trip was made by the geologist and myself to the Sturgeon River, which flows into this lake from the south. It is about two chains wide at the outlet, and on it are a number of rapids, which we portaged over at intervals for the first six miles, when we came to a fall about twenty-five feet high, where the river leaves a small lake. From the appearance of the portages this route is not much travelled. The timber up this river was the same as around the lake, being burned at the mouth, which is commonly the result of carelessness with camp fires, the mouth of a river being generally chosen as a camping ground. The north side of Upper Wahbahkimmug Lake extends into an arm about six miles long and empties into the Pike River. At the head of the river there is a long, flat and rocky rapid, which can be run with a light canoe. Near this rapid is an Indian cemetery. Pike River is a shallow and rocky stream about three miles long, which empties into a small lake. This we called Greedy Water as there are rivers running into it and apparently only one outlet, viz: Ogoki River. The timber close to Pike River (Kenogami) is chiefly pine, tamarac averaging ten inches at the base, a little spruce, poplar and birch. From Greedy Water Lake we proceeded north up the Palisade River, which flows south, and as its name implies is enclosed by bare steep rocks. The timber has all been burned for some years, with the exception of a few clumps, which have escaped the fire. From some high points we got a good view of the surrounding country, which was a brulé as far as we could see and continues so until we pass the thirty-eighth portage, which leads to Swampy Lake. The timber on the south side of this lake is burned, while it is healthy and green and somewhat larger on the north. Here there is more soil on the rocks and the country is more swampy, while the lakes, as far as Green Bush Lake, are shallow and muddy. There is another portage route from this lake i.e. Swampy Lake to Albany River, but from information we received, the route is more difficult. From Swampy Lake portage (39th) to Green Bush Lake the country is swampy and flat and the lakes are shallow, the bottom being covered by a soft mud, with only ten to fourteen inches of clear water, which made the canoeing difficult. The timber is chiefly spruce and tamarac. Some of the clumps of the spruce would be suitable for pulpwood, but there will be this drawback that the lakes, being shallow, will need a deal of improving to make them navigable. The timber is in a healthy condition, nor are there any brulés. This may be accounted for by the fact that the country is not very desirable to travel in. This country between Swampy Portage Lake (39th portage) and Green Bush Lake, (44th portage) has the best spruce, with the exception of the small area mentioned on Tamarac Lake, above the height of land. There is some jack pine and from its appearance the land has had no fire over it for quite a number of years. From Green Bush Lake the country is more rocky; consequently the timber is poor, consisting of small white birch, poplar, spruce and jack-pine, with some fair-sized tamarac near the water. This continued about the same during the remainder of our trip to the Albany River, the country being flat and slightly rolling.

At Green Bush Lake the muddy bottom disappears and the lake shores become more abrupt with granite sand or boulders instead of the accumulation of mud as on the lakes just passed over. From Green Bush Lake we pass into Pashkokogon Lake by a short portage. This is a large lake and is drained by the Pushkokogon River into the Albany. The timber here is the same as on the greater part of the trip. We had information from Indians that there was land suitable for agriculture along this river, which turned out to be sand drift. Lake St. Joseph is reached by the Clear Rapids portage about four miles up the Albany River from the mouth of Pushkokogon River. Here the Albany River widens into a lake expansion and is fed by another large river, which we explored for about eight miles, but from the appearance of the portages it was not much travelled. As the Albany River was the limit of our trip, we returned by the same route to Pike Lake or Greedy Water Lake, and went east down the Ogoki, the head of which is about a mile from where Pike River empties into the lake. Quite a number of rapids were passed, notably the Scotch rapids, which are somewhat difficult to run, but as there are no portages they have to be passed in this manner. The timber down this river is chiefly jack pine, most of which is burned, the country being a continuation of the burned district north of Smooth Rock Island and Wahbahkimmug Lakes. As we proceed we pass the mouth of a river running out of the north-east corner of Smooth Rock Island Lake, which the geologist and I explored from that lake. Passing over several rapids and portages we reach White Water Lake, which is a combination of two large lakes, the larger and more easterly called Highland Lake by the Indians, as they suppose it to be a height of land. The timber around this lake is all small and scrubby and the country rocky. As we proceed down the river, the country becomes low and flat, with here and there drifts of sandy loam until we reach White Clay Lake, which is a large lake and appears to be quite a hunting ground. From this lake we start south up a river over numerous portages, including the height of land. Leaving this portage, we go into a small lake about four miles long, and thence by Little Mud River to Round Lake, where our work of exploration ends. The timber is small and scrubby mostly burned. From Round Lake we continued down the Little Mud River to Lake Nepigon. This river is swift and crooked along its banks for about thirty-five miles. Here there is a good growth of spruce, ranging as high as twenty inches in diameter, also good tamarac and balsam, the latter as high as twelve inches in diameter, there is also some cedar of fairly good size. There were no brulés along this river and the timber looked healthy. This timber extends about twelve miles north of Lake Nepigon, but is not in the territory which was allotted to us for exploration.

Taking the territory as a whole, there is little or no timber, with the exception of the tracts mentioned on Tamarac Lake, and the lakes between the thirty-ninth and forty-fourth portages, but on account of the location and smallness of area, it is practically of no commercial value.

The whole country travelled is very rocky and barren, and is of no use for agriculture. The fires, I would judge, are altogether due to the carelessness of Indians with their camp fires. Evidences are to be seen that a second growth of timber has been burned.

The thanks of our party are due to you for your admirable arrangements for our well-being and comfort on the trip.

I have the honor to be, sir,

Your obedient servant,

JAS. A. SHARP,
Land and Timber Estimator.
Exploration Survey Party No. 7.

To H. B. PROUDFOOT, ESQ., O.L.S,
Surveyor in charge of Exploration Survey Party No. 7.

I, J. A. Sharp, of Sudbury, Land and Timber Estimator, make oath and say, that the foregoing report is true and correct, to the best of my knowledge and belief.

Sworn before me at Sudbury this 28th day of December, A.D. 1900. } JAS. A. SHARP.

T. J. RYAN.
A Comm'r H. C. J.

GEOLOGIST'S REPORT OF EXPLORATION SURVEY PARTY NO. 7

TORONTO, Dec. 10, 1900.

Sir:—In accordance with instructions received from Mr. Archibald Blue, Director of the Bureau of Mines, to accompany Survey Party No. 7, under supervision of Mr. H. B. Proudfoot, O. L. S., on a track survey from Wabinosh Bay at the northwest of Lake Nepigon, up Wabinosh River, across the Height of Land to Albany, and sections of the Albany and Savant Rivers, I left Toronto, June 23rd, and arrived at Nepigon the day after. My instructions required me to accompany Mr. James Sharp, timber estimator, in side trips from the track, and to examine and report on the geological features of the district explored, its land and timber, and mineralogical possibilities.

The party seven in number, left Helen Lake en route up the Nepigon River on July 5th, 1900, and arrived at Big Flat Rock Portage on the shore of Lake Nepigon two days later. On Monday, July 9th, we left the portage, and after encountering adverse weather for three days, we arrived at Nepigon House on the 12th of July. After a few days' preparation for our trip, we left Nepigon House, and arrived at Wabinosh River on July 16th.

WABINOSH RIVER.

Wabinosh River is small, about one chain wide, flowing at the first portage over boulders and ledges of hard Huronian trap, having a fairly well marked schistose appearance. Its shores are rocky, bearing scrub spruce, tamarac and birch. Where the river empties into Wabinosh Bay, would be a good place for a water power as the fall of the river would furnish ample motive force. At the end of the Bay, is a little plot of ground, with two or three huts, where there is a little surface soil, sandy in composition, in which potatoes are grown to good advantage during the summer. This deposit is of small area, only an acre or so, being but few a inches in depth, and has probably resulted from the abrasion and disintegration of the country rock.

LOWER WABINOSH LAKE.—During our stay on Lower Wabinosh Lake several trips were made inland. The first, in a north-easterly direction found the country hilly with high ridges, running about north and south, from one hundred to three hundred feet high. The rock formation of these ridges is mostly Huronian trap, but a talus of diorite, granite and crystalline quartz is often found at the base. Outcrops of this same trap occur at many places on Lower Wabinosh Lake, but in every case where these were examined they proved destitute of mineral. The timber of the locality is birch, tamarac, spruce and poplar, with a good deal of alder and other underbrush. The shore line of the lake has many angular boulders of trap and granite, while in a few places occur beaches of white, angular quartz sand.

At the northern end of the lake there flows in a small creek about fifty feet wide, ten feet deep, having a swift current. For two miles up the stream is tortuous; its bed is composed of mud, its shores of a sandy soil, with a slight intermixture of clay. The soil for about one hundred yards on either side of the river is level, and if cleared might be utilized for pasturage or root culture. The rock formations, noted at the north boundary of the lake, were of the same type as seen inland, being eruptive Huronian diabase, with numerous boulders from six to eight feet through, much weathered, often striated, and covered with moss, lichens and liverworts, the characteristic flora of the district.

From July 21st to 25th was occupied chiefly in portaging goods from the Wabinosh River, through Upper Wabinosh or Round Lake, over Sucker Lake, and a small chain of small lakes and rivers, across Tunnel Lake to Rocky Island Lake, at the end of the fourteenth or Height of Land portage.

The country passed over is well wooded with small spruce, tamarac, balsam, poplar, birch and jackpine, but the trees are quite small. The surface of the land is hilly; the rock formation of the district is, for the most part, Huronian trap, large outcrops occuring on the rivers and lakes, but apparently possessing no mineral. Between the eighth and twelfth portages, these outcrops possess a distinctly schistose structure, and bear all appearance of being of the Huronian age rather than the Laurentian. On July 24th, after passing the twelfth portage, which is quite high and rough at the outset, we began to meet less trap and more granite, and after the Height of Land was passed, the rock formation appeared to be all hard, flinty Laurentian granite, very silicious. The trap, in many places, extends over the granite in quite large intrusive sheets, and in all cases, the contact between the two formations is hard to discern, as the two shade so perfectly, one into the other, both in color and texture. This circumstance gave the impression at first that the trap was of the Laurentian age, the same as the granite, both having been laid down at the same time, and mingling one with the other : but later, because of its marked schistose structure, it was thought better to classify this trap as Huronian rather than Laurentian. Mineral may exist near this Laurentian-Huronian contact, but our examination failed to show any traces or reveal any contact veins at all.

At the fifteenth portage from the Wabinosh Bay, called Prairie Portage, situated at the northern end of Rocky Island Lake, there are indications of there having been a Hudson Bay Company's post. Here there are a number of charred bottom logs of a log house, and around the place for an acre or so, a clearing has been made, and during the summer blueberries and raspberries grew there quite plentifully.

Fawn Lake.—On July 26th, a side trip was made to Fawn Lake by canoe. The route lies first south east from Prairie Portage about half a mile, then east about half a mile into a bay of Rocky Island Lake. Near the foot of this bay is a very poorly marked portage into Fawn Lake, about twenty chains long, and winding so much in its course that it was necessary to blaze a path across before the canoe could be portaged. From this portage we paddled over Fawn Lake, which is long and narrow, somewhat like a high boot in shape, and about six miles long. The lake is deep, has a rock bottom, and abounds in pike, pickerel and white fish, as the large catches of several Indians demonstrated.

The landscape, as far as the eye can reach, may be considered as flat, although a few small hills do occur. The rock formation around the lake is hard, pink Laurentian felspar granite, outcropping in many places. Some trap was met but only as angular boulders. At the portage between Rocky Island Lake and Fawn Lake, running through the granite, are numerous joint or gash veins from one-eighth of an inch to two inches in width, filled with milky quartz and orthoclase felspar. This gangue is very glassy and hungry-looking and presents no indications of ore or mineral. The timber on Fawn Lake has nearly all been burnt, but there is a little small spruce and birch.

At the twentieth portage on the shore of Tamarac Lake, the rock formation is granite. Along the shore line and inland, large outcrops and boulders of granite were noticed. In a number of places the shore is sandy, especially at the south-western part of the lake, where the sand presents a stratified appearance, having crystalline quartz, hornblende grains and amethyst grains arranged in layers. Here at the south-western part, large granite outcrops occur, weathered to a dark brown color, but bearing no economic mineral. The timber around Tamarac Lake consists of small spruce, birch, tamarac, poplar and jackpine, the spruce being good at the south-western end of the lake.

At the twentieth portage is a rapid on a small river running from Tamarac to Smooth Rock Island Lake ; also at the twenty-first portage, where the river empties into the latter lake is a rapid and fall, having a good volume of water flowing over, and being about a chain in width. Both these rapids and falls would be adaptable for furnishing good water power. At the twenty-first portage, which is about twenty-three chains long, the rock formation is Laurentian granite. In it some hungry-looking amethystine quartz was noted in gash veins but bearing no mineral.

Over the twenty-first portage, we came into a long deep bay of Smooth Rock Island Lake. This lake according to the survey, is some twenty-five miles long, shaped very much like a huge devil-fish. At the southern end of the lake, the shores are steep, having rugged granite cliffs over hanging the water's edge. The surrounding country is flat ; the soil is scanty, the trees for the most part growing on the bare granite and deriving nourishment from the thick mat of moss surrounding their roots. The timber is small spruce, tamarac and jack pine.

While encamped at the southern end of the lake, several trips were made inland. The rock formation of the district is Laurentian granite, but we travelled northward from the twenty-first portage, we noted that their occurred more hornblende granite than the usual variety, having crystalline quartz and orthoclase felspar as inclusions. On the shore which has been rendered slanting and slippery by wave action, the granite is nearly every where exposed. The lake bottom is of solid granite like a floor, the boulders are angular and composed of granite and quartz. At the north end of the bay in which we first encamped, the country has been badly burned and blueberries grow there abundantly. Across this brulé is a rough portage, about twenty chains long between two bays of the lake. About one half a mile south of this portage, on the west shore, are two immense granite boulders, at least twelve feet through, lying high and dry about a hundred yards from the water, evidently placed there by ice action.

On a trip made round the lower part of the lake, under the guidance of two Indians, the same general features, noted above, were observed. The country is hummocky in appearance, and sloping down to the water's edge, and looks quite desolate on the east shore, where the timber has been burnt. Numerous gash veins were noted in the granite, filled without exception, with glassy crystalline quartz, being about two inches in width, showing no appearance of mineral and being very crooked and irregular. At a point on the east shore about ten miles north, numerous outcrops were examined for mineral without success. These outcrops were very much weathered, and in some places running through the granite, were small veinlets of hornblende schist, likely Huronian.

CARIBOU OR DEER LAKE.—While encamped in Lone Breast Bay at the north west of Smooth Rock Island, several exploration trips were made. Chief among these, was a four days' trip to Caribou Lake, also called "Deer Lake," in Indian termed "Atikosahgahegau" In company with the timber estimator and an Indian guide ; this trip was made from Lone Breast Bay on August 10th. The portage route to Deer Lake ran out of a long bay at the north eastern end of Smooth Rock Island Lake. There are five portages, all good and well marked to Deer Lake, the route following a river about three chains in width, with a swift current.

Deer Lake is about five miles long, and from one to four miles wide. The rock formation along the lake and river is dark colored, Laurentian hornblende granite. The shores are rocky and wooded with short spruce, tamarac and jack pine. Away to the south the height of land is distinguishable as a high ridge, while the surrounding country is flat. The flora is typically northern comprising lichens, liverworts and mosses. From Caribou Lake there is a portage route to Round Lake, but this we did not follow. Instead we pursued a course from Caribou Lake to Fawn Lake, which has been previously described. This route to Fawn Lake runs out of a south western bay of Deer Lake, and continues in a series of muddy lakes, in a south westerly direction until it enters a little bay at the south east corner of Fawn Lake. The first portage is fifteen chains long, running from Caribou Lake to a small creek, which empties into a muddy lake. Across the lake, a small creek is followed up the second portage, about half a chain long. From here the course is through a small muddy pond, almost entirely over grown with sphagnum and rank water plants, to the third portage, twenty-five chains long The next portage is about the same length, while the fifth is a good mile, long and rough. From here. through another mud lake, we came to the sixth portage which leads to Fawn Lake, being forty-six chains long and well marked. The mud lakes travelled over are all small, not one being more than ten acres in area. The vegetable growth is about six feet deep, carbonized quite extensively near the bottom, and might possibly be of value eventually for peat and moss litter.

The rock formation betwéen Deer and Fawn Lakes is the same as between Deer and Smooth Rock Island Lakes. Between Deer and Fawn Lakes, fewer outcrops occur, the country being more level and abounding in muskeg swamps and mud lakes. From Fawn Lake, the return trip was made by way of Rocky Island Lake and Prairie Portage, over the same route as travelled by the party on the way north.

SMOOTH ROCK ISLAND LAKE.—While encamped on the western side of Smooth Rock Island Lake,an exploration trip was made to a good sized river flowing out at the northwest part of the lake This river flows in a northwesterly direction and has a very swift current. At the first portage there is a series of rapids, about a quarter of a mile long, with a total drop of about thirty feet, which would furnish a good water power The portage is rough, over burnt fallen timber, about a quarter of a mile long, and runs from the lake to the river below the rapids. At this first portage some samples were taken, including several pieces of rock containing (1) white quartz mixed with grey rock matter, (2) pyrite impregnated within the quartz, also segregated, (3) pyrrhotite in small quantity, and (4) a little hornblende. These samples were sent to the Assay Office at Belleville, and were assayed by Mr. J. Walter Wells for gold, silver, copper and iron. The analysis resulted as follows :—

Gold	traces present
Nickel	"
Silica	24.45 per cent
Met. Iron	15.84 "
Sulphur	3.47 "

The country rock of this locality is Laurentian granite. Running through this at the first portage are several joint veins of milky quartz, from one inch to one foot in width. Also a vein of dark-colored ferruginous quartz in which pyrite was noted in minute quantities. Below the portage the river is from one hundred to one hundred and fifty feet wide, about twenty feet deep ; the water is dark colored and flows to the Albany River.

At a point in Smooth Rock Island Lake opposite the Hudson Bay Company's post, some mica was reported by members of the party, but on being examined it proved valueless. It is of the muscovite variety, in chunks from one-third of an inch to three inches square, very much twisted and dipping in all directions in the rock, which is hard Laurentian granite. Here and there in this granite are large inclusions of white crystalline quartz, while two small seams of hornblende schist were noted running parallel, slightly iron-stained but valueless, having a strike of north 20 degrees east, dip 70 degrees vertical, width 8 inches and 10 inches.

From Smooth Rock Island Lake to White Earth Lake ('Wahbahkimmug' Lake), the portage route runs up the Wahbahkimmug River, over five portages, which are all well marked and level. This river was explored on Aug. 6 by Mr. Sharp and myself, and later travelled over by the whole party. Its shores are rocky in many places, the rock being granite with here and there large inclusions of hornblende schist. At the first portage, the river empties into Smooth Rock Island Lake by a water fall about ten feet high, flowing through an almost perpendicularly walled channel of granite about twelve feet wide. This fall and another and the fifth portage, could be very economically utilized for water power.

While encamped on Wahbahkimmug Lake, several exploration trips were made to the surrounding country. On August 18th, a river at the south west of the lake was explored. Along this river is a portage route used by Indians during the hunting season. The first portage is about ten chains long, well marked, and cuts off a shallow rocky part of the river. Beyond this portage the stream averages nine inches in depth, a chain in width, has a mud bottom but rocky banks. About a mile and a half up the river we came to a small lake about a third of a mile long and six to eight chains wide. Above this lake the river flows by a series of rapids which cannot be followed up by canoe, but are passed by a long, difficult portage.

The rock formation along the river is Laurentian granite. At the first portage, there is an iron-stained outcrop with a band of red jaspery mineral, running through it about two feet wide, strongly impregnated with a greenish mineral, probably olivine. Samples of this mineral and rock were sent to Mr. J. Walter Wells for analysis and showed reddish jasper mixed with magnetic iron and very silicious. Exact assay :—

Silica	S 02	58.05 per cent
Met. Iron	Fe	14 35 "
Sulphur	S	00.076 "
Phosphorus	P	00.02
Titanium	Tn 02	None
Manganese	M	Traces

The country along the river is pretty level, and timbered with small jack pine, very little spruce, birch or poplar being seen.

Another trip was made to the main strait between Lower and Upper Wahbahkims mug Lakes, to examine some mica reported by Indians. This mica proved to be of les-value than reported, being of the muscovite variety, plates exposed on the surface being, at the largest only, about three inches square but mostly occurring in smaller sheets. The surrounding minerals are white quartz and felspar, very much twisted and distorted.

A high rocky ridge about two miles inland on the north shore of Wahbahkimmug Lake was explored for mineral indications This ridge is about three hundred feet high, running nearly east and west, composed of pink granite, which showed no trace of mineral. At the foot of this ridge northward, is an old dried up lake bed; further away about two miles northward, are three small lakes which empty by a little creek into Upper Wahbahkimmug Lake. On the south side of the ridge is a deep valley in which there lies a little lake about half a mile long, which empties into Lower Wahbahkimmug Lake by a small creek. During low water time this lake is reached by a well marked portage, a third of a mile long, used by the Indians during the trapping season.

From Upper Wahbahkimmug Lake, a day's trip was made up the Sturgeon River route. The mouth of this river lies at the southern part of the lake at the bottom of a long bay, where the river empties into the lake there is a rapid and water-fall named Sturgeon Rapids, and to pass this, is cut the first portage about ten chains long and well marked. The next four portages are rough and hard to follow, while the sixth is exceptionally good. This Sturgeon River route was at one time used by the Indians, but is now almost entirely abandoned. The river was followed up for seven miles or more and may be said to run in a northerly direction, being about on an average two chains in width, and having numerous swift rapids, passed by portages going up, but run on the return trip.

LITTLE STURGEON RAPIDS. About three miles up the river is a rough rapid known as Mahmansobahwetig, or Little Sturgeon Rapids. At the south end of the sixth portage is a good-sized lake about a mile long and sixty chains wide. Where this lake empties into the river is a fine waterfall with about twenty-five feet fall, called Black Beaver Falls, which could be utilized for water-power very economically.

The country around this lake and river is hummocky in appearance, fairly well timbered with jack pine and birch, but little spruce. The formation is Laurentian granite, outcropping all along the river and lake. At the third portage, running through the granite is a vein of jasper about ten inches wide, but containing no minerals of economic value.

WAHBAHKIMMUG LAKE. August 24th to 28th was occupied in a canoe trip around Wahbahkimming lake. The primary intention was to find a portage route to Saramie River, which was reported to run out of this lake but, as was afterwards found out, does not. Instead of hunting for this route we made a systematic exploration of the south and east shore line and surrounding country, so as to arrive at a general idea of the wealth of the territory.

The country may be said to be flat, much like it is around Smooth Rock Island Lake, and the timber has been burnt in many places The rock formation is Laurentian felspar granite, and although numerous brulés were examined no economic mineral or indications were seen. The timber is small jack pine, spruce and birch.

From Wahbahkimmug Lake we travelled to Osnaburg House. The portage route thither runs from a long bay at the north-east of Wahbahkimmug Lake, follows Pike River, running out of the above lake, through Pike Lake, up Palisade River and then by a series of lakes and rivers to the Albany River and Lake St. Joseph. There are twenty-three portages, all of them well marked and good, with the exception of a few muskeg portages, which were very wet and hard to walk over. Between Pike Lake and these muskeg portages the scenery is grand, granite cliffs from fifty to one hundred feet high rise up on either side of the Palisade River, and it is because of this feature that the river gets its name. The country inland is much more hilly than around either Smooth Rock Island Lake or Wahbahkimmug lake, where it is level and tiresome to the eye. The timber noted along the river consists of small birch, poplar and jack pine with very little spruce. On both sides of the river for a considerable distance the country has been fire-swept and presents a very desolate appearance.

The rock formation north of Pike Lake is Laurentian granite, gray in color and very hard and brittle. On Palisade River a piece of white quartz was picked up which looked as if it might contain some mineral of value and was sent to Mr. J. Walter Wells, of the Assay Office for analysis. It might be more exactly defined as a piece of white quartz containing a few specks of black, non-magnetic rock matter with a gray streak, probably hornblende, also a little pyrite. The analysis showed:

Gold	Traces present
Silver	None "
Copper	" "
Nickel	" "

On the south shore of a small lake called Swampy Portage Lake, a considerable extent of Laurentian gneiss was observed occurring with the granite.

Farther north, at many places on the shores of Green Bush and Pushkokogan Lakes, there are outcrops of grey granite, while inland this same formation occurs in the form of huge angular boulders. On these two lakes and along Pushkokogan River, are numerous evidences of Indian occupation, such as old wigwam poles, old camp grounds and storages of canoes and birch bark. As we approached the Albany River, these indications became more marked, and have probably been made by the Indians from the vicinity of Lake St. Joseph.

On a point opposite where the Pushkokogan River empties into the Albany, a fissure vein of dirty yellow quartz was noted running through the granite with a dip of 80 degrees from horizontal, strike north-west and south-east, width 14 inches, traced for about ten feet. This quartz was examined well and carefully for economic mineral but was proved barren.

OGOKI RIVER.—From Albany River back to Pike Lake we travelled by the same route as going northward. From Pike Lake we branched off eastward and followed the Ogoki or Okokeesibi River, which runs in an easterly direction out of Pike Lake and is studded with many granite reefs and abounds in bad rapids. At the first portage there is a waterfall about six feet high, with a very strong current, which could be utilized as a good water power. The Ogoki River empties into the White Water Lake by a bad rapid, passed by a short portage.

From the Ogoki River our route lay through White Water Lake, Highland Lake and River to White Clay Lake. From Pike Lake to White Clay Lake, the rock formation is pink Laurentian granite, which outcrops on all sides. On a trip inland from White Water Lake some boulders were noted with magnetite disseminated through them in small quantities, not enough, however, to be of economic value. The country round these lakes may be described as hilly, and the shore line rocky, but on Highland Lake and River the high hills and ridges begin to disappear, and the country becomes much flatter. In this locality numerous sand banks and sand beaches are seen until, when we get to White Clay Lake, we find high sand drifts at many places along the shore. The timber noted was small birch, spruce, poplar, tamarac and jack pine.

The canoe route from White Clay Lake to Lake Nepigon is well marked but in a few places quite rough. There are 17 portages, some of them long and one or two rough. At the ninth there is a fine waterfall, having at least a twenty foot drop with a large volume of water flowing over; at the tenth portage there is also a waterfall, about half the size of the last. These falls could be utilized for water-power without great expense.

From White Clay Lake to Round Lake, the shore line of the rivers and lakes is high and rocky and composed of Laurentian granite. Near Round Lake the formation changes from granite to trap and the contact is very hard to discern. From Round Lake to Lake Nepigon, Little Mud River was followed. This river has a swift current, flows through a sandy channel and winds very much. The soil in the river valley is a sandy loam, which would be well adapted for pasturage and the growth of hay and roots. At the third portage from Lake Nepigon there has at one time been a large camp ground, the land being cleared for an acre or more and covered with a thick sod. The timber along Little Mud River is spruce, tamarac, poplar and jack pine. The spruce and tamarac are good and could be easily floated down the river to Lake Nepigon.

SUMMARY OF EXPLORATION.

The region covered by this report might be considered as bounded on the north by Lake St. Joseph and the Albany, on the east by the portage route from White Clay Lake to Lake Nepigon, on the south by Lake Nepigon and on the west by Wahbahkimmug

Lake and the portage route to Osnaburgh House. A detailed description of the trip has been given, but it may be well to sum up briefly under various heads.

Soil and Agriculture.

At Nepigon House, Wabinosh Bay and Osnaburgh House, potatoes and hay are grown. The soil is sandy and of no great depth. At the head of Lower Wabinosh Lake and along Highland Lake and River are small areas of tillable land, already described. Along Little Mud River the soil is well adapted for roots and fodder, but not for other branches of agriculture. On this river near Lake Nepigon there is a Hudson Bay winter trading post, and here potatoes are grown during the summer. The soil here is a good deal like fibrous peat, and has no great depth or area. Some of this soil was sent to Belleville for analysis and resulted as follows:

		2nd test
Moisture	5 65 per cent.	5 48 per cent.
Vol. Comb	27 73 " "	25.87 " "
Carbon (fixed)	6 09 " "	6.77 " "
Ash (reddish)	60 53 " "	67.38 " "
Sulphur	None	None
Phosphorus	Traces	Traces
Quartz grains	....	....
	100.00	100.00

Mineral Resources.

Outcrops of rock are seen everywhere along the lakes and rivers, in fact the country may be said to be one continuous outcrop, so little soil is there. The rock in almost every case, except as around Smooth Rock Island Lake, Wahbahkimmug Lake and Palisade River, is barren. Samples of all likely-looking rock, mineralogically, were taken and forwarded, on return, to Mr. J. Walter Wells, B.Sc., at Belleville for analysis, and resulted as shown in this report. Along the very irregular contact between the Laurentian and Huronian formations, mineral may exist, but if so, this fact did not come under my notice. The mineralogical outlook is not very bright and only when Lake Nepigon was reached was any ore of value seen, that at Poplar Lodge. This ore is red jasper, mixed with a silicious iron ore, having a red streak and non-magnetic. Samples of this ore were sent to the Assay Office for analysis and resulted as follows:—

Silica	SO^2	22 02 per cent.
Met. Iron	Fe	31.08 " "
Sulphur	S	00.20 " "
Phosphorus	P	00.013 "
Titanium	TO^2	None
Manganese	Mn	Traces.

I believe some prospecting is being done at present at Poplar Lodge.

Fur and Game.

Small game, as partridge and rabbits, are fairly numerous in one or two localities especially around White Clay Lake. Duck are quite plentiful on the muskeg lakes *en route* to the Albany River, but are quite wild. No moose, caribou or bear were seen on the trip, and Indians report them to be scarce. As a whole, the country might be considered as a fair producer of the smaller fur-bearing animals, as the fox, otter, mink and musk rat, but very little larger game exists, which is largely accounted for, I believe, by the barren, desolate, burnt state of the country.

Water Powers.

At numerous places, already described, the volume of water and the vertical drop is fufficiently great to develop water power. As special instances, Black Beaver Falls, the salls on Wahbahkimmug and Little Mud Rivers, and on the Ogoki River may be noted.

2 years old heifer. Ox gave 1,121 lbs. of beef.

Some of the Bishop's cattle, Moose Factory. Party No. 6.

Bishopscourt and store, Moose Factory. Party No. 6.

Windigo Bay, Lake Nepigon. Party No. 7.

Nepigon House, Lake Nepigon. Party No. 7.

Rapid, portage No. 1   Party No. 7.

Between portages Nos. 5 and 6, Sucker Lake.   Party No. 7.

Smooth Rock Island Lake. Party No. 7.

Smooth Rock Island Lake, showing spruce timber. Party No. 7.

GEOLOGICAL FEATURES.

The chief rock formation of the district explored is Laurentian granite. In some localities, however, containing hornblende, giving a dark-coloured hornblende granite. From Wabinosh Bay nearly to the Height of Land, and again from Round Lake to Windego Bay, at the northern end of Lake Nepigon, the formation is hard, Huronian diabase and greenstone. So there is probably a belt of Huronian rock running across the northern end of Lake Nepigon, from fifteen to twenty-five miles wide. The rocks, especially between the sixth and twelfth portages from Wabinosh Bay, possess a distinctly schistose structure. This feature was also noted near Round Lake and so is somewhat confirmatory to the belief that the formation is Huronian. The contact between the Laurentian and Huronian formations is irregular and hard to follow; in places the Huronian flows over the Laurentian as large intrusive sheets of hornblende schist or trap, while in the Huronian area, the granite often out-crops in the form of hummocks. In many places the Laurentian granite has cracked and these fissures have been filled with quartz and hornblende schist, possibly Huronian. In the northern region, in some localities there occurs with the granite much gneiss, very much twisted and contorted. The hummocky appearance of the country in general, seems to point to the fact that at one time the whole district was swept and grooved out by glacial action. The hills and ridges may be a continuation of the Laurentian range.

Taken as a whole the entire district explored may be said to be one barren, desol ate stretch of country. In the northern district the muskeg swamps and ponds contain a good deal of dead matter, slightly carbonized, which may be eventually of value for peat and moss litter. The country if protected from fire for a number of years, may become valuable as regards timber. Mineralogically the country may be valuable, but careful search this summer failed to show it, while along the Laurentian-Huronian contact no mineral was noted at all.

Accompanying this report is a map, colored to show the geological formation, as directed by the Bureau of Mines.

I have the honor to be, Sir,
Your obedient servant,
F. J. SNELGROVE.

"B"

This is the Report marked "B" referred to in the Declaration of Frederick James Snelgrove, made before me this 10th day of December, A. D. 1900.

G. GARDNER,
A Notary Public for Ontario.

H. B. PROUDFOOT, ESQ., O.L.S.,
In charge of Exploration Party No. 7.

SURVEYOR'S REPORT OF EXPLORATION SURVEY PARTY NO. 8.

PARRY SOUND, ONTARIO, December 31st, 1900.

Sir:—

Under your instructions dated June 11th, 1900, placing me in charge of Exploration Survey Party No. 8 to make an exploratory survey of the country lying westward from Nepigon River and Lake to the height of land, and round Dog and Black Sturgeon Lakes, I left Collingwood on June 28th per steamer Majestic for Nepigon via Port Arthur. Our party included Mr. A. H. Smith, geologist, and Messrs. Claude Bryan, Allan Magee and Sidney Hurd, cook, Mr. John Piche, timber and land estimator, joining us at Sault Ste. Marie. We arrived at Nepigon on July 2nd, and left there on the 4th for the mouth of Gull River, on the west shore of Lake Nepigon, where I arrived on the 14th and hired two Indians who knew the Gull River country, as guides and canoemen, and commenced the survey of said river and exploration of that part of the country on the 16th. When we had worked up the river about twenty-five miles the Indians said that it would soon get very rough and rapid and would be slow work moving up with supplies and advised taking a canoe route, which went off to the west through a number of lakes and came into Gull River about fifty miles farther up. As this route swung off from ten to twenty miles from the river, and through a part of the country that it was necessary to explore, I made it a base of exploration and made a survey of it through to Gull River.

At Whitefish Lake where the route turns southward to the river, Mr. Piche went west with an Indian guide through a chain of lakes to the height of land, exploring the country on each side of the route as he went, and sketching in the several lakes that he passed through, estimating distances, while I continued my survey on to Gull River at Squirrel Lake. At the head of this lake I found the river coming in over a small rapid about twenty feet wide and not much water, and was told by the Indians that it was but a small stream from there to its headquarters and could only be ascended farther with small canoes, the water at that time being very low, and that there was nothing but small timber for a considerable distance and then burnt country on to the height of land. This route is used by the Indians in going light to and from Savanne on the Canadian Pacific Railway.

After surveying Squirrel Lake I surveyed Gull River down to where I had quit work coming up, where I went off on the portage route before mentioned. When coming down the river from Squirrel Lake we were able to run over many of the rapids with loaded canoes, where we would have been obliged to carry if we had tried to work up, and I was satisfied that the Indians' advice was good. I then moved down the river to within about fifteen miles of the mouth where I went northward through a portage route, surveying and exploring the country as I went. On Obugamiga Lake I turned eastward coming out into Wabinosh Lake which had been surveyed by Mr. Proudfoot, thence down to Wabinosh Bay on Nepigon Lake. From the west end of Obugamiga Lake, there is a canoe route westward, and turning southerly it comes into the canoe route west of Whitefish Lake over which Mr. Piche travelled to the height of land. Mr. Piche went west from Obugamiga Lake one day's travel with an Indian guide, and passed through all the timber that was worth anything and returned.

The country for ten to twenty miles east and south of Gull River, and north along the two exploration routes, is well timbered with spruce, jack pine, tamarac and poplar, in most places large enough for pulpwood; that immediately along Gull River is the best, in many places running as high as fifteen and eighteen inches on the stump. There is a small grove of good red pine on the east side of Portage Lake about ten miles up Gull River, of about twenty acres in extent. The timber in the above described country can all be brought into Lake Nepigon through the waters mentioned. Gull River is an exceptionally good stream to float timber down. I think about seventy-five per cent. of the above described country is stony, rock and swamp, the balance being sandy soil, with, in some places, a sandy loam suitable for agricultural purposes.

From Wabinosh Bay I went down Lake Nepigon and over a canoe route to Black Sturgeon Lake making a survey as I went. I only surveyed the east side of Black Sturgeon Lake, as Professor Bell, geologist, reports the west side to have been carefully surveyed by Mr. McKellar as well as the Black Sturgeon River down as far as Nonwatten Lake. I therefore again commenced survey at the head of Nonwatten Lake and surveyed down to the foot of Little Nonwatten Lake when I went in to the northwest angle of the township of Purdom and produced the north boundary of said township west to Black Sturgeon River, intersecting about half a mile south of the foot of Nonwatten Lake. I then moved up Black Sturgeon River to the mouth of Eagle Head River, which empties into Black Sturgeon Lake on the west side about two miles from the outlet of said lake, and surveyed the said river and the several lakes it passes through, up to the head of Eagle Head Lake and returned down the Lake to within about half a mile of its outlet, and surveyed a portage route from there through to Clear Water Lake. After exploring the country west and north of said lake I moved south through a number of small lakes to Dog Lake (I did not survey this route) and thence down through the said lake and Kaministiquia River to the Canadian Pacific Railway at Kaministiquia, and thence by the Canadian Pacific Railway to Sprucewood Siding on Black Sturgeon River. Thence I went up the river to the north boundary of the township of Lyon and surveyed Black Sturgeon River, from said boundary to the foot of Little Nonwatten Lake, connecting with the survey I had made before going into the north-west angle of the township of Purdom, when I commenced the survey of Black Sturgeon River at the Lyon boundary. Mr.

Smith, geologist, went with two Indians on an exploration through to the head waters of Wolf, Nonwatten and Eagle Head Rivers, it having been reported to me by some Indians that there was a large quantity of good white pine in that country, but, after a thorough exploration, Mr. Smith returned, having found a few scattered green trees over a large area. There had been a considerable quantity of pine there a few years ago, but it had been nearly all burnt.

The country lying between the Purdom township boundary produced, Black Sturgeon Lake and Lake Nepigon is well timbered with poplar, spruce and some jack pine and tamarac, poplar prevailing. There is a narrow belt from two to ten chains deep of red pine on the east shore of Black Sturgeon Lake commencing about half way down the lake and extending to the foot. There is very little good agricultural land in this section excepting about five miles square in the vicinity of the Indian Village at the bottom of McIntyre bay on Lake Nepigon. The Black Sturgeon River Valley, which is about a mile and a half wide, is timbered with spruce and poplar; but on either side after reaching the mountain ranges, the timber is too small to be of any use. The country lying west of Nonwatten and Black Sturgeon Lakes is timbered principally with poplar but some spruce and jack pine are found back to the mountainous country which is about ten miles from Nonwatten Lake and about five from Black Sturgeon Lake.

After getting into the mountainous country, which extends to the height of land between these waters and lake Superior, the timber is principally spruce, but not so large as on the Gull and Black Sturgeon waters. On the south shore of Young Sturgeon Lake, which is on the Eagle Head River, there is an area of about twenty acres of fairly good white pine and another small grove of scattered good pine farther up Eagle Head River before reaching Arrow Lake. Around Clear Water and Little Pike Lakes the country is timbered with spruce, poplar, jack pine and tamarac, extending from two to five miles on the north and west, and east to Eagle Head River and south about five miles. Outside of this to the west and north-west the country is all burnt over as far as I was able to see, and by report of Indians is burnt to the height of land, and continuing south through the Dog Lake country, the greater part of it is burnt.

I do not consider that any part of the country is adapted to agricultural purposes, the greater part being stony, rock and swamp with occasional small areas of light sandy soil.

Estimating roughly the area of timbered land which we explored, there are about six hundred square miles drained by Gull River, sixty per cent. of which timber is spruce, twenty per cent. poplar, and twenty per cent. jack pine and tamarac. The area of timbered country lying between the Gull River area and the height of land to the north and drained by a stream passing through Obugamiga Lake and into Wabinosh River, and thence into Nepigon Lake, is about six hundred square miles; about fifty per cent. of the timber is spruce, thirty-five per cent. poplar and fifteen per cent. jack pine and tamarac.

Eagle Head River, which is the principal branch of Black Sturgeon River, drains an area of about three hundred and fifty square miles of timbered country, about fifty per cent. of which timber is poplar, forty per cent. spruce and ten per cent. jack pine and tamarac.

Black Sturgeon River proper drains an area north of the township of Lyon of about three hundred square miles of timbered country, sixty per cent. of which timber is poplar, thirty per cent. spruce and ten per cent. jack pine and tamarac.

There is an area of about two hundred square miles between Black Sturgeon River country and Nepigon Lake and River north of the township of Purdom and the north boundary of said township produced west, which is fairly well timbered, seventy-five per cent. of which is poplar, twenty per cent. spruce and five per cent. jack pine, tamarac and cedar. Although there is some scattered pine in the several areas mentioned, the percentage is so small that it is not worth while classifying.

The several areas above mentioned only include country producing timber large enough for pulpwood, there being areas of considerable extent in all the above mentioned sections that have no merchantable timber, and areas from which the timber has all been burnt off.

I did not explore the country drained by the Kabitotiqua and Pushkokogon Rivers but was told by Indians, whose hunting ground it is, and who seem to know the country well, that it was timbered about the same as the Gull River country.

When I had finished exploring the Gull River and northern country and came down into the Black Sturgeon River country in September, I found that it was going to be impossible to get over all the country assigned me in my instructions without hiring more Indians and buying another large canoe so that my supplies might be kept abreast of my work, and therefore increased my party to ten men, and for a few days had eleven men, and bought a large canoe.

I was not able to use the Log supplied me by the Government through some defect in the instrument, and was therefore obliged to make all my measurements with micrometer which considerably delayed my traverse work.

I have the honor to be,

Sir,

Your obedient servant,

DAVID BEATTY, O. L. S.

Surveyor in charge of Exploration Survey Party No. 8.

The HON. E. J. DAVIS,

Commissioner of Crown Lands,

Toronto.

LAND AND TIMBER ESTIMATORS' REPORT OF EXPLORATION SURVEY PARTY NO. 8.

SUDBURY, Nipissing District,

Ontario, Dec. 17th, 1900.

DEAR SIR,—In connection with Exploration Survey Party No. 8, under your charge, I beg to submit my report as follows :—

Block No. 1.—North and east of Gull River. This part of the country is five miles long by about four miles wide ; it contains of spruce seven or eight cords, of poplar six or seven cords, of jack pine two or three cords per acre. There are about fifteen tamarac which would make about five ties to the tree ; about three piles grow to the acre. I did not find any red or white pine.

The soil is good for farming, being a sandy loam. Fifty per cent. of the country is good for farming, the remainder is rocky and mountainous. Size of timber—spruce eleven inches, poplar nine and one-half inches mean, jack pine nine and one-half inch mean

Block No 2.—Between Gull River and Obabicong Lake. This country lies south of Gull River and extends to Fairy Rapids then north to Obabicong Lake. It is well wooded with about seven to eight cords of spruce, two to three cords of poplar, and seven or eight cords of jack pine. There are about thirty tamarac trees per acre, which would make five railway ties per tree ; there would be about six piles per acre. The soil is divided as follows : Twenty per cent. sandy loam, marsh and swamp twenty-five per cent., and fifty per cent. rocky and mountainous. Size of timber : spruce, seven inch mean, poplar eight inch mean, jack pine ten inch mean.

Block No. 3.—South from Fairy Rapids, by way of Sturgeon Falls. The country from here east to Gull River, and south from Gull River for a distance of about ten miles, is extensively wooded with jack pine, there being between twenty-three and twenty-four cords per acre, four or five cords poplar, five or six cords spruce, and sixty tamarac trees, five ties per tree, six piles per acre. There are no red or white pine. The soil is divided as follows : Fifty per cent. sandy loam and fifty per cent. marsh and swamp.

Size of timber : spruce eight inches mean, poplar ten inches mean, jack pine twelve nches mean.

Block No. 4.—West by Round Lake and Fairy Rapids to head of Jackpine Lake, thence north of Gull and Jackpine Rivers and lake.

This section is about ten miles long by four miles wide. The country is well wooded, especially with jack pine, there being twenty-two or twenty three cords per acre. One or two cords spruce, one or two cords of poplar, and sixty tamarac trees grow to the acre.

Five railway ties per tree and ten good piles grow to the acre. The soil is divided approximately as follows: Twenty-five per cent. sandy loam, twenty five per cent. marsh, fifty per cent. mountainous. In this section spruce averages six inches mean, poplar seven inches mean, jack pine ten inches mean.

Block No. 5.—West of Sturgeon Falls and Fairy Rapids, along south bank of Gull River for a distance of fifteen miles by ten miles south from Gull River.

There are between three and four cords of spruce, one or two cords of poplar, and between four and ten cords of jackpine per acre. Sixty tamarac trees grow to the acre, and there are five railway ties per tree, eight good piles grow per acre. Only twenty-five per cent of this area is good for farming, being a good sandy loam; seventy per cent. is swamp and marsh and ten per cent. is rocky and mountainous. In this section spruce averages six inches, and poplar and jackpine averages eight inches.

Block No. 6.—From a small chain of lakes going north and west from Jackpine Lake thence to Little White Birch Lake, an area of twenty miles.

This section is well wooded, especially in jackpine; between thirty and thirty-one cords grow per acre, one or two cords of spruce, one or two cords of poplar; there are ten tamarac trees per acre and five ties per tree, and four piles per acre. There would be about three hundred white pine trees on the whole area averaging about two hundred B. M. feet per tree. There is scarcely any farming land excepting about ten per cent. which is sandy loam, the remainder is forty per cent marsh and fifty per cent. rocky and mountainous. The timber in this section averages seven inches mean.

Block No. 7.—South of Little Birch Lake extending to Whitefish Lake a distance of seven miles by two miles wide.

The country is wooded as follows:—spruce two or three cords, poplar two or three cords, jackpine two or three cords per acre, twenty tamarac trees grow per acre, one tree would make five good railway ties. The timber averages seven inches mean.

In this district there is no farming soil. Fifty per cent. is marsh, twenty-five per cent. swamp, and twenty-five per cent. is rocky and mountainous. There are no red or white pine.

Block No. 8.—Rice River height of land.—The country through which Rice River flows, lies south of Whitefish Lake, the lake to which Rice River is tributary.

There are a few miles of good timber on the south-east of Majata Lake, which extends to Whitefish Lake, the land of the west and north of Majata Lake is burnt over. The country south-west of Majata Lake which is drained by Rice River, is flat and sandy for a distance of ten miles. This flat is traversed by rocky ridges, which do not rise more than nine or ten feet above the level of the flat. This flat extends south-east from Rice River for a distance of five miles There is no timber of any value until Stone House Lake is reached, when we find large timber consisting of poplar, cedar, spruce and tamarac; there is no difference in the timber until Cedar Lake is reached. For a distance of fourteen miles along the river and lakes of the above described route, the timber extends for about two miles south-east. South east of this is burnt and blue or prairie grass is plentiful. The soil is sandy loam. On the north side of the above described river and lakes, the timber extends for eight miles, making in all an area of fourteen by ten miles that is valuable for timber; the timber averages as follows:—sixty spruce trees nine inch mean measurement, jackpine and poplar the same as spruce; tamarac trees, thirty per acre, five ties per tree, and ten piles per acre. The land around the west and south end of Cedar Lake is burnt through to Savanne waters. About five miles of the region lying between Cedar Lake and Rice River is timbered with spruce, poplar, cedar and scattered white pine, about three hundred in all, of good quality. The soil is divided as follows:—twenty-five per cent. sandy loam; twenty-five per cent. marsh; twenty per cent. swamp and thirty per cent. rocky and mountainous.

Block No 9.—Squirrel Lake. This region lies south of Squirrel Lake and of about two miles of Gull River, making a total of seven miles by six miles south. The country is covered with small shrubs, which are of no commercial value. There is a long marsh running east and west of this section.

The soil is divided as follows: Sandy loam ten per cent., marsh fifty per cent., rocky and mountainous forty per cent.

Block No. 10—South from Sixth Rapid down stream from Squirrel Lake a distance of eleven and one half miles This region is about ten by twenty miles; it is of no use for farming, the soil all being sand hills, boulders, pot-hole lakes, and a small percentage of swamp. The soil is divided as follows: swamp ten per cent., rocky and mountainous ninety per cent. The country is fairly well timbered, there being between nine or ten cords of spruce and nine or ten cords of jackpine per acre. Ten tamarac trees grow to the acre and five railway ties to the tree and ten piles per acre. There is no red or white pine.

Block No. 11.—Otter River. On the 28th of August I went exploring the Otter River. About six miles of this river is suitable for floating logs down, that is, from the mouth to the second lake—Otter Lake.

Otter Lake is walled on both sides by high mountains. Up to the head of Otter Lake there is no timber, but after the head is reached there is a strip of timber four miles wide by nine miles long. There the spruce grows to enormous sizes, some measuring twenty-four inches at the stump. Tamarac also grows in abundance, averaging one hundred and fifty trees, which would make five ties per tree, and eighteen piles per acre. Poplar grows one hundred per acre, the stump measuring from nine to ten inches. Otter River takes it rise from a small lake on that river. As to farming land, there is none.

Block No. 12—This section is north of Obabicong River and Little Obabicong Lake, east to Wabinosh Lake, a distance of eight miles deep. In this sixteen square miles of country there is no farming land. It is all rocky and mountainous. Fifty spruce, eight inches at the stump; fifty poplar, eight inches at the stump; fifty jackpine, eight inches at the stump, grow to the acre. Tamarac grows twenty five per acre, and five ties per tree, while eighteen piles grow to the acre. There is no red or white pine.

Block No. 13.—On the west side of Black Sturgeon Lake a creek flows into this lake about half way down, where there is about five square miles of comparatively level country, which is full of boulders, and is wooded with jackpine, spruce and poplar. I explored about five miles frontage on the lake. This area was timbered with about one hundred spruce trees averaging eight inches, thirty poplar trees averaging ten inches, and fifty jackpine averaging fifteen inches, per acre. Thirty tamarac trees grow to the acre, and would make five ties per tree, and eighteen piles per acre. About fifty per cent. of this section would make good farming land, it being a fine sandy loam; the remainder was rocky and mountainous. I did not find any white or red pine.

Block No. 14—The section is south of Black Sturgeon Lake. Between Black Sturgeon Lake and Nonwatten Lake there is an area of thirty-six square miles. Seventy-five per cent. of this area is sandy loam, good for farming purposes; the remainder is rocky and mountainous.

The whole area is timbered with spruce, poplar, and jackpine and tamarac in the following proportions: Thirty spruce per acre, averaging ten inches at the stump; jackpine and poplar are the same as spruce; thirty tamarac per acre, five ties per tree, and eighteen piles per acre.

Block No 15.—Is west of Nonwatten Lake and Little Nonwatten Lake, and down Black Sturgeon River to the last lake on the river. An area of seven by thirteen miles is timbered with a heavy growth of poplar and white birch; the poplar averages from twelve to fifteen inches at the stump. There would be about eighty poplar per acre. The white birch is good for nothing but white cordwood. Fifty per cent. of this area is good for farming purposes, it being a good sandy loam, the remainder is stony and mountainous, but timbered.

Block No. 16.—Is the country west of McIntyre Bay and across the country to Black Sturgeon Bay on Lake Nepigon. There is about four miles frontage, the land for two miles back from which is heavily timbered, the growth per acre being 100 spruce, averaging eight inches; one hundred and fifty jackpine, averaging eight inches; fifty poplar, averaging eight inches; thirty tamarac, five ties per tree. There are eleven piles per acre. The soil is all sandy loam, suitable for farming purposes, such as growing grain and root crops. There are no red or white pine.

Block No. 17.—Is south of McIntyre Bay. There is about eight miles frontage on the bay, by about five miles deep, which is timbered with the following trees: Seventy-five spruce, eight inches at the stump; twenty-five poplar, eight inches; forty jackpine, eight inches, and forty tamarac grow to the acre; three white pine per acre, which would make about two hundred B. M feet to the tree. The tamarac would make about five ties per tree, and three piles grow per acre. The soil is all sandy loam, valuable for farming.

The Indians had some potatoes that weighed one pound each, and an average crop of one hundred bushels per acre.

Block No. 18.—Is from the mouth of Sturgeon River on Black Sturgeon Lake to the Mountain Portage, north of Rat Root river and lake, and south of Little Sturgeon Lake, an area of four by eight miles. About fifty per cent. of the country is good sandy loam, the remainder is rocky and mountainous. The timber is from twelve to fifteen inches at the stump. The land is timbered as follows: sixty spruce, nine inch (mean measurement); seventy-five poplar, nine inches; fifty jack pine, nine inches; ten tamarac trees, five railway ties per tree, three piles per acre; one white pine tree per acre, fourteen inches mean measurement and about two hundred B. M. feet to the tree.

Block No 19—From Mountain Portage to Burnt Pine Portage there is a very swift river and a strip of timber on each side of the river about one and one half miles wide; the distance between the two portages is five miles. There is no good land, it is all swamp and wooded with tamarac, spruce and poplar.

Block No. 20.—On the south side of Eagle Head Lake there is four miles frontage on the lake by five miles south from the lake. It is timbered with a thick growth of spruce, poplar and tamarac; the land consists of sand, swamps and rocky hills.

Block No. 21.—From Hook Lake to Kaministiquia, the country is all burnt around Dog Lake. Down Kaministiquia river there is a small strip of timber, viz: spruce and jackpine.

The above is as close an estimate as could be obtained from the system and means employed by me.

I remain,

Your obedient servant,

DAVID BEATTY, Esq, JOHN PICHE,

O. L. S. Land and Timber Estimator Exploration Survey Party No. 8.

Parry Sound, Ont.

District of Nipissing,
Province of Ontario,
To wit:

I, John Piche, of the Township of Waters, in the District of Algoma, Explorer, do solemnly declare:

1. I personally made the explorations referred to in the foregoing pages of this book.

2. That the facts set forth in the preceding pages are true.

3. That the estimates given in the foregoing pages are faithfully made, and are as close and accurate as I could form.

And I make this solemn declaration, conscientiously believing it to be true, and knowing that it is of the same force and effect as if made under oath, and by virtue of the Canada Evidence Act of 1893.

Declared before me on the 17th day of December, 1900, at the town of Sudbury, in the District of Nipissing.

JOHN PICHE.

J. K. MACLENNAN,
A Commissioner in H. C. J., etc.

DESCRIPTION OF BLOCKS OF LAND EXPLORED BY J. PICHE WITH EXPLORATION SURVEY PARTY No. 8, IN CHARGE OF D. BEATTY.

No. 1. North and east of Gull Bay, a point forming the northeast shore of Gull Bay.

No. 2. Lies west of mouth of Gull River, extending to Fairy rapids, and north to Obacogaming Lake.

No. 3. Bounded on the east by the Gull Bay, on the north by Gull River, on the west by exploration 10 miles south from Sturgeon Falls and Fairy rapids.

No. 4. West of Round Lake and Fairy rapids to head of Jack-pine River and lake 10 by 4 miles.

No. 5. Lies west of Sturgeon Falls along south bank of Gull River a distance of 15 miles, and south from Gull River a width of 19 miles.

No. 6. First chain of lakes going northward, west from Jack-pine Lake, thence to foot of Little White Birch Lake, a distance of 5 miles by 4 miles.

No. 7. South of Little Birch Lake, extending to Whitefish Lake, a distance of 7 miles by 2 miles in width.

No. 8. Rice River of Piche's height of land route to Savanne, English and Gull River waters, a distance of 40 miles by 4 miles.

No. 9. South of Squirrel Lake and Gull River, a distance of 7 miles by 6 south.

No. 10. South from sixth rapid down stream from Squirrel Lake, 10 miles by 20 miles south.

No. 11. On Otter River, west of Obacogaming Lake, and 6 miles up Otter River where timbered land commences.

No. 12. North of Obabicong River and Little Obabicong Lake, east to Upper Wabinosh Lake, 83 miles.

No. 13. West of Black Sturgeon Lake, five miles frontage on lake and five miles west from lake.

No. 14. South Black Sturgeon Lake, between Black Sturgeon and Nonwatten Lakes, a distance of 6 miles by 6 miles.

No. 15. West of Nonwatten and Little Nonwatten, and down Black Sturgeon to last lake, 7 by 13 miles west.

No. 16. Is west of McIntyre Bay, across the country to Black Sturgeon Bay, 2 by 4 miles in area.

No. 17. Is south of McIntyre Bay on Lake Nepigon, eight miles frontage on the bay and five miles south.

No. 18. From the mouth of Sturgeon River on Black Sturgeon Lake, west to Mountain portage.

No 19. From Mountain portage to Burnt Pine portage, a distance of five miles by three miles.

No. 20. South of Eagle Head Lake, four miles front on lake and five miles south.

No. 21. From Eagle Head Lake and river to three miles west of Jackfish Lake and Clearwater to Hook Lake.

REPORT OF TIMBER, ETC., ON BLOCKS 1 TO 21.

—	Area in miles.	Per cent. of sandy loam.	Per cent. of marsh.	Per cent. of swamp.	Per cent. stony and mountainous land.	Acreage of timbered land.	Number of spruce trees per acre.	Number of cords of spruce pulpwood per acre.	Size of spruce trees.	Number of poplar trees per acre.	Number of cords of poplar pulpwood per acre.	Size of poplar trees.	Number of jackpine trees per acre.	Number of cords of jackpine pulpwood per acre.	Size of jackpine trees.	Number of tamarac trees per acre.	Number of tamarac railway ties in tree.	Number of tamarac piles per acre.	Number of white pine per acre.	Number of feet b.m. to pine tree.	Size of pine trees.	Number of pieces of board timber per acre.
No 1	20	50	....	....	50	12,800	40	7.35	11	50	6.84	9½	20	2.65	9½	15	5	3	....	....	....	....
" 2	200	25	12½	12½	50	64,000	100	7.43	7	25	2.42	8	50	7.58	10	30	5	6	....	....	....	....
" 3	200	50	25	25	....	1000	60	5.83	8	30	4 83	10	100	23.21	12	60	5	10	....	....	....	....
" 4	40	25	20	....	50	1900	25	1.36	6	25	1.85	7	150	22.76	10	60	5	10	....	....	....	....
" 5	150	20	50	20	10	48,000	60	3.27	6	60	1.46	8	100	9.73	8	60	5	8	....	....	....	....
" 6	20	10	40	....	50	6,400	20	1.48	7	20	1.48	7	200	30.34	10	10	5	4	7	200	14	....
" 7	14	....	50	25	25	4,480	40	2 97	7	40	2 97	7	40	2.97	7	20	5	4	....	....	....	....
" 8	160	25	25	20	30	24,320	50	7 37	9	60	7.37	9	60	7.37	9	30	5	10	11	200	14	....
" 9	42	10	50	....	40	....	....	....	....	....	....	....	....	....	....	....	....	....	....	....	....	....
" 10	200	....	....	10	90	8000	150	9.11	8	50	3.03	8	150	9 11	8	10	5	10	....	....	....	....
" 11	18	....	....	....	100	11,520	100	12 29	9	....	....	....	....	....	....	150	5	14	....	....	....	....
" 12	16	....	....	....	100	10,240	50	2.73	6	50	2.73	6	50	2.73	6	25	5	18	....	....	....	....
" 13	25	50	....	50	....	6,000	100	5.45	6	30	2 92	8	50	4 86	8	30	5	18	....	....	....	....
" 14	36	75	....	....	25	23,040	30	2.92	8	30	2.92	8	30	2.92	8	30	5	20	....	....	....	....
" 15	91	50	....	....	50	58,240	....	....	....	80	12.13	10	....	....	....	....	....	....	....	....	....	....
" 16	8	100	....	....	....	3,120	100	9 73	8	50	4.86	8	150	14·59	8	30	5	11	....	....	....	....
" 17	40	100	....	....	....	25,600	75	7.29	8	25	2.43	8	140	9.73	8	40	5	3	3	200	14	....
" 18	32	50	....	....	50	20,480	60	7.37	9	75	9.22	9	50	6.16	9	10	5	3	1	200	14	....
" 19	15	....	....	100	....	9,600	60	7.37	9	60	7.37	9	....	....	....	25	5	12	....	....	....	....
" 20	20	....	....	50	50	6,800	60	7.37	9	60	7 37	9	....	....	....	25	5	5	....	....	....	....
" 21	36	25	25	....	50	7,208	75	9.22	9	75	9 22	9	75	9.22	9	30	5	10	....	....	....	....

To DAVID BEATTY, ESQ., O.L.S.,
Parry Sound, Ont.

GEOLOGIST'S REPORT OF EXPLORATION SURVEY PARTY NO. 8.

TORONTO, Ont., Jan. 7th, 1901.

DEAR SIR,—In accordance with instructions from the Commissioner of Crown Lands issued last May, I beg to submit my report as Geologist of Exploration Party No. 8, of which you had charge.

The territory explored between 1st July and 9th of November consisted of country west of Lake Nepigon and the Nepigon River, Gull River, the territory to the north and east of Gull River, as far as the northern Height of Land, Black Sturgeon Lake and River, and the country between this river and Dog Lake.

The Nepigon is the largest river flowing into Lake Superior. It is about thirty-six miles long and flows nearly due south, but has a slight bend to the westward. The water is beautifully clear and teems with speckled trout, which often weigh as high as eight and nine pounds.

The fall between its head at Lake Nepigon and Nepigon Bay is over 300 feet. What with its mighty volume and steadiness of flow, this river would afford almost unlimited water-power. While in Nepigon Station a survey of part of the river near the Canadian Pacific Railway bridge was in progress for the purpose of building a power plant. Soundings were taken for a situation to build a dam, but the flow was not calculated, as the engineer in charge said it was unnecessary, owing to the tremendous volume.

The Nepigon River empties into Nepigon Bay through high banks of sand and gravel, which, as far as I could judge, form an old lake terrace. The Canadian Pacific Railway have built a bridge just at the mouth of the river, which is about seventy-five feet high. This bridge abuts on both sides on sand and gravel banks. Deep side cuts in the gravel have been made, forming the approaches to the bridge. I should judge the top of the gravel deposit to be one hundred feet above the level of the river.

About one mile from Nepigon Station, Lake Helen is reached, the first of four that occur on the river. The river enters this lake about half way up on the west side. Entering the river at this place, it is found to flow in a bed of alluvial sandy clay. Inland on both sides of the river above this point Laurentian gneiss was noted.

Columnar diabase (trap) starts at Camp Alexander, the first portage on the river, and forms the country rock on the west side of the river and Lake Jessie. Gneiss occurs at the narrows between Lake Jessie and Lake Maria, the second and third lakes on the river. At the head of Lake Maria the rock on both sides is a columnar diabase. On the east side, cliffs over three hundred feet high rise almost sheer from the water. At Island portage, a fine grained garnetiferous gneiss is found, but this seems to be only an outcropping, as the diabase extends north to Lake Nepigon without a break.

GULL RIVER REGION.

From Flat Rock Portage, which leads from the Nepigon River into the lake, to Gull Bay half way up the west side of Lake Nepigon, the whole rock formation seems to consist of dark greenish or black diabase, both massive, coarsely crystalline and columnar in structure. This rock evidently contains a considerable amount of magnetic iron, as the compass was deflected in many places.

Gull River, the first part of the district to be explored thoroughly by our party, empties into the head of Gull Bay through a low sandy flat. To the north and east of Gull Bay high, diabase hills are seen inland. These hills approach the river about five miles from its mouth, where they form a bluff about seventy-five feet high on the right hand bank. The river from this point, for twelve miles, to the third rapid, flows with a fairly good current in a bed of alluvial, sandy clay. Banks of clay and sand, in some places fifty feet high, form the sides. At the third rapid on the Gull an exploration to the north was taken to explore Round Lake. Immediately on the shores of this lake a peculiar granitic rock, containing a considerable amount of hematite, was noted. This rock is evidently of the Laurentian formation, the country to the north of this lake being of the same formation, but between Gull River and this lake only diabase dikes running north and south were noted.

An exploration to the south, and a mile up the river from the third rapid, was through a very hilly country. Two diabase hills were measured by the aneroid. The respective heights were found to be 340 and 450 feet above the level of the river. No other rock, except diabase, was noted in this district.

At the fourth portage the canoe route leaves the river, passes through a small lake and back again to the river, thus avoiding a number of bad rapids and falls of twenty feet. In Portage lake a small gneiss island is found about the centre; to the north of this point the Laurentian area is met with. A traverse of ten miles to the south from the sixth portage showed the country to be about three-quarters muskeg, with ridges of diabase running east and west. On the left bank of Gull River above the sixth portage a mountain rises 310 feet (aneroid). It is of columnar diabase, and on the side that faces the river a vertical cliff of one hundred feet rises majestically into the air. The base of the mountain is strewn with talus. Immediately across the river a range of hills of the same formation approaches from the northwest.

About thirty miles from its mouth the river branches. Taking the right hand branch and with a couple of portages Jackpine Lake is reached. On the last portage on Gull River, leading into Jackpine Lake, I found a hard, fine-grained, red rock, which is evidently an impure dolomite of the Keweenawan series of rocks. A pinkish grey sandstone, with large quartz pebbles and containing a considerable amount of calcareous matter, and also a reddish shale containing minute crystals of dolomite, was noted on the northeast of Snake Lake, the next lake immediately north of Jackpine Lake. The rock to the north of this area of shale and sandstone, and also immediately across the lake, is Laurentian gneiss. This gneiss contains a considerable amount of plagioclase and epidote.

From Jackpine Lake a portage route leads off to the southwest to Squirrel Lake at the head waters of Gull River. Following this route, the fixed rocks consisted of diabase till Devil's Crater Lake is reached. Here the Laurentian gneiss comes in contact with the diabase formation of the south. This gneiss is light grey in color and contains a considerable amount of plagioclase and epidote, but very little quartz.

A peculiar little lake ten chains wide, which was named Devil's Crater Lake, is found a couple of hundred yards from the side of the portage route. It lies in a round chasm with vertical sides. The depth of the lake below the surface of the country was estimated at about five hundred feet. This height was tested by a falling stone, which took from five and a half to six seconds to cover the distance. The lake appears to be immediately on the contact between the diabase and gneiss rocks. The gneiss near the contact was found in a very much altered and weathered condition.

From this point to Squirrel Lake, the portage route passes over the Laurentian gneiss and granite. The hills have a low, rounded appearance and nowhere do they rise more than two hundred feet above the surrounding country. In this district many little lakes and ponds are found.

White Birch Lake, a lake about three miles long, is found on the Portage route between Jack Pine Lake and Squirrel Lake. From this lake two explorations were made; one overland to the south to Big Whitefish, another lake on the same route, and another to the same lake by a circuitous canoe route of thirty miles. On both these explorations gneiss was the only fixed rock noticed. On the thirty mile exploration, fourteen small lakes and their connections were passed through. On Manitou Peewabic Lake—a narrow sheet of water six miles long—peculiar epidote syenite was noted, consisting of fresh colored orthoclase felspar and a little quartz.

Big Whitefish Lake is roughly circular in form, and about four miles in diameter. The formation is entirely Laurentian, gneiss being the prevailing rock. The outlet of this lake flows into Gull River about twenty miles below Squirrel Lake. From this lake Mr. Piche, our timber estimator, and one Indian travelled for about fifty miles in a southwesterly direction. He informed me that the formation was Laurentian, granite and gneiss as far as he could judge, but that at the end of his journey he encountered an area of slaty rock, which he thought belonged to the Huronian age.

Following a small creek, which flows into Big Whitefish Lake, and crossing four portages and three lakes, Squirrel Lake is reached. This is a large lake from which the Gull River takes its rise. The last portage leading into Squirrel Lake is very rough, being thickly strewn with drift, composed of granite, gneiss and hornblende and mica schist boulders. On this portage, about a quarter of a mile from Squirrel Lake, a contact between Laurentian gneiss and Huronian porphyroid occurs.

A piece of float pyrrhotite was picked up on this portage, but I was unable to locate where it had come from, but by its appearance and ragged edges, I should judge its location to be somewhere close at hand. The country being very low and covered so thickly with drift, and the time at my disposal limited, I was forced to give up the search.

The average of two assays of the mineral, which was made at the School of Practical Science, Toronto, gave the following result:

Copper	0.125	per cent.
Nickel	0.77	"
Iron	28.06	"
Sulphur	26.87	
Silver	Trace.	

Area of Huronian.

Squirrel Lake at the head of Gull River is about six miles long, running north-east and south-west. It contains a number of very deep bays. The whole shore line and the islands in it, are of Huronian age, with the exception of the bottom of a deep bay on the east side, where Laurentian granite is seen in contact with the Huronian. The rock is a dark olive-green porphyroid, containing light green inclusions, the strike of the formation being north, twenty six degrees east, and dipping at a high angle towards the west. At a quarter of a mile to the south of the outlet of this lake, a contact between diorite and porphyroid is met with. This diorite area cannot be very large, as it is only met with at the one point.

Passing down the Gull River from Squirrel Lake beds of Huronian porphyroids and diabase dikes are noted for a distance of twelve miles, the Huronian rocks having a strike of north, 30 degrees east. A twenty mile exploration to the south at this point proved the rock formation to be diabase, both massive and columnar in structure.

Twelve miles down from Squirrel Lake, the rock formation consists of a light green sandstone, with occasional out croppings of a red shale. High hills of diabase were noted on both sides of the river and quite close to its banks The sandstone extends down the river to the first forks, where our route branched off on the way up. The river runs for a considerable way below Squirrel Lake in a bed of sandy, alluvial clay, but as it nears the forks, a long succession of rapids is met with.

Obugamiga Lake to Wabinosh Bay.

After completing the survey and exploration of the Gull River, and getting fresh supplies, our party moved across country from a point about twenty miles up the Gull River, into Lake Obugamiga. This lake is nearly twelve miles long, running in a north easterly direction to Wabinosh Bay. The portage route between Gull River and this lake leads through four lakes. The second lake is called Big Mountain on account of the high diabase hills on the east and west sides. A number of islands occur in this lake composed of gneiss, which contains a considerable amount of magnetite. From the appearance of the diabase ridges, which are columnar in structure, I would judge them to be intrusions through the gneiss and forming part of the Keweenawan series. Slept-too-long Lake and the next lake are also surrounded by diabase hills. On the last portage going into Obugamiga Lake a small outcropping of hornblendic slate was noted, almost an edge, and with a strike south 80 degrees west. This slate was overlaid by the diabase, which forms the chief country rock of this section.

The north shore line of Obugamiga Lake is low and rolling, as compared to Big Mountain Lake, and in many places low sandy flats are seen covering a considerable area. On the south shore the country is considerably higher, with prominent hills of columnar diabase to be seen in many places.

From a point at the north-east end of this lake, where a small river comes in, I made an exploration to determine the northern height of land between the Hudson Bay waters and the waters running into Lake Nepigon. Compass readings and estimated distances were taken. The canoe route led considerably to the north-west through a country that had been burned over a good many years before. No great tract of large timber was noted, as the fire had only left small clumps of spruce, tamarac, poplar, jack pine and birch, but a healthy second growth is springing up, which in a few years should be valuable. The rock on both sides of the creek up to the fifth portage is a dark diabase, but at this portage a slaty rock, with a high dip and strike nearly east and west is noted. This slaty formation, which is evidently Huronian, cannot be of very great area, as it only appears on this one portage, the rock changing again to diabase.

The approach to Mink Bridge Portage, the seventh on the route, lies between high walls of pinkish gneiss, capped with a coarsely crystalline diabase. This gneiss has a very distinct, horizontal cleavage, which gives it a bedded appearance. From this point the diabase disappears and we enter into a Laurentian area of granite, which extends along the route to Onamakewash Lake, a large lake whose waters flow into Hudson Bay. This is approximately the northern height of land, as the second last portage into this lake leads from the waters flowing into Lake Obugamiga and Lake Nepigon. The shore line of this lake is low and has a rolling appearance. The gneiss, the fixed rock, is very red, containing a great deal of orthoclase and very little quartz or dark mineral.

As I had to meet my party again at Wabinosh Bay on Lake Nepigon, instead of going back over the same route, I determined to take a new route, which my Indian guide told me about. After leaving Onamakewash Lake and passing through a small pond and over two portages, another large lake was met with, which the Indians called Nanamego. This is a beautiful stretch of clear water nearly six miles long, with a low, rolling shore line, the prevailing rock being Laurentian, gneiss and syenite. Leaving this lake, the route passes through a number of small lakes, whose waters flow into Lake Nepigon. I met Mr. Proudfoot's exploration work at his 12th portage from Lake Nepigon, and from this point down to the lake will be reported on by his party. About a mile from the juncture of Mr. Proudfoot's exploration and my own, the gneiss stops and high diabase hills of the Keweenawan series appear.

Lake Nepigon, from Wabinosh to Black Sturgeon Bay.

On joining Mr. Beatty again at Wabinosh Bay, the whole party moved down the lake to Nepigon House, the Hudson Bay Company's post on Lake Nepigon. Here we were warmly received by Mr. Patterson, who was in charge of the post. After remaining there for two days on account of high winds, the party moved down to Black Sturgeon Bay for the purpose of exploring Black Sturgeon Lake and River, and the country between that lake and Dog Lake.

As the west side of Lake Nepigon had been explored and surveyed by Professor Robert Bell, of the Geological survey of Canada, in 1869, and as the season was growing short and a vast unexplored territory yet to be covered, it was not deemed advisable to make exploration of the shore line from Wabinosh Bay to Black Sturgeon Bay, but the general shore line of the lake observed, has that wild, rugged character formed by the high, trap mountains, which rise to five and six hundred feet in places.

Black Sturgeon Lake to River.

Leaving Lake Nepigon at the foot of Black Sturgeon Bay, a rough passage over water-worn boulders leads to a small lake about a quarter of a mile long. Another portage brings us from this lake into Black Sturgeon Lake, which is only separated by about a mile from Lake Nepigon. Both these Lakes appear to be on the same level. A little stream flows into the Portage Lake from the direction of Lake Nepigon, and another small stream enters Black Sturgeon Lake at the end of the portage from Portage Lake. The whole area is covered with boulders of gneiss, schist and diabase.

Black Sturgeon Lake, lying north and south, is thirteen miles long by about two miles wide. A high range of diabase hills, which runs up the south side of Black Sturgeon Bay, bends southward and strikes down the east side of Black Sturgeon Lake. The country immediately to the west of the lake is low and undulating for a considerable distance inland. A traverse of ten miles west showed the fixed rocks to be coarsely crystalline diabase, and the country heavily covered with Laurentian, Huronian and trap drift boulders. Four miles down from the head of the lake a gray sandstone hill is met with. The sandstone appeared to be slightly tilted to the east. Immediately overlying the sandstone, a dark and light red impure dolomite rested. This was covered by a capping of diabase.

A mile farther down the lake Laurentian gneiss, containing a considerable amount of epidote is met with. Between the Laurentian area and the Keweenawan, and a little inland, some beautiful ribboned green and pink sandstone were noted, but no considerable area of these rocks was found. A high ridge passes from this point down to the foot of the lake. It forms the contact between the Laurentian gneiss, which is so fine and dark in color that it almost approaches a hornblende schist, and an impure ferruginous dolomite.

At the lower extremity of Black Sturgeon Lake, and a little inland on the east side, a number of iron locations have been surveyed. Although I searched for two days I was unable to locate the outcropping, but I believe it occurs on Mining Location A L 1. As far as I could find out, the mineral is hematite iron. I was informed by one of our Indians that more and better iron was known to them in the same district, but that it had not been staked out. This district is well worthy of a more complete examination, and from what the Indians told me, a considerable deposit of iron is to be found there.

About two miles further south of these claims, a number of other locations are surveyed, but again my search for the exposure was without avail, and I was force to relinquish the quest.

Leaving Black Sturgeon Lake at the southern end where the Black Sturgeon takes its rise, the river flows southward in almost one continuous rapid for a distance of six miles to Nonwatten Lake.

At the first rapids down a diabase exposure was noted, but from this point to Nonwatten Lake, the banks are composed of red, horizontal shales, with light greenish layers. These shales, which appear to be a very impure dolomite, contain a great number of yellow-colored concretions. The country on both sides of the river is flat and level.

Nonwatten Lake, a beautiful little stretch of water, is about two miles in diameter. A high range of diabase (trap) hills rises on the east side of this lake and continue down the west side of Black Sturgeon River southward for a distance of about fifteen miles.

The river which runs in a south-westerly direction passes from Nonwatten Lake through Nonwattenose, one mile farther south and at two and three quarter miles below this, the last lake on the river occurs, namely, Esh quanonwatan. From this point to Black Bay on Lake Superior, the river flows in a very tortuous channel of alluvial clay and sand a little to the east or south.

About six miles below Esh-quanonwatan Lake high hills approach the river from the west. These hills, composed of columnar diabase, continue down to about three miles from the Canadian Pacific Railroad.

The high diabase hills on the east bank of the Black Sturgeon continue about fifteen miles from the Canadian Pacific Railway from this point to within six miles of the railway, Laurentian gneiss is noted high hills of this rock forming the banks of the river. From a point five miles below Esh-quanonwatan Lake, a portage route leads inland to Sturgeon Lake a lake at the head of Nonwatten River, which flows into Nonwatten Lake. With two Indians as guides, I explored the country along the route, both for timber, soil and mineral. The first portage off Black Sturgeon River is four miles long to Sucker Creek, the first half mile being over a sandy flat, but a good deal of the way being swampy, with small spruce and tamarac, and occasional bunches of poplar and birch. Sucker Creek flows south east into Black Sturgeon River. Passing up Sucker Creek about two miles, Sucker Lake is reached. This lake is roughly circular and about a mile and a half in diameter. Another portage of three miles and a half to the west brought us to Sturgeon Lake, this portage passes through good spruce. Some very large trees were to be seen on the higher ground.

From Sturgeon Lake we continued up the Nonwatten River for over a mile in a south-westerly direction, from this point portaged into a small lake, going a mile and a quarter west. Passing through this lake and over another portage to the westward, two miles long, Pine Lake is reached. A number of large scattered white pine were seen in this territory. About three hundred were counted, but the region is so isolated with poor water communications, and the timber is so badly scattered, that it would be impossible to take them out economically. On the south side of Pine Lake, a high trap mountain rises to a height of nearly five hundred feet. On the mile and a quarter portage, a large split erratic of diabase was noted, being thirty paces in diameter and fifteen feet high.

With the exception of the high hill on Pine Lake, the country is very flat, with rock exposures of red shale and pinkish sandstone. Instead of going back by the same route, I determined to go down the Nonwatten River to Black Sturgeon River, where I was to meet our party. The Nonwatten River flows between low banks nearly all the way to Nonwatten Lake. Occasionally the banks become steep and fairly high, but the country drained by the river is very flat and uniform. It is about thirty miles long from Sturgeon to Nonwatten Lake, and through nearly its whole length is very rapid. The general course is to the north-east. At the upper end for about six miles, a good tract of large spruce is noted. Twelve miles down from Sturgeon Lake, a good-sized creek comes in from the south, which drains five small lakes. From this point to Nonwatten Lake, the timber is large, but mixed spruce, poplar, birch and tamarac being the prevailing trees.

Production of the Purdom Line.

From a point a little below the rapids south of Nonwattenose Lake, on the Black Sturgeon River, a canoe route leads to the eastward to Sharp Mountain Lake, which lake touches the north-west corner of the township of Purdom. From this point the northern boundary was produced westward to the Black Sturgeon River, meeting at a point about half a mile below Nonwatten Lake. This line was nine miles and thirty chains long. Two lakes were met with on the line, namely, Frazer and Magee Lakes. From the corner of the township to the west of Magee Lake, diabase was the only rock met with; beyond this lake to the Black Sturgeon, a Laurentian gneiss, containing a considerable amount of muscovite and a few garnets was the prevailing rock. The gneiss, capped with coarsely crystalline diabase, formed a hill close to Black Sturgeon River.

Between Black Sturgeon Lake and Dog Lake.

A canoe route leads across country from the south-east corner of Black Sturgeon Lake, where the Eagle Head River comes in, to Dog Lake. After exploring and surveying the Black Sturgeon River as far south as Eshquanonwatan Lake, we determined to finish the rest of the river from the south and turn back through the country to Dog Lake. This route runs south-westerly through Young Sturgeon, Arrow, and Eagle Head Lakes, and the Eagle Head River; thence by way of Clear Water and a long chain of lakes to Dog Lake. The Eagle Head River proved to be almost unnavigable for our canoes, as it was very rapid, the party having to pole and drag the canoes for nearly eighteen miles against the rapid current. Three miles up the river branches, one section coming from Rat Root Lake—this we called Rat Root River—the other coming from Young Sturgeon Lake. Passing up Rat Root River a couple of miles, and crossing a short portage, another portage of two miles and a quarter leads to Young Sturgeon Lake. At the short portage before this, a brine spring is found near the foot of the first rapids on Rat Root River, where it bubbles up through diabase boulders. The country from Black Sturgeon to Young Sturgeon Lake is low and flat, with ridges of diabase found here and there. Where the Eagle Head River continues from Young Sturgeon Lake, a high range of diabase hills is encountered. Four miles up from this lake the river plunges in a series of rapids and falls through a gorge in the hills. A portage of a mile and a quarter was made to escape this. About a mile from this portage, further up the river, exposures of red shale, lying horizontal, similar to these observed on the Black Sturgeon River were noted. This formation continues for about five miles further up. At this point we had to strike overland, cutting a new portage, as we found the river too rapid to work against.

This portage, which was two and a quarter miles long, passed over a burnt country with fairly high ridges of diabase running south-east. The surrounding country is generally flat, except for these ridges. A few exposures of red shale were noted on the portage near the upper end. From the flat appearance of the surrounding country, and these exposures, I should judge the fixed rock to be red shale, with numerous intrusive dikes of diabase.

Continuing up the Eagle Head River, Arrow Lake was reached about six miles from this portage. This lake is about three miles long and surrounded by hills of columnar diabase. A few miles farther up the river Eagle Head Lake is reached. This proved to be quite a good-sized lake, about five miles long The east shore of this lake and the islands in it are of Laurentian gneiss, while the west shore, which is flat, is of the Keweenawan formation. Passing over a mile and a quarter portage, a small lake, and an eight chain portage, Clear Water Lake is reached. A mining claim, HP 709, is located on the north-west side of this lake, a fine red sandstone suitable for building purposes is exposed here. The rock is capped with a great thickness of diabase. This rock appeared to be a red dolomite, but undoubtedly impure ; red shales were also noted, dipping at a low angle under the trap capping.

Leaving Clear Water Lake, Hook Lake is the next lake of any consequence met with on the portage route to Dog Lake. On this lake Laurentian gneiss, with a strike of north 50 degrees east appears. From this lake down to the Kaministiquia River the whole formation is Laurentian, the gneiss in places being very fine grained and containing a considerable amount of muscovite. Dog Lake and the surrounding country, as far as I could judge, being surrounded by the Laurentian formation, passing through Dog Lake to its southern extremity, a portage leads into Little Dog Lake, which is on the Kaministiquia River. From this point down to the Canadian Pacific Railway only gneiss was observed, till quite near the track a ridge of Huronian slate with a high dip was noted.

Summary of Exploration.

This region covered by this report may be roughly stated as the country lying immediately to the west of Lake Nepigon and the Nepigon River. The two principal formations found in this region are the Keweenawan series of rocks and the Laurentian gneiss.

The Keweenawan Series.

This series of rock is known as Logan's Upper Copper-bearing series. In the Nepigon region, vast areas of diabase are to be found, generally coarsely crystalline, massive and columnar in structure, and containing along with the augite and felspar, a considerable amount of magnetic iron and iron pyrites. In a number of places, notably Black Sturgeon Lake, deposits of magnetic sand were found in conjunction with a black sand, which evidently represents the decay and grinding up of this trap rock. The diabase rock is not uniform in character, for while some is found hard and tough and impervious to the weather, other diabase rock becomes exceedingly friable and can be easily picked to pieces with the hand, and eventually crumbles into sand and gravel. The general outline of the hills of this rock are to a certain extent uniform ; flat tops, vertical sides, forming cliffs sometimes hundreds of feet high, which will be flanked by a terrace of talus that slopes away at an angle of about forty-five degrees.

The other rock of this series noted, were the sandstone and shales, together with a certain amount of impure dolomite. The sandstones sometimes contain small quartz pebbles and are generally friable in character. Numerous small clay-like concretions, are found in the shales. The shales contain a considerable amount of dolomite, and are called by Logan red indurated marls. A peculiar ferruginous, impure dolomite was found on the south west shore of Black Sturgeon, which evidently belonged to this series.

Laurentian Areas.

The chief rock found of this formation is gneiss. Vast areas of gneiss, containing a considerable amount of epidote were noted. One small area of diorite in Gull River was the only sample of this rock noted by me in the district explored.

Huronian Areas.

This areas of this formation are small and seem to occur in strips, surrounded by the Keweenawan. The largest area noted was that surrounding Squirrel Lake. The chief rocks of this formation noted were porphyroids, hornblende and mica schists and a peculiar greenish slate.

On Smooth Rock Island Lake. Party No. 7.

Ogoki River, rapids at portage 29. Party No. 7.

Rapids at portage 29. Party No. 7.

Rapids, west end of White Earth Lake. Party No. 7.

Indian canoemakers. Party No. 7.

Indian visitors. Portage 31. Party No. 7.

Shore of Palisade River. Party No. 7.

Burnt Hills, Palisade River. Party No, 7.

Economic Minerals.

No outcropping or exposure of any economic mineral came under my notice during the summer; but undoubtedly a considerable iron deposit occurs on the east side of the Black Sturgeon Lake, and which richly deserves a thorough investigation. I was informed by an Indian that a vein of iron, about fifteen feet wide, ran for a considerable distance somewhere to the east of the lake. This deposit has never been explored or staked out. The pyrrhotite float, found by me near Squirrel Lake, indicates that a deposit must be close at hand; it is also interesting, as it was found near the contact between the Huronian and Laurentian formations. Numerous small stringers of quartz were seen in the porphyroid rock on Squirrel Lake, but none of any size were found.

A number of brine springs are found in the district. The Indians in years gone by used to get their supply of salt from these springs. One is found at the foot of the first rapids on Rat Root River, another on the left bank of the Black Sturgeon River immediately below Nonwatten Lake.

A piece of native copper was shown to me by an Indian, who said it came from the west of Wabinosh Bay, but I am afraid no reliance can be put on a report of this kind.

The region north east of Dog Lake is reported to be rich in minerals. A number of iron claims north of Little Pike Lake have been surveyed, and other deposits of iron in this district are known to occur.

Undoubtedly the country explored by Party No. 8 is by no means barren of all mineral wealth, which will some day be found, and add considerably to the value of the territory explored.

Water Powers.

The rivers in the district are rapid as a rule, with many falls. The volume of water in the Nepigon river, as stated before, is tremendous, and could furnish a large amount of power. On the Kaministiquia River, near Dog Lake, there is a very high falls, besides which the volume of water is large and not subject to very great variations. Again, the Black Sturgeon, Gull, Wabinosh and Eagle Head rivers all could be harnessed to produce almost unlimited water-power. The rivers have a considerable volume and steadiness of flow, and if the time comes no great difficulty would be found in finding numerous places where power plants could be installed.

Indian Occupation.

A number of Indian villages are found on the west side of Lake Nepigon, namely, at Wabinosh bay, Nepigon House, Gull river and MacIntyre bay. In these places they have managed to clear small plots of ground, where they grow a few potatoes and other hardy vegetables. A number of families have built fairly substantial houses, but as the villages are only inhabited in the summer time the Indians prefer to live in tents. The shiftlessness of the Indian is plainly seen in numbers of half-constructed log huts, some with only a few tiers built and others roofless, or with the roof on but not chinked. The largest Indian village is situated near the Hudson Bay Co.'s post on Lake Nepigon. At this place they have a resident schoolmaster—Blae by name—who lives on Jack Fish island, where he has charge of quite a fine frame school-house. This building is used for service a number of times during the year by the Roman Catholics.

An English Church mission is situated on McIntyre bay. Here the Indians seem seem to be more prosperous than any place else in the district. They have a pretty little church, a graded street, a few head of cattle, besides which the houses are well built and the land cultivated, the Indians raising enough potatoes and turnips to last them all winter. The land in this district appears to be very good farming land and not much difficulty is found in clearing and cultivating it.

Our party, on arriving at the Gull River village in July and a few days before the annual payment of the treaty money, found about twenty families there. The inhabitants appeared to be strong and healthy, the men able to stand any amount of work. They also appeared to be free from consumption and other diseases which carry off so many Indians nearer civilization, but I was informed, both by white traders and the Indians themselves, that they are decreasing in numbers and gradually dying off, owing to the scarcity of game and the hardships they endure during the winter.

The Indians and Halfbreeds occupy themselves in the winter by trapping, and in the summer a large number of them find good employment as guides and packmen for the tourists that fairly swarm on the Nepigon river. For this work they sometimes get as high as three dollars a day. I believe now that quite a few work in the pulpwood camps, but I was informed by a foreman in one of these camps that they were never to be depended on to work any length of time, as they would suddenly get tired of the work and would go home, but while in employment proved, as a rule, to be good workmen.

The Indians in this district are either Roman Catholic, English Church or pagans. The Roman Catholics have the most converts and have a very fine church and school situated on the Nepigon river about three miles from Red Rock. Here a priest lives all the year round, who teaches the school. This settlement, which is highly prosperous, is composed mostly of Halfbreeds, who have built for themselves substantial frame dwellings, and cultivate the land to a considerable extent.

Fauna.

The region immediately to the west of Lake Nepigon appears at first sight to be a good game country. What with its many small streams and lakes and plentiful supply of food both for large and small animals, a more ideal hunting ground could not be found, and undoubtedly it has been a magnificent country in this respect, but the Indians have been so diligent in their pursuit for furs that it has left very few animals over this vast territory. Nevertheless a great number of pelts are yet procured from the region, but as a fur country it is nothing compared to what it was in former years, and the Indians have to work very assiduously to make trapping a paying business.

Large Animals. Moose, caribou and black bear are found in this region, but very few are to be met with. Five caribou, four moose and nine black bears were seen during the summer, which proves to a certain extent that they are not at all plentiful. An Indian who hunts all summer only manages to get a couple of caribou, and then he thinks that he is fortunate.

A few years ago the caribou used to be fairly plentiful, while moose were not to be found at all. About five years ago a couple of moose were killed on Wabinosh Bay. Since that date the moose have been growing more plentiful, while the caribou have been disappearing. Why this is so, it is impossible to answer, as the country affords splendid grazing ground for all these large animals. Hazel, crab, maple and small shrubs are plentiful, while the moss for the caribou is found covering large areas.

Red deer and wolves first made their appearance near Port Arthur about three or years ago. They are still very scarce, but still a number of them have been killed. Mr. Hodder, Indian Agent at Port Arthur, showed me the skin of the first wolf seen near that place. An Indian had killed the beast and had asked Mr. Hodder what kind of an animal it was. A number of men I met expressed the belief that the red deer had been driven into the district by the forest fires that had raged in the Northern States, and this theory appears quite feasible, as they were not found in the vicinity until after one of these great fires.

Black bear appears to be more plentiful than caribou and moose, for not only were more of them seen than any other large animal, but their tracks were noted throughout the district explored.

Small Fur-Bearing Animals. The principal small fur-bearing animals found in the district are mink, otter, fisher, beaver, martin, muskrat and fox, and these are the animals that the Indians trap in the winter. A number of varieties of fox are to be found, namely, the red, black and cross foxes, but the black fox is exceedingly rare, and only a few are killed each year.

The beaver is still found and numerous fresh workings were noted on the small inland lakes and rivers. The smaller and commoner animals are represented in the rabbit, squirrel, chipmunk, groundhog and the different species of field mouse.

The rabbit is trapped for both food and its skins, the Indians relying a good deal on the flesh of this animal for food. They also cut the skin in strips and make blankets for themselves. Sometimes as many as two hundred skins are used in this process.

Birds. Wild duck are found in great numbers on the small lakes, where they find abundance of food. The mallard, black, and sawbill, together with a number of fish-eating ducks, were the principal varieties. Other game birds noted were the tufted and Canadian grouse, the baldheaded eagle, various kinds of hawks, sandpiper, loon, burgomaster gull, owls, besides an occasional blue heron and bittern, and a large variety of small birds found in Southern Ontario.

The country is singularly free from snakes and lizards, and the large bull frog, whose voice is so well known in southern Ontario, is not to be found anywhere in the district.

FLORA. A few specimens of flora were collected by the party for identification on our return, but the district explored has a vegetation very similar to that of Georgian Bay. The timber consisted of spruce, jack pine, white and red pine, tamarac, balsam, poplar, birch, cedar and white wood. All these grow to a considerable size, some of them being found twenty-four inches thick at the butt. A few dwarfed elm and white ash were noted in the sandy flats along the river, but they were exceedingly small.

Alder, hazel, mountain ash, crab, maple, different kinds of willow, bear berry and high bush cranberry are the characteristic small trees and shrubs.

Where clearings have been made by fire, blue berries and raspberry bushes are found, while in the muskegs and the moss that covers the ground cranberries and caribou berries grow.

The small shallow lakes met with are sometimes almost completely clogged up with a luxuriant growth of mares-tails grass and other water plants. A few patches of wild rice were noted in a couple of places, but it is not plentiful.

ALEXANDER H. SMITH,
Geologist,
Exploration Party No. 8.

To D. BEATTY, Esq., O. L. S.,
In charge of Exploration Survey Party No. 8.

SURVEYORS' REPORT OF EXPLORATION SURVEY PARTY NO. 9.

GLENCOE, Ontario, January 1st, 1901.

SIR,—Having been appointed by the Government of Ontario to take charge of Exploration Party No. 9 and instructed to make an exploration of the land and timber lying between Dinorwic on the west, the height of land between the waters running into Hudson's Bay and Lake Superior respectively on the east, and Lac Seul or Lonely Lake Root River and Lake St. Joseph on the north. I beg to submit this, my report.

I left Glencoe on Monday, the 18th day of June, 1900, and went *via* Toronto and North Bay to Ignace Station on the Canadian Pacific Railway, being joined on the way by the other members of the party, which consisted of seven all told. Capt. McPhee, of Port Arthur, being Timber Explorer and John E. Davison, of Toronto, Geologist. Two other members from Toronto and two from Nipissing Village completed the number. These all remained with the party till it disbanded at the close of the season's work.

A part of the work allotted to me was a micrometer survey of Sturgeon Lake, and this I decided to perform first, exploring the country along the route between it and the railway. Before setting forth the results of our work I will briefly outline the course travelled by us in our surveys and exploration.

We left Ignace on the 22nd day of June, and made a track survey of the canoe. route to Sturgeon Lake, reaching it on the 2nd day of July. We made a careful micro meter survey of this lake with all of its numerous bays, which we could enter from the main lake without the necessity of portaging. The surrounding country for many miles was also explored by the timber explorer and the geologist.

This survey occupied our attention till the 13th day of August, and on the following day we took our departure from it, descending the Sturgeon river, which forms the outlet of Sturgeon lake. This outlet leaves the lake from near the southwest corner of what might be described as the north-west bay of the lake, by two channels about two miles apart ; but which unite about three miles below. This river has its course northwesterly and southwesterly through the country lying between Niven's Fourth Base Line run in 1897, on the south and Rat Root River and Lac Seul on the north, and enters Abram's lake by the long arm shown on the map. Extending northeasterly from the base line referred to, we then followed this chain of lakes, which I presume is known as Sturgeon or English river, in a northwesterly direction to within a few miles of Indian Reserve No. twenty-eight. At this point although the chain of lakes would seem to continue westerly, the channel of the river turns more northerly and this latter we followed by its very irregular course to Lac Seul. Continuing along its southerly shore westward we connected our survey with the Indian Reserve No. twenty-eight, at where its northern boundary intersects the lake, on the 31st day of August. Here we took an observation and erected a flag as per instructions.

We then turned eastward following the south shore of Lac Seul, and ascending Root River to near its head, then crossing over the height of land into Lake St. Joseph and on to its outlet into the head of the Albany river which we reached on the 21st day of September. Finding the post planted by D.L.S. Thomas Fawcett in 1885, as marking his magnetic station, we took an observation for latitude at it as per instructions. We started on the return trip the same day, following practically the same route as far as Abram Lake, and thence by Lakes Minnietakie and Big Sandy, and wagon road to the Canadian Pacific Railway, reaching Dinorwic on the 10th day of October and home a few days later.

This route is more particularly shown on the plan which I have prepared. We made a track survey of the whole course, and checked the same by observations taken for latitude almost every day when the conditions were favorable. Throughout the trip we gathered specimens of the flora which have already been forwarded to the Department. I also kept careful and regular readings of the barometer and thermometer from day to day as well as notes of the weather. The timber explorer and geologist made frequent trips inland as shown approximately on the map.

Along Root River and Lake St. Joseph we were not able to make many excursions inland, but from high elevations at favorable intervals, with the aid of a pair of field glasses, we could frequently get an extensive view of the surrounding country, for many miles and thus gain information regarding it. Having thus outlined the work I will now proceed to describe more particularly the route and the nature of the country explored.

From Ignace we went by train with our canoes and supplies to Osaquan Siding, five miles west, where we had arranged to have a team meet us and take our outfit to Camp lake, over a road one and one-quarter miles in length, cut out by Mr. W. H. Cobb of Ignace in a north-easterly direction. Here we put our canoes in the water, tested our instruments and proceeded with our work.

At Camp lake Mr. Cobb has built a small store house and other buildings for convenience in storing supplies &c. for shipping up the lakes to the mining regions. He had also made a small clearing at the edge of the lake and had only recently planted some potatoes. The soil was a whitish clay. Camp Lake is almost round in shape and about one mile in diameter. Crossing over we pass out of it by a shallow stream about one chain in width, with low grassy shores, and one and one quarter miles long to Indian Lake. This lake lies north and south, is about nine miles long and varies in width from five chains to two miles. The surrounding country is for the most part rough and rocky, with boulder hills in places covered with shallow, sandy soil. The timber is chiefly jackpine, spruce, poplar and birch. It is all small. The lake abounds in fish such as pike, pickerel, and it is said, white fish.

We saw evidences at deserted Indian camps of moose and bear having been taken in the locality. Passing out of Indian lake at its north end by a shallow stream four to five chains wide with slow current, a distance of one and one quarter miles in a north-easterly direction to Bear Lake, we encounter two falls about half a mile apart, necessitating portaging. The first portage is two and one-half chains long, and the fall in the water about four feet. The second one is eight chains long and the fall from twenty to twenty-five feet. A few red and white pine are seen in the locality.

Bear Lake lies in a north-easterly and northerly direction and is about eleven miles in length from portage No. 2 to portage No. 3, which latter is at its outlet. It varies in width from a quarter of a mile to two miles English River enters it from the east and this we explored for a distance of about twelve miles. The first ten miles has numerous lake expansions, after which it turns sharply to the south and continues in a stream of fairly uniform width of five to six chains, through a low flat with marshy, grassy shores. This flat is not much above the water-level of the river, for a mile or so from it, then it rises and falls again as far as we could see. The timber along both sides of this river is small and scrubby willows, alders, tamarac, and scattering spruce and poplar, none of it being of value. Traces of bear and moose are to be seen. The timber around Bear Lake is chiefly pitch pine, poplar and spruce, with birch and cedar in smaller quantities, the latter being only of a scrubby kind, fringing the shore in places. All are for the most part small. There are occasionally small patches of red and white pine. The land is rough and rocky, with a covering of sand and boulders in patches, and a thick coating of moss.

Portage No. 3 is four chains long. Here the stream for a short distance is split into three channels by islands, the portage being on the most westerly one. The fall is about eight or ten feet. Following English River from portage No. 3 for about three-quarters of a mile, brings us to Portage No. 4, which is about five chains long. The fall in the water at this place is about three feet. Continuing in a north-easterly direction for about three miles down English River, which is now from four to six chains wide, we reach Otter Lake, and turning north-easterly pass through it for a distance of about four miles to the mouth of what we here call Grassy River, which we now ascend. It is a shallow, and for the most part a sluggish, weedy stream, from fifty to twenty feet wide meandering in a north-easterly direction through a grassy flat from ten to 15 chains wide near its outlet, and contracting to three or four chains at a distance of three miles up. A portion of this was so shallow that it was necessary to tow our canoes for a considerable distance. Here it expands into Hut Lake of variable width and three miles long in a north-easterly direction, contracting again to a narrow, shallow, winding, weedy channel, for a distance of half a mile to portage No. 5, where the stream has a cross-section of from three to four feet and a fall of from twelve to fifteen feet in eight chains, which distance is the length of the portage.

Near this portage and south-easterly from it is a block of about fifty or sixty acres of fair red pine running about sixteen inches in diameter, which would probably cut two hundred and fifty thousand feet board measure. The country in this locality is more level and less rocky, there being few exposures. The soil is boulders covered with coarse sand and moss near the lakes, and boulders covered deeply with moss further back. This mossy covering is frequently two feet deep and makes travelling through it very tiresome. The timber is chiefly small spruce and tamarac on the low land, with jackpine, birch and poplar on the higher land. About half a mile above Portage No. 5, the stream again expands into Pine Lake, which is about two miles across in a north easterly direction to portage No. 6, which is eight and one-half chains long over an elevation about fifteen feet high to White Rock Lake. The river channel here is a narrow, rapid stream through boulders and having a fall of six or seven feet between these two lakes. The shores of these lakes are chiefly boulders and sand. The timber and soil are much the same as that already described, the timber being valuable only for fuel and the land not adapted for cultivation. Across White Rock Lake north-easterly is about two and three-quarters miles to the "narrows" between it and Young's Lake. This is a narrow winding channel which contracts to about six feet between boulders at one-quarter of a mile from White Rock Lake, and again expands into Young's Lake, which is crossed in a south-easterly direction two and one-half miles to Sturgeon Lake portage, which is over the height of land between the waters draining to Sturgeon Lake, and these draining to Otter Lake. The surface of Sturgeon Lake is probably fifteen feet higher than Young's Lake, but their waters find a common level in English River at Abram Lake.

This portage from Young's Lake to Sturgeon Lake is forty-seven and one-half chains long in a north-easterly direction, and is a new one cut out during the past winter as a sleigh road by a party developing some mining propositions on the lake. At the Sturgeon Lake end of this portage Mr. Cobb has put up a couple of small log buildings. The soil at this end of the lake as seen on the portage is heavy white clay with boulders and covered with moss, which latter seems general in this section of the country. The timber here is rather larger and more healthy than a good deal of that previously passed through. It is of the same mixed classes of spruce, jack-pine and tamarac with some birch and a few scrubby cedar near the water's edge. This timber runs as high as ten and twelve inches in diameter, but the average will be considerably less.

From the railroad at Osaquan Siding by this route is fifty miles, in which distance there are eight portages aggregating two and one-quarter miles, which includes the distance from the railroad to Camp Lake. This is the route most travelled by persons going to and from Sturgeon Lake. I might here mention that the water in these lakes and streams is said to be about two feet lower than usual.

The timber along the course is green, for the most part, of the kinds already mentioned, and is perhaps of better quality in proximity to Sturgeon Lake, and while there are a few pine, spruce and tamarac fit for logs, pulpwood and ties, they are not in sufficient quantity to be valuable. Fish of the kinds before mentioned seem to be plentiful in the lakes and streams passed through. Vegetation is scarce and game is not plentiful, but there were evidences in many places of bear, moose and caribou.

As before intimated we made a careful micrometric survey of Sturgeon Lake. This lake lies north-easterly and south-westerly and is a beautiful sheet of clear water containing many islands. Its extreme length is about thirty-six miles, of which one third is west of O.L.S. Niven's meridian line, run in 1890, between the Districts of Rainy River on the west and Thunder Bay on the east. This part is from two to four miles wide, and on its northerly side are several beautiful land-locked bays, surrounded with high rocky hills covered with green timber. West of Niven's meridian are the narrows, which vary in width from half a mile to a mile and are about twelve to fourteen miles long. The north-easterly part consists of two large deep bays with very irregular shore line. The rock is exposed on most of the shore line of the lake and bays as well as around many of the islands. I might mention that in my survey of the lake I did not make a survey of the numerous islands with which it is dotted most profusely in parts, as this would have occupied much more time than we had at our disposal. I did, however, where I could conveniently do so, tie many of them to the survey and sketched in the position of others, but there are several parts where they may not even be approximately represented.

A good deal of prospecting for gold was being done around the shores of the lake and a considerable number of mining claims seem to have been surveyed. These lay chiefly to the west of the meridian mentioned. In King's Bay, which is a narrow bay about three miles deep, extending westerly from the north end of the narrows, Mr. A. E. Shores is doing considerable work in developing a promising prospect. He had two shafts sunk to a depth of about sixty-five feet each and was cross-cutting on both. He was also doing considerable surface work, and had taken out considerable ore which looked very rich in free gold, and which he was saving. He kept from twenty to twenty-five men employed, and had neat log buildings erected for their accommodation and for the various purposes required in connection with his work. Mr. Shores shewed us considerable kindness in various ways during our stay upon this lake, which was much appreciated by all the members of our party.

In many places surrounding the lake and on the islands where the moss is burnt off, the rock is exposed and this makes prospecting so much more easy, that I am afraid the prospectors are not altogether guiltless of starting fires for thus facilitating their work, and the rich green color of the unburnt parts is fast giving place to the brownish red and dirty black of the scorched and burnt, but standing trunks, which in a few years will be blown down and another fire clear the country. Fires were smoking in various places during our stay upon the lake; these would often die down during the night to be fanned into flame by the breeze during the day. On account of the great depth of moss common to this section of country, these fires make comparatively slow progress and the recently burnt areas are of small extent, and are confined chiefly to near the lake shore and to islands.

Lying to the south and east of the lake is an extensive tract of older brulé. On many parts of this there is little or no soil, and the underbrush that has started in places is yet quite small. Around Sturgeon Lake there is very little vegetation; at the portage, however, I picked some stalks of wheat, oats and barley which had grown from seeds dropped from the horse feed taken in during the previous winter. These were strong and healthy in appearance, and were in head at the time. On asking Mr. Shores at the mining camp if he had made any attempt at growing potatoes or other garden stuff he replied "that when the land is cleared of moss and timber there is no soil to grow anything." And this is the condition of the greater part of it. The lake itself abounds in fish such as trout, pike, pickerel, black suckers, and, it is said, whitefish as well as other common kinds. As in other parts spoken of, game did not seem to be plentiful although about deserted Indian lodges at the outlets of some of the small rivers, would be found evidences of bear, moose, caribou, otters and fishers, and in certain localities other traces of them would be seen. There are a few Indians about this lake, but not a great number. The Hudson's Bay Company has a winter trading post here, and in close proximity to it is one conducted by a Mr. McLaren. The former was entirely closed when we were there, but the latter was in charge of an Indian. I might say that the original site of the Hudson's Bay Post on this lake was on the north shore, about two miles west of Niven's meridian, but it is now removed to the easterly side of the lake, seventeen or eighteen miles north-east of the old position, and is situated nearly opposite King's Bay, and near the entrance to the two long bays into which this end of the lake is divided. As already described, the outlet of this lake is by two channels about two miles apart, and which unite about three miles below.

On the most northerly channel the water leaves the level of the lake by a fall of from twenty to twenty-five feet in eight chains, and varying in width between rocky shores, from fifty feet to ten or twelve feet. This could be converted into a good water-power. On the south channel the level of the lake is left by a rapid about eight chains long and from fifty to sixty feet wide and having a fall of four or five feet. About half a mile below is another fall of about eighteen or twenty feet in five chains, and the channel is between rocky shores, and from twenty to sixty feet wide. This also would make a good water power. In the early part of August when we saw it, the water was low and there was probably three times the amount of water discharging by the south channel than there was by the northerly one, but the whole of the water from Sturgeon Lake could readily be turned through either channel if desired, and a good power obtained for reducing the ores of the region, or for other purposes. From Sturgeon Lake to O. L. S. Niven's fourth base line of 1897 at Abram Lake by Sturgeon River is about sixty five miles, in which distance there are fifteen portages most of which are short and fairly good. These aggregate a trifle more than two miles of carrying distance. The longest is thirty-one chains, seconded by another of twenty-five chains. These portages are all past rapids or falls in the river, the total fall of which between Sturgeon and Abram Lakes will amount to about two hundred feet. Several of them are not high, but the greater number of them have sufficient fall to create good water powers. The river between the different portages is generally of fair depth, and fairly uniform in width for considerable stretches, varying from two to eight chains, but there are many lake-like expansions on it. Except near the rapids or falls and a few other places, the current in slow. This seems to be the route travelled by the Hudson's Bay Company's carriers is transporting their goods to and from their post on Sturgeon Lake. The first thirteen portages occur within about twenty-four miles of Sturgeon Lake, and the country along this distance is for the most part green, except in the vicinity of some of the portages. The land is still rough and rocky, the shores in many places rising abruptly from the water, and in other places the shores are sand and boulders rising to rocky elevations at some distance back. There is no particular change in the timber, which consists of small, scrubby jack-pine and spruce upon the rocks with poplar and birch in places where there is more soil, and a few tamaracs in lower places. It is nearly all small and of poor quality, much of it being from four to six inches in diameter, but in occasional small patches of better soil it is larger. The water, although clear, on leaving Sturgeon Lake, gradually turns to a dark brown, as it receives water from its tributaries draining the lower land further back. The volume increases considerably too, and there is much more water passing over the falls at portage No. 13 than at No. 12. The former would make a particularly fine water power. Between portages twelve and thirteen the river expands into deep bays, and a river of considerable dimensions enters from the north This was ascended for quite a distance by the timber explorer and geologist, whose report of this section of the country shows that it is not materially different from that already seen. While near Sturgeon Lake there seemed to be quite a number of fish in the river, yet further down they seemed to be scarcer. Along this part there was very little indication of game, and the Indian lodges and trappers' dead-falls, common in other localities were here missing.

From portage No. 13 to portage No. 14 is about twenty miles, in which part there are numerous bays which receive many small tributaries. The country gets less rough and the shores lower and more sandy as we leave No 13, although rocky exposures along the river are yet numerous. The timber is of the same mixed kinds, with probably less jackpine and more poplar. In places it is larger, but nowhere is it sufficient in quantity and quality to be of commercial value except for fuel. The land between the river and the fourth base line is less stony than before. It is chiefly sand with boulders in places and certain limited areas of fair clay soil. The same deep coating of moss covering the ground keeps it very cold and there is very little undergrowth or vegetation. There is still little indication of game or fur-bearing animals, although some six or eight miles south of portage No. 14 we saw some recent beaver works along a creek. Portages fourteen and fifteen are a little over a mile apart. The former has a fall of about twenty feet and the latter about five feet, which brings us practically to the level of Abram Lake, as the current is light till it is reached. For a short distance below portage No. 15 the banks of the river are clayey, but soon get into more rocky country along the shores and it is also more rough and rocky further back. Deserted Indian wigwams are again appearing, and about a mile below portage No. 15 is apparently a winter trading post of the Hudson's Bay Company, while near by is a miner's prospect, but not much work had been done on it. For some distance the river expands at various places and then assumes a more uniform width and a straight course in a south-westerly direction, till it expands at Abram Lake near Niven's base line of 1897. From small patches of burnt country along the river below portage No. 15 a brulé of not long standing increases in extent, both north and south till it covers a considerable portion of this locality. Much of it, where the fire has probably been over it several times, is burnt clean, in other places the burnt and dried trunks of the jackpine and spruce remain, making a condition of affairs almost impassable and dangerous for travelling. There are some high, strong and rocky hills near this lower part of the river, and from them we got a view over a considerable extent of country, and it appeared to be burnt for a long distance northward.

The soil, where there is any, is chiefly sand among boulders, and not suitable for agriculture, except in occasional small patches. Rolling and rocky ridges extend east and west. About five miles above the fourth base line is a narrow, wet flat along the river, and here we saw two moose cross. There was also some evidence of red deer in this locality. In this brulé were blueberries of a fair sample, which were practically the only ones of a reasonable size met with during the season. We connected our survey of the river with the base line at Abram Lake, which last we crossed westerly along the north end, and passing through a shallow rocky channel about three chains wide, entered Pelican Lake. At this narrows there is quite a current, but it can be canoed without portaging. Crossing Pelican Lake northwesterly the channel again contracts around an island and Pelican Falls are reached, the portage being found by following the east side of the island. Around Abram and Pelican Lakes are many rocky exposures, which show recent marks of the prospector's hammer, and in a bay across Pelican Lake we noticed what was evidently a miner's camp, with a considerable pile of waste rock, but we did not get over to it on account of the high wind which was blowing at the time. The timber around these lakes is chiefly green yet. There are some small bunches of red and white pines, the other timbers are of the same mixed kinds, some being as large as ten and twelve inches, but not in great quantities. In the locality of Pelican Falls is some rolling land, sandy and stony, with fair clay in patches, but mostly having a hard-pan bottom. So far this is the best land we have seen. From the top of Souix Outlook on Pelican Lake, which the barometer showed to be about two hundred and fifty feet high, a magnificent view of the country was had. On the summit was a prospector's location stake with a flag attached.

Pelican portage is twenty-eight chains long in a northerly direction past a series of rapids and falls in the river, amounting to about fifteen feet, through a channel from three to four chains wide between rocky shores. Grass, golden rod and other weeds common in the older parts of Ontario were growing rankly here. From Pelican Falls to Lac Seul the river is a series of expansions of irregular shape, with deep bays extending inland from ten to twenty miles. These we followed, and from them got a good idea of a large section of the country, but found it similar in land, timber and general appearance to that we had previously passed through. It is yet, for the most part, in an unburnt state. Sturgeon or English River enters Lac Seul almost directly north of Pelican Falls but the course between the two is far from being a straight one. From a short distance below the falls there are two canoe routes to Lac Seul. One by a chain of lakes through Indian Reserve No. 28, is the shorter and more direct, and reaches the lake almost opposite the Hudson's Bay Company's Post. It has, however, one portage on it. This one we did not travel by, but followed the river route. The weather during this part of our work was very unsettled, with thunder storms and much high wind, which made travelling very dangerous in the more exposed parts. When wind-bound about midway between Pelican Falls and Lac Seul, we cut some trees and counted the growth of rings which showed as follows :

One jackpine nine inches diameter growing on the rocks and having considerable depth of soil on it had seventy-two rings.

One jackpine near the first but with less soil seven inches in diameter, seventy-two rings.

One jackpine near the first still less soil four inches in diameter, fifty-two rings.

One poplar nine inches in diameter in sandy flat, seventy-two rings.

From this it might be inferred that the fire which removed the previous growth of timber took place some seventy-five or eighty years ago, and this would probably apply to a large section of the country which we have seen.

On reaching Lac Seul there is more burnt country, and a good many of the islands are also burnt almost bare. Near the mouth of the Sturgeon or English River is a range of high burnt hills for a mile or so along the shore on the west, and from this to the Indian Reserve the timber is green. This part is probably less rough than much that we have seen, and there is more long poplar and birch near the shores. The soil is sand and boulders.

At the Hudson's Bay Post, which is situated on the opposite side of the lake from the Indian Reserve and about five miles from it, there are, in addition to the Hudson's Bay Company's buildings, several houses, including a parsonage and an English church, and the tidy, neat and comfortable appearance of the whole was quite pleasing. Here are kept cows, which are fed on wild hay, and do well Potatoes are said to be grown in patches on the mainland, but a sandy island in front of the post provides the chief planting ground, where it is said they do well. We were told that there was better land on the Indian Reserve, and that good potatoes were grown there, but nothing else was attempted in this line. Quite a number of apparently comfortable houses were seen near the waterfront on the reserve, and the Indians met with were well dressed.

Lac Seul, or Lonely Lake, is a large body of water with very irregular shore line, and many islands, rocky reefs and deep bays. The water is a dirty yellow color, and contains much sediment, and is much colder than that of Sturgeon Lake. The surface of the lake was some eight or ten feet lower than a very distinct water-line upon the rocks. It contained fish of the same kinds as spoken of in other lakes. The surrounding country is flatter than the land further south, but broken with many rough elevations. The soil and timber is not changed. Having completed the western portion of our survey, and explored the deep bay eastward from the mouth of the river, which we thought was the main lake till we reached the bottom of it, we continued along the south-easterly shore. We found that instead of running easterly or nearly so, as we had expected from the maps in our possession, it ran almost north for some eight or nine miles, then north-easterly for a similar distance to another high burnt hill on the east, where the lake contracts to a width of eight or ten chains. At this place we were able to locate our position on the map which we had, and which from this point is a fair representation of the course to the head of Root River.

Before leaving this part of the lake we ascended a river, which has its outlet into a bay about five miles south-east of the burnt hill last mentioned, for a distance of eight miles, till we came to a fall, past which the portage showed no travel unless by trappers. A short distance below this fall, a large creek enters, and this we ascended for several miles further. This river is about two chains wide with slow current through low, flat land with clay banks. In places the soil was of fair quality, but limited in area, and that near the river too low, being six or eight feet above low-water mark. The country was burnt to the north and green to the south. The timber was the same mixed poplar, birch, tamrac and spruce from four to ten inches in diameter. Returning to the burn hill, and continuing northerly for a distance of about twenty-four miles, through expan sions and channels of greater or less width, and among low islands, we reach the firs rapids on Root River. The country along the last stretch travelled does not improv upon that already described. It is perhaps flatter, but still broken, rough, and rocky ir parts, rolling boulders and sand in others, and low and swampy in other parts. A grea deal of this stretch was recently burnt over, and the new growth has not much of a start yet. On patches not yet burnt the timber is smaller and would seem to represent a growth of twenty-five or thirty years.

From a high hill about three miles below this rapid we obtained a splendid view o the country for fifteen or twenty miles, and as far as we could see it was burnt and deso late looking. There were no signs of game or mineral.

At this rapid there is a fall in the water of about four feet, which is no doubt con siderably reduced when Lac Seul is at its normal level. The channel here is two chain wide over rocky bottom, and the portage, six chains long, is swampy and bad. From this portage along Root River to the one over the height of land and into Lake St. Josep is about forty miles, in which distance there are ten portages including those mentioned The first nine of these aggregate three-quarters of a mile, and the last one is one-half mile in length. The total fall in the stream of these portages will be about sixty feet This river for the most part has a slow current, with very dark, dirty water and weedy marshy, and in places, willowy shores, with swampy flats of greater or less width on eithe side, which rise to rocky or sandy hills further back. There are a few expansions of the river, but the width is fairly uniform, being about five chains above Portage No. 1, and gradually contracting to about one and one-half chains between Portages Nos. 6 and 7 where the main channel turns to the west, and a smaller one carrying about one-quarter the amount of water, continues the canoe route to Lake St. Joseph. This channel is about fifty links wide, and contracts to about twenty links. The current in it increases as we ascend, and is greater than that of the main stream. The course becomes much more tortuous, and the overhanging willows sweep the canoe in passing. This route, however, is travelled a good deal by the Hudson's Bay Company's carriers, who find it necessary in this part to abandon their paddles and resort to poles for propelling their large canoes. As the stream was considerably swollen by recent rains at the time, we experienced no difficulty, but would think that with low water there might be more trouble in getting through with loaded canoes. The country along the river continues burnt, unless in places where a fringe of green timber remains near the shore. The brulé is probably two years old, and any attempt to travel through it is soon challenged by the now dry, dead timber, chiefly jackpine and spruce, which being small before the fire, having the small branches burt off, leaves a mass of sharp, hard, claw-like points which obstruct one at every move.

There are many high, rocky exposures in parts, with no particular change in the soil. There was some evidence of caribou. Fish in this stream did not seem plentiful. In the grassy expansion below the first portage were quite a number of wild ducks, and in other places a lesser number. The portage from Root River into Lake St. Joseph is over a slight elevation, a considerable part being through a tamarac and spruce swamp, over which is a pole-walk built by the Hudson's Bay Company. Lake St. Joseph is entered by a creek similar to the one just left, but which in a short distance expands into a wider channel, with considerable marsh for some distance along it. The map is a fair representation of it for about fifteen miles, which part is comparatively narrow, but with some bays not shown; beyond this distance to Osnaburg House it is difficult to recognize, as it is a large body of water very irregular in shape, and containing many deep bays and islands some of which are quite large. We found on several large areas of this lake a strong local magnetic disturbance of current which affected our compass needles and rendered them of little service in guiding us. The large area extended over probably ten miles, and we were not able to account for it by any mineral bearing rock which we came across. An observation taken at noon on an island on the 18th day of September shewed a magnetic variation of the needle amounting to about sixty-five degrees, while the variation in the undisturbed parts is about four or five degrees east or north

The water of the lake is fairly clear and very cold. The country along the westerly half of it is practically all burnt over to the south, and much of it is very clean. Continning easterly there are some patches of green timber, consisting of small jack pine, spruce, poplar and birch. The easterly half is not so much burnt, although there are numerous patches of comparatively recent brulés near the shore. The timber does not vary materially from that further south. There is probably less jack pine and all is more scrubby. The country is practically the same, rough and rocky in places, with rolling hills of boulders and sand, falling to low, swampy spruce land further back. Some of these swamps, however, are deeper than those seen further south. Where not burnt over, the same deep coating of moss covers the ground. There are a few sandy bays and some sandy islands, but the general aspect of the shore line is rocky, either fixed rocks or boulders.

At Osnaburg House, the Hudson Bay Company's fort on Lake St. Joseph, at the head of the Albany River, we obtained some potatoes which had been grown in the sandy land in the locality. They were said not to be as good as those grown in previous years, but were a fair sample. No attempt had been made to grow anything else. The manager, Mr. Wilson, at this post informed us that quite a number of furs were brought in, such as bear, otter, muskrat and sable, some foxes, martens and fishers, and very few mink or beaver. There are some caribou, but only very recently have moose been known on the lake. The fish are chiefly pike, pickerel and suckers.

Mr. Wilson had not heard of there being any valuable timber in this direction, on the Ontario side of the lake, nor any mineral lands. Some little prospecting had been done about two years ago, but so far as known no discoveries had been made. I might mention that we saw one wolf on the shore of a deep bay of this lake. There were some gulls on most of the lakes, and a few loons. Birds were comparatively scarce.

Returning from Osnaburg House, which we left on the 21st of September, we experienced very rough, stormy weather, with high winds, rain and snow, which made travelling on the larger waters very dangerous. The weather during July and August might be considered a good average, with cold nights and warm in the day-time. Throughout September there was a good deal of wind and the weather was stormy towards the latter part.

A very noticeable feature at Indian camping-grounds on the portages and other places is the great number of jackpine trees whose trunks are barked. It is said that this was done by the Indians, who make use of the pulpy matter found below the bark for food during certain seasons of the year.

Summing up, I might say that in the territory explored we found no timber in quantities fit for commercial purposes except fuel, and no tract of land fit for agricultural purposes.

The Sturgeon Lake district shows some promise of becoming valuable for mining purposes. Game and fur-bearing animals, though not abundant, are through parts of it in considerable numbers. About five or six bears, a similar number of moose and one wolf were seen by the party. Photographs were taken of objects of interest during the season, and the exposures made have already been forwarded to the Department. All of which is respectfully submitted.

I have the honor to be, sir,
Your obedient servant,
JAMES ROBERTSON,
Ontario Land Surveyor.

To the HONOURABLE E. J. DAVIS,
Commissioner of Crown Lands, Toronto, Ontario.

I, James Robertson, of the Village of Glencoe, in the County of Middlesex, Ontario Land Surveyor, make oath and say that the foregoing report is true and correct in all particulars.

Sworn before me at Glencoe, in the County of Middlesex, this third day of January, A.D. 1901.
W. D. MOSS,
A Notary Public for Ontario.

JAMES ROBERTSON.

LAND AND TIMBER ESTIMATOR'S REPORT OF EXPLORATION SURVEY PARTY No. 9.

PORT ARTHUR, ONTARIO,
DISTRICT OF THUNDER BAY,
December 5th, 1900.

DEAR SIR,—In connection with Exploration Survey Party No. 9, under your direction, I beg to report as follows .

June 20th.—I joined the Government Exploration Party No. 9 at Port Arthur, and proceeded by Canadian Pacific Railway to Ignace.

June 21st.—We laid over this day at Ignace fixing up our supplies, and planning our journey north.

June 22nd.—We left by train for Camp Lake, which is seven miles west of Ignace, and made a journey of two miles inland from Camp Lake. The timber in the neighborhood is small, second-growth jack pine from two to three inches thick. The soil is coarse sand and white clay not adapted for agricultural purposes. We saw a patch of potatoes planted on Camp Lake which promised no growth or yield. We continued from Camp Lake to Indian Lake, which is connected by a stream fifty feet wide, running north one and one-half miles into Indian Lake. The land along this stream is low and swampy, covered with small spruce from three to four inches thick. The land tributary to Indian Lake is rocky and sandy and no good for cultivation. The timber is of no commercial value, consisting of small jack pine, spruce, poplar and birch of recent growth from two to three inches thick.

June 23rd.—Camp No. 1. We explored the coast line of Indian Lake for twenty or thirty miles, taking advantage of high elevations to view the country back for several miles. The timber in the neighborhood of Indian Lake is small jack pine, spruce, poplar and birch of recent growth from two to three inches thick. The land is more rolling with rocky ridges, sand and boulders, and no good for cultivation.

June 25th.—Continuing from Indian Lake to Bear Lake, we put up camp 2 on the first portage on Sturgeon Lake route. The timber along this day's travel is small spruce and jack pine with occasional bunches of poplar and birch from three to five inches thick. The land along the river is low, flat and swampy, which slightly rises to rocky ridges of granite or gneiss. The soil is no good for cultivation.

June 26th.—Continuing from No. 1 portage down the river and over an expansion of the river into Bear Lake, we reached Portage No. 3, where we put up Camp No. 3. The timber along this neighborhood is small spruce, jack pine, poplar and birch with occasional patches of red pine from ten to twelve inches thick.

The land is rolling, rocky ridges with occasional flat patches of sandy soil, and no good for cultivation. It continues low and marshy along the rivers and streams. The surface of the country is mostly covered with a heavy coating of moss. Vegetation is rather scarce, game is not plentiful, but fish including pike, pickerel and whitefish appears to be abundant.

June 27th.—From Camp 3, Portage No. 3, we journeyed down the English River or Mattawan fifteen or eighteen miles from an arm of Bear Lake, in an easterly direction. This river will average from three to four chains wide, with a low, swampy, marshy country surrounding it. The timber is principally small spruce, with a few tamarac from three to four inches thick. The land in general along the river is marshy and wet covered with heavy coating of moss, and where a rise of land occurs is found rocky and sandy, and not suitable for settlement.

June 28th.—Continuing from Camp 3, on Portage No. 3 down the English River one mile to No. 4 Portage, we entered Otter Lake further down, and continued four miles across to Grassy River into Hat Lake, and from Hat Lake over a winding grassy stream to Pine Portage, where we put up Camp No. 4, on No. 5 portage. There is no change in timber, soil or rocks along this day's route. The land remains low, flat and swampy along the rivers, which are fringed with willows and rushes. Grassy River is difficult to canoe over in low water seasons.

June 29th.—We made a journey inland from Pine Lake, Portage No. 5. Camp No. 6, on a south-easterly course six miles, when a stream flowing north-east from thirty to forty feet wide interfered with our further progress and we returned to camp. The timber along this journey is jackpine, spruce, tamarac, popular and birch, and will average five inches thick. It is not large enough for pulp-wood, but the whole would average fifteen to twenty cords of fuel per acre. The land is low and flat with occasional rises. The soil is sandy with boulders covered with moss. There is no other vegetation and the land is unsuited for agriculture. Near this portage was found a bunch of red pine covering about sixty acres of land which will average sixteen to eighteen inches thick, and will cut about two hundred and fifty thousand feet of lumber, board measure. The soil is sandy. We also made a journey of two miles north-east of this portage and found no change in timber or soil.

June 30th.—From Pine Portage No. 5, Camp No. 4, we proceeded up the river about sixty chains to Pine Lake, which is about four miles across, making No. 6 portage of five chains, into White Rock Lake, and crossing White Rock Lake into Young's Lake which enters a narrow channel of rocks. We continued our course across Young's Lake to Sturgeon Lake Portage No. 7, and went over the portage to Sturgeon Lake, and returned The timber along this journey is small, scrubby spruce, jackpine, poplar and birch, from five to six inches thick. The soil is white sand and boulders. The land is flat with low rocky ridges. Vegetation is scanty and game is also scarce, but the waters are clear and well stocked with fish.

July 2nd.—We broke up Camp No. 4, Portage No. 5, and traversed the same route as that on June 30, but found no change in land, timber, or streams. The portage from Young's Lake to Sturgeon Lake is three-quarters of a mile long, with a rise of thirty or forty feet which falls back to Sturgeon Lake. The timber along this portage consists of tamarac, spruce, jackpine, poplar and balsam, which will average seven inches thick.

July 3rd—We crossed Sturgeon Lake portage back to Young's Lake, and proceeded south six or seven miles, took up a south-easterly direction inland and found a new lake which is about one and one-half miles wide. The timber along this route is small spruce, jackpine, and tamarac with an occasional bunch of poplar and birch averaging six inches thick. The soil is sand and gravel. The shore line of this lake is granite, gravel and boulders. The timber along this route varies in size as the land rises and falls, that on the ridges being larger than that in the swamps or low ground, and will average from five to eight inches. There is a little white pine among it.

July 4th—Crossing Sturgeon Lake to Young's Lake we proceeded south west one and one-half miles by canoe, turned north and found a deep bay at two and one-half miles. We journeyed by land five miles north-west, the first three miles being low swampy country, with water visible all the way. The land then rises gradually to the height of one hundred and fifty feet where a few exposures of rock show up at different points. The timber along this route is small spruce and jackpine, and will average from four to six inches thick, but is of no commercial value and would only be suitable for fuel. The soil on the high land is white sand with boulders, and no good for cultivation.

Timber in the lowlands is principally spruce, tamarac and some cedar but nothing of a merchantable character. The land is all covered with a heavy coating of moss. There is no vegetation, game is scarce, and there are no wild fruits.

July 5th.—From Sturgeon Lake Portage No. 7, Camp No. 5, we proceeded across the bay by canoe, one mile to the south shore, and journeyed inland eight miles, in a south-easterly direction. The land is nearly all swamp with occasional narrow ridges of rocks which immediately gives place to swampy lands again. Water is visible all through the swamps, they are heavily covered with moss underlaid with boulders and quicksand, and entirely unfit for agriculture. The timber on the ridges is of the original birches, poplar and balsam, mixed with a very thick undergrowth, and is very difficult to travel through. This bunch of timber will average twelve inches thick. The timber on the low lands is tamarac, cedar and spruce, and will average from six to seven inches thick, and is very thin and scattered. There is very little appearance of vegetation, as the soil is covered with a heavy coating of moss. A small stream was crossed about a mile from the lake, running north east, but it is not large enough for any driving purposes. A few exposures of rock are found but no mineral so far.

July 9th.—Heavy rain and windstorms detained us in camp for two days. The weather permitting we broke up camp 5 on Sturgeon Lake, portage No. 7, and proceeded by canoe to a point ten miles north-east of the portage, and put up camp No. 6 on an island. The timber along the shores is small, scrubby poplar, jackpine, birch and spruce. from four to six inches thick. There are a few exposures of rock along the shores. The land is sand and boulders covered with thick coatings of moss. We returned back by canoe three miles to a point south-west and journeyed inland two miles north-west. The land gradually rises to a height of fifty feet for a mile from the shore, and falls suddenly into a spruce swamp about four chains wide, then rises to a height of one hundred feet above the lake level, and again falls into low, swampy ground. The timber along this route is small jack-pine, spruce, poplar and birch, and averages from four to six inches thick. I also saw at this point a small patch of potatoes which did not promise any growth or yield. The soil is white sand and boulders not adapted for farming. We explored along the shore line which showed a few rock exposures and returned to camp.

July 10th.—We journeyed by canoe from camp 6, which is situated on an island north-east of Portage No. 7, Sturgeon Lake, or Sturgeon Lake Portage, and ran a south-easterly course for two miles, continuing down an arm or bay of the lake three miles west. The land rises from the bay from fifty to one hundred feet. Numerous exposures of rock occur along this shore, and also on the high ridges. The timber is small and of no commercial value, and consists of spruce, jackpine, poplar and birch. There are a few white pine trees of poor quality along this coast. The soil is still sand and boulders covered with moss. Some small streams enter into the bay. We met with no minerals although the formation is somewhat favorable. On ascending a hill I could view the country south and west for eight or ten miles. The country shows no change and appears to be low and swampy. No water is visible from this high elevation, and there is little or no vegetation to be seen.

July 11th.—We left Camp 6 and proceeded south-west four miles crossing Sturgeon Lake in a south easterly direction for three miles, when we entered a lake two chains east of Sturgeon River, which is a fair-sized stream, and flows southwest into Sturgeon Lake. The timber on the south-east side of Sturgeon Lake is similar to that on the south-western side, small poplar, birch, jackpine and spruce, from four to six inches thick. The shore line is fringed with scrubby cedar. The soil is sandy and rocky. Entering this small lake and canoeing around its inlet, coming in from the south-east end of the lake, we canoed up the stream for about fifteen chains. The water is sluggish and dark, about twenty-five feet wide and from three to four feet deep, and then becomes narrower with a swift current. The lake is about a mile long by half a mile wide. The timber surrounding this lake is all burnt by recent fires, but a small undergrowth of poplar is now springing up. The soil is sandy and rocky, and worthless for cultivation. We proceeded up the south arm of this lake one and one-half miles by canoe, and journeyed inland. The first half mile brought us to an elevation of one hundred feet above the lake level. We crossed a small marsh covered with wild grass; three miles farther on we came to a lake about two miles long by thirty chains wide. The timber along this region has been all destroyed by recent fires, with the exception of occasional low swamps. The soil being quicksand and boulders is no good for agricultural purposes. Caribou and moose are plentiful in this neighborhood. We camped on this lake over night, July 11th, and called it No 3 Lake.

July 12th.—Continuing our journey of the 11th, No. 4 lake was reached five miles further south. It is one mile long by fifteen chains wide. No change in timber or soil was observed with the exception of occasional white pine trees. The land surface is uneven and broken up but not high. The timber is close and small and will average five inches thick. Lake No. 5 we reached a mile and a quarter further on in the same direction. It is eight chains long and five chains wide. There is no change in timber or soil. The land rises and falls with uneven surface, and no exposures of rock. The next lake met was No. 6—six and a half miles inland, and is thirty chains long by twenty chains wide. These last two lakes has an old Indian portage between them six or seven chains long. The land is still rolling and the timber and soil remain as last described. Moose, caribou and bear appear to be plentiful. The waters of this lake appear to run north. We arrived at Lake No. 7—seven and one quarter miles inland, twenty chains long by eight chains wide, running in an easterly direction. The waters of this lake run south to Lake No. 8, which is eight miles inland and is two and one-half miles long by two and one-half miles wide. The timber surrounding this lake is jackpine, spruce, poplar and birch, and averages from four to six inches thick. The soil is white sand and rocks covered with moss. Vegetation is scarce. The waters of this lake flow in a south-easterly direction, possibly into Mattawan or English river. The water of all these lakes is clear, and apparently full of fish of different kinds. Moose and caribou appear to be plentiful in these regions, but other game, such as rabbits, partridge and other birds are scarce. We travelled two miles further back, which is ten miles inland, over a level, flat country, very thickly wooded with balsam, spruce, birch and poplar, but the timber is of no commercial value. At the end of ten miles the land fell into a low cedar swamp, mixed with tamarac and spruce, not in a healthy condition. The soil is still rocky and sandy and no good for cultivation. There is very little vegetation and no wild fruits of any description. We returned to camp 6 at 10 p.m.

July 13th.—We left Camp 6, and crossed the bay two miles west from camp by canoe, and proceeded by foot four miles inland on a west course. The first one-half mile from the lake is covered with jackpine, poplar and birch in a fairly healthy condition which will average ten inches thick, but timber of this size only occurs at intervals in small bunches. The next one-half mile the land falls to a small spruce swamp, and then rises to low, rocky, mossy ridges of jackpine. There are a few exposures of rock. The soil is sand and boulders, unfit for cultivation. At half-a-mile inland a stream was crossed flowing south, apparently draining from the swamps. There were no other changes to note on this journey.

July 14th.—Leaving camp 6, portage No. 7, we proceeded east by canoe four miles exploring the shore line, and landing at different points. We found (four miles east of camp) an old Hudson's Bay Post. The timber along this shore and islands is small spruce, poplar and birch, and will average five inches thick. The soil is sandy and rocky covered with moss. Continuing the journey three miles further east, and turning west down a long bay two or three miles, we landed at the head of the bay, and proceeded inland three miles west. The timber along the first mile, is birch with a few poplar and jack-pine of fairly good quality, and in a fairly healthy condition, and will average about ten inches thick, but would be of no other use than cordwood. Along the next mile and one-half, the timber changes more to jackpine and spruce with an average of ten inches thick, but it is of poor quality, short and scrubby. The next one-half mile we travelled over was a bare, rocky ridge covered with white moss, and blueberry patches. From this ridge I could view the country back for six or eight miles. The soil was still unchanged and valueless for agriculture. There was no vegetation, and no signs of game. From this point we returned to camp.

July 16th.—We broke up Camp 6 at 6 a.m., and proceeded six miles further up Sturgeon Lake to a point about sixteen or seventeen miles from Sturgeon Lake portage and made Camp 7, on an island. We started south-west up Six-Mile Bay. The channel of this bay is winding and shallow. Reaching the end of this bay we entered into a small river, on which we continued one-half mile by canoe, and entered into a small lake, No. 9. We found two mining claims here located, on which some excavation or stripping had been done on the vein which shows up at the water line. Continuing still further and making a portage to the river, we found a marshy lake, which led us to a larger lake (No. 10) about three miles long and averaging thirty chains wide. The timber tributary to this lake is of a poor quality, being small spruce, birch, poplar and jackpine, averaging from four to five inches in diameter. The soil is also of poor quality, composed of sand and boulders, and is useless for farming purposes. The land is flat with occasional rocky ridges. This bay is full of small islands, which are all covered with small timber, birch, poplar and jackpine. Vegetation and game is scarce.

July 17th.—We left camp and proceeded on the same course as that on the 16th. Continuing up six miles we entered into a small lake (No. 11) situated on the south-west end of No 10 lake. We continued our journey by land in a westerly direction and found a lake two miles long by thirty chains wide. The water in this lake is darker than that in Sturgeon Lake. The land rises gradually for the first mile to fifty feet, then drops suddenly forty or fifty feet. There are a few exposures of rocks. The second mile is low, swampy land, with some cedar, spruce and tamarac trees of poor quality and scattered, which will average eight inches thick. The soil is sand and boulders, and is no good for cultivation. Vegetation is scarce and there are few or no signs of game. We did not return to camp this night but camped on the shores of No. 11 lake.

July 18th.—Leaving Lake No. 11 we proceeded two miles farther west, until we came to Lake of Bays, No. 12. This lake is west, ten degrees north, of Sturgeon Lake, about ten miles long, and will average from one-half to one mile wide. Part of the lake runs north and south and part east and west. The country on the south side and north end appears to be mountainous ranges of rocks, from one hundred and fifty to two hundred feet high, covered with small jackpine, poplar and birch. The north-west side appears to consist of low land. The land between No. 11 and 12 lakes for two miles rises and falls, with rolling rocky ridges and low mossy land, with no appearance of vegetation The soil is sand, gravel and boulders, with a light growth of spruce, jackpine and occasional tamarac trees of no commercial value. The Lake of Bays is a beautiful sheet of clear water with numerous sandy beaches. There are a few exposures of rock. We did not find any minerals. We returned from Lake of Bays to camp. The land along the shores of Six-Mile Bay is very uneven, with some jackpine and poplar that averages from six to seven inches.

July 19th.—The weather being unfavorable for work in the woods, as it was raining very heavily, we explored along the shore line north-east of the camp. We entered a river one mile north-east of camp, which is situated at the mouth of Six-mile bay (sixteen miles north-east of Sturgeon Lake portage). We canoed up this river one mile, where the water became rapid. The river will average one chain wide up to this point. We snagged our canoe, and had to return for repairs, after which we continued our course along the coast line for five miles. We then made a journey two miles inland. The timber along these two miles will average twelve inches thick, and is composed of spruce, tamarac, birch and poplar, but this size of timber only occurs in small bunches according to the state of the land. About thirty-one and one-quarter miles from Sturgeon Lake there is a small lake thirty chains long and ten chains wide. The water is clear and the stream runs south-east to Sturgeon Lake. We returned to camp at 7 p. m.

July 20th.—We left camp at the mouth of Six-mile Bay and crossed the lake in a southerly direction to prospect the shore line. The timber on the south shore at this point is small spruce and tamarac, with a little birch and poplar mixed. It will average from four to six inches. The shore line is fringed with small, scrubby cedar. There are a few exposures of rocks. The soil is white sand with somewhat fewer boulders than in other parts, but entirely unfit for settlement. The land is low and flat along the shores. We crossed Niven's meridian line, which runs north and south across the lake, two miles east of our present camping place. There is no change in the timber or land at these points. The woods in this section are very dense and hard to travel through, and consist mostly of small balsam and spruce, the soil remaining as previously noted. North-east of the meridian line two miles, the land rises for half a mile from the shore to a height of about fifty feet. There is no change in timber or soil characteristics. The surface of the land is covered with a thick coating of moss. Continuing still further north-east and making short trips inland from one half to one mile, we found similar conditions. There is no vegetation, and very little signs of game.

July 21st.—We left Camp 7 by canoe on a north-east course for two miles, and entered the river. We canoed up one mile to the rapids and portaged across six chains north-west. We entered the river again and went up one half a mile to the next portage, which is about seven chains long. We crossed this portage and proceeded one mile, when we reached another portage twelve chains long. We then entered a large lake, No. 14, eight or ten miles long, and averaging two miles wide, with numerous low, flat islands, some nearly one mile long. The lake appeared to run north-east and south-west. Coasting along the shore on the south-west side for six or seven miles and continuing up an arm or bay for fifty chains to an inlet, we found a portage fifteen chains in length which we crossed, and came upon another small chain of lakes, apparently an Indian canoe route to some larger lakes, but not running in our direction. On returning and coasting along the south-east coast of this lake to the portage leading to Sturgeon Lake, we found that the elevation between Sturgeon Lake and No. 14 lake is about fifteen feet, with a gradual fall. There is no change in soil, timber or rocks on this route.

July 23rd.—We broke up Camp 7 and proceeded north-east ten miles, and put up Camp No. 8. After lunch we continued south-west two miles by canoe, and then journeyed inland four miles west. Three chains from the shore line the land rises to

Palisade River. Party No. 7.

Rapids and Falls. Portage 34. Party No. 7.

Dinner time on Portage No. 38. Party No. 7.

Natives, Burnt Rock Lake. Party No. 7.

Green Bush Lake. Party No. 7.

Pushkokogan Lake at Outlet. Party No. 7.

1st rapids below Pushkokogan Lake. Party No. 7.

Fish trap at 2nd rapids north of Pushkokogan Lake. Party No. 7.

a height of about one hundred and fifty feet, with bare, rocky cliffs. Five chains further the land declines to a swamp, and then rises and falls in ridges, on which are jackpine, spruce and poplar, averaging from four to six inches thick, with mossy surface. The soil is sand and boulders, and not of an arable nature. There was no vegetation visible.

July 24th.—Leaving Camp 8 by canoe, we proceeded seven miles south-west to the south shore of the lake and entered a river, up which we canoed for half a mile. This river is from two to three chains wide. The waters are flowing east into the lake, with a fall at this point of about nine feet. We crossed a portage about half a mile long, in the course of which the land gradually rises to an elevation of about thirty feet, showing few exposures of rock. The timber is small, dense spruce, from two to four inches thick. Along the river there are a few black ash, birch and poplar, which will average from four to five inches in diameter. The soil is still sand and boulders, except along the margins of these rivers, which would be more fertile. We paddled back to the mouth of the river and proceeded one mile south-west, when we entered another river, and continued up the same for one mile. There was no portage from this river. It had a gradual fall of ten feet. We went inland five miles from this point, and found the land more rolling and travelling more easy. The last mile was low, swampy, mossy country with dead, falling timber. There was no change in soil and no vegetation.

July 25th.—From Camp 8 we proceeded by canoe six miles down King's Bay or Shores' Mine, and about a mile further up we found a small river running a westerly course of about twenty chains. We portaged over six chains to a small lake, about five chains wide by twenty chains long, and then journeyed by land in a westerly direction. The land rises for the first half mile to about fifty feet, then suddenly falls into a spruce swamp, which extended for six miles. The timber is fairly lofty, but would only average from four to five inches thick, and is rather scattered, and of no commercial utility. The soil is not adapted for agricultural purposes. It is covered with from two to three feet of moss, beneath which is a bed of boulders and quicksand. There is no vegetation. Frost is found in many places along these swamps, and the water is icy cold, but on an average very good drinking water. There were no signs of game, and birds of any kind were scarce. On visiting Shores' Mine we were shown some very rich samples of gold ore, but the prospects of the mine depend on much further development. They have at present a shaft sunk sixty feet deep, where they are doing some cross cutting and drifting. They have another shaft down twenty or thirty feet deep, apparently on an extension of the same vein, which is also under development. Shores' Mine is situated thirty or thirty-five miles north-east of Sturgeon Portage Lake, in a bay six miles deep, known as King's Bay.

July 26th.—We moved Camp 8 and proceeded ten miles north-east up through the narrows, six or eight miles north-east of King's Bay, or Shores' Mine, or one mile south-west of present Hudson's Bay Post. We made Camp 9 on a small island, half a mile from the main shore. After lunch we travelled by canoe three miles to the south-east shore, landed, and journeyed inland three miles. The country is burnt all along this region from two to four miles back from the lake, and along the coast line for twenty or twenty-five miles, with the exception of an occasional green spruce swamp. The fire has apparently run over the same ground twice, as the land is burned clean and free of falling timber. A four or five years' growth of young birch and poplar is springing up. The land is still rocky and sandy and rises from ten to one hundred feet, and falls back again to low, swampy ground. The soil is worthless for cultivation.

July 27th.—Departing from No. 9 Camp we went east by canoe two miles, landed and crossed a portage thirty chains long to a small lake, No. 21. This lake is two miles long by one mile wide. The water is clear and apparently deep, and has an elevation of twenty feet above Sturgeon Lake. The country around this lake is burnt, but a small growth of birch and poplar is replacing the former timber. The topographical features of the surface outline are uneven, as it falls and rises at intervals to one hundred and fifty and two hundred feet high. Continuing the trip further by land five or six miles we camped over night. This section is still burnt country with land low and flat, ranging in rolling, rocky ridges. The burnt country is still covered with a growth of young poplar and birch with their leaves turning to autumn tints. The soil is sand, boulders and gravel without vegetation and there is little or no sign of game.

July 28th.—Extending yesterday's journey four miles further inland from where we camped over night, our course still being easterly, we travelled over a barren, burnt country, with no timber. The soil is but a bed of boulders and rocks. The land is low and flat with occasional ridges of rocks. We carried our climbers which we used as often as we found timber large enough for that purpose. Near the end of this ten mile journey the land fell into a low spruce swamp, which the fire had not reached. The timber is small and scrubby, being only from three to four inches thick. There is no vegetation except Indian or Labrador tea. Travelling through these regions is very difficult work owing to its numerous obstructions, such as fallen timber and deep moss. One almost wants snow shoes to prevent sinking in the spongy soil. There is water in almost all the low swamps, and frost retained by the moss keeps the waters icy cold. This journey was taken from a point eight miles east of King's Bay or Shores' Mine. We returned back to camp No. 9, which is situated on a small island about one mile south of Hudson's Bay Company's Post or McLaren's trading post.

July 31st.—We were detained one day in camp by a heavy storm. We broke up camp 9, continued east eight miles to a point on mainland seven miles east of Hudson's Bay Company's or McLaren's Trading Post, and put up camp 10. Travelling east three miles by canoe we entered a stream, which empties from a chain of lakes known as the Nepigon Canoe Route from Sturgeon Lake. We continued over these small lakes, making portages, and entered larger lakes from four to six miles long and averaging from one-half to one mile wide. The country along these lakes is more or less burnt. The only timber the fire has left is principally in the low, wet swamps, and is composed of spruce, tamarac and balsam, averaging from four to five inches thick. The land is low and swampy, with rocky ridges, and there is no vegetation and game is scarce

August 1st.—We left camp 10 and extended the journey of July 31st over the Nepigon Route to locate the divide (which is the base of our exploring division) which we located about twenty miles east and ten miles north of Sturgeon Lake. We crossed the portage of about forty or fifty chains over the divide, to another chain of lakes running east. These waters also flowed east from the divide. We continued our journey south after crossing back from the divide over these chains of lakes, until we covered thirty-six miles from camp 10. There are numerous portages along these chains of lakes. When we reached thirty-six miles from camp 10 we found the stream too small to navigate any further. We were out three days and two nights on this trip. The country is all burnt along this route as far as we could see from the highest elevations, from which we could view the country for fifteen or twenty miles. The fire is of a recent date, as the timber is yet standing. The land is more rolling with high ranges of rocky hills of gneiss or granite from fifty to two hundred feet high. There is no soil in these regions, as the land is entirely composed of rock, gravel and boulders. The water in the lakes is fairly clear, and apparently well supplied with fish of different kinds. These lakes are not often frequented by Indians, as there are very few signs of any camping grounds to be seen. These waters are beautiful stretches of a continuous chain of lakes with numerous islands and long bays. It is rather difficult to locate the channels where the lakes close up to narrow passages. The shore lines are bold and rocky, with numerous reefs, which often surprise the navigator, as there is nothing to indicate their presence. This region is a vast, barren, fruitless waste. There is no vegetation and the only signs of game we saw were a few traces of bear. There are no minerals. These waters all enter the north-eastern arm of Sturgeon Lake.

August 2nd.—On our return from the Nepigon route trip back to Camp 10, we had no other experience than morning fogs, and the soft side of a boulder for a bed. We camped this night near the divide or height of land. For the first five or six miles on this route from Sturgeon Lake there is more or less green timber of a poor quality, jack-pine, spruce, birch and poplar mixed, which averages from four to six inches thick. The elevation above Sturgeon Lake to the end of this journey would be about twenty-five or thirty feet. There are also to be seen at different points along these lakes traces of ice or glacier marks. There are numerous changes in the rocks on the coast of these lakes. We found no minerals.

August 3rd.— Resuming on our journey back to Camp 10 we found the wind against us with a very choppy sea. We had yet twenty miles to travel to camp. These waters are very difficult to travel over without a guide, as the most experienced men may get lost. I pulled through the whole trip without once making a mistake. We had to find the channels from the lay of the land at a distance. We arrived back safe to camp.

August 4th.—This being moving day we were all called up at 5 a.m. and proceeded on our way shortly afterwards. We pitched Camp 11 nine miles north-east of Camp 10, or four miles east of King's Point, on the south-east shore of the lake. We had lunch and proceeded to Shores' Mine. Mr. Davidson and myself were courteously shown around the mine. They have two shafts under development which show up fairly well. No. 1 shaft is down sixty feet and No. 2, twenty or thirty feet. The strike of the vein appears to be north-east and south-west, with a quartzite dike running parallel with the vein at a distance of fifty feet from the vein. As the weather was threatening, and a storm was evident, we pushed back for camp and joined the surveyor's party, who were continuing the survey of the lake. We all started for camp but the storm caught us and we had all we could do to get back safe.

August 6th.—From Camp 11 which is situated on an island on the south shore of Sturgeon Lake, or four miles southeast of Shores' Mine or King's Bay, we proceeded south-west to King's Point. We explored along the shore line of this point until we reached King's second gold claim, which is about two miles north-east of Shores' Mine, and is known as King's Point. Its vein is from two to four feet wide with a strike north-east by south-west. Continuing half a mile further along this point we reached another vein, belonging to the same firm, about three feet wide with a strike south-east by north-west. The timber along these shores is small and scrubby. It consists of jack-pine, spruce, poplar and birch, and averages from three to five inches thick. The land is rolling, rocky ridges and of no use for farming purposes. There are numerous mining claims taken up in this neighborhood, but no development has been done. We traversed the coast line for ten miles north-east from this point. There are numerous islands along the coast, but no change in timber, soil or rocks. The shore lines are bold and rocky, and easily explored. There is very little indication of iron in this region, although we met some attraction at certain points. There are very few Indians on this lake, although we saw numerous old lodges along the shore.

August 7th.—We left camp 11 on a journey towards Savant Lake and travelled ten miles down the coast when the weather got too rough to venture any further. We camped with an explorer named John Craig on the lake near his claim, of which he showed us samples of very rich gold ore. The shore lines in this part are very rocky and full of reefs dangerous to navigation. The land rises and falls to rocky ridges along the shore. The timber is all mixed and from four to six inches thick. The land is sandy and rocky covered with moss. There are numerous islands along the course which are all covered with small mixed timber. No other vegetation was noticed.

August 8th.—We left John Craig's camp, where we had stayed all night, and continued two miles to the end of the lake where we were surprised to find a four mile portage. We had spent the best part of the forenoon looking for this portage but could not find it. We continued our journey along the coast four miles further when we entered a deep long bay, which we followed three miles north and three miles north-west. We then entered a river up which we travelled two miles until we reached a portage running west one mile. We camped here overnight. The timber along the north east part of the lake does not show any change from that of other parts of the lake. The shore line is rocky and full of reefs and sand bars. The land is rolling, rocky ridges and destitute of arable soil or vegetation.

August 9th—Continuing our journey north east by east for eight miles to our base line or boundary, which is the divide or height of land, and crossing our boundary we reached a large river which runs north-east. This river will average twenty feet wide and is rapid in places. The timber along our course, which is about eight miles inland, is small jackpine and spruce, varying in size from three to five inches thick. The land is rolling, rocky ridges or granite rocks without soil or vegetation. There are some signs of moose and caribou and we found some moose horns. Other small game is scarce.

August 10th.—Breaking up camp 11 we proceeded to cross the lake to the south-west shore and the mouth of Sturgeon river, where we put up camp 12, our operations being somewhat retarded by rough weather. The land is rolling and rocky along the shore of the bay from which the river flows. The bay is about three miles long to the mouth of the river or first portage. The river has two branches about a mile apart, which unite at the second portage, about two miles from the first portage. The timber along the bay up to the first portage is small jackpine, spruce, birch and tamarac. The rocks are covered with moss and there is neither soil nor vegetation.

August 11th.—We left camp 12 at Sturgeon River, down which we proceeded by canoe for seven miles, crossing three portages. The first of these at the mouth of the river is about seven chains long and has a gradual fall of about twenty feet. The second portage is only about three chains long, and about one and one-half miles from the first portage. The third portage is about twenty chains long and about one and one-half miles from the second and has a gradual fall of fifteen feet. The fourth portage is three miles from the third on Sturgeon River, with a gradual fall of about twelve feet. The timber along the river from the third to the fourth portage is small jackpine, spruce, tamarac and poplar and averages from four to five inches thick. The land is boulders and rock covered with moss with complete absence of soil. The land is undulating rocky ridges and low swamps. We made a journey of six miles inland south from the fourth portage. The country for three miles inland is burnt and has neither timber nor soil, being rocky ridges and low mossy swamps. One mile south from the fourth portage inland we found a peat bed of about two hundred acres in extent; the first genuine peat bed seen on this expedition. On ascending a high elevation we could view the country back for eight or ten miles and it appeared to be burnt and barren. There is no vegetation except patches of blueberries which occur occasionally.

August 14th.—We broke up Camp 12 and portaged our stores down the Sturgeon River establishing Camp 13 at the fourth portage.

August 15th.—We broke up Camp 13 and proceeded several miles inland on a journey due north. The land runs in low ridges of granite rocks, and falls to low spruce swamps The ridges are covered with a small growth of jackpine, poplar and birch and the timber averages from four to five inches thick. There is no soil or vegetation except Labrador tea and no game or signs of beaver or other furbearing animals. We returned to Camp 14, on No. 6 portage.

August 16th.—We abandoned Camp 14 on Six Portage and proceeded two miles to Portage No. 7, which is thirty-two chains long with a gradual fall of twenty-five or thirty feet. We portaged across and loaded up for Portage No. 8, which is about two miles from Portage No. 7; crossed Portage No. 8 which is about four chains long and proceeded to Portage No. 9 where we put up Camp No. 15, for the night. There is no change in timber or soil along this river so far. The shore lines are rocky and full of bays and reefs.

August 17th,—On breaking up Camp 15 on Portage 9, we proceeded one mile to Portage No. 10, which is ten chains long with a fall of ten feet, thence one mile to Portage No. 11, which is seven chains long with a fall of twelve feet. We crossed Portage No. 7 from which we continued our course half a mile to Portage No 12 which is eleven chains long with a fall of ten feet. Two miles further we reached Portage No. 13.

We separated from the party at this point and proceeded three miles up an arm of the expansion of the river running north, and entered a large stream flowing from the north into Sturgeon River. We made a portage of three chains to an expansion of the river, (twenty chains wide and one and one-half miles long) to the second rapids where we camped overnight. There was no change in the timber, soil or rocks during this day's journey. The timber along the river is small spruce and tamarac from three to four inches thick. The land is low and swampy and we saw no vegetation or marks of game. There are no exposures of rocks at this point.

August 18th.—We left our camping-ground at the second rapids on North River and travelled up the river for two miles. We could not find a portage, so continued north by land for about two hundred feet high, where we could view the country for fifteen miles north and west The land travelled over rises in places to one hundred feet high of bare rock, and then falls into flat land covered with small spruce and jackpine from three to four inches thick. The soil is sand and boulders covered with thick moss. The view we obtained from this high range of mountains enabled us to report on the country for twenty miles from Sturgeon River. There is apparently no change in the country viewed from this point, except that a few miles distant there is a vast, burnt, barren waste, with rolling ranges of rocks. There is no vegetation or soil.

On returning to our camp we found an old portage or Indian trail leading up the river. When we reached our canoe we proceeded over this portage sixty chains long. It led us into another expansion of the river, up which we proceeded for seven miles, making one more portage. This river is difficult for canoeing, as the portages are from sixty chains to over a mile long and very rough. We explored along the river for twenty miles, and found it very rapid and rocky in places. The timber and land along this part of the country is valueless. The land is rough and rocky. We returned to where we had left our blankets, and proceeded to overhaul the rest of the party, who had a full day's start of us, and we feared we would have some difficulty in finding their camp, and as we had no food, and a storm was threatening, we were in a very unpleasant position. We arrived at Portage No. 13, which we crossed, continuing over the great expansion of Sturgeon River in an easterly direction. Not knowing where the channel or other party was, we concluded to land and camp for the night.

There was no change in the timber or land along the river and its expansions. The timber is all mixed and small, from four to five inches thick. The land is rolling and rock ridges, without soil or vegetation, and no game excepting a few rabbits.

August 20th.—Having rejoined the other party we left Camp 16, which was situated sixteen miles west of portage No. 15 on Sturgeon River, and proceeded north three miles by canoe, down an arm or bay of the river, and thence eight miles inland. We found the country low and flat, but not swampy. The soil is light and sandy, with occasional patches of clay, but too closely resembling hard pan to be of any use for cultivation. The timber consists of spruce, jackpine, balsam and poplar of poor quality, averaging about six inches thick. There are no exposures of rocks or boulders. Water was scarce, although we passed a small lake in the first mile.

August 21.—We broke up Camp 16, and continued eighteen miles down the river to a point near Abram Lake, or five miles east of Niven's meridian line. The country along the river is more or less burnt on both sides for twenty miles, or as far as we could see with the aid of the field glass. We obtained a good view of the country from the barren hills. There is very little soil in these regions. Eight miles west of Portage No. 15, we saw a Hudson's Bay Post, which is only kept in winter. Near this post we saw where there has been a little mining done in the shape of a small drift run under a cliff close to the shore. This region does not promise any great future for either timber, soil or mineral productions. Game and fur-bearing animals are scarce, as the country is all burnt, which leaves them without either shelter or food.

August 22nd.—From Camp No. 18 we proceeded down the river five miles by canoe to explore and locate Niven's meridian line. The land along from Camp 18 to Niven's line is rocky, with an uneven surface The timber that escaped the fire is of a poor quality, and has no commercial value. The soil is poor and incapable of cultivation. We also made a short journey of one mile over a rise of land from a point two miles east of Niven's line, which appeared to be an old portage or trail. This led to a small lake, one mile long by twenty chains wide, which was most likely some Indian canoe route, running south-east from Sturgeon River. The land surrounding this lake is low and swampy, with high ranges of bare rocks a few miles to the south. We noted no change in timber or soil. Vegetation is scarce.

August 23rd.—From Camp 18 we proceeded eighteen miles down the river or part of Minnietakie Lake and Abram Lake to Pelican Falls, and put up Camp 19 on the west end of the portage, which was twenty-eight chains across, with a gradual fall of twenty feet. The falls are divided into two channels entering a basin a few hundred feet below and forming part of Sturgeon River.

This river is practically a chain of lakes from twenty chains to three miles wide, with numerous islands and deep bays.

The timber along the route from Camp 18 to Camp 19 presents no new characteristics. The shore line for a few chains back is lined with poplar and birch with jackpine and occasional patches of red and white pine. Back a few chains from the shore the land is rocky and rolling with occasional patches of sand and gravel soil. The shore line is bold and rocky. On making trips inland we did not find any change in timber or soil and found no minerals.

August 24th.—From Camp 19 we proceeded down the great expansion of Sturgeon River about twenty miles and put up Camp 20 on an island in a bay or northeast arm of the river, to further explore the country. The timber along this route is not of a merchantable quality and is composed of small jackpine, spruce, poplar and birch, with occasional patches of white pine of a poor grade. The shore line is covered with a small growth of birch and poplar. The soil varies from whitish clay to sand and hard pan and gravel, and runs in small patches between rocky heights.

The surface outline is uneven and rolling and covered with moss. We landed occasionally and ascended high elevations, which enabled us with the aid of the field glass to view the country for twenty miles north and south of the river.

The country does not appear to improve from these points. We secured a view from a point twelve miles north-east of Pelican Falls or Portage No. 16. With the aid of a field glass we observed the interior for fifteen miles. The land, having been burned over, appears to be a barren, desolate waste. There is no vegetation, except a few patches of blueberries that are occasionally found on the burnt land amongst the rocks and boulders. We did not see any signs of game of any kind. There are but few Indians, as their hunting grounds have been destroyed by fire.

August 26th.—We were weather-hound in the arm of the river running north-west, twenty miles from Pelican Falls. The timber around this bay consists of jackpine, spruce, balsam, poplar and birch, averaging from four to six inches thick. The soil is of a whitish clay and hard pan, not adapted for farming. The high water mark of this bay is ten feet above its present level. The glacier or ice marks are common along the shores of Sturgeon River. There are no minerals, vegetation or game.

August 27th.—From Camp 20 we proceeded to explore the east arm of the river. In returning we went five miles out of this bay in a south-easterly direction for four miles. While returning out of this second bay the wind increased to a gale, which compelled us to land. We were wind-bound here for the rest of the day and put up Camp 21. There is no change in soil or timber on these bays.

August 28th.—From Camp 21 we continued our journey to Lake Seul by a channel, which was very difficult to follow, owing to its winding around in different directions and passing over some small rapids. The wind increased to a terrific gale, which compelled us to land. While storm-bound we put up Camp No. 22. The timber, soil and rocks are still unchanged.

August 29th.—From Camp 22 we continued down the expansion of the river twelve miles to Lake Seul, and put up Camp No. 23, on the island at the mouth of the Sturgeon River. The timber and soil presented no new features, except occasional patches of white clay and hardpan. We paddled three miles down a bay in a north-easterly direction, about twelve miles from Camp 23 on Lake Seul, and found similar conditions.

August 30th.—From Camp 23 we proceeded two miles west by canoe, and entered a small stream, whence we travelled inland a westerly direction. At the end of five miles, with the aid of climbers, we were able to view the country for three miles further. The land rises and falls from rocky ridges to low spruce swamps. The timber on the ridges is mostly jackpine, birch and poplar from three to four inches thick. The swamps are heavily covered with moss underlaid with boulders and quicksand. The stream where we landed is flowing from the south-west running out of a lake (shown on the map) crossing Indian Reserve 28.

The timber along the shore is small and scrubby. We counted the rings or years' growth on some of the trees, and found that although from four to nine inches thick they are from seventy to 80 years old. The timber is partly burnt along the shore and islands near the entrance to Lake Seul. The country does not appear to improve as we go north. There is no vegetation, and game is scarce.

August 31st.—We proceeded from camp No. 23, six miles east to the entrance of a small stream up which we paddled for three miles south. This stream is from four to ten feet wide, with only water enough to float our canoe. The water is sluggish and the land low and flat. The river is fringed with grass and willows with clay banks, but this soil only extends from three to four chains back, where it gives place to sand and boulders, and rocky ridges. We extended this journey four miles inland, to a point where we used the climbers, which afforded us a view of the country for three miles further. The land rises for a short distance from the stream, then runs in rocky ridges and falls into large, wet, spruce swamps, covered with thick moss. We also crossed a bed of peat twenty acres in extent but not of good quality. The rock is still unchanged. The timber is small and stunted. Moose and caribou are plentiful along the river.

September 1st.—On leaving Camp 23 we proceeded fifteen miles east along the south shore of Lake Seul to explore the coast line, bays, and rivers, making short journeys inland at intervals. There is no change in timber, soil or rocks.

The land is rolling ridges of rock covered with moss. The soil is still sand and boulders, unfit for cultivation. The shores of Lake Seul are full of reefs, deep bays, long narrow points and numerous islands, which bar the view of the lake. Some of the islands are barren. We did not visit the Hudson's Bay post, but have learned from Mr. Robertson that they only raise a few potatoes, as other crops would not be a success, as the soil is too sandy. They keep cows and cut wild hay for them. Game continues scarce. There are no ducks as there is no wild rice or anything else for them to feed on.

September 3rd.—We were wind and storm bound at camp 23.

September 4th.—From camp 23, after storing part of our supplies, we proceeded down the eastern arm of Lake Seul about fifteen miles (partly over the same journey as that on September 1st) to explore farther down the bay, and also to find the channel to Root River, which is very difficult to follow. We reached the end of this bay or arm, and followed the north shore of it back to a point two miles north of camp No. 23 and put up camp 24. There are numerous small streams entering into these bays, where grass and other small vegetation is found. We made short journeys inland at intervals which gave us a general idea of the soil, timber and country at large. We often obtained magnificent views of the country for many miles back from high elevations. Nowhere during this day's journey of thirty miles has the timber or land shown any change. The bay is dotted with islands, varying in size from five to thirty feet high, composed of rock covered with moss, and a small growth of jackpine, birch and poplar. The shores are bold, with reefs and shoals. Vegetation is scarce. There are a few old Indian lodges in this neighborhood.

September 5th.—From camp 24 we journeyed down another arm of Lake Seul west to explore and locate the channel leading to Root River. We reached the end of this arm, which is fifteen miles long by sixty chains wide and coasted along the south shore of the bay back to a point not far from camp 24, where we put up camp No. 25. There is no change in the timber or the land, except an occasional low spruce swamp where frost and ice are often found. There is nothing of any value in this part.

September 6th.—From camp 25 we proceeded down another arm of Lake Seul in a zigzag direction, first north about five miles, then south five miles, and south-east five miles. We reached the end of the bay and paddled back eight miles to a point fifteen miles north-east of camp No. 25, or mouth of Sturgeon River. This bay will average fifty chains wide and is dotted with islands. There are numerous small streams, and bays fringed with willows and rushes. The soil, which is not arable, changes from clay to hard pan and gravel. The timber is irregular, some trees being lofty, others short and scrubby, marking the changes of the soil. There are occasional red and white pine trees found but of poor quality.

September 7th.—From camp 26, on the north shore of the lake at a point two miles north-east of Patterson's trading post on the north side, or twelve miles from camp 23 at the mouth of Sturgeon River, we continued down the fourth arm of Lake Seul, five miles to the end of the bay, and explored along the shores, landing at different points, and then returned five miles and put up camp 27 near camp 26 on the north shore of the lake. There is no change in timber or soil.

September 8th—From Camp 27 we returned two miles to Patterson's trading post and obtained some information as to the channel to Root River. We left Patterson's and continued down a channel, which led us into a bay that we had already travelled over. We entered into a large river flowing from the south-east, which we continued up four miles, when we came to an expansion one mile wide and one mile long. When we came to the first rapids we could find no portage, so turned back eight chains, and proceeded up a smaller stream for five miles to where it was obstructed by driftwood and fallen timber. This stream flows from the south-east. We then turned back to the expansion of the river and put up Camp No. 28. This river is from two to three chains wide.

The land along the river is low and flat. The soil from ten to fifteen chains back is of darkish clay and would be suitable for cultivation. Some distance from the river the land rises to rocky ridges, with sand and boulders covered with thick moss. As the land varies so does the timber in size and quality, averaging from four to eight inches thick. It consists of jackpine, birch, spruce and poplar. Game and vegetation are scarce, except for occasional patches of cranberries and mountain-ash berries. The weather is getting much cooler as we proceed north. There are a few old Indian lodges around these bays.

September 10th.—From Camp No. 28, we proceeded six miles down the river to the bay, and continued five miles north to the end of the bay, making a short turn into the channel of Root River. We continued for fifteen miles on the Root River route and put up Camp No. 29 about ten miles south-west of the entrance. The country is still burnt and barren without any signs of soil, vegetation or game. The land in places consists of bare rocky hills which fall again into wet swamps.

September 11th.—From Camp No. 29 we paddled twelve miles to the entrance of Root River, and seven miles up the river to the first portage, where we put up Camp 30. The river along this journey will average from three to four chains wide and is very winding. The waters are dark and muddy. The land along the river is low and swampy, and rises to rocky ridges a short distance back. The timber is spruce and tamarac of small, poor quality from three to five inches thick. We viewed the barren hills with the field glass, and saw that the country was burned back a distance of from fifteen to twenty miles. This makes the country not only dangerous but almost impossible to travel over, as the fallen trees are piled four or five tiers deep. We found no signs of soil or vegetation.

September 12th—From Camp 30 on Root River, sixteen miles from the mouth, we made a short trip of two miles inland south-east of the camp or First portage. The land rises slightly from the river over a sandy slope and then to rocky elevations, and falls to low spruce swamps, which give place to sand and hardpan with small patches of green poplar, followed by a stretch of burnt country for fifteen or twenty miles. We get good views of the country from burnt hills, which show no change as far as we could see. There is no vegetation, soil, timber or minerals. We returned and broke up Camp No 30, and proceeded six miles to another expansion of the river, where we got wind-bound, and put up Camp No. 31.

September 13th.—From Camp No. 31 we proceeded up the river for eighteen miles to portage No. 4 and put up Camp No. 32. We made three portages averaging six chains, twenty-one chains, and two chains long. The river gets narrow as we near Lake Joseph or the divide. The land is low and flat along the river.

The country is still burnt except a fringe along the bank, where it is low and wet. We observed no noteworthy alteration in timber, soil or vegetation.

September 14th.—We left camp No. 32 on portage No. 4 and proceeded east seven miles up a small river, to where it was blocked with fallen dead timber. We made a journey of three miles on foot and got a view of the country for ten and fifteen miles south and east We saw no change in timber, soil or rocks. The land along the river is low and swampy. We returned and broke up camp No. 32 and started up the river for seven miles making two portages, five and six, and camped on portage No. 7.

September 17th.—After being detained in camp some time by rough weather we started from camp No. 33, on portage No. 7, travelling two miles up Root River to portage No. 8, which is one chain long with a fall of six feet. We continued one mile further to portage No. 9, which is four chains long with a fall of five feet, and thence half a mile further to portage No. 10, which is forty chains long and crosses the divide in Lake Joseph waters. We proceeded two miles down a winding stream to the opening of Lake Joseph, and thence down an arm or channel of the lake about eleven miles, and put up camp No. 34. The country still consists of barren, burnt land, and rolling rocky hills, alternating with low wet swamps. Vegetation continues scarce with no signs of game.

September 18th.—From Camp No. 34, we continued down Lake Joseph thirty miles, [illegible] p Camp No 35 at a point 40 miles from portage No. 10, or entrance to Lake Jr ph. After leaving Camp No. 34, we encountered a magnetic current which rendered outcompasses useless. We could not account for this as there were no local attractions [illegible] borhood. The rocks were free from magnetic iron or other materials. This current extended for ten miles on our journey. The lake is full of islands and long bays, which render the channels difficult to find. The land is generally flat with occasional ridges of bare rock. The land is sterile, consisting of sand and boulders. The shore lines are fringed with a small growth of poplar and birch, which runs back to burnt, barren country.

September 19th.—From Camp No 35 we continued down the lake for twenty-eight miles, and put up Camp No. 36, on an island about five miles from the Osnaburgh House or Hudson Bay Post. The land in this region is low and flat with occasional ridges of rock. The shore line is rocky with many reefs and shoals. The timber is like that on other parts of the lake, and of little value, being spruce and tamarac, with small patches of poplar and birch from three to five inches thick. The soil is composed of gravel, sand and boulders, and bed rock covered with moss. Vegetation is sparse and there were no signs of game.

September 20th.—We broke up Camp No. 36 and put up Camp No. 37, on a sand point two miles from the Osnaburgh House, or Hudson's Bay Post. We missed the channel and went several miles out of our course, when we met some Indians whom we hired to guide us to the Hudson's Bay Post. The timber, soil and rock present no new features and the country is more or less burnt.

September 21st.—From Camp No. 37 we proceeded to the Hudson's Bay Post where we secured the services of an Indian to guide us down six miles to a post planted on an island in the first rapids on the Albany River. We located this post and took observations, carved the names of our party on the post which is eight inches square and photographed the post and party. This was the end of our division, according to our instructions which we very faithfully executed all through. This post is planted with a stone mound built around it six miles north-east of Hudson's Bay Post. The land in this region is rocky and rolling. The soil is sand and gravel, no good for farming. The timber is small and scrubby wilth less jackpine than in most other parts and averaging from three to five inches thick.

September 22nd.—The Hudson Bay people keep no animals except dogs which they use in winter as sleigh dogs. There are five buildings at this point, including a church built by the Church of England, and apparently not completed yet. These people do not raise any crops except a few potatoes, which give a very poor yield and are a poor quality, small, tough and hard to cook. The soil in which they are grown is sand with a little black soil mixed in. There is no mineral or timber that would be marketable.

Mr. Wilson, manager of the Hudson Bay Post, told us that fur is not so plentiful as in former years. This can be accounted for by the fact that the country has been burnt over for years.

The Hudson Bay Company carry a fairly heavy stock here, so we added a few more groceries to our list of supplies and started on our homeward journey, over the same route as we came out, except that part from Sturgeon River or Abram Lake to Dinorwic, on the Canadian Pacific Railway.

October 8th.—From Camp No. 46 on Pelican Portage, we continued twelve miles to the entrance of Minnietakie Lake, and then up Minnietakie Lake for fifteen miles, where we put up Camp No. 47. This lake is thirty miles long, running in a south-easterly direction, with large expansions, deep bays and numerous islands. The land runs in ridges of rock and sand. The country as far back as we could see is burnt, with the exception of a few green bunches of red and white pine, which show along the shore and islands, and average about ten inches thick, but are of a poor quality. There is no soil or timber except a second growth of poplar and birch where the fire has run over.

October 9th.—From Camp No. 47, situated on Minnietakie Lake, we journeyed for fifteen miles until Big Sandy Lake was reached. The soil, rock and timber has not changed since yesterday's report. A few prospectors' camps are found along the shore. The water is fairly clear and well supplied with fish. There is a small mine or prospect being worked about six miles from Big Sandy Lake Portage, which we learnt was promising to become ore-producing. We continued on to the Big Sandy Lake Portage, which passes over a ridge sixty feet high and thirty chains across We left Big Sandy Lake Portage and paddled our way seven miles through heavy seas until we reached the

Hudson Bay warehouse. Big Sandy Lake is a beautiful sheet of water without any islands, and is seven miles long by five miles wide.

The country is partly burnt and rocky. Small bunches of jackpine, birch and poplar with occasional trees of red and white pine were seen, from eight to twelve inches thick. The soil is more clay, but useless for agriculture. We stored our canoes in the Hudson's Bay warehouse and packed up all our goods to be shipped by Hudson's Bay wagon to Dinorwic. We put up Camp No. 48, where we remained over night.

October 10th.—We broke up Camp No. 48, and left for Dinorwic. We travelled nine miles carrying our dunnage over wet rough roads and arrived at the Canadian Pacific railway station at 10 a.m., taking the local train to Ignace and travelling thence east by the main line.

GENERAL SUMMARY.

There is very little variation in the general characteristics of the territory explored by our expedition. One-half of the country we traveled over is burnt, and the other half contains no marketable timber except one block of sixty acres of red pine on Pine Lake Portage, which would cut about three hundred feet of lumber. The spruce is generally found in the low swamps with sand and boulders, the bottom covered with a heavy coating of moss, and too poor to grow heavy timber. The timber is more mixed on the rolling lands, which would have no other value than as cordwood or fuel, and the spruce in general is too small for any market use. The soil is not available for any agricultural purposes.

I have the honor to be, Sir,
Your obedient servant,

DANIEL MCPHEE.

JAMES ROBERTSON, Esq., O.L.S.,
In charge ot Exploring Party No. 9.

Ontario.
District of Thunder Bay }
To wit. }

I, Daniel McPhee, of the Town of Port Arthur, in the District of Thunder Bay, Explorer, make oath and say that the above entries in this book are correct and the result of my exploration in Exploration Survey Party No. 9.

Sworn before me in the Town of Port Arthur, in the District of Thunder Bay, this 6th day of December, 1900. } DANIEL MCPHEE.

(Sd.) FRANK H. KEEFER,
A Comm. in H. C. J. &c.

GEOLOGIST'S REPORT OF EXPLORATION SURVEY PARTY NO. 9.

TORONTO, ONT., Dec. 28th, 1900.

Sir: In accordance with the instructions of the Director of the Bureau of Mines, I joined Exploration Survey Party No. 9 that left Toronto on June 19th under your direction. The party was completed to its full number of seven by the addition of Mr. MacPheo of Port Arthur, who was acting as timber estimator.

Having completed our preparations we left Osaquan Siding five miles west of Ignace on June 22nd, and started northward with our equipment in three Peterboro' canoes.

Our work, very fortunately, lay along an ideal canoe route which, starting at Camp Lake lay down Camp River into Indian Lake, thence through Bear, Otter, Hut, Pine, White, Rock and Young's Lakes to Sturgeon Lake, separated from the former by a height of land forming a three quarter of a mile portage. A careful micrometric survey was made of the last named lake. From Sturgeon Lake our course lay down the Sturgeon River through Abram and Pelican Lakes, and in to Lonely Lake, or Lac Seul, thence easterly, and some forty miles up its chief eastern inlet Root River to the great height of land separating Lake St. Joseph from the Lonely Lake drainage systems. The former flows up the Albany River to James Bay; the latter, by the English and Winnipeg Rivers, to Lake Winnipeg, and thence by the Nelson River to Hudson's Bay. We made a very rapid trip down Lake St. Joseph, and arrived at the furthermost point of our journey, the head of the Albany River, on September 21st. After taking an observation at the post planted there in 1885 by Mr. Fawcett, we began our return journey on the same date. Being detained by strong headwinds and heavy rains, we reached Dinorwic by way of Minnietakie and Big Sandy Lakes, on October 10th and arrived in Toronto on the 12th.

Along all our course we made many exploring trips into the bush, at times on foot, when possible by canoe. These trips were of such lengths as were considered necessary, taking into consideration the nature of the country and the time at our disposal, but generally the whole surrounding area was explored. During the summer's work nearly two thousand miles of canoeing besides several hundred miles of walking were accomplished.

I should like to acknowledge the kindness of prospectors and miners who assisted us in every way possible, and also of Mr. A. E. Shores, of Sturgeon Lake, who very kindly brought in our mail for us during the few weeks we spent in his neighborhood. I would also especially thank Dr. Coleman and Mr. J. W. Bain for the kind and hearty way in which they have assisted me in the preparation of this report, more particularly in the determination of some of the rock specimens.

General Outline of the Region.

The whole country covered by our work seemed to have, for the most part, a north-east and south-west bearing, both in its drainage system and in the general direction of its ridges. The lakes, both large and small, usually lay in valleys surrounded by rocky ridges that run parallel to the shores and often reach a height of 150 to 200 feet.

The country may be divided into two great areas, the Lonely Lake including that of Minnietakie and Sturgeon Lake, and the Lake St. Joseph. Both the Minnietakie and the Sturgeon Lake systems strike north-easterly, and find a common level in Abram Lake, the former by Abram Falls, the latter by the Sturgeon River, which flows westerly from the northern part of Sturgeon Lake. These waters find their way to Lac Seul, which is separated from Lake St. Joseph by the height of land of northern Ontario.

The rocks of the region vary greatly, the lower part being almost entirely Laurentian but changing to Keewatin near Sturgeon Lake, which lies in a narrow belt of these rocks. Sturgeon River, Lac Seul and Lake St. Joseph lie in the Laurentian, although the rocks on the lakes show a very great diversity of appearance, as to suggest different geological conditions. Minnietakie Lake lies in a similar belt of Keewatin rocks in a hollow formed by the decomposition of the soft schists of that series.

The Laurentian area was composed principally of gneiss, usually very micaceous, and on Lake St Joseph particularly so. The Keewatin shows the characteristic schists, along with diorites, porphyrites, and diabases.

The whole district covered was well glaciated, and showed many good illustrations of ice action, here by low morainal hills, and there by rounded stone-worn rocks carryin striæ, distinct and clear. The soil in some places is scanty, the rocks being covered with a heavy blanket of moss. Parts are covered with sand, sandy loam, and in places by clay, and carry a good growth of spruce, poplar, balsam, jackpine, along with tamarac, cedar and balm of Gilead.

General Geology.

The country in the neighborhood of Ignace is well covered with drift material. A mile to the west granite biotite gneiss outcrops, that at the C.P.R. quarry has a good cleavage, and can be obtained in any sized blocks. The stone for the culverts and the bridge foundations of this section is obtained from here. When visited several men were at work.

Our real start, was made from Osaquan, five miles west of Ignace. The portage eading to Camp Lake is good, three-quarters of a mile long, and is used as a wagon road thus forming a base for carrying supplies to Sturgeon Lake. The country between the railway and the lake, is level with few exposures of granitic gneiss Camp Lake itself shows no rock exposures and has low sandy shores, and Camp River, its outlet, is similar. On approaching Indian Lake a few low outcrops occur, and the lake itself shows rocky shores of gneiss, which, rising abruptly from the water, show, in places, quite a marked cleavage north 20 degrees west, also one north-easterly. The lake is almost cut in two by a long sand-bar stretching out towards the eastern shore, and forming two nearly circular bodies of water about four miles in diameter.

Some three miles further on is the turn to the portage, around a high rocky buff of gneiss, tending to be granitic, and containing granitic inclusions.

Turning down Indian River two rocky portages are encountered, the first is short but rocky ; a second portage is necessitated by a very pretty fall, preceded by a slight rapid. A small island divides it into two cascades, where the river, dashing into a hollow in the solid rock, is thrown to the lower level. The total fall is about twenty-five feet. The banks of the river past the rapids are low, but several hundred feet back, hills of gneiss rise up above the water to a considerable height. These rocks are slightly granitoid, but about four and a half miles down Bear Lake on the north-west shore, become distinctly more gneissoid, and of a reddish color. A mile further down they are much contorted, being thrown into sharp bends, sometimes through 180 degrees, and at others, into a sharp wave-like form.

Near the first portage into Otter Lake, a hard gneiss was noticed with a strike north 10 degrees east and dip 45 degrees west. This changes very rapidly into the more micaceous variety at the end of the portage. On the neck of the lake leading to the Mattawa River the same rock appears in low outcrops. The river itself, as far as seen, showed very few signs of formation, but where observed, was of the same characteristic gneiss ; banks low, valley half a mile wide and bounded on either side by low-lying hills.

Gneiss outcrops on the portage leading out of Bear Lake into Otter Lake, and along the south shore of the latter. From Otter Lake we passed up Grass River, where there are no rock outcrops shown by the banks, which are low and swampy.

Hut Lake—the portage from which shows biotite gneiss with chloritic inclusions—is a small expansion of the river, which shows schists, as does also the portage into Pine Lake where hornblendic schist has a strike south 70 degrees east. The country to the south east of this portage is gently undulating, but on account of the thick coating of soil, showed no outcrops as far as we progressed—some seven miles.

Near the portage into White Rock Lake green hornblendic schists outcrop. The lake lies in a basin between high hills, those to the north and west being rough and ragged in appearance, and of a greater eminence than those to the east The inlet from White Rock to Young's Lake is very narrow, and at one place is entirely through boulders, thus showing the drift covered nature of the country. A winter road some three quarters of a mile long crosses the point near the inlet from Young's Lake, and might be used as a portage.

Young's Lake, the last lake of this series, forming an ideal canoe route between Osaquan and Sturgeon Lake, is a pretty sheet of water, with a shape like a "V," and, like the previous lakes, is surrounded by a chain of green hills untouched by fire. Hornblende schist with strike north 60 degrees east is presented near the portage into Sturgeon Lake, where there is a contact between that schist and a much weathered basic eruptive.

To the south-east of Young's Lake the country shows a good coating of soil, so that no exposures were found. At one and a half miles a small stream flowing northerly to Sturgeon Lake was encountered. A small lake at eight miles shows no exposure along its shores, which are covered with boulders of granite and gneiss.

This system of lakes forms an excellent canoe route, portages are all short, and including that into Sturgeon Lake, seven in number, making two miles of carrying. The Sturgeon Lake portage, from which the railway can be reached in two days, is well nigh three-fourths of a mile long, and crosses the height of land between the Sturgeon and Minnietakie Lake drainage systems. The former has an elevation of 15 to 20 feet above Young's Lake. The portage shows exposures of green diorites and quartzites.

According to Mr. McInnes of the Dominion Geological Survey, "Sturgeon Lake lies in a belt of Keewatin rocks but little wider than the lake itself and made up of the usually divergent types." The lake is a very beautiful expanse of water, about fifty-five miles long, and from one to seven miles wide The shores are green except patches that have been recently burnt, and which occur most frequently in the lower part of the lake, as do also the islands which are quite numerous enough to make the lake navigable in moderately rough weather.

On the long arm first encountered from Young's Lake, the rock consists of alternations of coarse hard diabases, quartzites with felspar and green schists, striking northwesterly. Felspathic quartzites occur in many places along the northern shores. A large bay about seven miles down the north side, is crossed by the contact about half way down where granite outcrops, and quite close at hand beautiful light colored chlorite schist. The bottom of the bay shows green schists of the same kind with small seams of quartzite.*

From the bottom of the bay land rises to a height of 100 feet and a mile inland a very fine-grained diabase outcrops. The western side shows alternations of quartzite and green schists, that continue for some distance. and about ten miles down the contact between them shows edges, with fragments of schists included in the quartzite, the whole having a banded appearance, and striking in all directions, making a peculiar patchwork appearance. The schists are slightly magnetic, due to small quantities of magnetite. The quartzite in places shows a granitic structure, where a boss of red Laurentian granite bursts through in an eruptive way. The same banded outcrops extend for some distance further till off the old Hudson Bay fort a contact of chloritic schist and sheared porphyry occurs. The porphyry carries many small bunchy veins of quartz which, on entering the schist, rapidly disappear.

Past the fort, light coloured schists, probably chloritic tending to talcose, are encountered, but on account of the open shore and heavy sea, we could not land. The schist, in large bay behind old Hudson Bay post, shows a strike of north 65 degrees east and a vertical dip ; to the westward the country rises into low hills. Hard chlorite and hornblende schists form the shore line of the next large bay, where a wide vein of white quartz, carrying chalcopyrite, occurs near the entrance. On the other side of the channel, a ferruginous quartz vein, with brown color, becomes cryptocrystalline and shows brown and grey colored chert.

On a small island off the mouth of the bay, there occurs a contact between a porphyroid, formed by large crystals of felspar in green schist, and hornblende porphyrite, showing needle-like crystals of hornblende ; the contact strikes north and south. The usual green schist is interrupted above the narrows by amphibolite, strike north 40 degrees east, and dip nearly vertical. An island in the narrows shows a band of micro-granite, red in color, and carrying small quantities of pyrite. Further on, above the narrows, the dark hornblendic schists rise abruptly from the water, and become slightly magnetic, although when powdered, no magnetite was found. This gives way to alternating bands of schist and quartzite, that occur all along the north shore, and continue down King's Bay.

At the lower end of the bay is a mine owned by Mr. Shores, who is working on a vein of ferruginous quartz, occurring in a contact between fine-grained greenish felsite and quartz porphyry.

The south shore of the bay is mostly of chlorite schist, but past the mine, quartz-porphyry changing to diorite on the north side, occurs. This gives place to a light-coloured weathered schist at King's Point. Past the point, the shores again show numerous exposures of green schist, striking north 70 degrees east, and containing a good-sized quartz vein, about a mile further. At two and a half miles, alternations of schist and diorite occur, also quartz porphyry ; a band of granite intrudes at four miles and extends some distance, and a vein of quartz striking north east occurs at the contact. Hard green schists again come into evidence, to be displaced by gneiss, varying from grey to pinkish in colour, on the much weathered surface.

A dark dioritic-looking rock carries a quartz vein with a felsite band running parallel to it, showing a strike north 36 degrees west. Northward, gneiss extends all along the lake, is fine-grained, and contains mica in small concentrations, but near the Sturgeon River the colour varies from pinkish to grey.

*On account of the similarity between some varieties of quartzite and felsite a few exposures taken in the field for the former are probably the latter.

According to Mr. McInnes, "The northern edge of the gneiss area lies about a mile to the south of Sturgeon Lake, at the furthermost end, and keeps nearly parallel to the shore line. When the lake widens suddenly to two and a half miles, the south shore is just at the contact, the rocks consisting chiefly of quartz porphyries that in places becomes quite granitoid, with abundant blebs of opalescent quartz, often crushed and sheared to a schist, and generally holding iron pyrites. Quartz porphyries that vary to quartzites and hydro-micaceous schists occur along the shore line. Diorites and green schists also appear."

On a large bay, some six miles down, a greatly weathered dike of diabase, much broken by cleavages, protrudes through the surrounding schists. About a mile to the south-east of this bay an exceedingly red Laurentian granite was seen outcropping, while a dark colored weathered granite intervenes between it and the lake. Quartz porphyries, varying to felsites, occur near Niven's Line between Thunder Bay and Rainy River. Inclusions of small masses of light-colored quartzite give the rock an anorthosite appearance.

Just east of the line, on an island, in the centre of the lake is a high hill composed of white quartz sand, and which can be seen for several miles down the open channel. The shore line becomes boulder-covered for some miles, but shows porphyry varying to hard green schists, at one and a half miles further. The schists near the narrows are magnetic, but no distinct veins could be found; probably the magnetite is finely disseminated throughout the whole mass. Near the narrows a peculiar rotten rock carrying large felspar crystals occurs. Mr. McInnes, in his report describes it thus: "An intrusive mass of porphyrite with a ground mass of quartz and felspar, abundantly speckled with pyrite, and with large crystals of orthoclase. The rock weathers deeply, and is rusty from the decomposition of the pyrite, and waterworn surfaces are thickly covered with protruding felspar crystals often of large size.

"Along the narrows, Keewatin diorites, quartzites and schists occur, strike north 50 degrees east. At the lower end is a small area of massive crystalline felsite, which becomes in places crypto-crystalline felsite, which becomes a closed grained or crypto-crystalline rock composed mostly of quartz"

Green schists, aud porphyries extend along the shore till near McLaren's Bay. A large vein of iron stained quartz occurs on an island near the entrance. The strike of the vein is north 80 degrees east and the dip 80 degrees north. Near McLaren's post green chloritic schists occur, showing a white colored quartz porphyry, with blueish quartz, on a hill behind the post.

The south side af the bay is composed for the most part of reddish quartzite, with alternations of schists, and shows beautiful examples of ice action in smooth, plainly striated surfaces bearing glaciations south 21 degrees west. In places the schists become sericitic, striking north 34 degrees east and dip 70 degrees south, 56 degrees east at four miles from the post.

On the north side green chlorite schists interbanded with quartzites, are occasionally exposed, till past the bay they become dioritic. Outcrops are in low rounded hills that occur quite frequently.

Alternations of chlorite schist, quartzites, and quartz porphyries extend from this point to the gneiss area near the foot of the lake. About ten miles from the northern end the quartz-porphyry carries large crystals and shows striæ south 17 degrees west at seven miles, chlorite schist and quartzite, in narrow bands run nearly parallel to the shore, while near the contact, quartz porphyry again occurs.

Along the northern shore gneiss occurs of a whiteish color but becoming pinkish and granitic a short distance inland, and on a deep northern bay. The gneiss area extends southward some miles beyond the Sturgeon River.

On account of the extent and irregularity of the shore line a ready access is obtained to a large part of this Keewatin belt, and Sturgeon Lake thus becomes a new and apparently promising field for the prospector. With the exception of the shore line exposures, little work of this kind has been done. Quite a number of claims have been surveyed on the lower part of the lake; some of which show free gold.

A great drawback to prospecting is the heavy coating of moss that everywhere covers the rocks, but this is being gradually overcome by fires which sweep moss and timber before them.

EXPLORATIONS NEAR STURGEON LAKE.

ON THE NORTH SIDE OF THE LAKE.—Entering near Cobb's house, some six miles we down, explored in a north-westerly direction.

In the rear of Cobb's house, a weathered gabbro outcrops in a small mound-like elevation. The country is rolling in low hills, with few outcrops; at two miles falls to a level spruce plain which extends for miles in the same direction.

Entering again, off the first large bay of any considerable size, in the same direction, the country near the lake is high, but it falls again at about two miles from a dike of diabase, into a similar wet plain of five inch spruce and balsam.

At fourteen miles, there is a hilly country, with few outcrops, but one of greenish diabase occurs four miles inland, beyond the surface is of a rolling nature, falling in places to a moss-covered muskeg.

Entering from the bay below the Hudson Bay post, the country was found to be much the same, falling to a level swamp-like area at two and a half miles, where a hard chlorite schist was struck.

From the western side of a large six-mile bay, sixteen miles down from the portage, we paddled up a river that widens into a small lake No. 9, at the foot of a rapid. The shores of the lake are composed of green schists with interposed quartzite bands, often quite large and veinlike in appearance. One at the rapids has been stripped for some distance.

A portage of six chains leads past the rapids, where the river becomes shallow; near lake ten, which stretches three miles in a northeast and southwest direction, it becomes very wide.

The shores of lake 10, although low, show many exposures of the same green schist, alternating with bands of sheared quartz porphyry, somewhat decomposed and striking north 65 degrees east. This continues under a thick mossy coating and is again seen near the shores of lake 11, which is two miles long with dark waters.

About 2¾ miles from lake 10, hard chloritic schist with pyrite rises abruptly with a strike north 20 degrees east. A dike of diabase rises out of the centre of this schist elevation about two hundred feet from the edge, but on crossing a gully the schist is again found in situ. This outcrops several times between here and Lake of Bays (or what we believe to be the Lake of Bays), and also near its southern extremity. At this point the lake No. 12 has fine sandy shores, with hills rising two hundred feet above the water, and giving it a wild, nestled appearance. The waters of the lake were found to be fresh and clear.

The shores of lake 13 show exposures of green Keewatin schists, but the country is so well covered with soil and moss, that outcrops are infrequent. Beyond and between it and Sturgeon Lake the ground is gently undulating. Lake 14 probably has an outlet into lake 13, thence a dark river flows into Sturgeon Lake.

The low-lying shores of lake 13, as well as the river flowing from it, show no exposures.

Lake 14 has a length of eight miles and lies to the west of King's Bay. It is separated from Sturgeon Lake by a river, with two rapids giving a drop of twelve feet. Its shores are composed mostly of schist striking northeast. A vein of calcite 60 feet wide and showing a red weathering rises through the schist about five miles from the portage, on the southern shore.

About thirty miles down we made a trip to connect with lake 14. Up to 2¾ miles the country is crossed by several ridges of chlorite schist, but at this point quartz porphyry stretches off to the northeast. The country gently falls to lake 17, which apparently is surrounded by muskeg in a low-lying country. Near its southern shore quartzite is exposed.

A small river, flowing into King's Bay, expands into a small lake, 18. The shores of the lake are low and moss covered, with no exposures. A mile to the west of 18, a weathered diorite, with pyrite, was met with. A shot hole was observed in a quartz vein, 10 chains further on. The surface is covered with sandy soil and boulders, that make outcrops difficult to find. After several slight elevations, a fall occurs at two and a half miles, and from there inward a spruce muskeg extends; at two and three quarter miles holding a small swampy lake, 19; at five and a half miles, an outcrop occurs of chloritic schist, striking north 75 degrees east, dip perpendicular, and at six miles, a second lake, 20, is marshy and without any visible outlet.

On the south side, we made a trip one and a half miles down this shore, and then inland.

The surface is undulating, rising for the first quarter of a mile, then follows a series of depressions and elevations for seven or eight miles. The whole is thickly moss covered, while the lower parts tend to be muskeg, with a growth of spruce, poplar and balsam. Diorite outcrops at four miles down.

Near the shores of lake 1, the country was recently burnt, and now is covered with a low scrub. The outlet is on the western side, through a narrow three-foot channel, extending to a length of one hundred feet.

On the western side, a sluggish river, twenty to twenty-five feet wide comes in and flows westerly. On the first bend, felsite carrying pure white quartz veins occurs, and near by chlorite schist striking north 20 degrees east. An arm of the lake stretches southwest. To the southwest of this arm, a rise occurs near the shore; beyond which is a beaver meadow, and at about one mile a greatly weathered red Laurentian granite outcrops.

On a ridge at one and a quarter miles, a quartzite band carrying iron pyrites runs northeast and at three miles on lake 3, chloritic schist outcrops. This lake stretches two miles north and south by half a mile wide and is at a considerable elevation above Sturgeon Lake. The country near is fairly level, and well wooded, with no rock outcroppings. Lake No. 4, is low-lying. Chlorite schist occurs on the north shore of lake 7. Lakes 5 and 6 lie in a hilly moss covered country. Past lake 8, the country falls to a low plain, covered by spruce, and only slightly above water. This level country probably extends to the Mattawa River, without any variation.

On the bay running off McLaren's Bay, a band of chlorite schist is exposed, but the coast line down and around the bay is principally of reddish quartzite, with striæ south 21 degrees west shown on smooth well glaciated surfaces.

The river flowing northwest into this bay has a fall of ten feet. Two miles up weathered green quartz diorite outcrops, and at the first portage, greenish diabase was observed cutting across the end of the lake in a wide dike.

The river is composed of a series of lakes extending some 35 miles, and joined by short channels, or in some cases by only a swift current. The first lake stretches for a mile southwest, then follows a smaller lake, separated from the third by a swift current. This last lake is nearly circular, a quarter of a mile in diameter; its shores show outcroppings of quartzite.

The next lake stretches 4 miles to the south west and then several miles south. The country around the two preceding lakes is rather low but becoming higher on this, shows at one and a half miles up, a granite outcrop bearing hornblende, and approaching mica schist, strike north 24 degrees east and dip nearly vertical. Most of the western side of the lake is burnt for at least ten miles inland, but the south east side is still green.

At one mile along the south west shore shows there is a boss of red Laurentian granite and close at hand mica schist, tending to be gneissic. At three miles, quartzite outcrops in a wide band.

On the south side of the channel connecting this and the next lake is an outcrop of weathered diorite, 100 feet high, and burnt free of all vegetation. The surface shows protruding crystals of hornblende; the strike runs north and south.

About 15 miles back fine-grained gneiss tending to be sericitic, outcrops in the burnt country, striking north 66 degrees east, and with a nearly perpendicular dip. Half a mile further, this shades off into a greenish-colored granitic-looking rock—in reality, a syenitic gneiss which extends to the next lake.

By means of a short river we reach the last lake of this series, which extends some fifteen miles in a course varying between east, south and west. The country is mostly green, with burnt patches. At four miles, Laurentian granite was seen, and a mile further, it becomes fine-grained, and much like quartzite in appearance, becoming a granitoid gneiss four miles further, but gradually losing its schistosity and extending down the remainder of the lake in low rounded hills which are free from all soil and vegetation. The river flowing into the end of the lake shows mica schist striking north 6 degrees east and dipping 65 degrees north 82 degrees

Rapids at portage No. 46. Party No. 7.

Tebähkewinne and another Indian. Party No. 7.

Pushkokogan River between portages 46 and 47. Party No. 7.

Falls and rapids at portage 47. Party No. 7.

Bare hills, east side of White Water Lake. Party No. 7.

White Water Lake in windy weather. Party No. 7.

Falls at portage No. 53. Party No. 7.

Portage No. 53, Ogoki River. Party No. 7.

-west at the first portage. Granitic bands run parallel to the river along its northerly course, and can be traced back through the country running in a similar direction. Our trip ended at a point some 35 miles to the south west of Sturgeon Lake.

The waters of these lakes are dark and cold, the country generally wooded to the east, and south east, but to the south west has been burnt, especially along the latter portion. About 18 miles from Sturgeon Lake—on the first large lake—a cross trip was made to a lake to the north-east, one mile distant. This lake seems to be across a height of land, as we judged by its clear waters, and although we did not follow it to its outlet, it forms, in our opinion, one of the Nepigon drainage system. On account of the sandy shores no exposure were seen near the lake, or between it and the former.

Two small swampy lakes off the north east corner of Sturgeon Lake show many exposures of gneiss with a reddish weathering.

A large bay runs northward from the north-west corner, at four miles narrowing to a river, with a slow current. Both bay and river lie in the gneiss area. From the first rapid on the river we proceeded ten miles to the north-east, to the height of land between Sturgeon and Savant Lakes, as we supposed. Here a river flowing north-easterly, and full of rapids, was struck, and as no inflowing stream had been observed on Sturgeon Lake, it is probable that this ridge of red granitoid gneiss forms the height of land between these two lakes. All the surrounding country was well covered with trees and moss.

Sturgeon River.

At its outlet, the Sturgeon River is wide, flowing through clear-cut shores of grey gneiss which extends down the river till past the third portage. The river proper begins at the first falls, about two miles west of the lake, where the river separating around a large island, has a drop of 25 feet. A mile further is the second portage, with a fall of 10 feet. The river is united beyond the second rapids. One and a half miles further, is the third rapid dropping twenty feet.

Grey gneiss near the third portage often rises 30 to 40 feet perpendicularly from the river, which in places is a mile wide

The fourth portage is at about seven and a half miles, and shows outcrops of syenitic granite, which has been exposed for some two miles along the banks. The country to the north is green and well wooded, but south bank has been burnt.

Half a mile to the south of the fourth portage, two hills of reddish syenite are exposed on account of the burnt nature of the country; they extend for half a mile, and gradually drop till at one and a half miles there is a fall of 40 feet into a peat bog, with peat of good quality. This extends for about quarter of a mile, where the land rises again into the same red syenitic granite, covered by a troublesome brulé.

At two and a half miles a lake was seen stretching out to the west. The country falls to the valley of the lake, but rises at four miles to a range of long low hills, striking east and west and with an elevation of 200 feet. From here on, extends a level spruce region, mossy and low, forming the beginning of green timber at five miles. Entering at three miles, on the north bank, a ridge was found to run along the river, then falling to a spruce and moss country, with gentle ridges of syenite, intervening at short intervals; the country wholly green.

The same granitic rock extends along the river, till at one and a half miles past the fourth portage, gneiss striking north 70 degrees east and dip 75 degrees south, 20 degrees east. This becomes interbanded in pink and grey, but at ten and a half miles disappears under a sandy beach, that extends some miles along the river, with occasional exposures of gneiss.

At the 5th portage, which is over a hill of drift accumulation, the river drops 15 feet and the 6th portage shows granitic outcrops, tending to be gneissic. To the north of the river between the 6th and 7th portages, granitic gneiss was found exposed in many low ridges, which rise above the surrounding country in gentle elevations. After rising to half a mile, a drop takes place, interrupted by an outcrop of the same gneiss at one mile, but falling at one and a half miles before being interrupted again by an outcrop of granitic gneiss. At five miles a stream flows north-east, and at seven level spruce country continues, with a fall towards the north-east.

Granite and gneiss, with syenite, extend along the river, past the ninth portage, where glaciations are south 28 degrees west. Near the 13th portage, a very biotite gneiss is exposed.

Entering on the north side just above the 13th portage, the gneiss was found to continue for several miles. A river, after passing some distance through a low-lying country, empties into this bay. It shows several portages, past rapids well filled with boulders, and may be navigable at high water.

North of the second rapid a bold outcrop, 20 feet high, shows pink granitoid gneiss, but beyond this the country falls, but rises again into a low hill at one and a half miles distance, gently falling from this point to the north east. At three miles it rises into a bluff of gneissic granite, 150 feet high, from which the country can be seen for ten miles —it is partly burnt, but green to the eastward.

Immediately above the second rapid the river widens into a lake-like expansion, ending in a narrow short rapid flowing through solid rock. A short lift out of 15 feet is all the portaging necessary. A mile further is the 4th portage, three-quarters of a mile long, and partly uncut. On the upper end is a second expansion, stretching three-quarters of a mile to the north-east and lying between red perpendicular banks of granitoid gneiss that rise, bare, from the water.

A 13 mile lake, in places 3 to 4 miles wide, lies between the 13th and 14th portages on Sturgeon River, and is partly in the same gneiss area. Some ten miles above the 14th portage, syenite is exposed, and near the portage shows striæ, south, 28 degrees west.

About a mile to the east of this portage the country seems well covered with drift, and no exposures were found. One mile to the north, a small marshy lake was observed in an area covered with a fine, sandy clay loam, and which extends some eight miles, where it slightly declines in a mossy, spruce country. The syenite, extending along this part of Sturgeon River, is displaced by a band of mica gneiss, striking north 44 degrees east and dipping perpendicularly. A fifteen foot rapid, through a narrow channel, cuts through green Huronian schist, with the same strike and dip, 80 degrees north-west. A band of gneiss crosses the river and at one and a half miles further becomes much contorted, becoming a schist near the Hudson Bay post, where a quartz vein breaks through the hornblende schist. The vein shows a narrow edge of serpentine. The borders of this schist form the contact. Green Huronian schists extend down the remainder of the river, becoming chloritic till within three miles of Abram Lake, where a belt of grey sericitic schists striking north 80 degrees east and dipping 65 degrees south is noticed. At one mile from the lake is an intrusive mass of weathered quartz porphyry. A portage leads over a high elevation into a lake to the south that probably forms one of a series, as it appears to be higher than the river. Near the lake chlorite schist is exposed. Hard, green schists displace the hydromicaceous variety as Abram Lake is reached.

The schists tending from chlorite to sericite, and with narrow bands of altered diorite, become on the north shore of Abram Lake, just past Niven's line, a schist conglomerate, with large inclusions, often a foot in diameter, of quartz diorite, and quartzite, and also small lenses of quartz.

On an island near the lake, a series of four pot holes was seen, the largest cut away to a semicircle, but the rest are nearly circular and still contain the pebbles. A large bay, on the east side near Pelican Lake, shows a fine-grained agglomerate. Bands of green and white schists are exposed as you approach the entrance into Pelican Lake, and also on the arm leading to Sturgeon River.

According to the report of Mr. W. A. Parks to the Bureau of Mines for 1898, "the south bay of Pelican Lake is largely of fine grained green schist. In the second bay of this lake the rocks are fine green micaceous and quartzose schist with pyrite rising into a bluff of considerable height known as the Sioux Outlook, notable as being the site of the last battle between the native tribes and their powerful enemy to the south. The shores northward from this point are practically the same till they reach the Laurentian contact near the outlet of Vermilion River, at which point fine granitic and felsitic bands occur, all much contorted but becoming distinctly gneissoid at the head of the bay."

The northern shore of Pelican Lake is composed almost entirely of gneiss with bands of light and dark and becoming in places a biotite gneiss. The remainder of Sturgeon River lies in the gneiss area, and is often very swift, making canoeing somewhat dangerous. Nine miles past Pelican Falls—the 16th and last interruption on the river—the country to the north was found to rise for one mile, continued fairly level till 3, but at that point becomes burnt. A ridge of gneiss runs north and south, and extends to five miles. The surface is well covered with drift material. A long bay running off from the river about 9 miles from Pelican Falls, shows a very weathered gneiss, with a water mark ten feet above the present level. Off the mouth of the bay, the gneiss is contorted.

Black and red colored gneiss extends along the banks and at 13 miles shows granitic inclusions. As Lac Seul is approached, the gneiss is thrown into a gentle wavelike contortion. The lake shore opposite the river mouth shows a framework structure of light and dark gneiss; the light pink and more granitic variety forming a sort of net work, with the intervening spaces filled by the darker and more micaceous gneiss.

To the west of the river mouth the surface is fairly level and well covered by drift, that shows occasional exposures, with a strike north 56 degrees east, and dip 65 degrees north 34 west, and showing striæ south 44 degrees west.

Gneiss forms the only rock of the south shore; half a mile east on 26 mile bay it becomes very contorted, and rises perpendicularly 75 feet; and in places is garnetiferous. The north side becomes green, granitic gneiss with hornblende, and in places very coarse-grained.

The river flowing north into the bay has low banks with few exposures. Three miles up it widens into a lake, on whose borders, and also at the rapids above, are seen examples of the same rock. Near the narrows on the south shore is a high, burnt, rounded hill of granitic gneiss stretching north and south.

At the narrows it assumes the black and pink banded structure, in places granitic, which extends to the mouth of Root River, where a fine garnetiferous gneiss outcrops in a high elevation. The country is all burnt to the east as far as visible. The lower part of the river is rather swampy. Two miles up is the first rapid where garnetiferous gneiss is seen with a northerly strike and westerly dip, and this same rock was found to continue to the south of the river in frequent exposures.

From 6 to 12 miles the country is rather low, but the same rock is in evidence, becoming much decomposed, at 12 miles. The gneiss here loses its pinkish tint and becomes greyer. At twenty one miles is the second portage with a fall of 10 to 12 feet showing exposures of gneiss. The country to the south of the river becomes higher and full of rocky ridges.

The third portage, 31 miles up, cuts off a bend of the river having two rapids. The trail is over a level, sandy plain covered with jack pine, is a quarter of a mile long, and forms an excellent camp ground. Shortly past this portage a claim is surveyed on the east bank. The gneiss, in places garnetiferous, is very much decomposed, crumbling in the hand.

The last ten miles of the river is interrupted by seven short portages, with a fall of many feet. The stream becomes narrower and has a rapid current for the last three miles. There is little change in the geological character of the district—gneiss, varying between the fine-grained schistose variety and the coarser, more granitic. Near Root portage several high outcrops of grey granite occur, rising about 150 feet above the river. The region adjoining Root River is for the most part burnt, although many green patches extend along the banks. The trees for the most part are still standing in the burnt area.

A river, entering Root River immediately below the 4th portage, shows low banks for the first three miles, where large masses of gneiss are exposed. At four miles this becomes quite granitoid, with small seams of the fine-grained variety. From an elevation a view of the surrounding burnt country, streching as far as the horizon, was had.

The portage over the height of land separating Root River from Lake St. Joseph is very level. A Hudson Bay Company's trail leads over the lower muskegy portion, and so makes the portage a good one. Both ends show grass, and loamy sandy soil, and no exposures were noticed in the half mile across it.

Lake St. Joseph.

Lake St. Joseph is a splendid sheet of water, about 80 miles long, and lying almost entirely in the gneiss area. The shores are regular, and singularly free from bays. Many islands scattered through the lake add greatly to its beauty. Its outlet is by the

Albany River, at the north east extremity and the first few miles from the portage are through a low lying country. At 2¾ miles down the south shore is an exposure of biotite gneiss, banded white and dark in narrow bands, where the lake widens, hills of glacial sand and gravel 40 to 60 feet high, run parallel to the shore line. Half a mile further, gneiss strikes north 86 degrees east and dips 85 degrees north, containing small blotches of pinky quartz. Grey muscovite gneiss outcrops again, on the south shore of the first bay. The north shore of the bay shows augen-gneiss with sheared planes containing thin layers of sericite.

At about eleven miles is the contact with the Huronian where phyllites, striking north 76 degrees east and dipping 80 north 14 west outcrop. These become hornblendic, carrying lenses of quartz. At 16 miles chlorite on hornblende schist occur along the shore. An area of great magnetic variation extends for 10 miles down the schist ; no apparent cause was found, as the adjoining schist showed no trace of magnetite. At one point the variation was 65 degrees. These green schists, with a little diorite, extend for about 20 miles along the southern shore ; in places they show very weathered surfaces, and near the contact become sericite. The eastern end of the schist area shows few exposures along a boulder-covered shore, till at 33 miles a greenish granitic rock is seen, becoming a fine grey gneiss, which at forty miles turn to the darker variety, that soon gives place to the more granitic, with inclusions of dark mica gneiss.

An island half way down the lake exposes a decomposed hornblende granite, that breaks through in a boss, with veinlets of quartz.

At fifty miles, although gneiss was observed along the shore, green schists occur at one and a half and two miles to the south, in low outcrops, in a level country. At about 60 miles a band of grey granite crosses the lake and becoming coarser, shows large crystals of protruding felspar, with striæ south 43 degrees west. Fine-grained gneiss, with well-shaped garnets, is exposed several miles to the west of Osnaburg House. Near the house it becomes granitic and coarse-grained, but there is once more fine-grained dark gneiss at the foot of the lake, where the Albany River escapes. A second great compass deviation was noticed towards the eastern end of the lake. The country to the south of the lake is greatly burnt, but displays large areas of green timber.

On our return, the south shores of Abram Lake were found to show exposures of the same green schists as occur along the remainder of the lake. Abram Falls, a short rapid, joins the former to Lake Minnietakie.

Lake Minnietakie.

On account of our hurried passage through the last named lake, I had little opportunity of observing the character of its rocks, and with the exception of a few exposures along the northern shore, no further information was obtained.

The shores of Lake Minnietakie were found to be very similar to those of Abram Lake. Hard green schists, in places tending to be sericitic, extend along the north shore. A quartzless porphyry, or porphyrite, was noticed some distance down, lying between green schists. About half way along the shore is a high outcrop of weathered quartz diabase. The schists become hard and slaty as the long arm extending south west is approached. On the long bay, the schists carry small veins of quartz The Golden Rod Mining Co. of New York, is developing a property on the north shore near the portage leading to Big Sandy Bay ; but time would not permit our visiting it.

Referring to Dr. Coleman's report on this region, printed in the 1895 report of the Bureau of Mines, a fine description of the lake can be obtained. "Sandy Lake empties into Minnietakie with a fall, as measured by aneroid, of 20 25 feet. However, the canoe route does not follow the creek, which flows from the north end of the lake, but crosses a steep portage of a quarter of a mile at the point where the two lakes approach one another most closely.

" Lake Minnietakie is more than 27 miles in greatest length, and at some points several miles wide, but is very irregular in shape, having long narrow bays towards the south-west and wider stretches with many large and small islands towards the north-east. The water of the lake is beautifully clear, and its shores are usually rocky, though stratified sand rises to a height of 20 feet about the middle of the southern shore, and is washed into beautiful beaches.

"The entrance to Lake Minnietakie is by a long narrow bay running first northeast, then curving to the east, about 17 miles long before the lake widens. A long point separates this bay from a shorter one with more varied outlines to the south. The whole shore line of this part of the lake consists of Huronian rocks, of great variety, chiefly the green schists described by Dr. Lawson in the region to the south, as Keewatin. In general the strike of the schist is parallel to the direction of the bay. Just after crossing the sandy lake portage, the schists are hard and folded. Half way down the projecting points consist of yellowish sericite schist; evidently altered quartz porphyry, and at some places a true porphyry with large blebs of quartz. In the inlets along shore, one finds the green schists; so that apparently the bay has been hollowed out of a band of the softer, yellowish, altered quartz porphyry, the harder points projecting still as points. A small outcrop of rather fine-grained granite or gneiss occurs on the shore not very far from the portage perhaps in connection with the gneissoid rock of Sandy Lake. A somewhat grey clay slate or phyllite is found in the point near the opening of the long bay into the lake. At several points along the southeast shore of this bay there are veins of quartz more or less charged with sulphides. Fahlbands, i.e., beds of schist containing much sulphide such as copper pyrites, are also found widely extended along the shore.

"The long bay stretching 7 miles along the southern shore of the long point just described, has on its shore rocks of a quite similar geological character, sheared porphyroids and green Huronian schists, and need not be described in detail.

"The north-west shore presents chiefly green schists with some veins of quartz and bands charged with sulphides, but two assays of rusty quartz showed no signs of gold. Going north-westerly along the southern shore much of the beach is found to consist of Laurentian boulders at first, but the only rock found in place is hard, grey-green, and scarcely schistose, and is probably an eruptive rock of Huronian age. This was found on a small island.

"Clay slate of a dark grey color, and showing two directions of cleavage is found west of the bouldery shore, and contains some small bedded veins of quartz with sulphides; then follows a beautiful sand beach consisting almost entirely of garnet and magnetite, derived no doubt from the adjoining sand cliffs which rise 20 feet from the lake, and are being undermined and re-arranged by wave action.

"Near the point where the shore bends to the south, green chloritic schist is found, but a dark grey slate, interbedded with sheared porphyrite occurs at the south end of this arm of the lake. From this point round most of the eastern shores to the outlet of the lake, Huronian rocks of green or grey color, sometimes very schistose, and at others, massive looking are found. The islands which are many and often large with narrow channels between, present more variety of constitution. On one of them a grey green porphyrite with crystals of white striated felspar one inch long was obtained; and on another an eruptive mass of greenish-grey quartz diorite, like some of the so-called protogene of the Shoal Lake region. The same granite-like rock was found at the eastern end of a deep bay on the mainland; but for lack of time, its outlines were not completely traced."

Sandy Lake.

Referring to the same report, "Sandy Lake is a fine body of water, six or seven miles from north to south, by four miles from east to west. It contains few islands, and these small; so that a sea dangerous to Peterboro' canoes can easily arise, as we had occasion to discover. The water of the lake is beautifully clear, shores not generally high, and often consist of drift materials covered with second growth woods. The Hudson Bay Company have two large York boats on it to transport supplies, and bales of fur from their storehouse at the north-east end of the long portage to the next portage into Minnietakie Lake.

"About two miles east of the landing at the long portage, a rather coarse-grained reddish grey granite is the only rock found. Here a point which projects displays a small mass of grey schist, seemingly included in the granite, having a strike 40 degrees east of north. A third of a mile to the north east, contorted greyish gneiss occurs in the granite, and beyond this, grey Huronian schist with a strike of 70 degrees. In a deep bay on the east of the lake, green banded schist with small bluish quartz veins occur, having a strike of thirty degrees, and on the north-east side of the bay similar schists with more or less contorted bedding show a strike of 20 degrees or 25 degrees.

"The point that projects to the south of the portage to Lake Minnietakie, consists of coarse grained porphyritic syenite of a light flesh color. At the portage contorted green schist occurs again, so that the syenite is apparently an isolated boss. An island west of the point is formed of the same rock."

"A waggon road of about nine miles in length, connects the town of Dinorwic with the Hudson's Bay storehouse on the south side of Big Sandy Lake. About half way across the portage is found the contact of the two formations." "The portage shows drift material, with silty clay on the lower parts, also black loam. Two stony ridges are crossed, probably moraines. Boulders are chiefly gneiss and granite with a few green schists. Beyond one finds clay again and sometimes sand, barren looking for the greater part, but covered with black loam at the bottom."

Surface Geology.

From Dinorwic a covering of sandy clay extends till beyond the southern shores of Camp Lake, which is of a light grey color, and shows occasional exposures. On the Camp Lake end of the portage is a small clearing of a quarter of an acre, where potatoes were being planted on the day of our visit, June 22nd. Later, we heard that on account of the wet season, they had not been a success.

The shores of Indian Lake are of a rocky nature as is also the country in the vicinity which is covered with a scanty soil of sandy clay but deep enough to give support to a thick growth of jack pine, spruce and poplar, some of which reach a considerable size, but the average diameter is somewhat low. A sand bar projecting from the west shore nearly cuts the lake into equal portions; beaches of fine sand are also quite numerous along the northern portion of the lake. Sandy soil mixed with clay, extends around Bear Lake, with its rocky shelving shores. The country is rather level, and in places tends to be low, but everywhere is seen a thick growth of moss. The first portage shows a covering of rather sandy clay loam, which extends with varying thickness up to Otter Lake, with occasional outcrops.

As you approach the Mattawa River from Bear Lake the surface gradually becomes more level, and lower, with few hills. The river itself flows through a low lying area, that is swampy for some distance back from the banks. From Otter Lake up the Grass River many large boulders were observed in the soil, that extends to Sturgeon Lake without much change. Grass River and Hut Lake show few exposures in a country composed of drift sand. Hut Lake shows exposures with striæ south 12 degrees west.

The shores of Pine Lake are covered with boulders of gneiss and granite with few exposures, Soil, somewhat sandy, that on the portage between White Rock and Pine Lake supports a good growth of cedar and poplar, with many spruce and balsam intermingled. White Rock Lake is of much the same character as the preceding lake. It is connected with Young's Lake by a sluggish river, ending in a rocky channel with a good current.

Young's Lake.—The sandy soil about Young's Lake contains many boulders. Half a mile to the south east of the lake, extends a narrow belt of scattered white pine of good quality. Beyond the surface becomes level with an occasional low hill. No massive rock was seen for eight miles, where a small lake one and a quarter miles by half a mile, was encountered. Many animal trails, especially those of the bear, were observed around the lake. On the northern side of Young's Lake the soil is inclined to contain gravel (probably the result of decaying rocks); the trees although small on the lower parts, reach a diameter of 10 and 12 inches on elevations.

The portage into Sturgeon Lake shows fine white clay in considerable thickness, with several outcrops of the country rock. A small clearing with two log houses, greets the eye on the eastern end of the portage. Some five miles up, stands a prospector's cabin; and a small clearing on fine grey sand, mixed with gravel, lies to the rear. Here potatoes were growing, but on account of the time, July 9th, they had only reached a heighs of some four inches. About 10 miles up the north shore the soil is of a fine sandy nature, with occasional areas carrying a perceptible quantity of clay. An island off the shore at this place, has a thin covering of black loam formed from the decomposition of leaves and woody matter.

Entering the bush some fourteen miles down this shore, a considerable elevation was found near the lake, covered by a fine growth of poplar and extending half a mile inland, where the country becomes lower and more level. At three and a half miles a cool stream runs south, draining a rather low-lying spruce area that exists to the west. The soil is of a fine sandy nature, that extends along Sturgeon Lake to lake 9 which receives its waters from a stream flowing from lake 10. The banks of the stream are low, while its bed is composed of soft muck of unknown depth that gives out evil smells on being disturbed.

Lake 10 extends three miles north and south with rocky shores that show but a scanty covering of sand. Between lakes 10 and 11, two miles to the west, are several ridges showing sheared quartz porphyry. The waters of lake 11 are dark, and the soil in the vicinity is of red and white sand, with many boulders. Several ridges of green schist separated by deep gullies lie between lakes 11 and 12. Lake 12 is a very pretty body of water with bays stretching in all directions, the clear water beating upon sandy shores. The southern part of the lake showed a soil of sand covered with a thin layer of dark loam.

Near Niven's Line, a wide, deep river with low banks flows into Sturgeon Lake. A mossy portage through tall, slim spruce, leads to the upper part of the river. After a second portage the river widens into a small lake, which a short portage separates from Lake 14, that stretches 7 to 8 miles to the north-east and is surrounded by high hills. A ridge separates Lakes 14 and 15. Sand forms the predominant part of the soil in this area, which is everywhere coated with moss.

At the narrows, a rise of 150 to 200 feet is made a quarter of a mile back from the lake in a rapid incline, beyond the country gradually sinks to a level, covered with moss and grey sand with many boulders. Lake 16 is a depression in the muskeg, as is also Lake 17.

King's Bay receives the dark waters of Lake 18 by a small river. The soil to the west of Lake 18, is sandy, and contains much gravel. Between Lakes 18 and 19 the country is level, with the same soil, that lowers into a muskeg completely surrounding Lake 20.

One and a half miles from the portage the south side of Sturgeon Lake shows a rather level country. At a quarter of a mile to the south of the lake, a stream flows northerly. After crossing several small ridges, the country drops to a spruce area, that extends many miles to the south-east.

A narrow channel leads from the first large bay on this side of Sturgeon Lake to Lake No. 1, and a river 25 feet wide flows into its northeast corner. Veins of white quartz in green schist were found on the banks, and a short distance to the south of the lake, lies a small beaver meadow. Soon the Laurentian area is reached at an elevation of considerable height. Around Lakes 3 and 4, the soil is sandy and well covered with a heavy layer of moss; many small marshy lakes lie in this vicinity. Near Lake 5, the ground is well trodden by bear and other animals. The soil is a sandy clay with boulders. A high knoll separates Lakes 5 and 6, around the end of which a stream, now dry, flows to Lake 4.

A canoe portage, showing that these lakes form a means of communication, was noticed between Lakes 5 and 6, the latter having its outlet into the former. Steep declivities that show a scanty covering of sand surround both these lakes. Lake 7 is low-lying, with muskegy shores; and the soil changes slightly, becoming more sandy, though overlayed with moss, around Lake 8, whose shores rise into low sandy hills. Beyond the country becomes lower, with dark surface soil, and extends unchanged as far as the Mattawa River. No exposures were observed for several miles in the latter part of the distance

A hill of some hundred feet elevation rises along Sturgeon Lake to the south of McLaren's Bay. The country is covered by brulé and makes difficult travelling. The recent fires have swept away the coating of moss, and have left the rocks in many places exposed. Many fruits grow in profusion, including wild cherries, raspberries and blueberries. Some five miles back a spruce country exists.

A river, forming the outlet of a long series of lakes, flows into McLaren's Bay. Its shores show reddish quartzite with good glaciations striking south, twenty-one degrees west. A very pretty system of lakes extends some thirty-five miles to the south and southwest.

The shores of those near Sturgeon lake are green, but to the southward they become more and more burnt, till at last, for the last ten miles, the country is entirely burnt in all directions.

The river empties into the bay by a slight rapid, followed at two miles by a second and then the system of lakes occurs separated usually by swift currents, but few portages necessary. The first three lakes are small, with green shores. The fourth lake is quite long, extending some nine or ten miles to the southwest. The western side of the lake is burnt in places but the eastern shore remains green. The contact of Laurentian and Huronian crosses near this lake.

A short river joins this and the succeeding lake, which is very irregular in direction, continuing some fifteen miles in a course varying between east, south and west, A good sized stream flows into the last long lake, but after making two portages, one of which we had to cut out, the stream became impassable. We proceeded as far as possible on foot, but the fallen timber in the burnt country made travelling useless and well nigh impossible.

The soil of all this region is sandy and somewhat scanty, especially around the last lake, where the moss is also missing. These waters are dark in contrast to the waters of a lake to the northeast of the first large lake of this series, where clear water was found. A sandy ridge, green and well covered by bush, separates these two lakes, the latter of which probably forms one of the Nepigon series.

The lower end of Sturgeon Lake shows many sandy beaches in enclosed bays. Two small lakes off the northern extremity occur in muskeg that extends back some distance, and is too soft to travel on. A long bay, a little to the west, extends seven miles north; its northern end receives the waters of a small river, probably the outlet of some inland lake. On the first bend of the river stands a deserted trading post. A short distance beyond a rapid necessitates a portage of nearly a mile. Across the rapids are the remains of an old fort, once the scene of an Indian visitation, as its timbers are cut and seared by many shot holes. To the northeast of the portage the country continues fairly level with deep soil of sand; covered with deep moss. At three miles a river flows to the northeast with a rapid current; as it continues in the same direction for several miles, it probably flows along one side of the height of land between Sturgeon and Savant Lakes, flowing into the latter. Syenitic gneiss outcrops in this region in quite frequent exposures of considerable height.

STURGEON RIVER.—The wide entrance to Sturgeon River is through a gneiss region of grey color. It is a grand river, very often lake-like in character without any perceptable current, while in places the flow is rapid and exciting, though perhaps somewhat dangerous. From Sturgeon Lake to Abram Lake is about sixty miles. The first rapid is some two miles down the river on the north branch; it has a fall of twenty-five feet and is suitable for power development. The river unites after the second rapid, Beyond the third rapid, with a drop of twenty feet, there is a lake-like expansion three-quarters of a mile wide. Green country extends nearly to the fourth portage, seven and a half miles down, where the south sides shows a brulé, with blue-berries and raspberries in great abundance. The soil along the river shows a fine grey clay that becomes sandy a short distance back.

The burnt country to the south of the fourth portage extends some distance inland. Several high exposures rise in the vicinity of the river; these fall at two and a half miles to a long peat bog of good quality. At four miles hills of gneiss rise two hundred and eighty feet, stretching to the east and west, with a similar range at six miles, separated from the former by a valley of moss-covered, low country.

The north side of the river, rises into a sand and moss-covered ridge, running parallel to the river, and stretching out into a rolling green country of poplar, spruce and balsam. A row of green stumps, cut by an axe, shows that man has been there before.

Pushing in from nine miles—the land to the north rises near the river, showing a short stretch of clay land, that gives way to sand as the elevation in the vicinity of the river is ascended. The surface remains fairly level with the exception of a few granitic gneiss elevations that run parallel to the river, and are separated from each other by a level, sand and moss-covered country. At five miles a small stream flows to the northeast through a very similar country.

Between the 4th and 6th portages the south bank is usually higher than the north; the banks are of clay, with many boulders lining the shore; that shows outcrops of syenite. The 8th portage is somewhat difficult, being over the rocky unused bed of the river Shortly past the 8th portage the river narrows, and assumes a ess lake like appearance. The 9th rapid is full of boulders and very rocky; the syenite on the 10th shows striæ south 28 degrees west. The 10th and 11th are separated by a short rapid with a fall of ten feet and a very swift current. Beyond the 12th, the river spreads out into a lake about a mile wide, off the eastern side of which is the channel to the 13th. The shores of the lake are sandy in places. The country is mostly green, with only occasional patches of brulé. Groves of poplar extend along the shore, and back for a quarter of a mile, where spruce and balsam supplant them, showing however a sprinkling of red pine near the lake expansion.

A large river flows into the Sturgeon River on the north side, near the thirteenth portage, and separated from it by a short rapid. For some distance the river is 2 to 3 chains wide, with low-lying shores. Some four miles up a second rapid occurs. This portage was found to be about half a mile long and the river to have a drop of 40 feet. About a mile of river intervenes between it and the 3rd portage, a mere lift out, in turn separated from the fourth by a mile of slowly moving water. On account of low water, this last portage was three quarters of a mile long, only partly cut out. Above, a long narrow lake extended to the north east between hills of gneiss that rise perpendicularly from the water to a height of 60 feet. The tops were denuded of soil but a thick coating of moss seemed sufficient for sustaining a thick growth of scrub spruce. The same spareness of soil was noticed along the river and on the portages.

A rather level, sandy soil extends to the north of the second portage on this river. About five miles north is a high hill from which the surrounding country was seen to be green to the east and west, but burnt trees were standing to the north. Many of the spruce and poplar reach 12 inches diameter. Tracks of bear, moose and caribou were seen in great abundance.

Gneiss near 13th portage shows striæ south 20 degrees west. Between this and the 14th portage, extends some 13 miles of lake. The soil is quite thick in this region, sand beaches occur frequently along the shore. Near the 14th portage there is presented a level, sandy soil, extending almost unbroken, with the exception of a small marshy lake at one mile. Beyond this, a light clay soil extends for several miles to the north, and is covered with a thick growth of large poplar with spruce on the lower parts. At eight miles a spruce swamp appears extending far inland. The 14th rapid is through green Huronian schist that is exposed for a short distance near the portage and beyond the 15th rapid a low country with clayey soil extends, with no high elevations. Past the Hudson Bay post a sandy point stretches out into the river. Green schists extend from this point to Abram Lake, with many bold exposures near the river, carrying striæ south 36 degrees west. This schist area is denuded of timber, but is covered with a quickly growing scrub, while the hill tops are thickly covered with raspberries and blueberries.

A small island at the entrance to Abram Lake showed a fine clay soil, that extends in patches along the river, having near the lake a good growth of poplar, jack pine and spruce. Sand seems to obtain around Abram and Pelican Lakes.

Fine clay occurs on Pelican portage, which is half a mile long, and is the last portage on the river. A brownish clay mixed with sand extends down the river, showing few outcrops on the banks. About nine miles down the country becomes very rocky and is covered by spruce with odd jack pine, poplar and tamarac. To the north of the river the land rises for three quarters of a mile then falls to a mossy low-land drained by a stream flowing into the Sturgeon River. A high ridge at three miles gives a good view of the country, which shows a small brulé extending to a ridge at five miles.

The same clayey soil extends down to Lonely Lake, in places containing much sand and many boulders. The trees continue much the same, spruce holding the preponderance over five inch poplar, with a few balsam and jack pine. Several sand bars along the river show fox-holes still inhabited and gneiss off the river mouth on Lac Seul shows striæ south 42 degrees west. The sandy soil, with clay, extends westerly towards Canoe Lake reservation, 28, a light-colored sandy soil covers the gneiss near the river. The Hudson Bay post, and the reservation, have well kept houses, while onions, corn, potatoes and other vegetables were growing well in the gardens.

To the east of the river a level sandy soil extends to a small creek that flows through a low muskegy country. The upper end of the creek cuts through clay. Many moose tracks were seen on the soft banks, while the stream bed was cut, and the waters made muddy by their continual crossing. Beaver work was also noticed. Wild currants grew in abundance in the bush, where partridge were quite abundant. The trees, tall poplar and spruce, were large and healthy looking.

Lac Seul.—To the eastward the shores of Lac Seul present a rocky appearance with a sandy soil that in places contained more or less clay. A river flowing into a large bay near the narrows shows clayey banks. The soil is dark in parts, yields blackberries in abundance, while birch, poplar, spruce, with some balm of Gilead, grow thickly. Numerons moose tracks were observed. On the lake expansion of the river many fine 20-inch spruce, now dead, once grew. Above the lake is a rapid, but no portage could be found. A stream joining the larger one immediately below the rapid, became blocked some two or three miles up.

Along to the narrows, and beyond, the south side is much burnt. To the south-east a high outcrop of reddish gneiss gives a good view of the surrounding country, burnt for ten to fifteen miles; the trees are still standing, and, before the fire, had not reached a very large size; spruce and jack-pine predominate. The soil continues sandy, and in places is somewhat scanty. A green belt runs along the upper portion of the lake whose waters grow darker and darker from the inflowing Root River. Near the river, the tops of many high gneiss exposures are seen, having lost their earthy covering. Moss is very thick, and scrubby spruce and jack-pine grow thickly near the river.

Root River.—Root River empties through a low sandy mouth, in a wide bay of reeds, where ducks were very numerous, flocks of 40 and 50 being frequently seen. The country near the river continues low to the first rapid some two miles up. This shows sandy, slightly argillaceous soil. A fringe of green extends along the river. Back of the river at six miles, the land rises gently for a quarter of a mile, then falls slightly, and then rises into a long rolling ridge of gneiss, covered scantily with burnt moss and sand. From this ridge the same forest of poles extended for miles. An isolated green ridge extended some seven miles away. Several Indian camps, now deserted, were observed along the river. The second portage is about 21 miles up. The country becomes higher near the river, and exposures more frequent. High sandy banks occur about 28 miles up. The third portage, which cuts off two or three rapids, is across a level jack-pine plain of sand, well grass covered at the ends. It makes a good and much used camp ground.

The last stretch of the river shows seven rapids, and in places a swift current. The portages are usually grass covered and show sandy soil. Most of the country near the river, here very winding, is low, covered with alders, and shows high exposures of grey granite.

Just below the fourth portage where the gneiss shows striæ, south 44 degrees west, a good sized stream flows into the river from the south side. Continuing up this stream about six miles, we found the country to be burnt, with trees, consisting of small spruce and jack pine, still standing. A scanty covering of sand and burnt moss covers the rocks. The gneiss in places tends to be granitoid.

The Root portage leading over the height of land between Lake St. Joseph and LacSeul, is some half a mile long. A muskeg crossed by a Hudson Bay Company pole walk, forms the central portion. The northern end shows a thick covering of sandy clay, and is all grass-covered.

The waters of the upper end of Lake St. Joseph are dark and cool. The land rises gently behind the lake, and on the narrow arm and along the shore is still green. Some 3 miles down, hills of sand 30 to 40 feet in height running parallel to the shores, are perhaps moraines. These were noticed also, along the north side. The gneiss is exposed in low and frequent outcrops, that on the first deep bay increase in height, and rise perpendicularly from the water to a height of 15 feet. Farther down the lake they reach a height of 75 to a hundred feet in a high sea wall. The country around the lake seems to have been burnt at different times, and so has a patchy appearance, brulé alternating with young second growth trees.

About 24 miles down are some islands composed entirely of sand cliffs and showing much Indian occupation, followed at 29 miles, by low sandy shores, with spruce, jack pine and poplar growing thickly. The lower portion of the lake is mostly green, while the soil becomes thicker near the Osnaburg House, close to which a sand bar stretches out into the channel. The fort buildings are enclosed by a fence. Near the enclosure is a garden, where potatoes were growing in grey sandy soil. An unfinished church gives the place a certain air of civilization. The well kept Hudson Bay store showed a stock of skins of beaver, otter, bear, muskrat, fox, etc.,—the remnant of last season's catch. Several families of Indians encamped there enjoyed the novelty of seeing so many pale faces. The Albany River, some seven miles past the fort, escapes from the Lake in a rapid, flowing through rocky shores of gneiss and boulders. The lower portion of the lake showed a large rounded outcrop of grey, coarse-grained granite in the gneiss area, and covered with a scanty layer of sand. The Albany River formed the farthermost point of our trip.

On account of our rapid trip to the railway the nature of the country around Minnietakie and Big Sandy Lakes was not observed. These lakes we covered by Dr. Coleman's report in 1895.

Economic Geology.

No quartz was found in the Laurentian area of Indian and Bear Lakes. According to Mr. Parks' 1898 report to the Bureau of Mines, "mineralized fahlbands occur in the vicinity of Hut and Otter Lakes, but no quartz veins were found crossing these belts." Sturgeon Lake presents many quartz veins in its green, felsitic and dioritic schists, that often carry pyrrhotite. Felsitic schists outcrop along the north shore, along with narrow bands of quartzites.

Some miners at the camp some six miles down showed pannings of free gold, but where the vein was situated, we do not know.

The first large bay carries out crops of granite and quartzite, at the contact. This should be a likely place for gold, but none was observed. Some ten miles along this shore quartzite is associated with a narrow band of green schist carrying small quantities of pyrrhotite. Above this, bunchy quartzite bands spread out in all directions.

Near the old Hudson Bay fort, the junction of green schist and quartzless porphyry shows many small quartz veins, that disappear in the schist. In the rear of the fort Mr. Parks found a deposit of quartz into which a shot had been put.

On Lake 9, a quartzite vein has been stripped for some distance, and has, I think, been recently located. It was found to contain iron pyrites.

The south eastern shore of lake 14, shows a wide 40 foot vein or band of calcite, carrying iron pyrites. A smaller 18 foot vein lay quite close to it. The surfaces were deeply iron stained; on panning no trace of gold was found. On a large bay some 16 miles down, a fifteen foot quartz vein carrying small quantities of chalcopyrite is associated with the green schists. Several shots had been put in. On the mainland, on the south side of the entrance to the bay, a five foot quartz vein carrying lenses of pyrite and associated with chert, had also been attacked by the prospector. An assay made at Belleville, gave gold to the value of $1.80 a ton.

An island in the narrows carries micro-granite with small quantities of pyrite. Near the narrows, on the north side, a bold cliff of hornblendic schist fronts the lake. Although a crushed sample revealed no traces of magnetite, the compass variation is very great, on land, but on water, no effect is produced.

Quartzite bands alternating with green schists, occur along King's Bay. At the foot of the bay is a mine, being developed by Mr. Shores. Two parallel veins 70 feet apart are on the property. The smaller, 3 foot vein, occurring in felsite and quartz porphyry and with a strike south 80 degrees east is being opened up, and the intention is to cut across to the larger one. At the time of our visit the shaft had reached a depth of 60 to 70 feet. A vein of pyrrhotite is carried by the quartz, that showed a trace of gold and also of copper. The quartz porphyry forming the contact, is rich in iron pyrites that assayed $2.20 gold. The vein itself is of rusty quartz that carries pyrite and free gold.

Some very rich specimens of the latter were shown us by Mr. Shores. The earth on the top of the vein is also very rich, even a pinch panning gold. About 18 men were employed during the summer. Upon the success of this mine depends to some extent the future of the Sturgeon Lake region, as a field for mining operations.

Near lake 18, a quartz porphyry shows a shot hole, and close by a quartz vein with pyrite also bears mark of the prospector. Below King's Bay many small quartz veins were observed, not many of which had escaped the hammer. Some 2½ miles below, a large quartz vein occurs in diorite (?) and schist striking north-east. This is supposed to be a new and rich discovery. A sample taken showed no colors on panning.

On the inlet of lake 1 to the south of Sturgeon Lake, felsite with veins of quartz occurs, while about a mile to the south of the same lake, a high elevation of reddish quartzite carries iron pyrites. The green schists to lake 6. show no quartz veins.

The very rich decomposed crystalline rock along this part of Sturgeon Lake carries iron pyrites. Further down the shore, reddish quartzite is associated with felsite carrying small quantities of pyrrhotite. At 20 miles, quartzite and a granitic rock are associated with green schists that carry numerous quartz veins. The green schists on the south shore display magnetism similar to that on the upper part of the narrows, on the opposite side of the lake.

A large vein of ferruginous quartz cuts across an island near McLaren's Bay, and is associated with quartzite. The bay exposes much quartzite, as does also the first lake of the series, that empties into the bay.

Alternating bands of green schist and quartzite extend down most of the remainder of the south shore. Some twelve miles from the foot of the lake two large veins of iron-stained quartz striking north sixty degrees east and dipping seventy-five degrees north-westerly have been staked out. At seven miles a quartz vein shows signs of work and a short distance further on narrow bands of schist and quartzite running parallel to the shore line have also been subjected to the prospector's hammer. A prospector named Craig with whom we camped, showed us some samples of quartz carrying free gold. A sample of syenite found back from the southern shore was found to contain magnetite. From various camps along the lake, samples of galena were found, and although no vein or ore body of this nature was seen by the party, the probability is that some exists in the region. The foot of the lake, being in a granite and gneiss area, showed no signs of minerals.

Sturgeon Lake area may be well worth prospecting, as the irregular shore line gives easy access to the surrounding country. Although the lake shores have been pretty closely observed by the prospector little or practically no work has been done in the surrounding area. This is perhaps due to the well-wooded nature of the country, that makes prospecting difficult.

The gneiss and syenite of Sturgeon River was found to be barren, and no mineral was encountered until the Huronian area of Abram Lake was met with. Here, near the old Hudson Bay fort, a vein of quartz rises through the green schists. A tunnel had been driven for about twenty feet before the work was abandoned. The schist at the vein contact carried pyrrhotite, that showed a gold assay of $2.60 per ton. The quartz carried small quantities of pyrite. About ten miles from the 15th Sturgeon River portage, a three foot vein of magnetite occurred, running parallel to the schist which here shows a strike north, seventy-two degrees east. Striæ here point south, thirty-six degrees west.

A lense of quartz near Abram Lake shows a shot hole but no gold was observed. Abram Lake seems to have been well prospected. Many small quartz veins occur in the schists, but none of any size came under my notice. The schists of this region are well mineralized, and are of the same nature as those of Sturgeon Lake. Mr. Parks examined the Abram and Pelican Lake regions and reported on them to the Bureau of Mines in 1898.

As is usual, the gneiss area of lower Sturgeon River and Lac Seul, displayed no sign of economic mineral, and it is not until you reach the Huronian area of Lake St. Joseph showing mineralized schists, that anything of value was seen. Here phyllites and hornblende schists contain lenses of quartz, but no gold bearing pyrites. These schists become very magnetic, so much so that a variation of 65 degrees was observed at one place. The schists showed no signs of magnetite when tested with the magnet, and the only other cause, is that there may be a large submerged vein. This may be probable, as the area of disturbance extending some ten miles in length shows as great a variation on land as on water. The schists show an occasional quartz vein, and towards the contact becomes sericitic, with lenses of quartz.

A second great magnetic variation lies near the foot of the lake in the gneiss area. I can give no reason for it, as no schists appear anywhere in the vicinity.

Time would not allow a close examination of Lake Minnietakie, so I would refer to Dr. Coleman's report of 1895 on this, and also on Big Sandy Lake.

At Dinorwic we saw some good samples of galena, with pyrite found somewhere in the Minnietakie Lake region. A mine is being opened up by the Golden Rod Mining Company of New York on the long arm of Minnietakie, but its character was not ascertained in our speedy trip down the lake.

It may be noted that gold is distributed in small quantities through the Huronian areas, and that it occasionally becomes more valuable. In places it is in the free state and in others is carried by sulphides usually pyrite. Geologists who have studied the region closely say that these rocks resemble Lawson's Keewatin series, to the south of the Lake of the Woods, and that there is no geological reason why they should not be equally as valuable.

Timber and Agriculture.

From Ignace to Sturgeon Lake jack pine, birch and poplar, with spruce, extend over a green country. The larger trees reach from 15 to 18 inches in diameter, but the average is not above 8 or 9. Occasional stretches of clay lie in this area, and may be suitable for agriculture. On the south-east shore of Young's Lake is a belt of scattered white pine. Some of them reach 18 to 20 inches in diameter.

On the higher land, near Sturgeon Lake, jack pine, poplar and spruce grow to a good size—often to 15 inches—but on the long, low stretches, balsam and tamarac take the place of jack pine and poplar. As the ground becomes still lower, and increases in moisture, spruce alone grows, with occasional tamarac. All around the lake, at a distance of four to eight miles from the shore, there seems to extend a mossy spruce swamp, apparently endless. In these areas the trees have an average of from four to six inches. Occasional areas of sandy clay were found in this region but none of any extent.

The green country of Sturgeon Lake extends westward down the Sturgeon River with but little interruption from burnt areas that occasionally occur along the river. Similar trees grow along the river, and on the clayey banks, are healthy, and of good size, often 15 inches in diameter. A large brulé exists on the river near Abram Lake, and extends north and south as far as the eye can reach. Abram and Pelican Lakes show good sized poplar and spruce, with jack pine and a scattering of tamarac. The soil is clayey, but sand areas far exceed these tracts in extent.

Lower Sturgeon River and the western end of Lac Seul are still green, with spruce and poplar predominating, towards the eastern end the country becomes largely burnt, in an area that extends along Root River, and includes a large part of Lake St. Joseph. On the latter lake the trees are mostly poles from 1 to 4 inches in diameter, and have grown very closely together, making travelling very slow and difficult. It is probable that this burnt area extends through to the northern, or at least nearly as far as the northern extremity of Sturgeon Lake.

The thin soil of the Lonely Lake region gets scanty on Lake St. Joseph, making but a poor foothold for trees of any size. The lower end of the lake shows shores green with poplar and spruce, but very few jack pine and no pine, white or red. Small balsam and tamarac are scattered through the bush.

As a whole, this region can never be a great agricultural country, as the areas of good land are too small in extent. No pine of any account was met with, only a few scattered clumps, but spruce was found to be thickly scattered throughout the whole district, while the jack pine of the southern portions was displaced by poplar in the north. If the logs could be gotten to the front, an almost inexhaustible supply of pulp wood, and common timber would be assured.

Your obedient servant,

J. E. Davison

I, John E. Davison, of the City of Toronto, in the County of York, Geologist, do solemnly declare that this report is true and correct, to the best of my knowledge, information, and belief, and I make this declaration conscientiously, believing it to be true, so help me God.

Declared before me at Toronto, in the County of York, this 28th day of December, 1900. J. E. DAVISON.

GEO. B. KIRKPATRICK,
A Commissioner.

SURVEYOR'S REPORT OF EXPLORATION SURVEY PARTY No. 10.

RAT PORTAGE,
RAINY RIVER DISTRICT, ONTARIO,
January 2nd, 1900.

SIR,—I have the honor to submit the following report of Exploration Party No. 10, of which I was in charge, and which, acting on your instructions of June 22nd, 1900, explored that part of the Rainy River district lying north of the Canadian Pacific Railway, between the west boundary of the Province and the Indian Reserve on Lac Seul.

I had six assistants, including the geologist and the land and timber estimator, and on June 29th shipped my outfit, including canoes, from Rat Portage to Margach station on the Canadian Pacific Railway, eight miles east of Rat Portage, and on June 30th, the porty went out and crossed to Black Sturgeon Lake, ready to begin our allotted task on July 2nd. By consulting the map of the territory to be examined, I laid out a scheme of exploration to which I was able, substantially to adhere, and which was as follows: We began at the outlet of Black Sturgeon Lake on Winnipeg River and went easterly up that lake and through the chain of lakes and connecting streams and portages which terminates at the southeast end of Lacourse Lake, a mile and three-quarters north of a point on the Canadian Pacific Railway track, which is one mile east of Jackpine Siding. Thence we portaged out to Jackpine Siding and took the train to Eagle River station. On the above run we made a subsidary trip from the north-west corner of Favel Lake through a chain of lakes to the south-end of Grassy Narrows Bay on English River. We also traversed the chain of lakes leading south from Canyon Lake towards Hawk Lake. We turned back, however, when about two-thirds of the way through because we found the whole region to be a brulé having been burned over in 1886. We also made an exploration and survey from Rice Bay and the north shore of Black Sturgeon Lake as far as to Ant Lake, and inclusive of it. We followed the north-east bay of Canyon Lake to its outlet through Canyon River.

According to the map our route should have led from Canyon Lake through a chain of lakes that would have brought us out at Vermilion Bay on Eagle Lake. The only lakes we saw however were those we followed out to Jackpine Siding as already mentioned.

From Eagle River station we went down the Eagle River and down the Wabigoon River to the north boundary of Mutrie Township at which point, being the limit of surveyed territory; we resumed our explorations and surveys.

We went up Canyon River from Wabigoon River to Canyon Lake, and up Blackwater Creek to the lake of the same name.

Arrived at the mouth of Wabigoon River we ascended English River to Lac Seul, surveying en route two bays omitted on the survey of this stream by D. L S. Fawcett in 1885, viz, Oxalis Bay on Fishing Lake and South Bay on Maynard Lake. We continned eastward as far as the Hudson Bay Company's post on Lac Seul at the Narrows opposite Indian Reserve No. 28. We learned here that O. L. S. Robertson in charge of Exploration Party No. 9, had preceded us by only a few days. Procuring an

Indian guide here we proceeded to the northwest corner of Indian Reserve No. 28, and worked westward through Burntwood, Route, Williams, McIntyre, Islands and Farewell Bays and from the bottom of Bay of Islands across the long portage into Wabuskong Lake, and northerly through this lake and Cedar River to English River again. The timber estimator and the geologist explored and sketched the chain of lakes extending from the northerly end of Wabuskong Lake to Oak Lake on English River. From Wabuskong Lake we took the route through to Clay Lake on Wabigoon River, and then down that stream to its mouth and then on down the English River to Grassy Narrows where we connected our survey brought through from Favel Lake with known points on English River. Thence down the English River to Winnipeg River, surveying and exploring Roches Moutons Bay, a bay on the north side of Separation Lake and Gong Bay farther down. We then went up Winnipeg River to the head of the south-west bay of Swan Lake and thence southerly through a chain of lakes and streams to Malachi Lake, shown on existing maps. We then turned northerly through Malachi Lake and creek, Otter Lake, Mink Creek, and Cygnet Lake into Swan Lake again. This part of the territory was examined more minutely than other parts because of the reported existence of an extensive tract of white and red pine, which, however, we failed to find. Ascending the Winnipeg River we went ashore for a day or so on the north shore of Sand Lake, at the portage on the route leading north to English River. From Sand Lake we came on up the Winnipeg River to Rat Portage, the point from which we set out, arriving on November 12th. Thus, by following the numerous waterways that intersect the territory, we had the best chance to examine it The geologist and the land and timber explorer were able to use their canoe to great advantage and to gain a knowledge of the country without so many very long trips into the bush, although these were made at points where it was necessary. As a result of our work, I feel justified in saying that we have obtained a fairly good estimate of the character and resources of the territory examined.

We had no time to devote to that part of our territory lying along the railway, nor to the country between Dinorwic and Lac Seul. However, the country along the Canadian Pacific Railway is pretty well known, and the same is true, although in a somewhat lesser degree of the country north of Dinorwic, where O. L S. Niven's surveys, the local surveys of mining locations and the explorations of O. L. S. Robertson in charge of Exploration Party No. 9, during the past season, have contributed to a knowledge of this tract.

The phenomenally dry weather experienced in the district in the spring and early summer was succeeded by showery weather with thunderstorms about the time we took the bush, and there was a very heavy total precipitation by the 7th of October, after which the weather was fine to the end of the trip. We had the usual equinoctial winds in September, which made canoeing laborious and even hazardous at times.

We made a track survey with compass and Rochon micrometer of the routes followed except where a survey already existed. On English River and Lac Seul we surveyed the bays that had been omitted by D. L S. Fawcett in his survey. On Lac Seul there was no survey between the Indian Reserve and Stony Point, nor any between Manitoba Point and Shanty Narrows. I surveyed two bays west of Shanty Narrows. On the copy (reduced) of D L S. Fawcett's map, furnished me, there is a gap on the south shore line east of Grassy Narrows, Indian village, and I was not sure whether this meant another hiatus in Fawcett's survey. But as I was in doubt about it and as there are a large number of islands in the bay in question, and as the season was already far advanced at the time I was there, with a considerable extent of territory still to be examined, I concluded to let it alone On the first route through Black Sturgeon Lake, etc, I connected my survey with the point where the north boundary of the township of Jaffray meets the shore line of Black Sturgeon Lake on the north boundary of Lot 16, and at the easterly end of this route I connected with Mileboard 170 on the Canadian Pacific railway, and by means of O.L S. Stewart's survey of the Canadian Pacific railway track, and the surveys of Jaffray and Haycock townships, I was able to check my work. The closing error was 52.00 chains, being about two per cent. of the direct distance between the tie points, the error having a negative sign, *i. e.*, my survey made the distance too ,reat.

Some of the error is no doubt due to local attraction of the magnetic needle, as this source of disturbance was detected at certain points where observations were made. The magnetic declination was determined from time to time by observation of polaris, and the values used in the different localities are given in the plotted field notes. In the field the compass bearings were recorded, and the astronomical bearings are given in the copied notes. The difficulty of finding the true declination at any point, unaffected by local attraction, is a notable source of error in extensive surveys of this kind, made with the compass.

Except on the few narrow streams which we meandered, our surveys were combinations of triangulation and telemeter methods. From the main stations bearings were taken to subsidiary points, which were thus located by intersections. Thus in the case of narrow lakes the telemeter survey would be carried along one shore whilst the salient points of the opposite shore were located concurrently. Islands were located by the same method. This method made the field notes more complicated and slower to plot, and is not so good for an accurate delineation of the coast line, in most cases, as the direct method of survey, from point to point along the shore by bearing and distance, but we were obliged to employ it as much as possible in order to reduce the amount of paddling and the number of landings to a minimum, laden as our canoes generally were with outfit and supplies and opposed by head winds and heavy seas, making the paddling very laborious and going ashore on rocky points dangerous to the canoes. In some situations, however, we were able to work with light canoes whilst the camp remained stationary for one or more days. In a few instances deep bays were only sketched in as we could not spare the time for their complete survey, especially as it was not imperatively needed. These omissions were very few, however. Numerous islands, especially in Lac Seul, had to be passed the same way.

Messrs. Johnston's and Nash's reports will deal in detail with the mining and the land and timber resources of the territory gone over and incidents of the topography, so that my remarks on these points will be supplemental to theirs and otherwise general in character.

The slope of the territory is toward the north and west, and it is drained to the Winnipeg River and its great tributary the English River. This point has an important bearing on the question of getting out the timber. It is noteworthy that the natural drift of the logs when put into the streams would be away from the existing railway.

The rock is practically granite throughout and is in broken ridges which have in general a north-east and south-west trend, where any trend can be recognized. This broken, corrugated character belongs more especially to the country along the Winnipeg River and easterly to the Wabigoon River, and some of the country east of the southerly portion of the Wabigoon River. It may be said that the whole territory explored is broken and rocky, with scant patches of clay here and there in the gulches and at low levels around certain of the lakes and streams, excepting that portion of the territory lying between Wabigoon River on the west, English River on the north-east and Lac Seul on the east. This major and westerly portion from the Wabigoon River west to the Provincial boundary is, generally speaking, a rough, rocky tract of country with little soil, a rather noticeable absence of swamps and no marked elevations. The whole tract is densely wooded with jack pine, poplar, birch, spruce, balsam, cedar, etc., and has a few unimportant areas of red and white pine. There is a total large amount of poplar and spruce pulpwood, and also a large amount of jack pine and tamarac for railway ties, mine timbering and other similar purposes, with an immense quantity of all sorts for cordwood, the whole tract being a typical Laurentian area. Forest fires causing great loss of timber have swept over large areas, especially in the Winnipeg River region and eastward along the Black Sturgeon Lake and Canyon Lake chain of waters, as described by the timber explorer.

The jack pine has its station on the rocky ridges and the rocks generally, and also on sandy and gravelly spots. The poplar, birch and spruce grow on the slopes in the glacial debris of boulders, gravel, &c., and in the gulches and on the flats where there is soil. Sometimes it is found on the rocks among the jack pine, but then it is mostly small and stunted.

As regards the floating out of the timber when cut into logs the whole region is well served by the numerous water routes The country adjacent to Canyon Lake and its tributaries would send its logs down the Wabigoon River whilst the country west would employ the Winnipeg. The English River would, of course, be used for its adjacent territory.

The second division of the region explored by us, viz: the portion east of the

Against wind and current. Ogoki River. Party No. 7.

Falls at portage 65. Party No. 7.

Pitikigouching River. Party No. 7.

Indians travelling. Party No. 7.

Nepigon Trou Party Nc 7.

Nepigon Trout. Party No. 7.

Approaching rapids, north of Kenogami Lake. Party No. 7.

northerly part of Wabigoon River is distinguished physically from the first division just described by its comparatively flat and low-lying surface with its extensive areas of clay soil and consequently heavy growth of large timber, this feature being apparently due to the more effective glaciation of the rocks, by which they were planed down, and left with a comparatively even surface. The Keewatin shore also, from the mouth of the Wabigoon River, north-easterly, is of the same flat character, and is also, I may add, densely timbered with poplar, spruce, etc., a circumstance, by the way, which adds to the value of the Ontario forest, because it will permit lumbering operations to be carried on in a larger way, and, therefore, at less cost per unit of output. As we go down the Wabigoon River this level tract, being our second division, is met with on the east shore of Clay Lake, an expansion of the river at the portage on the route to Wabuskong Lake. There is quite a tract of clay land at the west end of this portage. At Blackwater Lake, a little further north, there is another tract of level clay land, supporting a heavy growth of poplar, spruce, etc. The same clay is met with again lower down stream at the first falls below Clay Lake. We meet it on the lower reaches of the Wabigoon, then up the English River to Lac Seul, and along the shores and on the islands of that lake. The entire surface is not soil-covered without any protruding rock; the rock is never many feet below the surface, and is constantly showing itself in small knobs on low ridges, but these are often covered except on their very highest points with a thin layer of soil. The result is that the heavy forest covers nine-tenths of the surface. From the bottom of the Bay of Islands, Lac Seul, the flat country extends to Wabuskong Lake, and reaches its maximum development on Cedar River and the chain of lakes extending from near Wabuskong Lake to Oak Lake on English River. But it may be said to prevail with greater or lesser interuptions over all the territory lying north of the canoe route from the Bay of Islands through to Clay Lake and east of the Wabigoon River. This large tract is well timbered throughout, with a large percentage of the surface clay, but the most extensive belts of heavy timber and the largest tracts of good land are in the portion about Cedar River with the chains of lakes to the west of it, between these waters, the English River and Lac Seul. For further details I must refer to the timber estimator's report.

The country along both sides of Wabigoon River from a point a few miles below Mutrie Township, down to Clay Lake and thence along the west shore to Canyon River is of the same rough character as the country to the west, already described, the timber being small with considerable brulé country. Between Mutrie Township and the nearest falls on the Wabigoon River down stream there is some good clay land near the river.

This north-eastern portion of the territory explored by Party No. 10 contains large areas of clay land and extensive forests of valuable timber, poplar, spruce, jack-pine, tamarac, etc., and the timber could be readily taken out along the different waterways, whilst the waterfalls on the English River would supply the power for any manufacturing that might be undertaken. The arable land could be profitably cultivated by any settlers that might come in owing to the opening up of the timber industry.

I will now add a few words about the southerly portion of our territory lying along the Canadian Pacific Railway, being acquainted with it to some extent through repeated surveying and exploring expeditions.

The country between the Manitoba boundary and Rat Portage is in general of a rough broken character, being chiefly a Laurentian area with the usual jack pine ridges and poplar gulches. Lumbering in ties and cordwood has been carried on at many points, but the timber now standing is important chiefly for cordwood. Much of the original timber has been killed by fire. There are numerous lakes.

East of Rat Portage the same Laurentian rocks prevail with the characteristic topography, timber and soil and patches of clay, of gravel or of sand. The country is generally a brulé. Jack-pine and poplar are the prevailing timber and the tract has great resources in cordwood. At the west end of Eagle Lake the clay land begins and extends easterly to Dinorwic. The Pioneer Farm in Dryden is in this belt, The timber does not merit special mention further than to say that here have been fine forests of jack-pine, most of which have b[illegible] ..ed by fire.

North of Dinorwic towards Lac Seul a belt of Huronian schists cross the country striking north-east and south-west. A number of mining claims have been located [illegible]ere and some development work done.

The clay soil found at various points of the region examined, and notably in the north-east, is apparently the same as that occurring at other points of the Rainy River District, its greatest development, perhaps, being around Wabigoon and Eagle Lakes, as in the Township of Dryden. It is also seen mantling the rocks on the islands and around the shores of the Lake of the Woods. Excellent brick is made from it at Rat Portage and at Wabigoon. It appears to be a pure levigated clay, very free from sand, and such as would be deposited from suspension in water and thus can form only a clay soil, but not a loam. It was deposited on the floor of the vanished glacial Lake Agassiz, and might appropriately be called the Agassiz clay. It is found only at low elevations, but would appear to occur at high elevations as we go north. At the mouth of Mattawa River, on English River, it covers the plateau to a height of twenty five feet above the stream. Perhaps the most salient feature of the topography of the Rainy River District is the great amount of water-surface and the extremely irregular shore lines. These characteristics are well sustained by that portion under discussion. In giving a brief description of these it will perhaps be as well to go over them in the order in which we took them up in our field work.

The Black Sturgeon Lake discharges into Winnipeg River over a rapid which can be run in a canoe at high water, but at low water a portage must be made. This is a long, narrow lake, with dark-coloured water, containing pike and pickerel and some sturgeon. On the south shore it is nearly all brulé, but on the north shore, towards the east end, there is some fine timber, and from this part a good many railway ties have been taken. From the east end of the lake we pass through a stream that would float sawlogs into Silver Lake. There are two short portages, and the ascent to Silver Lake, sometimes called Manitou Lake, is about twenty-five feet. It is thus a water-power of considerable value. Silver Lake is, perhaps, the most beautiful sheet of water seen on the entire trip. It is approximately circular in plan, with an arm extending towards the east. The surrounding scenery is quite interesting in this generally flat country, especially on the north side. Members of our party sounded the lake off the south shore of Swallow Nest Island and found a depth of over one hundred and fifty feet. The lake abounded in lake trout, so that it is quite a resort for Rat Portage sportsmen. The water is clear and sweet.

From Silver Lake we passed through a series of six small lakes into Favel Lake, the last of the six being East Lake. Long Pine portage, between John and Mark Lakes, is over 40.00 chs. in length; the rest of the portages are short. Between East Lake and Favel Lake is a height of land, East Lake discharging towards the west and Favel Lake towards the east. The country along this route, from Black Sturgeon Lake, is rocky and broken, with very little soil and not much timber, being largely a brulé. Favel Lake is a narrow sheet of water, lying in a rough, rocky country, and discharging into Canyon Lake through a creek meandering through a heath muskeg. Logs could be driven down this creek at ordinarily high water. There is about 23.00 chains of this narrow, crooked creek, after which it becomes wider and deeper and would float logs all the year round. From the north-west arm of Favel Lake a canoe route goes northerly to Grassy Narrows Lake of English River, through a chain of fine lakes, connected by good portages, none of them long. The next lake to English River is a good-sized sheet of water, with the surrounding country rough and broken and partly burnt over. Not much timber of any value was seen here or at any other point on this route. Canyon Lake discharges into Wabigoon River, through Canyon River, from the north-east arm of the lake, the distance to the river being about four miles through a hilly country. The westerly part of Canyon Lake is in a deep ravine, with steep banks, towards the east, however, the land becomes flatter. Following up Canyon Lake through its south arm, we can pass by a chain of seven lakes to a point within a mile and three quarters of the railway track, about a mile east of Jackpine Siding. From the two lakes next the track the water goes south, but from the other five it goes into Canyon Lake.

On all the portages spoken of thus far there are good trails, except, perhaps, the one between Boulder and Augite Lakes, which needs clearing out.

Wabigoon River.—On Eagle River there are three falls within a couple of miles of the railway, constituting fine water powers. Between the north boundary of Mutrie Township and English River there are five notable falls, two above Clay Lake and three below, the descent at each, taken in order going down stream, being fifteen feet, eighteen feet, ten feet, four feet and eight feet, respectively. In most instances the head could be increased by a dam, so that at each of the four highest of these falls a large horse-power could be obtained.

For a few miles after leaving Mutrie Township the stream is from four to six chains wide, with very slight current; the water being extremely low, however, at the time of our visit. It winds about between clay banks over a clay bed, and widening out at intervals into marshy lakes, much of the country being brulé. It maintains its river-like character until it reaches Clay Lake, which acts as a settling basin for its yellowish, muddy waters. Issuing from Clay Lake the water is tolerably clear, and at the same time the river widens out irregularly here and there into lake-like expansions and conforms to the type of the Laurentian River. There is still not much current, except at narrow places. Our survey was made in August, when the water was very low. Passing down the lower portion of the stream again in the early part of October we found the water had risen about six feet, and was then bank-full.

Canyon River.—This stream of dark water, through which the Canyon Lake system discharges into Wabigoon River, is from one to three chains wide, and winds through a rough country throughout its course of about four miles. At the mouth is a fall and rapid making a total descent of about twenty feet, farther up stream is another fall of twenty five feet, while at the outlet of Canyon Lake there is a fall of about thirty feet. At an average stage of the water there is a valuable power at each of these falls, although at high water in the Wabigoon River the effective head of the lowest falls would be materially reduced, but probably the whole head of the stream could be utilized at the mouth.

Lac Seul.—The portion of the lake that came under our cognizance is remarkable for its deeply indented coast line and its numerous islands. The water is clear and the shores for the most part flat. The large islands are in general well-timbered. Almost from Indian Reserve 28 the lake country is well-timbered, and shows areas of clay land all the way to English River.

English River.—This stream receives the drainage of a large territory, and when it joins the Winnipeg River it is almost, if not quite, equal in magnitude to that stream. From Lac Seul to Barnston Lake it maintains the stream like character tolerably well, but below that it is a succession of lakes connected by narrower channels. Its principal tributary from the north in this quarter is the Mattawa River, whilst in the south it receives in the Wabigoon a most important addition to its volume. There is a very noticeable decrease in volume where it issues from Lac Seul, compared with its volume at the narrows below the mouth of the Wabigoon River. There are quite a number of falls between Lac Seul and Winnipeg River. The first are two or three miles below Lac Seul outlet, where are two falls near together, the first a fall of ten feet and the other of sixteen feet, the width of the stream being much contracted. Next come the falls near the mouth of Cedar River, where there is a total descent of twenty feet, divided among three cascades and a rapid. The portage here is forty-three chains across. Next come the falls, about a mile east of the east end of Barnston Lake, where the stream is contracted to a width of about four chains, the fall being ten feet, forming a beautiful cascade, the snow-white mass of foam being visible far out on the lake. The next falls are at Oak Lake, where there is a descent of about twenty feet in a distance of half a mile, one fall with a drop of eight feet, where it pours over the rim of Oak Lake. The upper fall is six feet, and the fall in the rapids about five feet. The next fall is at the narrows between Maynard and Tide Lakes, where there is a fall of ten feet or over, while the river is very swift for half a mile above, so that the total descent there must be as much as fifteen feet. There is a portage of about eight chains here on the north shore, and also one across the peninsula opposite the falls on the south shore. The next falls are some four or five miles below Separation Lake, where within three miles there are three falls, the first, Sunset Falls, of ten feet, and the other two of about twelve feet each, making a total fall of about thirty-five feet, and the largest power on the river. The last, or lowest, fall is a little below the outlet of Deer Lake, and the drop is about eight feet. From the Wabigoon River up to Lac Seul the country along the English River is well-timbered with poplar, spruce, jack pine, etc., on clay land, as mentioned elsewhere in these reports; the water is clear, and whitefish and other kinds of fish are found in it.

WINNIPEG RIVER.—We made no track survey of this stream except to run a tie traverse from a traverse post on an island in the mouth of English River to the south-west bay of Swan Lake, and a survey from this point around to the south-west arm of Swan Lake. The country along Winnipeg River from Swan Lake to Sand Lake, and thence to Rat Portage is for the most part rough and broken, the timber being chiefly jack pine; there are areas of burnt country also. At White Dog Indian Reserve—Islington—there is a tract of good clay land, and a smaller tract of the same at the Dalles, about ten miles from Rat Portage.

At High Falls, on Winnipeg River, there is a fall of about twenty-five feet, constituting a most valuable water-power. This is about thirty miles from Rat Portage. At the Dalles, about ten miles from Rat Portage, there is a strong rapids, but a small steamer makes trips up and past it, using a line, however, in going up stream. From the south-west corner of Swan Lake we made an exploration survey southerly up a chain of lakes and streams to Malachi Lake (shown on existing maps), and then northerly again through Malachi Creek, Otter Lake, Mink Creek and Cygnet Lake to Swan Lake again. Logs could be run out along the Scot Lake route to West Swan Lake east. There is, however, not much timber to come out from this region. We examined this part of the country between Winnipeg River and the Manitoba boundary, in order to discover the area of white and red pine and hemlock, reported to have been seen in this particular section. We could not see anything of the kind, however, I regret to say.

ZOOLOGY.

MOOSE.—This noble game is quite plentiful in this northerly part of the Rainy River District, and is killed in summer, autumn and early winter without any regard to game laws. A good many are killed in midsummer, when the flies drive them to take refuge in the water and marshy lakes and streams, where also they feed on a certain plant. An old Hudson's Bay Company's employee at Lac Seul Post remembers when the moose first began to come into that country, about forty years since.

CARIBOU.—This deer is seen in large herds on the ice and around the shores of lakes in winter, and less frequently and in small herds or singly in summer. It would appear that they migrate slowly southward after the lakes have frozen over, the return movement toward the north taking place before the ice goes in spring.

WOLVES.—These are not numerous, still there are a few large timber wolves in the country, and occasionally one is killed. We saw the carcase of one on the shore of Sturgeon Lake, and one was ambushed and shot by a trapper on Eagle Lake last spring, They hunt the caribou successfully, although I think infrequently. I know of two well authenticated instances where a wolf ran a caribou down and killed him, the wolf being afterwards shot by a hunter. In both instances, my informant averred that the caribou would have escaped had he not gone repeatedly into the bush, where he fatigued himself in the deep snow, whilst his pursuer profited by the manoeuvre by alighting at every spring on the track left by the deer, and so avoided the stress of breaking a road through the snow. In this region, where rain in winter is of rare occurrence, and the snow is otherwise dry and feathery and seldom deep, it must seldom happen that there is a crust on the snow strong enough to carry a wolf and give him his proverbial advantage in running down his prey. The action of the Legislature in raising the wolf bounty to twenty dollars is highly to be commended. It appears that the wolves follow the caribou in their migrations. A person at the post on the Lac Seul informed me that the wolves disappeared about the time the ice went in the spring. I have not heard of an instance where wolves pulled down a moose.

RED DEER.—The Indians killed a number of red deer at Eagle Lake last winter, I believe, and in June last I saw red deer tracks on an inland in that lake. Red deer have been occasionally met with in the Seine River and in the Rainy River region, and farther north also, and it would appear that they are coming northwards. They evidently come from the State of Minnesota, where they are retiring before the advance of settlement. Were they protected, our woods would have in a few years a most important addition to their game animals, but otherwise the Indians will promptly kill them off as fast as they appear.

FUR BEARING ANIMALS.—These are the black bear, the Canada lynx, fisher, red, cross, black and silver-gray foxes, marten, skunk, weasel, otter, beaver, mink and muskrat.

There is also the ground hog, porcupine, red squirrel, flying-squirrel, hare, dormouse and field mouse. The domestic rat is as yet unknown. The beaver has become very scarce; only one was seen on the trip, although fresh cutting was seen in three places. Even at the Hudson Bay Company's post at Lac Seul, only a few beaver skins are obtained in a year: whilst formerly one of the sub-posts would send out a thousand. The numbers of Indians constantly trapping over the country from year to year keep the fur-bearing animals from becoming plentiful:

BIRDS.—Wild geese fly over in spring and autumn, but they seldom alight and not many are killed. At Goose Lake, however, west of Cedar River, Messrs. Nash and Johnston saw a number of geese which had apparently located on the rice beds there.

DUCKS.—These are to be found in great numbers at the rice beds that are scattered all over the district. Amongst the different kinds the mallard is well represented. The ducks become very fat after having fed on the rice for a time, so heavy as to be unable to get on the wing in some instances, I believe. There is a story of an Indian who in late autumn came upon a large flock of ducks which, being unable to fly, took refuge in the bush and were there slaughtered by the dusky sportsman, assisted by his faithful dog, to the number of several hundreds. The Rat Portage sportsmen make large bags at the rice grounds in the autumn.

LAND GAME BIRDS —The other game birds of the region are the gray partridge, spruce partridge and the prairie chicken or pintailed grouse. Specimens of the latter were seen down the Wabigoon River and down the Winnipeg River, and the species was quite plentiful one time about the Indian reserve on Lac Seul. This bird, as is well known, will colonize a tract of burned woods and remain a number of years, and then leave when the new forest appears. Prairie chickens are to be found almost anywhere along the Canadian Pacific Railway track, along which they have now penetrated from the west to Sudbury and beyond. Ptarmigan have been seen at various points in this district in times past, having come south to obtain subsistence when the prevalence of deep snow over their habitat in the north deprived them of access to the shrub willow, on the buds of which they usually feed.

GREBES.—These are represented by the loon or great northern diver, and the helldiver.

There is a species of rail, I believe, but specimens are rare. Other aquatic birds are the white gull, the grey gull, the heron, the bittern, the crow-duck and the king-fisher.

BIRDS OF PREY.—Of birds of prey we have the bald eagle, the fish eagle, grey owl, horned owl, snowy owl, and various species of hawks. Crows were seen everywhere, but ravens are not numerous; they remain all winter.

SMALL BIRDS.—Smaller birds which may be mentioned are therobin, swallow, martin, night-hawk, bluejay, whiskey-jack, chick-a-dee, three species of wood-pecker, the wren and other warblers, one or two species of climbers, and two or three species of ground sparrow. One humming bird was seen on our trip. There has been a colony of English sparrows at Rat Portage for a number of years, and we saw one on Deer Lake, on English River, and others on Winnipeg River, north of Swan Lake.

FISH —The pike is found in all waters, and the pickerel is about as common, although his favorite haunts are in waters where there is a current. Whitefish is found in Lac Seul, Wabuskong Lake, English River and Winnipeg River, and Sturgeon River, in the larger waters. Lake trout are found in Silver Lake, in Linklater and Daniels Lakes, lying north of Hawk Lake, in some of the lakes on the Grassy Narrows route in Lac Seul, English River and Winnipeg River and in Wabuskong Lake. These latter trout are rather rank-tasting and are said by the Indians to be too rich. Some years ago fishing was carried on for one or two seasons in Williams Bay, Lac Seul, the catch being hauled out to the Canadian Pacific Railway by dog-team. Sturgeon and pickerel in quantity have also been taken out of Canyon Lake, at the east end, within the last few years.

BOTANY.

TIMBER TREES.—The timber trees are jack pine, poplar, white birch, tamarac, balsam, cedar, red pine, white pine and willow. The jack pine station is on the rocks, or on sandy and gravelly plains, and in scattering trees amongst the poplar, etc, where there is good soil. It is the great resource for cordwood in this country, and is beginning to be taken for railroad ties as well.

SPRUCE.—This tree is about as common in these western forests as any other species, and has its station anywhere from the rocky ridge to the muskeg, but it is only where there is soil that the tree attains a useful size. The clay soil covering the country around Cedar River accounts for the fine spruce forests of that locality.

OTHER TREES.—The species to which allusion will now be made are not of much, if any, economic interest. Black ash and soft elm were seen at a few points, as at Black Sturgeon Lake and on the upper reaches of Wabigoon River. A single specimen of the western or ashleaved maple was seen at White Dog Reserve on Winnipeg River. On low points along the shore of Winnipeg River, near Swan Lake, and at points on the shores of lakes south of that, stunted trees of oak and ironwood were seen. On Wabigoon River a stunted or dwarf oak was met with, having an edible acorn about the size and shape of a thirty-two calibre conical pistol bullet. On the Wabigoon River was also found a species of hawthorn with slender and smooth stems, bearing a large fleshy haw. The mountain ash was in evidence on the cool, shady north slopes of steep ridges. The specimens seen were of slender growth, none of them being more than two inches in diameter at the butt, and when bent over with the weight of their clusters of scarlet berries formed a beautiful object.

The hazel grows almost everywhere, and nuts appear to have been a good crop last year. There are two kinds of cherries—the red cherry and the choke cherry.

WILD FRUITS.—The wild plum tree was occasionally met with. It is found in the woods around Lac Seul, but I could not learn that the fruit was gathered or even produced. Wild plums are gathered, however, at Indian House on English River and at White Dog Reserve on Winnipeg River. Of small fruits, which are, however, common throughout the Province, I may mention raspberry, strawberry, dewberry, black, red, and skunk currants, blueberry, saskatoon, gooseberry and highland cranberry. The cranberry is not plentiful owing to the generally infrequent occurrence of the muskegs where it grows.

WILD RICE.—This is found throughout, as on Winnipeg River, Black Sturgeon Lake, Wabigoon, English and Cedar Rivers, Lac Seul, Wabuskong Lake, etc. Its favorite station appears to be in a muddy bay or channel where a small stream comes in, causing a slight current. Wild rice is a staple article of diet with the Indians, and it is for sale at the Hudson's Bay Company's posts. The crop was a failure in 1899 owing to the extremely high water at a critical time in the season. The crop in 1900, however, was very good and one family on Lake Wabuskong gathered as much as ten bushels. In old times an allowance of wild rice used to constitute the entire ration dealt out by the Hudson Bay Company to their employees at certain points, sometimes a little pickerel oil was added as a luxury, so I am informed by ex-employees yet living at Fort Seul. This shows that wild rice is a nourishing article of diet. As already mentioned wild ducks in great numbers frequent the rice beds. The Indians gather the grain in September and obtain it by bending the stalks over the gunwale of the canoe and then beating the grain from the straw with a stick, It is prepared for food by heating it on an improvised scaffold over a slow fire to thoroughly dry it, after which it is treated in a sort of mortar which is sometimes no more than a hemispherical excavation in the stiff clay soil. Being pounded in this receptacle, the hull made brittle by the previous heating over the fire is readily detached from the enclosed kernel. The mass is then treated on a sieve and the separated grain finally cleaned in the wind to get rid of dust. One way of preparing it for eating is to parch it in a pan over the fire ; it is then like popped corn.

GRAINS, GRASSES AND VEGETABLES.

A little Indian corn is raised at the Indian Reserves on the Winnipeg River, but it is of no importance. The valuable blue-joint grass grows along sheltered places of rivers and lakes, and in general where there is a moist soil containing humus. A good deal of it grows along the Wabigoon River, although not in large patches as a rule. Wild pease and vetches, so valuable as fodder, are found in birch and poplar woods or in brules, where there is soil, their favorite station is a bank of clay soil in open bush or brulé with a southerly aspect. At Lac Seul we noticed the plantain and white clover ; these plants are usually found where there is a settlement. A similar phenomenon is the distribution of red clover and timothy along the Canadian Pacific Railway track. Canada thistles grow at the Indian Reserve on Swan Lake.

Potatoes are cultivated at all the Indian villages where the same plot is sown year after year to the manifest detrioration of the tubers. The Indians do not appear to know that the best potatoes are raised on new land or they are too indifferent in the matter to undertake the labour of clearing a fresh potato patch. The Colorado beetle has not yet appeared on the Indian Reserves in the north, although it has long been a pest at the settlements along the railway.

Wild Flowers and Plants.

The following flowers and plants are indigenous in this region explored and in Rainy River District generally. The white water lily, yellow water lily; marsh marigold; tiger lily; wild rose; wild primrose, blue bells, arbutus, anemones, violets, asters, heliotrope, golden rod, lady's-slipper, species of the mint and of the mustard families, potentilla, sarsaparilla, dock, prince's pine (a specific for kidney trouble among the lumbermen) partridge berry, bear-berry, gold-thread, dandelion, lambs quarter, small burr, flag and various species of heath, including Labrador tea. A single specimen of the yellow oxalis was seen in Oxalis Bay, Fishing Lake, English River.

Second Growth Forests.

From the occurrence of charred stumps and the general absence of fallen trunks, it would appear that the present forest is a second growth in that it has succeeded one destroyed by fire many decades ago. These evidences of ancient forest fires, in the presence of charcoal, and the general second-growth appearance of bush are observable all the way from Winnipeg River and Lake of the Woods to the shore of Lake Superior, and are, I think sufficient to verify the tradition existing amongst some of the Indians. According to this there was, more than one hundred years ago—the hottest and dryest summer that had ever been known, or that has been since, when fires broke out in the bush and raged until the whole country was burnt over. Remnants of the old forests are occasionally met with, chiefly in swampy places where the moisture checked the fire.

Climate.—It will make this sketch more complete to add a few words about the climate of the Rainy River District. Navigation on the Lake of the Woods lasts for about half the year, on an average. The snow fall is not great, and the snow is light and feathery all winter, as rain during this season is all but unknown, and even a thaw is very rare. The winters are of course long, and are generally cold, but there is a noticeable absence of storms, and some winters are so favourable in this respect and in the absence of extremely cold weather as to constitute a winter climate, which for wholesomeness and general agreeableness is not excelled in any part of the Dominion. Winter and spring are dry, whilst the summer and sometimes the autumn are moist and even wet. The plentiful rainfall in a land of forests and lakes ensures a cool, pure atmosphere so that the summers are never hot, although there are, nearly every year, a few hot days when the wind is coming off the heated prairies of the south and south-west. The nights however are almost invariably cool, and it may be said that a blanket is an agreeable wrap every night in the year.

Finally, I would point out the great resources of the territory prospected by us as a sporting, summer and general health resort, for it has attractions in these respects which are not surpassed by any tract of country on the continent. It numerous water-route through the forest make it simply a paradise for canoeists. The position of Rat Portage, a town on the main-line of a continental railway, as a point whence voyages can be taken by canoe over so many different routes to points near and far, over the main water-ways through the inland lakes, is quite unique. There is no doubt, I think, that in time this section of the Province so accessible from the railway and possessing all the qualities desired by the tourist or summer settler in search of sport, varied recreation, or for a haven in which to rest and recuperate will become widely known, its advantages appreciated, and its benignant resources exploited. In anticipation of that time, one improvement suggests itself, which would also be a good thing *per se*, viz.: The gradual re-stocking of the region with beaver.

I further respectfully submit that the Government would be fully justified in setting aside a portion of this territory for a national park and game preserve, viz., that portion bounded by the Wabigoon River, English River, the western boundary and the Canadian Pacific Railway track.

To sum up in a word: The portion of Rainy River District examined by Exploration Party No. 10, during the past season, contains immense quantities of timber for cordwood, railway ties, piles, mine timbering and other construction purposes, and a great quantity of spruce, poplar, and birch for pulp-wood, the timber being accessible and easy to get out along the numerous water-ways intersecting and abounding in the territory in question with ample water power for its manufacture if necessary.

The agricultural areas, though circumscribed, would yet furnish an important source of supplies when the timber resources came to be exploited, although the present crop growing on this land, viz.: the timber, is probably as valuable a one as any that could be substituted for it.

The mining capabilities of the region are at present not very great apparently, but further and more detailed exploration and the progress of geological science may yet reveal substantial wealth in these Archean rocks.

All of which is respectfully submitted.

I have the honour to be, sir,

Your obedient servant,

JOHN MCAREE, O.L.S.

The HON. E. J. DAVIS,
Commissioner of Crown Lands,
Crown Lands Department, Toronto.

LAND AND TIMBER ESTIMATOR'S REPORT OF EXPLORATION SURVEY PARTY NO. 10.

RAT PORTAGE, RAINY RIVER DISTRICT, ONTARIO,
December 24th, 1900.

Sir,—I beg herewith to submit my report as land and timber estimator on Exploration Party No. 10 which, in your charge, examined during the past season, under instructions from the Crown Lands Department, that portion of the Rainy River District lying north of the Canadian Pacific Railway, and extending from the western boundary of the Province of Ontario, to the Indian reservation on Lac Seul.

July 2nd. I arrived at camp on Black Sturgeon Lake with the party, having travelled by canoe from Margach station on the Canadian Pacific Railway.

July 3rd. The party started west to outlet of Black Sturgeon Lake, which is the Winnipeg River. Mr. Johnson and I continued through the bush in a northerly direction, until we reached the banks of the Winnipeg River. The country through here is all brulé. We returned on the southern side of Black Sturgeon Lake and climbed a high, bare hill, from which was obtained a good view of the surrounding country. For seven or eight miles either north, east or west to the Winnipeg River, all was burnt excepting in the deeper ravines and just on the lake shore, where the timber has escaped the fire. The timber is in no place either thick, or of good size. It is composed of 40 per cent. poplar, 30 per cent. jackpine, 10 per cent. spruce, 10 per cent. tamarac, with a scattering of white birch and balsam on the lake shore, also some odd red and white pine. The country is rocky and barren and of granite formation.

July 4th. We proceeded east to the central portion of Black Sturgeon Lake. On the north shore there is considerable green timber, which extends for seven or eight miles along the shore and for one and a half to three miles back. There is but an occasional red pine, white pine or tamarac, the timber being jackpine, spruce, poplar, white birch and balsam, in all forty-five cords per acre. Of this 10 per cent. is spruce of medium size. This timber could readily be put in Black Sturgeon Lake, but would have to be taken out by way of the Winnipeg River. The water here courses north. The balance of this section is burnt bare.

July 5th We proceeded from Black Sturgeon Lake up Rice Creek. Here on the west side of Rice Creek there is a belt of good land, extending back from the creek about half a mile and along the creek one mile. The land, which is in a valley, through which runs the creek, is covered with fair-sized spruce, balm of Gilead, tamarac and poplar, in all forty-five cords per acre, of which 15 per cent. is spruce and 15 per cent. tamarac. This timber would have to go out by way of the Winnipeg River. The remainder of the land on the east and west sides of the creek in the valley is rough and rocky, with some green timber, about twenty cords per acre. The balance of the land on the Rice Creek is covered with second growth jackpine, spruce, poplar and tamarac, about twenty cords to the acre.

July 6th. We visited the northwest end of Ant Lake. The country has been burnt for some years and small second growth covers the whole surface. Travelling west from the northwest end of Ant Lake, about one mile, we discovered another small lake. On the east of Ant Lake there is a grove of red pine about eight thousand feet B.M., but no other timber of any size.

July 7th. On account of heavy rain we returned from Ant Lake to our old camp on Black Sturgeon by canoe.

July 9th. We travelled to the south-east of Black Sturgeon Lake by canoe. On the lake shore the timber consists of poplar, jackpine, small ash, scrub oak, elm, tamarac and white birch of little value. To the north of here, distant about three miles back in the woods by land, there is heavier timber, principally jackpine, poplar, white birch, spruce and tamarac, 10 per cent. of tamarac and 10 per cent. of spruce. I saw here moose, caribou and muskrat. We made our camp on Silver Lake.

July 10th. Silver Lake is twenty-five feet higher than Black Sturgeon Lake. The inlet into Black Sturgeon Lake is a rapid about one-eighth of a mile in length. The timber along this rapid is of no value. Silver Lake is clear, and two soundings made off the south shore of Swallow Nest Island showed one hundred and forty and one hundred and ninety feet of water. The timber is all second growth, the country having been burnt, excepting on Swallow Nest Island. There is here twenty-five thousand feet B.M. of red and white pine, taking thirty logs to the thousand. The country is high, rocky and of granite formation.

July 12th. We went round Silver Lake and landed on the north side, proceeding on foot inland to the north-east end. We found the country burnt for miles in every direction.

July 14th. We went north from Silver Lake through a chain of lakes, and made three portages in four miles. The country was still burnt and rocky. Here I saw a fresh beaver cutting. The rocks are of a granite formation.

July 16th. I left Camp on Silver Lake by canoe with Johnson we proceeded east and portaged into Bog Lake, traversed a shallow creek from Bog Lake to Clear Lake, a distance of a quarter of a mile, and then portaged into John Lake on the north. This lake empties into Clear Lake with a fall of over twelve feet, being twenty feet wide at the falls. There is no timber but small second growth on this route, excepting a little, not of much account, on the portages, which are flat. The country is hilly and broken. Leaving John Lake we crossed Big Pine portage, on the east side of which there is one hundred and eighty thousand feet B.M. of fine large red pine. This pine is old forest but green. We crossed over to Little Pine portage where we camped that night.

July 17th. The party moved camp to Sand-fly portage, at the north-west corner of Favel Lake, by an old Indian route, which goes to Grassy Narrows, on the English River. I left the party at Little Pine portage, and went about two miles east from the end of South East bay of Favel Lake. Here I found another lake which enters into Favel Lake by a fall of twenty feet in height, affording considerable power. Along the shore of the rapids above the falls, there is a big bunch of timber, half spruce and half tamarac. This belt of timber extends back from the creek on each side from one-eighth to one-quarter of a mile, and is at least two or three miles in length. I returned to Little Pine portage and proceeded with the party to Sand-fly portage, where we camped.

July 18th. Went north with the party by the lake route towards Grassy narrows. The timber in this district is all second growth, and does not include any pine. We camped on Shore Lake.

July 19th. Went five miles east of Big Lake and climbed to the top of a high hill from which I could see the surrounding country for a radius of nine miles each way. A number of points of the English River are visible from here. Fire has cleaned out all the timber in this district, leaving nothing but rock.

July 20th. We went north by canoe up the Grassy narrows route. I climbed another high hill from which I could see three swamps covered with small spruce trees from four to six inches thick. There is also considerable jack pine in places large enough for railway tie timber. The jackpine would cut fifty cords per acre. We moved back to Sandfly portage.

July 21st. Moved into Canyon Lake, some distance past the narrows. From here I explored both north and south. The lakes are of considerable extent, and surrounded by hills. I found several small bunches of red pine which would run about thirty-six logs per thousand. The rock formation is all granite.

July 23rd. We continued south, crossed a large clear water Lake, and camped on the south side of Daniel's Lake. Camp No. 9.

July 24th. We left camp and went three miles east. To the south and west the country is bare, but there is a small second growth of pine to the east. To the north east of Daniel's Lake there are ten thousand feet of red pine near Camp No. 9.

July 25th. Returned to Camp No. 8 on Canyon Lake, and thence went back to the north, through a narrow lake for two and a half miles east and north of Camp No. 8. The timber is very good for three miles. It runs about 7 per cent tamarac, 50 per cent jackpine, and 40 per cent spruce. The latter is from three inches to ten inches on the stump. The timber would cut sixty-five cords per acre. North of this the timber is small.

July 26th. We went north east of Canyon Lake to the outlet. Here we found a water-fall of thirty feet, but no timber of commercial value. The water-fall here mentioned is one of great value ; the immense area of water of the lakes empties through a gorge about twenty five feet wide, and the fall could be more easily and cheaply dammed than anything of the kind I have ever seen.

July 27th On the eastern arm of Canyon Lake I found one million five hundred thousand feet of red pine of fine quality which could be logged very cheaply, as it grows on the last arm of Canyon Lake. On the peninsula to the west of this timber is about one hundred acres of very good land, clay covered with small second growth, while to the east of this there is a big tract, extending over two miles, of timber half spruce and half tamarac, which would cut about forty-five cords per acre.

July 28th. We moved south from Canyon Lake into Forest Lake and made a portage, where there is a water-fall of seven feet. Here also there is a great deal of spruce, large poplar, birch, tamarac and jackpine. About 50 per cent of this timber was spruce and poplar. This timber would cut fully seventy cords per acre, being big and heavy.

July 29th. We continued southwest through the chain of lakes, and found one hundred and twenty thousand feet of red and white pine on the southwest end of Boulder Lake, also some spruce, poplar, jackpine, white birch and balsam.

July 30th. We continued south from Boulder Lake, principally through a burnt country. There are, however, some bunches of timber on Augite Lake, comprising large poplar, spruce, jackpine and white birch.

July 31st We reached Joe Lacourse's mining camp, two miles north of the track, east of Jackpine Station, on the Canadian Pacific Railway.

August 1st. We left Joe Lacourse's camp to find the watercourse to the Canadian Pacific Railway and located the same.

August 2nd We left Joe Lacourse's camp for Jackpine tank on the Canadian Pacific Railway between Parrywood and Summit stations.

Augast 3rd. The party moved out to Jackpine tank.

August 4th. The party moved to Eagle River station.

August 6th. We moved down Eagle River to Wabigoon River and located township lines, and made our camp near the north boundary of Mutrie township.

August 7th. We travelled down Winnipeg River over small rapids and continued down stream, where we came to a waterfall of about sixteen feet in one hundred yards, at which we camped.

August 8th. We left the fall and continued down Wabigoon River. The country through which we passed was all brulé.

August 9th. We went east by land ten miles and found nothing but brulé country in every direction. There were vast quantities of blueberries. We also saw some moose. The rock formation is granite.

August 10th. We moved by canoe down Wabigoon River to Clay Lake in a north-westerly direction. We explored the north-east side and found a creek running north-east to Blackwater Lake. We camped on Blackwater Lake.

August 11th I left camp by canoe alone, and entered and traversed the woods to the south-east about three miles. Here I discovered fine spruce and tamarac timber on good land. A creek runs through this valley. This is a large tract of timber in good condition and would cut eighty cords per acre. It could be readily placed in the lake. There were here two old hunter's camps, well fitted, where many moose and caribou had been killed, the skulls of which are strewn around.

August 13th. I went by canoe one mile to the north-east end of Blackwater Lake, where I landed and followed a creek for about three miles. I left the valley of the creek and went south-east about two and a half miles, where I discovered a lake of the chain of lakes shown on the blue-print map. These lakes lie between Wabigoon River and Lac Seul. I saw a large bear lying in the creek when I returned. All through this part is well timbered, spruce and jackpine in the valley and on the hill-side principally jackpine. The whole would cut from seventy to seventy five cords per acre, but is not quite such large timber as on the south-east side before described. An extensive fire has passed through south of the chain of lakes mentioned above, quite recently. The fire went through the second-growth stuff and missed the good timber.

August 14th. Mr. McAree sent Johnson and myself to find the outlet of Canyon Lake into the Wabigoon River, which we found.

August 15th. We tried to find the portage to Wabuskong Indian village.

August 16th. We went south into a large bay of Wabigoon River and found no timber but small jackpine. The party left to explore Canyon River where we camped at 6 p.m.

August 17th. At the upper chute of Canyon River there is a rapid two hundred and fifty yards long terminating in a fall, with a total descent of thirty feet. About one-quarter of a mile further there is another fall of ten feet and again another rapid about one mile distant, two hundred yards in length, terminating in a fall, the total fall being twenty-five feet. The current is very swift in many places. The next fall is about twenty-five feet over a smooth rock. The last fall drops into Wabigoon River. The total descent of this rapid is twenty feet. There is a great deal of timber growing in the Black Lake, Canyon Lake, and Wabigoon River country, chiefly pulpwood, all of which could be brought by water to the last fall, that which enters the Wabigoon River. By harnessing this fall a sufficient power can be developed to conduct any manufacturing process that might be required. The banks on each side of this fall are high, and the cost of damming would not be great.

August 18th. From camp at station (149) we travelled north-east by land all day. There is red and white pine through here to the extent of over two million feet, of which 20 per cent. is white pine. The timber is not large, and would cut at the rate of about twenty-four logs per thousand. East of this there is two hundred and fifty thousand feet of large spruce, averaging twenty-four logs per thousand, and there is also a large quantity of smaller spruce, poplar and tamarac fit for pulpwood.

August 19th. We saw four moose.

August 20th. We left camp at station (149) with Johnson, and went east about two miles into a large bay of Wabigoon River. Here are rapids with a large body of water running south. We continued down Wabigoon River over one chute with a drop of eleven feet, and down another which had a fall of four feet in seventy-five yards. At several other points there is a strong current, and the river flows swiftly for one-third of the distance we travelled that day.

August 21st. We left camp at station (176) in the morning and went east by land, passing through some spots of very good clay loam land, situated from one-quarter to one mile from the shore of Wabigoon River. Through here is a lot of very good poplar along the banks of the river, but back of that the country rises, and has been swept by fire.

We continued down the river in the afternoon, passing three waterfalls, the first about six feet, the second three feet, and the third fourteen inches. Below this again, towards evening, we came to another waterfall, where we camped. This rapid has a fall of twelve feet.

August 22nd. We portaged over the fall on Wabigoon River. This is the last fall on the river and is situated about two and a half miles from its mouth, where it enters the English River. We arrived at English River and travelled down the east side to Dahm's trading post. He can grow potatoes at his camp. Left Dahm's camp and went west by canoe about five miles on English River. The soil is clay loam in patches and the rock formation granite. We camped on English River at Observation Point. The timber from Dahm's camp to Observation Point is jackpine, poplar, white birch, balsam and spruce, all small. It will cut thirty-five cords per acre.

August 23rd. We left Observation Point with the party by way of English River to the Indian village at Grassy Narrows. The Indians grow good potatoes. The timber is stunted and of no value. An Indian had killed two moose that day. The west end of the lake is reported by the Indians to be splendid for whitefish. We returned to Observation Point.

August 24th. The whole party started east up English River. The timber is green on the south side of English River. It consists of poplar, white birch, spruce, jackpine and tamarac, and would cut thirty-five cords per acre, 7 per cent. being spruce. Johnson and I continued on to Maynard's Lake, where there is a good waterfall of six feet drop. We crossed to the east shore of Maynard's Lake, and went into the bush in a south-east direction four miles. The timber is all green, with a few clusters of red and white pine. The total would cut thirty-five cords per acre, of which only 2 per cent. is pine and 7 per cent. spruce. We camped on Maynard's Lake.

August 25th. We went east by canoe to the unsurveyed part of English River and saw a few scattered red pine on the shore. We went up to Oak Lake Falls and camped on the north side of Oak Lake at the fall. The river empties into Oak Lake at a fall of about twelve feet, having an island in the centre. The timber which we saw on English River this day was too small to be of value.

August 26th. We went to the south east arm of Oak Lake, a journey of seven miles by water, and continued by land to the south east, four miles. There is good agricultural land here and some fine poplar, spruce, birch, jackpine, etc. The whole would cut fifty cords per acre, of which 15 per cent. is spruce. There are two lakes which empty into the south east arm of Oak Lake I saw eleven very large red pine trees, the only ones seen away from the water-side. We returned to the camp on Oak Lake falls.

August 27th. We proceeded up the English River to Long portage. The first waterfall on Oak Lake drops eight feet, the second four feet, the third five feet in a distance of eight hundred yards. The timber on the first four miles is all poplar excepting 5 per cent. spruce, an odd tamarac and an occasional small jackpine ridge. The whole will average 40 cords per acre. We camped on Long portage.

August 28th. We crossed Long portage, about half a mile in length, of which the rapids run the whole distance, a large body of water with a descent of twenty-five to thirty feet in that distance. We continued by canoe one-quarter of a mile further, to another fall of three feet. All along this route the soil is good. The timber is poplar and jackpine with 20 per cent. spruce. The whole would cut forty cords per acre, which condition continues up to Mattawan River. At the mouth of Mattawan River there is a Hudson Bay Company's post, which is still used during the winter months. They have five houses and seven acres of cleared land, which has been cultivated in the past. The soil up to the mouth of the Mattawan River is good for agriculture on both sides. From the Mattawan River we continued up the English River nine or ten miles. The soil is clay. The timber is poplar, tamarac, spruce and jackpine, 6 or 7 per cent. spruce. The whole will average forty cords per acre. We camped at Owl camp on English River.

August 29th. We continued up English River past two falls. The timber is good comprising poplar, spruce, birch and balsam. The whole will cut sixty cords per acre, 10 per cent. of which is spruce. We reached and entered Lac Seul, the west end of which was burnt some four or five years ago. It rained all day. There are a few white pine on the Ontario side of Lac Seul at the last portage. We camped at West Bay.

August 30th. We went five miles inland from camp, and saw two large creeks and had to fell trees to cross the same. Here there is fine poplar, tamarac, jackpine and spruce, all large. It will cut eighty cords per acre, 25 per cent. spruce, 25 per cent. tamarac, with 40 per cent. poplar. This poplar is large, very sound and without blemish We returned to camp on West Bay.

August 31st. We travelled in an easterly direction on Lac Seul. The country is somewhat more rough, but not high. The timber is thick and of good quality, but not quite so large as the last described. We camped near Manitoba point, Lac Seul.

September 1st. We went by canoe four miles on Lac Seul and we were wind-bound at Stony point. The timber here is small poplar, etc, and the soil is sandy.

September 2nd. Starting by canoe on Lac Seul we passed around several large bays and saw scattered white pine and three clusters of red pine on the islands. The balance of timber is small and the soil sandy. We camped on an island in Lac Seul.

September 3rd. We continued by canoe on Lac Seul in an easterly direction and arrived at the Lac Seul Hudson's Bay Company's Post. The timber on the islands is small, principally jackpine and poplar. Some of the islands are burnt. We made a camp at Fort Lac Seul.

September 4th. We were wind-bound all day at Hudson's Bay Company's Post, Fort Lac Seul.

September 5th. We left by canoe to the north-west corner post of Indian Reserve. We found the timber small. We camped in Burntwood Bay.

September 6th. On leaving Burntwood Bay we went into another large bay. There are a great many islands in this part of Lac Seul and they are well wooded with poplar, jackpine, white birch and about 7 per cent. of spruce, which are very large. The islands are all green and the timber on them would average fifty cords per acre. We camped on Williams Bay.

September 7th. We went by canoe south two miles into Williams Bay and then travelled on foot fifteen miles, crossing Beaver River. The timber about one mile from Williams Bay is of good size, poplar, spruce and tamarac and will cut eighty cords per acre, 20 per cent. is spruce. After that to Beaver River the timber is small spruce. The soil is sandy with ridges in places.

September 8th. We continued south and crossed Beaver River by felling a tree. The river at this point is from twenty to thirty feet wide and about six feet deep. Higher up there are two lakes about five miles apart. There have been many beaver here and they have so dammed the upper lake that the river has taken another course. This particular dam is fifteen feet high and trees of many years' growth grow on top of the dam. There are still beaver here as evidenced by fresh cuttings and there are several other beaver dams on the stream. Here are also several other creeks running into this stream. The timber in the valley of Beaver creek is large spruce and tamarac. The valley is about one quarter of a mile in width and the timber will average one hundred cords per acre. Beyond the valley on each side the timber is small and second growth. I went seven miles south of Beaver Creek where the country has been burnt by recent by recent fires. From here west the country is also burnt. This is probably caused by the same fire which passed through Blackwater Lake, as it evidently came from the south. From the highest point I could see seven miles to the west, and five miles to the east along the valley of the river, of which about one half has been burnt and the balance is good timber. We returned to the north side of Beaver River. Beaver River is a small flowing stream which empties into Lac Seul at the foot of Route bay, where it falls some twenty feet.

September 9th. We returned to camp on Williams Bay.

September 10th. We moved camp about five miles west. The timber is mixed as before, poplar, jackpine,, spruce, birch and tamarac, and would cut fifty or sixty cords per acre, of which 15 per cent. is spruce of fine quality. We travelled across two peninsulas on which there is some large spruce and camped on McIntyre Bay.

September 11th. We went to end of McIntyre Bay about fifteen miles and inspected the south side. The timber here is not so heavy. Would cut about thirty-five cords per acre, 15 per cent spruce. We returned to camp on McIntyre Bay.

September 12th. We paddled westerly by Stony point and also touched at Manioba point and Poplar narrows, south of Big Island. The timber is chiefly poplar, large and heavy, and will cut sixty five cords per acre. It rained and thundered. We made camp on a large island in the Bay of Islands.

September 13th. Leaving this camp by canoe we travelled seven miles south and went inland about four miles in a southerly direction. Here we found two lakes, one to the south-east and the other to the south-west. The district through here is all burnt as also from William's Bay to Bay of Islands, excepting on the shore for about one mile back, which is rough and high and covered with small growth of the usual varieties, and a great deal of down timber. We returned to camp on Bay of Islands.

September 14th. We travelled by canoe in a westerly direction towards Wabuskong portage. The timber is green and of good size being spruce, tamarac, poplar, white birch and jackpine. It will cut fifty-five cords per acre, twenty-five per cent. spruce and twenty-five per cent. tamarac. We continued around the Bay of Islands in a south-westerly direction some twenty miles and found the timber small, principally poplar. The country is rough but not high. We camped at Scout Bay.

September 15th. We remained at Scout Bay owing to heavy rains. The timber is very good, comprising quantities of spruce, tamarac, jackpine, white birch, etc. It would cut sixty-five cords per acre, twenty-five per cent. spruce and twenty-five per cent. tamarac.

September 16th. Going south from Scout Bay by land we found an Indian trail running west and followed it four miles. We met an Indian in a canoe, hunting up a large creek. He informed me this creek came from lakes which lie to the south and he kindly marked them on my map. His information afterwards proved to be correct. I returned to the place where I first struck the Indian trail. The timber through this west Indian trail is medium in size, good, and sound, and fully half spruce. It will cut about forty-five cords per acre. We continued south on foot three and a half miles and reached a small lake named Florence Lake, about seven miles south of Scout Bay. The timber through here is exceedingly good, being large, fully half spruce, and will cut fifty-five or sixty cords per acre all told. We camped on Florence Lake.

September 17th. We continued on foot south from Florence Lake seven miles, crossed two creeks and several muskegs and came to a lake I called Lake Louise. This lake is ten miles in length and four miles wide at its widest part. We traversed the east end of Lake Louise and continued south three miles. We returned and camped on the south-east end of Lake Louise. The timber through this day's course is good spruce, poplar, tamarac, white birch and jackpine. It will cut seventy cords per acre, fifty per cent. spruce of good size, fit for lumber. This timber can be readily delivered at Lake Florence or Lake Louise and floated through a chain of lakes into Cedar River emptying into the English River. It can be handled cheaply. After three miles south of Lake Louise the country is brulé. Around Lakes Florence and Louise there is good clay soil.

September 18th. We returned to Scout Bay and camped after dark.

September 19th. I travelled round the west shore of Scout Bay and up four creeks, from half a mile to two miles each, looking for Wabuskong portage, which I failed to find. The timber was tamarac, spruce, poplar and white birch. It will cut sixty-five cords per acre, 30 per cent of spruce. I camped on a point on Scout Bay.

September 20th. I went by canoe along the west shore and joined the main camp at Shanty Narrows on Lac Seul.

September 21st. The whole party moved back to Scout Bay. While Mr. McAree was surveying here I went north on foot three-quarters of a mile, where I found a small lake. The timber through here is practically all spruce. about fifty cords per acre. It runs from four inches to ten inches on the stump, and the trees grow tall. We returned to camp on Scout Bay.

September 22nd. We crossed Wabuskong portage, which is one and three-quarter miles long, crossed two creeks, and came to a creek covered with wild rice and hay. This creek leads, after two miles, into Wabuskong Lake, where we camped. The timber through this portage is spruce one-third, the balance poplar, tamarac, jack-pine and balsam and will cut sixty-five cords per acre.

September 23rd. By canoe to the south-east end of Wabuskong Lake to an Indian village, where I procured potatoes and turnips. The Indians here are prosperous. They have good large houses and raise potatoes, etc. The timber will cut forty cords per acre, poplar, spruce, jack-pine and cedar and some red pine. We returned north to the

Cedar River, which is the outlet of Wabuskong Lake. We discovered a large body of quartz about an eighth of a mile from Cedar River. The timber was chiefly poplar, about forty cords per acre. We camped at Trout Falls on Cedar River.

September 24th. I accompanied the party down Cedar River to English River, about fifteen miles, and returned, making three portages on the way down, and four on the return trip. The timber is not large and will cut thirty cords per acre on the river side, principally poplar and tamarac. Back from the river the timber is better. We passed three waterfalls. The stream is a good body of water with a fairly good current.

September 25th. As it was blowing and raining all day we remained in camp.

September 26th. We went to north-east bay of Wabuskong Lake by canoe, intending to push through to English River but rain, snow and wind prevented. We camped on north-east bay of Wabuskong Lake. The timber is poplar, tamarac, spruce, birch and balsam, and will cut forty cords per acre.

September 27th. We left North East Bay of Wabuskong Lake direct north by land and reached West Bay on Lac Seul, crossing four creeks and six narrow ridges of rock and eight swamps. The timber is spruce, poplar, tamarac and birch and will cut fifty cords per acre, 25 per cent. of which is spruce. I passed two small lakes, the outlets of which are to the north of Lac Seul. The whole distance traversed was about ten miles. It snowed during the night.

September 28th We returned by the same course to Wabuskong Lake and camped. I could see a large swamp to the north west towards Cedar River. This swamp is three miles long and contains a great deal of good spruce.

September 29th. We travelled by canoe to the east side of Wabuskong lake to an Indian village. We went over to west side of Wabuskong Lake where the poplar is very large, and will cut forty-five cords per acre. There is also some small cedar.

October 1st. We left the Indian village by canoe to the south-west of Wabuskong Lake to the first water-fall, where we camped, the distance travelled being five or six miles. The timber is somewhat light and burnt, and is principally poplar, with a few red pine.

October 3rd. We went twelve miles to the south-west to the second portage on Perault Lake, and returned three miles. The timber in this part will yield fifteen cords per acre.

October 4th. We started north by canoe to Wabuskong Lake, and met Mr. McAree's party on Perault Lake.

October 5th. Left McAree on Perault Lake and travelled in a northerly direction, making one portage, and reached Cedar River. The distance travelled was twenty-five miles

October 6th. Went down Cedar River by canoe one mile and branched off west by a creek, half a mile long, into a chain of lakes. There is good land and spruce timber in this region. I continued west and made a portage at a waterfall. I went up Channel Lake, and thence south by land one-quarter of a mile to a small lake, where there is fine timber, spruce, poplar and tamarac. It will cut one hundred cords per acre, and there appears to be a considerable tract of it. I returned to the canoe and continued across Channel Lake to a jackpine ridge, which proved to have been burnt some years ago. I camped on Indian Lake.

October 7th. I travelled west by canoe to a large, clear water lake, eight or nine miles in length and about five miles wide, with a rough shore. The timber was not worth mentioning. I was informed by an Indian, Chief Perault, that there were eight lakes emptying into this clear water lake, around which there is a quantity of good spruce. I, however, did not see this, but have always found his information correct. I returned and camped on the west end of Indian Lake.

October 8th. I went north by land three miles. The country is very rough. The timber is not burnt and of the usual varieties, fifty cords to the acre, 20 per cent. of which is spruce. I returned to the canoe and crossed the portage into Duck Lake. On this portage and along that side of Duck Lake is timber that will cut seventy-five cords per acre, of which 20 per cent. is spruce. I continued to the west end of Duck Lake and portaged into Goose Lake. The timber on Goose Lake and portages will cut fifty cords per acre, of which 20 per cent. is spruce. I went down a creek into Rice Lake. The creek is about twenty feet wide, with a good current of four feet of water, and could be driven with any timber from the height of land at Duck Lake. The timber on the creek is small tamarac. I could see big timber in the distance. I camped on Rice Lake.

October 9th I continued west by Rice Lake three miles, and crossed the half-mile portage at the west end into Alder Creek, which is twelve miles long, I continued down Alder Creek to Oak Lake on the English River and camped. The distance travelled was fifteen or sixteen miles. The timber along this route is principally poplar, very large and clean, the finest I have ever seen in the west. It will make beautiful and valuable lumber. There are millions of feet and it will cut eighty cords per acre. The country is almost free from bushes. These trees will run from twelve to sixteen inches on the stump. They are tall and straight with very few limbs and they show neither rot or blemish. All the timber on this course is in a healthy state. There is also on the Rice Lake end, and on the lower end of Alder Creek, big spruce and tamarac, very thrifty and large. The ground is very good soil and level with very few ridges, and the timber can be taken out by way of Oak Lake on the English River.

October 10th I left Oak Lake and went down English River to the mouth of Wabigoon River, where we camped at Dahm's trading post.

October 11th I travelled to Grassy Narrows and met Mr. McAree and party at 1 p. m. We camped at the Indian village. The timber is all second growth and of no value.

October 12th. We went down the English River to Roches Moutons bay and found the country rocky and covered with timber second growth, consisting of scrub oak and jack pine. We camped on the bay.

October 13th I assisted in survey on Roches Moutons bay, and camped. The timber here is small second growth.

October 15th. I went west to the south end of Separation Lake and assisted on survey. The timber on this course is small jack pine, spruce, balsam, poplar and scrub oak, and would cut twenty-five cords per acre, all small stuff. I saw also some gray willow and hazel, as found in most of the country we have passed through. We camped on the west side of the mouth of Narrow bay leading south from Separation Lake en route to Sand Lake and Rat Portage.

October 16th. I left camp by canoe to the south shore of Separation Lake. Here I came on Vermilion Lake and crossed the winter route from English River to Winnipeg River. I went through the bush one and a half miles to the south-east and returned to the same place by a circular course. Here the country is flat and boggy for a distance of four or five miles. The timber is poplar, spruce, tamarac and jack pine, which will cut sixty cords per acre, 25 per cent. spruce, 25 per cent. tamarac. On the higher ridges it will only cut thirty-five cords per acre. The soil appeared to be black loam. The timber on the other side, that is to say, the south side, is very similar. I returned to camp on Separation Lake.

October 17th The whole party went north-west down English River and after seven or eight miles' journey came to a waterfall, which has a drop of seven feet. The banks are low and in the narrows on the way coming down the current is swift. The timber is principally small jack pine, about thirty-five cords per acre. The country is not high. There are two creeks running into the river from the south-west. We camped at the fall.

October 18th. We went south-east by canoe two and a half miles and landed in a large bay. The timber has been blown down, excepting some large spruce, which will average eighteen logs per thousand. We struck through the bush in a southerly direction, traveling five or six miles. We crossed a high ridge and three creeks, all of which have swift currents and flow from the south-west. The timber for the first three miles on this inland journey is spruce, balsam and white birch, and for the first mile the spruce is the only timber of any size standing, the balance having been blown down by storm. The spruce is not thick. The timber on the other two miles is good and all standing and will cut fifty cords per acre, 30 per cent. spruce. The balance of the journey was over higher land, which produces no timber of value. In the valleys of the second and third creeks, which join each other on the course, there are some very large tamarac which measure twenty-four inches on the stump. We returned by the same course and camped with the party at the fall.

October 19th. I went with the party down English River making three portages in three miles. At each portage there is a waterfall. There is no timber along this route. We continued in a north-west direction past the Indian House and camped.

Dawson Route across Kaministiquia River. Party No. 8.

Portage at south end of Obugamiga Lake. Party No. 8.

Kaministiquia River and Dog Lake, looking south. Party No. 8.

Dog Lake, looking south. Party No. 8.

On traverse of lake. Party No. 8.

Indians bathing. Party No. 8.

Indians bathing. Party No. 8.

A chute on Gull River. Party No. 8.

On Nepigon River. Party No. 8.

Indians at Gull River. Party No. 8.

On Gull River, Party No. 8.

A Cariboo Luncheon. Party No. 8.

Nepigon House. H. B. Co.'s post. Party No. 8.

Wabinosh Village, Lake Nepigon Party No. 8.

Inner barn, Wabinosh Bay, Lake Nepigon. Party No. 8.

Cliff, Wabinosh Bay, Lake Nepigon. Party No 8.

Echo Rock, Lake Nepigon. Party No. 8.

Bush fire starting, Sturgeon Lake. Party No. 9.

Bush fire, Sturgeon Lake. Party No. 9.

Fall at No. 7 portage. Party No. 9.

Indian wigwam, Sturgeon Lake. Party No. 9.

Rapids on Sturgeon River. Party No. 9.

Falls, centre channel, portage 13, Sturgeon River. Party No. 9.

Lac Seul. Party No 9.

October 20th. We continued down the English River to its mouth, and up the Winnipeg River a short distance. We did not see any timber this day. The country is rocky and rough.

October 22nd. Johnson and Keys went down to the Manitoba boundary, so I took Key's place in the surveying canoe, reaching South West bay of Swan Lake at night. The timber is small, not worth mentioning, the country being barren and rocky.

October 23rd. I journeyed south by land about two miles on a jack pine ridge. Here the ridge ends at a valley, which is about two miles across and of great length. For seven miles along this valley there is poplar of large size, which will cut seventy cords per acre. On my return journey I crossed over to another ridge on the north and reached the lake about two miles north of the camp. The timber is spruce, tamarac and poplar, all large and will cut eighty cords per acre, of which fifty thousand feet B.M. is spruce. There are also wild fruit trees, plums, cranberries and hawthorn. The soil not on the ridges is good.

October 24th. I went with the party to a bay on the south-west of Swan Lake and continued the journey up Scot Creek. There are two falls. The first has a descent of ten feet in one hundred and fifty yards, and the other six feet in fifty or sixty yards. We continued our journey until we reached another fall, at the head of Tom Lake. Near here I went inland to the west about one half mile to the top of a high ridge, where I obtained a good view of the surrounding country. The timber along Scot Cree k or half a mile on its west side and one mile on its east side is all good, principally poplar mixed with some jack pine and spruce, all of large size. There is also some elm and iron wood and oak in some of the low points. This is a large belt of good sized timber, and will cut in all sixty cords per acre. All this timber can be taken out by way of Scot Creek, through Swan Lake to Winnipeg River. The country is rolling with a clay soil. The water is dark in shade as also is the lake above. About a mile below the third fall a fire passed through some four or five years ago.

October 25th. It has been reported that in this country, between the Winnipeg River and the Manitoba boundary, there is a quantity of hemlock and pine which I have been endeavouring to locate. We camped on the east shore of Tom Lake.

October 26th. We went inland about five miles directly east from camp. For the first three miles the timber is green, but for eight or ten miles further on it has been destroyed by fire. The green timber is jack pine, spruce, tamarac, poplar and birch. It will cut thirty cords per acre and is very scattered.

October 27th I accompanied the party in a south-west direction until we reached a creek running in the same direction. I went up this creek until I reached a lake about three miles long running north and south. I left the canoe on the east side of the lake about half way up and followed an Indian trail running north east. Passing through a swamp of small spruce I reached a low creek running west. There are some scat re white pine around this lake. The remainder of the timber is jack pine. It will cut forty cords per acre. I camped at the south east end of North Scot Lake.

October 28th. I went north about one mile along the shore and thence east by land one and a half miles. I saw some scattered whitepine at the base of ridges, in all thirty thousand feet of sound timber in addition to much that was faulty. The remainder of the timber is jackpine, spruce, poplar and balsam, which will cut forty cords per acre. Crossing to the west of Scot Lake I went inland and saw a good sized lake with a creek joining it to Scot Lake. The country is high and rocky. The timber is jackpine, spruce, poplar and balsam which will cut thirty-five cords per acre. On the north side of Scot Lake there are two hundred and twenty good sized white pine, but most of them are faulty.

October 29th. I left with Johnson and travelled east by land about one and a half miles until we reached Malachi Lake which is seven or eight miles in length and running north and south. We continued up the west shore of this lake to the narrows, passing an Indian trail, we returned by this trail to camp. The timber is mixed and will cut sixty five cords per acre, 20 per cent. being spruce.

October 30th. We cleaned out the Indian portage to Malachi Lake in the forenoon. In the afternoon we portaged to Malachi Lake, went south by canoe to the narrows and camped. The timber on the portage is heavy but scattered and would cut sixty-five cords per acre. It comprises jackpine, spruce, balsam and tamarac, with 20 per cent. spruce.

October 31st. We went south on Malachi Lake two and a half miles and up a creek half a mile, when we reached a small waterfall of six feet. We continued south one and a half miles by the creek. We reached a place where railway ties have been cut in recent years. Between the tie cutting and the camp on both sides of the lake is the usual kind of timber, which will cut thirty-five cords per acre. The creek is swift and only four to six feet in width. There is no timber on the ridges but second growth and no soil. We left the creek and continued on foot south three miles. The country is rough and rocky, without timber. We returned to camp on Malachi Lake.

November 1st. We left camp and went inland by North East Bay on Malachi Lake and travelled east about one mile. We could see that the country was burnt here for six or seven miles. Thence we went north by Malachi Creek, which is the outlet of Malachi Lake. This leads into Otter Lake, so named because we saw an otter there. There is a river which empties into Otter Lake on the north-east shore at a waterfall of about eight feet. Just here there are twenty-seven good red pine trees. The rest of the country is burnt. We took the canoe down to the outlet of Otter Lake and made a portage into Mink Creek. Here there is a fall into Mink Creek of six feet. We went one-eighth of a mile further, and made another portage of two hundred and fifty yards at another fall of ten feet. Between these falls there are one hundred and thirty-five small red pine trees. The remainder of the country is burnt. We camped half-way down Mink Creek.

November 2nd. We continued down Mink Creek about two miles, and made a portage at a fall four feet high, and farther on made another portage at a fall of sixteen feet. We reached Cygnet Lake, which is about six miles long, climbed a high ridge, and got a good view of the surrounding country which is bare and burnt. Three miles further down we climbed another ridge on the west shore. The country here is also burnt, excepting a few small trees. I turned to the west and went about one and a quarter miles, arriving at the south east bay of Swan Lake. The land here is low with a clay soil. On this course there are ten thousand feet of spruce of good size, twenty-four logs to the thousand. The other timber is poplar, jackpine and birch, the whole of which will cut seventy-five cords per acre. We returned to camp on Cygnet Lake.

November 3rd. We went north by canoe about one mile, and made a portage at a fall of about four feet. We crossed a pond and made another portage at a fall of seven feet. Continuing north about one mile we made another portage into Swan Lake at a fall of eight feet. The country is low and rough, and the timber small of the usual varieties. It would cut twenty-five cords per acre. We camped on the south-west bay of Swan Lake.

November 4th. We went east by canoe about one and a half miles and then one and a half miles inland, south-east from a bay in Swan Lake. Then we continued two miles south, through a good clay soil country. The belt of poplar, described on October 24th, continues right through to Swan Lake, with an occasional ridge of jackpine. We returned to camp on Swan Lake.

November 5th. We left by canoe and went one mile west up to the boundary line at mile post No. 47. We crossed two creeks on this course, one of which is twenty feet wide. They both run north to the Winnipeg River. The country is rough. The timber will cut thirty-five cords per acre of the usual varieties. This section has been reported to Mr. McAree as bearing hemlock and pine, but we saw nothing of it. Having travelled a distance of about six miles we returned to camp on Swan Lake.

November 6th. We travelled with the party north-east down Swan Lake to Winnipeg River. We reached the White Dog Indian Reservation and the Hudson's Bay Company's trading post. The land along the shore is high on the Indian Reservation and the timber is of the usual kinds, but none of it is very large. Here there is a Church Missionary Society Mission and school well attended. Mr. Macdonald is in charge of the Hudson's Bay Company's post. The river current above the Hudson's Bay Company's post is very swift, until the first fall of seven or eight feet is reached. We made a portage here. A mile or so further up the river there are three falls, about one-half a mile apart. The three have a total descent of about twenty-five feet and one portage passes them all. Along this portage the timber is good and very thrifty, consisting of spruce, jackpine and poplar. On either end of the portage the timber is very small. The distance travelled was from twelve to fourteen miles. We camped at the portage.

November 7th. We continued up Winnipeg River with the party. We made one portage of about 100 yards. There is a very swift current here. We went up to the north-east corner of Sand Lake and camped at the portage on the route leading to English River, having travelled twenty miles. We passed some small red and white pine scattered on the islands, but most of the islands are burnt over.

November 8th. We attempted to go from Sand Lake to English River, but the ice was too thick on the first small lake. Howevpr, Linklater and I crossed this lake on the ice with one canoe and four hundred pounds of baggage. We saw that it was impossible to reach English River under the conditions and returned. I then thoroughly examined the north east end of Sand Lake. Here are six hundred and fifty thousand feet of red pine in good condition, other timber being scarce. The country is burnt.

November 9th. The whole party left Sand Lake and reached Winnipeg River. We encountered a heavy head wind and snow. We paddled against the current fifteen miles and camped. There are a few scattered pine along the island towards the east shore of Sand Lake and a grove of red pine on the Winnipeg River below the Dalles at the Indian fishing ground, which I did not inspect. Most of the timber on the river is small, scrubby second-growth of the usual kinds.

November 10th. We reached the Indian reserve at Dalles. Here Mr. McAree connected his Black Sturgeon Lake survey with the Indian village. We camped at the School house, by favor of Mr. Clark, the teacher. There are a few scattered red pine along the river up to the Dalles rapids. The other timber is small.

November 12th. We left the Dalles rapids and reached Rat Portage. Here again there is a scattering of red pine. The other timber is jackpine, poplar, etc.

SUMMARY.—TIMBER, AND MEANS FOR LOGGING AND DRIVING.

The timber areas examined may be divided into three sections, viz: That which would have to be logged and driven by way of the English River, the Wabigoon River, and the Winnipeg River.

ENGLISH RIVER AND TRIBUTARIES.—The most important timber in this district or section is that which is to be found on Lac Seul Beaver River, Cedar River, Alder Creek, and Florence and Louise Lakes.

That on Lac Seul would drive direct into the English River, to the mouth of the Wabigoon River, where there is a large and valuable water power. It is assumed that all the timber through here would be manufactured at this water power. That on the Beaver River would also come by way Lac Seul to this place, while that of the Cedar River and Alder Creek would reach the same point by way of the English River. The timber in the Canyon Lake and Blackwater Lake country on the other hand should be manufactured at the last fall on Canyon River, where it empties into the Wabigoon River.

To transport this lumber to the Canadian Pacific Railway track would necessitate a spur track of come forty miles in length, having its terminus at the mouth of the Wabigoon River. The timber through here could not be transported by any other means. The timber in this whole section, namely, Lac Seul, Beaver River, Cedar River, Alder Creek, Florence and Louise Lakes with their tributaries, is exceedingly valuable and comprises spruce, poplar, jack pine and tamarac, a great deal of which is of large size and extent. The most valuable are the spruce and poplar, which are sufficiently large to cut many million feet of lumber. The poplar especially is very valuable and good and in large quantities, while there is a vast quality of smaller timber suitable for the manufacture of pulp.

THE WABIGOON RIVER SECTION.—Takes in from the Canyon River country, south and west, to the Canadian Pacific Railway track. The logs would also be driven to the mouth of the Canyon River. The timber here includes three and a half million feet of red and white pine of good quantity and also large quantities of spruce and poplar.

THE WINNIPEG RIVER SECTION includes all the timber on the banks of the Winnipeg River and tributaries, lakes, etc. This embodies the Black Sturgeon, Swan Lake and Sand Lake countries. All the timber in this section would necessarily have to go down the Winnipeg River to Lac Du Bonnet. There is some good timber through this section, notably the poplar and spruce of Swan Lake and Black Sturgeon Lake and the red pine on Sand Lake, all of which have already been described. The banks of the Winnipeg River itself are very rough and broken and the timber has been for the most part burnt. What green timber to be found is small.

On the territory covered by Exploration Party No. 10, I estimate that there are 600 square miles of agricultural land; 1,500 square miles of pulpwood land yielding 4,000,000 cords of spruce and 9,000,000 cords of poplar; and another 1,500 square miles yielding 2,000,000 cords of spruce and 3,000,000 cords of poplar, or a total of 6,000,000 cords of spruce and 12,000,000 cords of poplar.

All of which is respectfully submitted.

I have the honor to be, Sir,
Your obedient servant,
JOHN NASH,
Land and Timber Estimator, Exploration Party No. 10

Mr. JOHN MCAREE, O.L.S.
Surveyor in charge of Exploration Party No. 10.
Rat Portage.

I, John Nash, of the Town of Rat Portage, Land and Timber Estimator, Exploration Party No. 10, make oath and say that the contents of this report are correct and true, to the best of my knowledge and belief.

Sworn before me at Rat Portage this 24th day of December, 1900.
H. LANGFORD,
A Commissioner.

JOHN NASH.

GEOLOGISTS' REPORT OF EXPLORATION SURVEY PARTY NO. 10.

TORONTO, ONTARIO, January, 31st, 1901.

Sir :—

In accordance with instructions received from the Director of the Bureau of Mines, I left for Rat Portage on June 19th and arrived at my destination on the morning of Friday, June 22nd. I reported immediately to John McAree, O. L. S., in charge of Exploration Party No. 10 and found that our departure would be delayed until the arrival of part of the outfit then on its way from the east. By June 29th everything was in readiness and a start was made the following morning.

The party travelled by rail to Margach and then by canoe northward on Rossland lake to Black Sturgeon Lake. A north easterly course was then taken through Manitou and Canyon Lakes, the country both to the right and left being explored by way of several canoe routes. Turning southward from the east end of Canyon lake we followed a series of lakes towards the Canadian Pacific Railway tracks, reaching the same at Jackpine Siding and there took train again to Eagle River. Eagle River was followed to its confluence with Wabigoon River, the latter then affording us an easy route to English River. The ascent of this fine river then began, the Hudson's Bay Post on Lac Seul being our objective point. On reaching the Post which was the most easterly point visited by Party No. 10, we began the return trip by following the Ontario shore of Lac Seul to the west side of Island Bay and then by portage to Lake Wabuskong. The outlet of this lake was followed to its entry into the English River. At Wabuskong the party divided, the surveyor and his assistants taking a southerly course to Wabigoon River and then down, the latter, the timber estimator and geologist, pursuing a westerly and more direct course, to Oak Lake. The party united again at the lake of the Grassy Narrows and at once began the descent of English River. On reaching Winnipeg River we explored it to the Manitoba boundary and then began its ascent. A halt was made eight miles below the Indian village of White Dog and two weeks were spent in a thorough exploration of the country between Winnipeg River and Manitoba. After which we again continued our course up stream. Our arrival at Rat Portage on November 12th enabled me to report in Toronto on Thursday, November 15th.

I desire to express my thanks to Mr. McAree and Mr. Nash for kindness shown me during the summer and to Dr. Coleman for the determination of many of the rocks described in this report.

Margach Station is some eight miles east of of Rat Portage and only a couple of chains distant from shores of Rossland Lake. It is situated in an area of basic, Laurentian rocks, containing hornblende, biotite, felspar and small crystals of bluish quartz. They differ considerably from the more typical rocks of the same age which are found just north of them and into which they gradually merge. The lake extends northward two and one quarter miles, and a portage one-third mile in length leads from it to Black Sturgeon Lake.

Black Sturgeon Lake.

Black Sturgeon Lake is a fine body of water, seventeen miles in length, a narrows two miles long dividing it into a western and eastern portion. At its northwestern extremity it empties over a short rapid into a bay of Winnipeg River. The western portion is surrounded by gently sloping ridges of granitoid gneiss usually light in color, but occasionally darkened by a higher percentage of biotite. A band of massive gneiss striking west south-west, crosses the lower end of the rapid. The rock in this locality is seamed in all directions by numerous narrow dikes of coarse pinkish or reddish pegmatite.

Rice Creek, which enters the north shore of Black Sturgeon Lake two miles west of the narrows, was followed to its source in Ant Lake. For one and a half miles above Black Sturgeon Lake it flows through flat country, the west side showing some good loamy soil, the opposite, however, is marshy. Above this the stream cuts frequent outcrops of granite, with swampy land intervening. Short rapids occur at intervals along the stream but the ascent is slight, the rapids giving the only evidence of current. The lake is surrounded by numerous and high exposures of coarse granite, whose bareness is their most conspicuous feature. The valleys are narrow and frequently walled by almost perpendicular cliffs. Toward the north the country apparently becomes much higher and presents a bare appearance.

A belt of Huronian rocks, one and a quarter miles wide, crosses the narrows of Black Sturgeon Lake and continues with decreasing width in a northerly direction. Both sides of this band consist of hornblende schist with a narrow central portion resembling a sheared diorite but which differs little from the former when examined under the microscope. Both cleave into thin laminæ and in each there are thread-like veins of white mineral probably feldspar. On the west contact the strike is north, twenty degrees east, with a dip of eighty-five degrees to the southeast. On the east side, where the hornblende schist forms the shore line for some third of a mile, the strike is north, ten degrees east, and a dip vertical. Fine-grained granite outcrops along the south shore of the lake east of the schist; and associated with and lying between granite and Huronian rock is a handsome mottled rock that crumbles easily under the hammer. It was found to contain dolomite, quartz, microcline, plagoclase, biotite and either talc or muscovite. An excursion was made from the north shore of Black Sturgeon Lake at a point three and a half miles east of the narrows for the purpose of examining the Huronian belt. A walk of one and a quarter miles in a northwesterly direction over three ridges of gneiss brought us to the east side of the band of schist. Its width was found to have decreased to half a mile, and the western strike changed to north, forty-five degrees east, dip vertical; on the east side strike north, twenty degrees east, with vertical dip. Toward the north the width of this Huronian area rapidly lessens and large inclusions of pink gneiss are frequent.

The gneiss lying east of the Huronian rock gradually assumes a greyish tone and at the east end of Black Sturgeon Lake gives place to fine grey granite, containing only a little quartz which is in small bluish crystals. The outcrops of this rock are numerous and high, and continue eastward to the west side of Manitou Lake where pink gneiss is found again. The small lake lying between Black Sturgeon and Manitou Lakes is only slightly higher than Black Sturgeon Lake a sluggish stream joining the two. The southwest end of this lake is bounded by a rocky wall some twenty five feet high, in which is found a soft, dark, greatly decomposed rock that is much worn by wave action. In passing eastward to Manitou Lake two portages are passed; the fall in the rapids at the lower one is twenty feet, in the upper five feet.

Manitou Lake.

Manitou is a beautiful lake of remarkable clear water about six miles long and three in breadth. Its shore is almost entirely rock and in the main quite high. The country around the north-western portion presents much the same appearance as that around Ant Lake, consisting of bare ridges of massive pink gneiss many of which become dark and with finer lamination for a narrow width along their apex. On the other hand but more rarely the schistosity becomes very indistinct, and the rock approaches granite. Several small lakes surrounded by swampy land were found in this district and drained by a narrow and very rapid creek into Manitou Lake.

An interesting feature of the rock on this lake is that in several places it is found in almost horizontal, long, flat, more or less lens-shaped pieces, lying one above the other as in masonry. These pieces vary in thickness from one to eight inches and separate distinctly from each other, and their composition is that of fine grained granite, the microscope showing nothing unusual in the mineral constituents. This structure is best shown on the south shore of Swallow Island, about the centre of Manitou Lake where it is exposed in the face of a cliff some seventy feet high, the upper portion being granite, the bottom also granite, but arranged in flat pieces as described. A similar structure is found on both the north and south shores of the lake east of Swallow Island. On the south shore it is exposed in a width of only two hundred and fifty yards and on each side gradually merges into the massive granite. Many of the flat pieces have been loosened by the waves thus forming caves into which a canoe can be paddled. From examination of these caves it would appear that this arrangement in the rock dips on the south shore toward the south-west at about twenty degrees.

On the opposite side of the lake the exposures are poor, the shore being low and well timbered. The only change is in the dip which on this side is toward the north at fifteen degrees.

From the north-east extremity of Manitou Lake we made a short portage into a small creek leading from Lake one. This small lake is grown up thicky with weeds and is completely surrounded by low marshy shores. Above it the creek is narrow and winding but deep and leads to a clear rock-enclosed lake, lying north-east of No. one. On this lake exposures of granite showing the horizontal-jointed arrangement as on Manitou Lake were noted. By a portage from the north side of this lake we reached the third lake on the route and then by a portage nearly two-thirds of a mile long we arrived at a small round lake named Lake 4. The last two lakes are both surrounded by low gently-sloping ridges of granite which, as is common in the rock of the district, shows some schistosity in places. A portage of two and a half chains over a low ridge leaves the west side of Lake 4. Lake 5 is a long narrow band of water followed for only a part of its length by the canoe route; it receives a fair sized stream at its north end just east of the portage leading to Lake 6. The sixth lake is reached by a short portage over a sandy slope and occupies a position on the watershed between Manitou Lake and English River There is a slight descent from Lake 6 to Lake 7, the water of the latter flowing eastward and then northerly to Wabigoon River. The shores of Lake 7 are high and very rocky; the flat, bedded appearance of the rock along the north shore is very similar to that on Manitou Lake, and it would appear occurs at intervals over the whole distance between the two lakes. The shores reach their greatest height along the narrow channel running to the north-west, where they rise in smooth rounded ridges one hundred feet above the lake. On Lakes 8 and 9 the rock, which is quite granitic, but in places showing schistosity is much seamed and broken frequently showing perpendicular walls, forty to sixty feet high. East of the ninth lake there is an area of very coarse pegmatite which extends northward past Lake 10. This rock which rises into splendid ridges, two hundred to two hundred and twenty-five feet above the lakes, partakes of the nature of a granite porphyry, the crystals of feldspar, which are exceptionally large, resting in a ground mass of granite. The feldspar crystals frequently show twinning. When seen in masses this is a very handsome rock and owing to the scarcity of soil and vegetation the exposures are very fine. The summit of these ridges presents an excellent view of the country northward in which the valley of the English River could be easily traced and beyond it the blue hills of Keewatin. Lakes 11 and 12 are surrounded by massive gneiss, which is greatly shattered and broken in places, forming heaps of angular boulders. On Lake 12, more notably on the south shore, glacial striæ, north, forty-five degrees east, are faintly visible.

From Lake No. 12 we returned to Lake No. 7 and followed the small winding creek that empties into a long narrow arm of water extending in a direction almost due east and west for some sixteen miles. This body of water was named Canyon Lake. A low ridge of granite follows along the north side of the creek; the south side is low and covered with light soil, upon which there is a fair growth of timber. The west end of Canyon Lake is surrounded by granite that rises in low hills, but both to the north and south the country rises rapidly. Two miles to the east the lakes become shallow and weedy, with more the appearance of a wide river. Opposite this on the north shore there is a shallow deposit of sand, stretching for half a mile along the lake, and here the monotony and bareness of the country is broken by a grove of red pines. East of this both shores of the lake are formed of one continuous exposure of granite until the east end of the lake is neared, when the north shore becomes flat and the rock covered with soil. Two deep bays are found on the north side of the lake, and from the most easterly one it discharges by a short and rapid stream into Wabigoon River. The last two miles of the north shore of the lake are formed of a shallow deposit of sandy soil and boulders, which supports a thick growth of small timber. East of the lake the sand reaches a considerable depth, and contains few boulders. On the ridges many fine red pine grow, while spruce and tamarac crowd the intervening valleys.

The country south of Canyon Lake was explored by way of a large creek, which enters the lake about four miles from the west end. This stream and the lakes drained by it, together with several draining southward, form a canoe route between Canyon and Hawk Lakes.

A short distance from Canyon Lake the stream becomes very swift and is choked with boulders, necessitating a short portage. A couple of chains above the portage the stream widens into a small lake, elevated some eight feet above Canyon Lake. A granite ridge fifty feet high forms the southern shore, and over it a portage of one-fifth of a mile leads to Bush Lake. The latter is long and narrow, with a regular outline, and bears a great resemblance to Canyon Lake, which lies half a mile to the north and in a general way parallel with it. The east end is surrounded by low swampy land, but to the west both banks are of rock, covered in places with light soil, upon which there is a scant growth of vegetation. Five miles west the lake widens to form deep bays to the south and north, the former being in a low level district. Beyond this the lake becomes narrow, sometimes not more than three chains in width, and continues in a westerly direction for some eight miles. The rock towards the west end shows slight schistosity. The canoe route leaves the south side of the lake at a point one and a half miles from the east end and follows the valley of the stream flowing from Daniel's Lake. Several rapids occur in the lower part of the stream, necessitating a portage one-third of a mile in length. A short distance above the portage the stream widens into Daniel's Lake. Unlike the two long lakes to the north this has its greatest length in a north and south direction, and as would be expected has a very irregular shore line with deep bays running to the east and west. The rock becomes distinctly gneissoid toward the southern end of the lake, and the exposures though quite low at the north end, become of considerable height to the south and east. Linklater Lake, the next on the route, lies just south of Daniel's Lake, from which it is separated by a narrow low ridge of gneiss. The portage here does not follow the stream flowing from the more southerly lake. There is scarcely any vegetation here to hide the light colored rounded hillocks of gneiss which glisten brightly under the unobstructed sunlight.

On the south shore of Linklater Lake high dark ridges rise abruptly above the flat smooth gneiss and these proved to be the northern boundary of a Huronian area. The contact is very well shown, the gneiss and hornblende schist butting one against the other. The Huronian area consists largely of hornblende schist, but what appeared to be a sheared diorite was found in several places, always however at some distance from the Laurentian. At the east end of Linklater Lake the strike is north, eighty-five degrees east, but to the west becomes due east and west. The dip is uniformly about eighty-six degrees to the north. Inclusions of gneiss, some quite large, are common, also dikes of very course pegmatite. The gneiss adjacent to the Huronian is quite garnetiferous and often stained by iron. Two small lakes lying west of Linklater Lake are along the contact; they drain toward the southwest.

From the east end of Canyon Lake we followed a narrow channel in a southerly direction until we reached a large creek which entered on the west side. We followed a portage some five chains in length along the south side of the creek into Forest Lake. For a third of a mile our route followed a westerly direction but after passing a narrows lined with large boulders turned southward. The northern end of the lake is surrounded by sand, but half a mile to the south exposures of granitoid gneiss become frequent. Here as on Daniel's Lake there is no definite line of separation between the granite and the gneiss, as they merge gradually one into the other. Both shores of the lake are low, but the elevation increases rapidly for some distance from the lake. A sluggish stream with marshy banks enters the west side of the lake near the southern end. This we followed for nearly a mile when Boulder Lake was reached. The southern part of this lake is studded with islands of all sizes and shapes and grouped in a manner that is very pleasing to the eye. Except on the south shore the rock is massive Laurentian gneiss.

On the south, however, there is an area of Huronian rocks whose northern boundary, as on Linklater Lake, is marked by a series of rugged bluffs rising thirty or forty feet above their surroundings. Finely laminated hornblende schist containing some chlorite was the only Huronian rock found in this area. In places the schist is cut by dikes of very coarse pegmatite containing immense crystals of feldspar and biotite; some fine crystals of tourmaline were found in one of these. Inclusions of gneiss are common and frequently bands of the schist are stained, probably by iron, the surface sometimes showing considerable decomposition. The strike of the schist is north seventy degrees east and dip seventy-five degrees north, twenty degrees west. Compared with Laurentian country the Huronian is rougher and the higher portions lack the smooth roundness common in exposures of the gneiss and granite.

The large bay at the west end of Boulder Lake lies in the Huronian area, but on its west side there is only a narrow band of the dark schist one-fourth of a mile wide. This is very contorted and thickly impregnated with garnets. In the gneiss and schist to the east and south of the bay garnets are very numerous, the rock in places consisting almost entirely of that mineral. Augite Lake is reached by a portage of a few chains from the south-west extremity of Boulder Lake. Its level is about fifteen feet above that of the latter. Split-Boulder Lake lies immediately west of Augite Lake and drains eastward by a very short stream. At the time of our visit there was not enough water in this creek to float our canoes, making it necessary to portage along the stream bed. Both of these lakes lie along the contact of the Huronian and Laurentian, the latter forming their northern shore. A large lake lies south of the west end of Split-Boulder Lake and occupies a position some thirty feet above the northern lake. It was reached by portaging one-fifth of a mile over a steep ridge. The lake is shallow with flat, smooth shores that are very irregular in their outline. It rests in an area of hornblende schist, which is cut in many places by bands of pegmatite. Lacourse Lake lies one-fourth of a mile to the east; it greatly resembles the lake previously mentioned having flat, low shores and an irregular outline. It drains into several small lakes east of it, the water of the latter flowing south past the railroad.

Over a large area here the rock presents a strange appearance, narrow bands of hornblende schist occuring alternately with pegmatite and sometimes granite. At the northern part of the lake the schist predominates but to the south, especially near the railroad, the areas of granite rock are much the larger. The schist apparently goes to no great depth and seems to be everywhere underlaid by pegmatite and granite. This was well shown on the west shore of Swan Lake, near the railroad track. Here an exposure of coarse granite or pegmatite is broken across the strike of the schist leaving a perpendicular wall of rock some twenty feet high. A band of schist about twelve feet wide is exposed in this wall, with the granite enclosing it on both sides, and below the schist merely filling a trough in the coarser rock. Similar exposures were seen elsewhere, in one the schist having a depth of not more than two feet. It would seem therefore that, had erosion gone only a score of feet deeper, much of the southern part of the Huronian area crossed in passing from Boulder Lake to the railroad would have entirely disappeared. As would be expected, the southern limit of the Huronian is quite indefinite but in the vicinity of Jack Pine, on the Canadian Pacific Railway line, the only place in which I had an opportunity for examination, the railroad forms a rough line of separation.

Eagle and Wabigoon Rivers.

From Jack Pine we travelled by rail to Eagle River Station, and then began the descent of the Eagle and Wabigoon Rivers. Below the Canadian Pacific Railway line the Eagle River has an average width of two and a half chains, and follows a northerly course for about six miles, when it joins the Wabigoon River; the united streams then flowing in an northwesterly direction. This lower part of the Eagle River and the upper part of the Wabigoon River cross an extensive area of white clay similar in appearance to, and probably identical with, that of the Wabigoon district. In many places the river banks are low and flat, stretching far back as great meadows of marsh hay, while in others they are higher and dry, with fine undulating country behind. Little rock was seen until we were north of the township lying west of Sanford (number 36 on our maps). Narrow, low exposures of gneissoid rock then became common, and as we continued down stream, this rock was found to gradually displace the clay.

Five miles north of Sanford the river contracts, and flows over a rocky ridge in a fine rapid, terminated by a perpendicular fall of six feet. In a distance of two hundred yards the river level lowers between fifteen and twenty feet, and, as the volume of water is large, this would make a valuable water power. Several miles down stream a second, and much larger rapid occurs. Here the river is crossed by a ridge of very massive gneiss, through which the river has cut a narrow gorge. Large boulders choke this channel and the water as it rushes through is dashed high into the air. Several rapids occur in the next two hundred yards, making a total fall of some twenty-five feet. This marks the northern limit of any clay deposit of any considerable area, the nature of the country now rapidly changing, the low undulating banks of clay giving place to bare, rugged hills of gneiss; each shore, however, continues to be skirted by a narrow strip of clay. As far as the first expansion on the river, Clay Lake, the country has no interesting feature, being a succession of rocky ridges separated by narrow, often swampy valleys. Small timber grows on many of these ridges near the river, but exploration at some distance inland proved the country to be quite bare over large areas. On one of these excursions (on the east side of the river and eight miles above Clay Lake) a large deposit of coarse sand resulting from the decay of a granite rock, was found. A small stream of beautifully sweet water flowed along a depression near its centre, and afforded a cool refreshing drink on a hot August day. These springs are said to be common in some districts, but few came under my observation.

Clay Lake lies in a shallow depression surrounded by low hills of rude gneiss that rise forty to fifty feet above its level. It is weedy and muddy, with water darkened by sediment and floating vegetable matter, and which has an odor that makes it disagreeable. Numerous islands are scattered through the lake; all are low and most of them small. A deep bay extends north-eastwards and receives a large creek, which drains Blackwater Lake and the country east of it. Blackwater Lake is also shallow and weedy, with water stained a deep red by the swamps and muskegs in its vicinity. To the north and east of this lake the country rises rapidly in bare, rounded ridges of coarse, pinkish gneiss. To the south-east there is a valley containing several square miles of good soil, on which grows splendid timber, mostly spruce and poplar. Along the south shore there are many evidences of glacial action, the striæ showing plainly for many feet in the smooth gneiss. Much low land lies west of Blackwater Lake, but the gneiss, where exposed, shows a distinct lamination, separating into pieces six to eight inches thick. The strike is north, forty degrees east. This gneiss continues along the north shore of the bay previously mentioned, and a little way past the narrows leading to the northerly portion of Clay Lake. Its color is much darker than the pinkish gneiss found on each side of it. This is caused by a high percentage of biotite. Considerable marshy land is found at the northern part of the lake, especially on the west side, where several small streams enter. Beds of clay have been deposited on many of the flats, and in these concretions are very common, the beach in places being paved with them. Glacial action is shown over a arge area here, the rocks being grooved and polished everywhere.

Canyon River, the outlet of Canyon Lake, enters the west side of Wabigoon River a mile below Clay Lake. Immediately above the Wabigoon River a rapid, two hundred and fifty yards long, with a total fall of about twelve feet occurs. Above the rapid the river has the width of about a chain, and a good depth, the current being very slight. This continues for a mile, when the river narrows and flows over a smooth mass of gneiss, causing a fall of five feet. This is followed by smooth water again, but a third fall, having a drop of twenty-five feet, occurs one and one half miles up stream. There are really two falls here, the lower twelve and the upper thirteen feet. The gorge is narrow and crowded with large angular masses of rock, broken from the cliff which overhangs the river bed. One and one fourth miles up stream another fall of twelve feet is met; and above this half a mile of smooth water is broken, just as Canyon Lake is reached, by two falls, the lower twenty-three and the upper eight feet in height.

The rock is gneissoid for two miles, but then gradually becomes granitic. The outcrops are low in the vicinity of Wabigoon River but rapidly become higher farther up the stream reaching a height of one hundred feet in places, notably between falls two and three. Soil is found only as shallow deposits along the bottoms of the valleys, which are all short and narrow.

Below Canyon River the Wabigoon River continues wide without noticeable current for two and a half miles, but is then suddenly contracted by a narrow strip of land. In times of flood the latter is made an island by the river filling a channel half a mile west of the main one. For a hundred yards the current is swift, but below this again the river widens with deep bays, both east and west. For four miles the river is unbroken by rapids, though bordered by almost continuous outcrops of gneiss. A ridge of rudely laminated gneiss then crosses the valley, and over it the water drops in a perpendicular fall of seven feet, and then rushes through a narrow chute, to break in a short rapid below, with a total fall of ten feet. A mile below a short rapid has a fall of four feet. The banks of the river are then flat for five miles, the level country continuing inland for a considerable distance. A loamy soil, through which boulders are thickly scattered, forms the greater part of the shore of this section of the river, but below exposures of gneiss are not infrequent. This flat country terminates in a low ridge of gneiss, which crosses the river and causes a fine rapid a chain or so in length. The remaining length of Wabigoon River is narrow and usually swift. Gneiss forms the shores which become high and often precipitous, their sides, at the time of our visit, were frequently brightened by profusely scattered mountain ashtrees richly loaded with bright red berries. The banks of the Wabigoon River gradually decrease in height as the valley of the English River is reached, the country being comparatively flat at the confluence.

The trip up the English River and along Lac Seul to the Hudson's Bay Post was made in the interval between August 24th and September 3rd, and as much time was spent in portaging and some lost on account of bad weather, my opportunities for examining the country were somewhat limited. I can speak, therefore, only in general terms of this part of the district. Its geology, however, is of slight importance, except for a deposit of clay east of the Mattawa River, the rest of the area being Laurentian gneiss.

Coarse pink gneiss is found along the river from the mouth of the Wabigoon River to Oak Lake. The strike varies between north forty-five degrees east, and north sixty-five degrees east. Between Oak Lake and Barnston's Lake the gneiss is dark and finer in its lamination, and in many places contains large-sized garnets of fair quality. Before reaching Barnston's Lake we passed a number of isolated areas of good soil; the shores in places showing clay six to eight feet deep. Two miles above this lake the shores of the river become low and flat, and are covered by clay on which grow meadows of tall grass. The banks continue to be of clay for some ten miles above the Mattawa River and for the greater part of this distance, are high and well wooded with poplar. Between Clay Lake and Lac Seul the river is bordered by frequent outcrops of gneiss and some small areas of clay.

Falls and rapids occur at the following places; One-third mile below Maynard's Lake rapid, fall nine feet. In first mile of river above Oak Lake, three falls, eight, three and one-half and four feet, respectively.

One mile above Barnston's Lake, falls, ten feet.

Two miles above same lake, a long rapid with three falls of eight, three and six feet.

Ten miles above the Mattawa River, rapid, fourteen feet.

One-third mile above this, falls six feet.

Lac Seul.

The Ontario shore of the western part of Lac Seul (that part west of Shanty Narrows) is composed mainly of pink gneiss. Sandy beaches are common, and on most of the bays clay is found in the valleys and on the flats. Beyond a mile and a half from the lake, however, little good soil is found, the country being merely a series of ridges, high and bare, with swamps of small spruce and muskegs in the valleys. At Shanty Narrows a long rocky point on the south shore contracts the lake to a width of only a few chains, but east of this it becomes quite wide, with deep bays both to the right and left. Gneiss is almost continually exposed along the shores, but is hidden sometimes by swampy land at the bottom of a bay, or by small deposits of sand along the beach. A narrow band of mica schist, running with the gneiss, crosses near the entrance to the deep bay immediately west of Poplar narrows. The gneiss on the east side of this band is quite micaceous for a short distance, but then resumes its former character. Three miles west of Williams bay the schistosity of the gneiss increases and in half a mile it becomes quite micaceous and finely laminated. Three narrow bands of mica schist follow the strike of the gneiss, which at this point is north, sixty degrees east. Amber mica occurs with the biotite and the gneiss at one place has small scales of iron prites scattered through it. Many dikes of pegmatite cut the gneiss in various directions, as do a few narrow irregular seams of quartz. Before Williams bay is reached the gneiss becomes coarse and massive and continues so as far as our visit extended.

The Hudson's Bay post at Lac Seul is situated in an area of sandy land on the Keewatin shore. It is nearly directly north of Dinorwic, and about thirty-five miles distant. The buildings and stock of the company gave all of our party considerable surprise, the former being large and built in a substantial and modern style, while the variety and quality of the articles on sale were much better than would be expected where Indians were the only purchasers. Several of the families living at the post had plots of ground under cultivation, in which grew various garden vegetables. We were fortunate enough to procure a supply of potatoes, which were of splendid quality and good size, though not yet at their maturity.

While returning along the south shore the timber estimator and myself made two excursions inland to explore the country at some distance from the lake. The first was made from the bottom of Williams bay, from which we travelled south by compass for seven miles and then five miles southwest. On leaving the bay the shore rises quickly to some twenty feet above the lake, after which the elevation slowly and gradually increases for two miles. Moss and vegetable mould gave a thin covering to the rock for a greater part of this distance, but where ridges occurred the bare gneiss was exposed. South of this the country rose rapidly with frequent outcrops of rock. Four miles from our starting point the slope became southward and remained so until we reached Beaver river, some seven miles distant from Williams bay. This stream, which flows in a north-easterly direction, has an average width of twenty five feet and is four to five feet deep. It drains through a winding valley of varying width, the swampy nature of which is indicated by the deep color of the water and the fringe of spruce along its center. In several places the stream has been dammed by beavers, and the fresh cuttings of these animals were seen at a number of points. Our course now turned to the south-west and following along the valley of the Beaver River crossed several broad ridges of massive gneiss. In this distance we reached an open country which afforded an excellent view in all directions except the north. A large area to the west and south had recently been overrun by fire, leaving the rocks exposed as bare, smooth ridges. On these hills generally, on the southern slopes, there were shallow deposits of sand and small boulders, while a little clay was seen at a few places. To the south-east the country rose in high ridges, covered by a scanty growth of jack pine, and to the east was low and wooded, though somewhat rocky.

The second trip was made from the shore of Wild Rice Bay. On the north shore of this bay the gneiss is quite micaceous and near the west end it becomes hornblendic. Leaving the west end of Wild Rice bay, we travelled south over rolling country for three miles, the higher portions showing outcrops of gneiss and many loose boulders. For this distance the ascent had been gradual, but now the slope became southward, and, after going a couple of miles through swamp land crossed by two low ridges, we reached a beautiful clear lake, to which the name Florence was given. The length of this lake north and south is about two miles, the breadth half of that distance, excepting the south-eastern part, where the shore is swampy, it is surrounded by sloping rocky hills covered by small timber mostly jackpine. A stream of clear water, indicating more lakes drained by it, enters at the north-eastern extremity and finds an outlet again at the west side three quarters of a mile to the south. Continuing south from Florence Lake we passed several swamps separated by rocky ridges crossing in a southwesterly direction. In three miles we reached a second and much larger lake which was named "Louise." The water of this lake is beautifully clear, its outline regular, and its shores formed almost entirely of rock. Two small lakes, draining into the larger one, lie in a depression one mile east of Louise Lake.

All outcrops of rock were gneiss in which garnets were of frequent occurrence. Small areas of clay were seen on both lakes and sand was commonly found in the higher places, associated with boulders of various sizes.

Lake Wabuskong.

On returning to Wild Rice Bay our course followed westward along a winding creek for two and one-half miles. A portage one and three-quarters miles long then leads southward into a second creek which flows through low country for a mile and then enters Lake Wabuskong. The latter is an irregularly shaped lake about ten miles long. On its northeastern shore an Indian village bearing the same name is situated on a high bank of loamy soil that lies close along the lake. A dozen comfortable log buildings, including a schoolhouse, have been built by the Indians, who, at the time of our visit, inhabited eight of the dwelling houses, while several other families, more true to their Indian nature, lived in bark wigwams. Each family had a small plot of ground sown with potatoes, and though these had evidently been poorly cared for, the yield was good, better in fact than many I have seen in southern Ontario. A small patch of turnips in front of the chief's house had received more attention, and these were an excellent crop. There can be no doubt that where land can be found in this region it is of a very fertile nature, and speaking from my experience during the summer, the climate is not unfavorable to the growth of vegetables or the hardier grains.

The rock on the northern part of the lake is all gneiss and this continues eastward past the narrow part of the lake to within one and a half miles of the Indian village. Here a belt of granite is found and is remarkable because of being arranged in horizontal layers, similar to that described on Manitou Lake. It is exposed along the shore for about a mile, when it is hidden by the deep deposit of soil upon which the Indians have built their houses. North and east of the village gneiss is found again.

Wabuskong (pronounced by the Indians O-bosh-kong) drains at its northwestern extremity into Cedar River, which follows a northwesterly course for some eighteen miles before entering the English River. The lake discharges over a fall of four feet. Below the lake the river passes through three miles of rolling-wooded country, with few rock exposures, where it descends a rapid seventy hards in length with a fall of seven feet. For several miles the valley remains narrow and the river swift, but below, where swamp land is found, the river widens and its course becomes tortuous. One and a half miles above the English River a sudden break in the river valley causes a rapid two hundred and fifty yards long, the river then remaining wide and sluggish for its remaining length.

The country between Wabuskong Lake and the western part of Lac Seul was explored by crossing in a northerly direction from the north side of the former to a point one mile from its eastern boundary. At this point a sand bank fifteen feet high formed the shore of Wabuskong Lake, but was immediately succeeded by rock and swamp, which continued northward for three and one-half miles, several low ridges of rock crossing it in a southwesterly direction. The swamp gave place to one-half mile of rocky, broken country, which rose in three distinct ridges and formed the height of land between the two lakes. The slope now changed to the north and remained gradual till Lac Seul was reached. This northern slope was fairly well covered with soil and boulders, upon which was a good growth of poplar, birch, spruce and tamarac.

The canoe route to Wabigoon River follows a chain of lakes extending in a southward direction from the northeastern extremity of Wabuskong Lake. Owing to lack of time I did not accompany the surveying party past the third lake, but returned, and, in company with Mr. Nash, the timber estimator, took a westerly course to Oak Lake.

South of the Indian village the lake narrows and is bounded by hills of greyish gneiss that often show perpendicular faces ranging from forty to fifty feet high. Beyond this narrows the lake widens and continues about three miles southeast. This part of the lake is thickly studded with rocky islands.

The river enters the lake at the south east extremity, and widens, one half mile up stream, into a small lake which received the inflowing stream on the south shore. A ridge of gneiss thirty feet high separates this from the next lake which lies one quarter of a mile to the south, and is some twenty feet higher. The upper lake discharges into the lower one by a rapid two hundred and fifty yards long, with a fall of twenty feet. This lake called Five Mile Lake has a length as indicated by its name, and a width of half that distance. Our course lay along its east side, which is one continual exposure of gneiss, except where cut by several narrow bands of mica schist, that cross near its centre. Near this mica schist the gneiss is fairly laminated and largely composed of biotite, associated with a small amount of hornblende. The strike is south-westerly and the dip vertical. Southward the gneiss becomes very massive and gradually loses its schistosity, showing a decided granitoid nature south of the long narrow channel, through which lake our course was directed.

The route followed through to Oak Lake, leaves the west side of Cedar River at a point one mile below the upper fall, and follows a remarkably straight creek one half to two chains wide, in a direction north, eighty degrees west, for one and one half miles. Here we reached a narrow lake about one mile long, which we crossed in a westerly direction, and then ascended the creek entering at its north western extremity. The country around this lake is low but broken with many exposures of dark grey gneiss striking north, seventy degrees east, and dipping seventy five degrees north, twenty degrees west. Above this lake the creek leads through a marshy pond one half mile in length. On the south shore the stream enters over a fall of two feet after descending one of six feet a few yards farther up. A portage sixty yards long follows the west side of the creek, and reaches the wider part of the stream above the falls. For one third of a mile the creek passes low, wooded banks and then gradually widens to form a shallow channel, varying from three to eight chains wide, and extending two and one half miles south, seventy degrees west. The south side of this channel is low and covered with a shallow deposit of sandy soil and boulders; the northern rises rapidly in a ridge of massive gneiss, the top of which affords an excellent view to the north and north-west, showing the country in that direction to be a series of similar ridges separated by narrow valleys. This narrow stretch of water terminates in Indian Lake, which is one and a quarter miles long and half as broad. It is surrounded by low shores of massive grey gneiss. The creek which we had followed now turned southward, while our route followed a portage to a lake to the west. Having been told of a large lake a short distance up stream it was decided to explore the same.

In one third of a mile the creek expands into a long narrow channel, that stretches one and a half miles to the south, into the wider portion of Clearwater Lake, so called because of the clearness and purity of the water. It is the largest of the lakes on the route having a length east and west of five miles and a width of about the same. Only the northern and western parts of the lake were visited. The broad point running southward into the lake ends in a cliff of granitic rock fifty feet high, which has a horizontal structure similar to that noted on Wabuskong and Manitou Lakes. Samples from this locality and from Wabuskong Lake showed both rocks to be common granite without any interesting features minerologically. The band of granite is exposed along the shore for nearly a mile, and is bounded by gneiss of a granitoid character. East of the lake, rocky ridges, often with precipitous cliffs, rise to a height one hundred and fifty to two hundred feet above the lake, but south and west the country is lower and rolling.

From Indian Lake we crossed a portage four hundred and fifty yards in length and entered the east end of Little Duck Lake, which runs south-westerly one and three-quarters miles. Our course left the north-east side of the lake and followed a portage in a westerly direction one thousand nine hundred and forty yards, and then entered Goose Lake which it followed its full length of one mile. The shores of this lake are low and covered by sandy soil, except for a quarter of a mile north of the portage where a granitoid gneiss outcrops in a hill thirty feet high. The lake discharges at the west end by a creek twenty feet wide, which flows northeastward one and a quarter miles. A rapid occurs one quarter mile below the lake and is passed by a portage two hundred and fifty yards long, but the rest of the creek is through marshy lands and shows little current. It enters on the south side of Goose Lake.

The latter has a length of three miles east and west, and a width of nearly a mile. South and east of the lake the country is high but the north shore is partly marshy with low outcrops of rock, that increase rapidly in height, however, at a distance from the shore. The lake drains from the north shore by a narrow, winding creek, which runs westerly to Oak Lake. Instead of following the creek from its head, we went by way of a one-half mile portage from the west end of the lake, and reached the creek at some distance below its source. The average width of this stream is about twelve feet. Heavy rains had swollen it at the time of our visit, and the strong current aided very much in the descent though our passage was greatly retarded by the over-hanging alder bushes, which in places almost blocked the way. For five miles the creek flows in a direction south, seventy degrees west, passing through a valley one and a half miles wide with banks that rise slowly to the level of the surrounding country; then its course is more southward for four miles and lies in flat level land, mostly swampy and often muskeg. Turning westward again it is constricted by rocky banks and flows down a rapid two hundred and fifty yards long, but soon widens to a width of one chain which increases gradually as Oak Lake is approached. The rock at the east end of Oak Lake is gneiss, striking north forty-three degrees east, and dipping northward at sixty degrees. A sample broken from a narrow band of dark rock cutting the gneiss, proved on examination to be gabbro.

In following English River from the lake of the Grassy Narrows to Separation Lake, one passes through a country which is remarkably level and low considering its rocky character. Rock is everywhere, but all exposures are low and smooth, glacial action having worn the district down to a monotonous flatness. Our somewhat hurried passage gave little opportunity for rock examination, but all exposures noted were gneiss, which in places partook of the nature of a granite. Clay is found in shallow beds over small areas. At Separation Lake a band of dark crystalline hornblende schist crosses English River. The eastern boundary of this belt is marked by the deep bay running southward from the lake. The strike on this side is north, fifty degrees east, and dip northwest, at seventy-five degrees, the rock cleaving easily into thin laminæ. Much coarse pegmatite is associated with the rock. This band of Huronian continues down the river for about seven miles, and then gives place to granite and gneiss of Laurentian age. A rapid is situated about the centre of the band and there the strike has become north, sixty degrees east, and on the western boundary is north seventy degrees east, dip northward at eighty-five degrees. The schistosity of this rock shows much greater distinctness near the borders than at the centre of the area. A few narrow irregular quartz veins were seen.

Immediately west of the Huronian rock, and along the south side of English River, a small outcrop of granite occurs. It appears to be of a pegmatitic nature, and is beautifully coloured by small crystals of phlogopite, this is succeeded by massive pink and grey gneiss, which becomes quite granitoid in many places. The country continues flat along the river, but now rises rapidly at the south, where it becomes rough and broken. Eight miles below Separation Lake the river is narrowed by a ridge of gneiss, and the water plunges down a rapid one hundred and fifty yards long. The fall was estimated at ten feet. One and a quarter miles down another and almost perpendicular fall of twelve feet occurs. A short distance below this again there is a third fall, the three occurring in a distance of less than two miles. A magnificent water power could be obtained by damming, so as to bring the three falls to a common head of not less than thirty feet. Below the third fall a band of dark, finely laminated rock, probably a hornblende gneiss is reached Biotite enters largely into the composition of this rock, which strikes north, eighty degrees east, and dips north at eighty-five degrees. In places feldspar gives the rock a pinkish color, but gray is the prevailing tone. Garnets are plentiful in it. As the river is descended the lamination loses its distinctness, the color becoming pinkish, and as Lone Man's Lake is reached the rock becomes massive and granitoid, in places splitting into irregular slabs varying from eight to twelve inches in thickness. Massive gneiss continues to the Winnipeg River, except for a band of rock one and one-half miles wide and similar to the grey gneiss above Lone Man's Lake. This crosses the river immediately below Deer Lake.

Winnipeg River, at the boundary between Ontario and Manitoba, passes between low exposures of massive red gneiss. One and one-half miles above the boundary the river is broken by a short rapid (fall four feet) with a mile of swift water above. Twelve miles above the Manitoba boundary a small stream flowing from the south enters a bay of Winnipeg River. This stream was followed in our exploration, and together with a slightly smaller one lying to the east, aided us greatly in our work in the western part of this district. From Winnipeg River we followed the creek in a southerly direction for about eight miles The width varies from a few feet to one and a half chains, the upper portion having the greater width. A long rapid occurs a mile from Winnipeg River, fall ten feet; a second one, falling six feet, is met a short distance above. A fall of thirty feet over a smooth, steep ridge occurs immediately below the first lake expansion on the stream. Between the Winnipeg River and the second rapid the river follows a channel between rocky banks that reach a height of one hundred and twenty-five feet. Much of the remaining part of the river has banks of good soil, mostly clay, but in many places the rock, which is always exposed at no great distance from the stream, is found extending to the water's edge. The country both east and west of the first lake was explored and found to be very rocky, with little or no soil except in the swamps. From this lake we followed a wide, sluggish stream to the second lake, from which exploration was carried on as far west as the Manitoba boundary, a d about five miles to the east. The country proved to be similar to that of the previous lake.

Malachi Lake.

By a portage of ninety-two chains we passed eastward to Lake Malachi. The valley of the small creek which enters the south end of this lake, was explored to within a short distance of the railroad, and found to be a series of rocky ridges with swamp mostly of tamarac in the valleys. East of Malachi Lake high bare ridges rise one hundred feet above the water's level. From these splendid views could be had of the country, and it was seen to continue bare and rough a long way towards the east. Malachi Lake drains from its north-east extremity by a winding, grassy creek, which flows north and east for about one and a half miles. It then enters a lake one and a half miles in length, from the north side of which it flows over a rapid sixty yards long. One-fourth of a mile lower down another rapid, with a fall of ten feet, has a length of one hundred and fifty yards. The portage here follows the brow of the cliff on the left side of the stream, and is very difficult to make, both ends being very steep. Below this rapid, smooth water is found for three miles, when two falls occur, the upper four feet, the other six feet within one hundred yards of one another. The stream here enters a large lake some six miles in length, around which the country is high and, except in the valleys, is quite bare. It empties at the north end over a rapid four feet high, followed one quarter mile below by a second one with a drop of four feet also. The last rapid on the stream occurs at the entrance of the latter into Swan Lake.

This whole area is Laurentian gneiss striking uniformly between north, sixty-five degrees east, and north, seventy degrees east. It is very massive and shows little variation in composition or even in color. In the southern part of the district it out-crops as broad high ridges that follow the strike in a general way, but to the north, especially in the vicinity of Swan lake, is found as low rounded hillocks.

Above our camp on Swan Lake the Winnipeg River continues quite wide to within a short distance of the Indian village of White Dog, when it suddenly narrows and becomes very swift. Two miles above White Dog, there is a fall of seven feet, below which is a swift, boiling current that makes canoeing both difficult and dangerous. One mile higher up, there is half a mile of very swift water in which there are three falls, twelve, eight and five feet respectively. The total fall cannot be less than thirty feet, making an opportunity for developing an excellent water power at this point. Above these falls the river widens and continues so until Sandy Lake is crossed. Above the latter, the river continues narrow and often very swift until the rapid at Indian Reservation 38 is passed, when it widens again and continues so to within a short distance of Rat Portage.

All the rock which came under my observation between Swan Lake and the Huronian area at Rat Portage, was massive Laurentian gneiss. It was exposed almost continuously along the river. On the north-east shore of Sandy Lake, there are three deposits of sand on which grow some fine red pine. The largest of these is three quarters of a mile in length and about two-fifths of a mile in its widest part.

Surface Geology

Surface deposits, worthy of note, were so uncommon, that I have mentioned them as well as the topography of the country in connection with the general geology of the localities in which they occur, making it unnecessary for me to again refer to them. The more important area of soil however will be briefly mentioned below.

Economic Geology.

As is usual in the Laurentian, nothing of economic value was seen in the granitic and gneissoid area mentioned in this report, except the accumulations of clay on the Wabigoon and English Rivers. It is improbable however that the clay is suited for other purposes than agriculture, except perhaps the manufacture of brick or material of a similar nature.

Several veins of white quartz were seen in the hornblende schist on the south side of Black Sturgeon Lake, and a number of others on the north side. They followed the strike of the schist and varied in width from one to fifteen inches. No gold was seen in any of these, but that metal is found in the locality and several mining claims are reported to be located in this strip of Huronian rock.

Several bands of schist on the south shore of Linklater Lake were well mineralized, in one of which quartz was present but only as a vein a few inches in width. Narrow veins of quartz were common and the locality seems one which will reward the diligent prospector.

In the Huronian area lying between Boulder and Lacourse Lakes, belts of the schist are impregnated with fine scales of iron pyrites. The best example of this is seen on the east shore of Lacourse Lake, where a prospector, after whom the lake was named, has sunk a pit about five feet deep. The schist here is stained deep red by oxide of iron, and the fumes from the oxidation of the sulphur in the freshly exposed sulphide, can be detected at a considerable distance from the shallow pit. Quartz veins are common along the north part of this area of schist, but few of considerable size were seen. Here again a prospector might be well repaid for a summer's work.

Timber and Agriculture.

The higher portions of the country lying along our course from Black Sturgeon Lake to Lake Lacourse, were generally bare, but sometimes timbered by jack pine and small spruce. The slopes usually were covered by a light growth of poplar, white birch, balsam and spruce, and the swamps with spruce and tamarac. The only timber of importance, however, in this district was seen along the east side of Canyon Lake, and on the north shore of Forest Lake. In the former place there are three ridges of red pine with swamps of good spruce and tamarac between them. On Forest Lake a mile or more of shore is well covered with poplar, tamarac, spruce, &c. The clay banks of Wabigoon River support a good growth of poplar in many places, but the rocky region between Clay Lake and the English River shows little timber except of a scrubby nature. East of the Blackwater River there are several square miles of good poplar and spruce, but farther east the country is bare over large areas. Little timber of importance is found along English River except on the clay area near the mouth of Mattawa River. The banks of the stream are well wooded, mostly with poplar, for seven or eight miles. In the Lac Seul and Wabuskong Lake districts the country is fairly well wooded but largely with scrubby jack pine or stunted spruce. On the flats it is better; these often having a growth of good poplar, white birch, &c. Good spruce and tamarac are found in the swamps along the Winnipeg River and the lower parts of the English River, but these are

Church at Osnaburg House, Lake S Joseph. Par y No. 9.

Osnaburg House H. B. Co.'s post, Lake St. Joseph. Par y No. 9.

Burnt jackpine country. Roo R ver. 'arty No. 9.

Three generations Lake St. Joseph Indians. Osnaburg House. Party No. 9.

Shore of Lake St. Joseph. Party No. 9.

Fawcett's post at head of Albany River, Lake St. Joseph. Party No. 9.

Falls on Eagle River, north of C. P. R. track. Party No. 10.

Clay flats, Wabigoon River. Party No. 10.

Bell's Lake from Keewatin shore. Party No. 10

Rapids on English River. Party No. 10

Fall on Winnipeg River. Party No. 10.

Rapid, Canyon River. Party No. 10.

The gorge, Canyon River. Party No. 10

Digging potatoes. H. B. Co 's post at Lac Seul. Party No. 10.

C. M. S. Church, H. B. Co.'s post, Lac Seul. Party No. 10.

York boat, H. B. Co.s post, Lac Seul. Party No. 10.

Camp at Lac Seul. Party No. 10.

Halfbreed woman and child, H. B. Co.'s post, Lac Seul. Party No. 10.

seldom of sufficient area to make them important. Three groves of red pine grow on the north shore of Sandy Lake. The largest has a length of about three-quarters of a mile and a width of half that distance.

The clay deposits on the Eagle and Wabigoon Rivers, and immediately above the Mattawa River on the English River, should prove valuable for agricultural purposes. There was nothing by which their fertility could be judged except the growth of timber and grass which they supported, but these would indicate that there is nothing lacking in that respect. The cutting of marsh hay on the Wabigoon River should, I think, be quite profitable, if it were harvested at the proper season and then taken to the railroad in the winter. The price of this article is very high in the locality, where the supply must be brought long distances by rail.

I have the honor to be, Sir,

Your obedient servant,

JOHN A. JOHNSTON

JOHN McAREE, ESQ., O.L.S.,
Surveyor in charge of Exploration Survey Party No. 10., Rat Portage, Ontario.

N.B.—All distances and areas are approximate except where otherwise stated.

I, John A. Johnston, of the City of Toronto, in the County of York, solemnly declare that the above report, dated Jan. 31st, 1901, and the facts and statements therein contained, made by me, are, to the best of my knowledge, true and correct.

JOHN A. JOHNSTON.

Sworn before me at Toronto,
this fifth day of February, 1901.

GEO. B. KIRKPATRICK,
A Commissioner, &c.

INDEX.

K

L.

Mc.

M.

N.

O

P

Q.

R.

S

T.

U.

V.

W.

CPSIA information can be obtained
at www.ICGtesting.com
Printed in the USA
LVHW01s2058251217
560732LV00012B/102/P

9 781331 918141